Advances in Donkey and Mule Research

Advances in Donkey and Mule Research

Guest Editors

Ana Martins-Bessa
Amy McLean

MDPI

Basel • Beijing • Wuhan • Barcelona • Belgrade • Novi Sad • Cluj • Manchester

Guest Editors
Ana Martins-Bessa
Department of
Veterinary Sciences
University of Trás-os-Montes
e Alto Douro
Vila Real
Portugal

Amy McLean
Department of
Animal Science
University of California
Davis, CA
United States

Editorial Office
MDPI AG
Grosspeteranlage 5
4052 Basel, Switzerland

This is a reprint of the Special Issue, published open access by the journal *Animals* (ISSN 2076-2615), freely accessible at: www.mdpi.com/journal/animals/special_issues/Advances_Donkey_Mule_Research.

For citation purposes, cite each article independently as indicated on the article page online and using the guide below:

Lastname, A.A.; Lastname, B.B. Article Title. *Journal Name* **Year**, *Volume Number*, Page Range.

ISBN 978-3-7258-3054-1 (Hbk)
ISBN 978-3-7258-3053-4 (PDF)
https://doi.org/10.3390/books978-3-7258-3053-4

Cover image courtesy of AEPGA–Association for the Study and Protection of Donkeys

Contents

About the Editors

Ana Martins-Bessa

Ana Martins-Bessa (DVM, PhD) holds a degree in veterinary medicine from the University of Lisbon. She completed her European PhD in veterinary sciences at the Veterinary Faculty of Madrid in 2005, specializing in semen cryopreservation technology. Her subsequent research has focused on sperm preservation in various species, including dogs, swine, horses, and more recently, the Miranda donkey breed.

She has contributed to an international project on male donkey reproduction, collaborating with researchers from UTAD, the Autonomous University of Barcelona, the University of Pisa, and the University of Illinois. More recently, she has been involved in a national project focused on female donkey reproduction, specifically addressing chronic endometritis.

Ana Martins-Bessa has published several articles in specialized journals, with her primary research interests including andrology, reproductive physiology and medicine, new reproductive technologies, and the use of animal models to study reproductive aging.

Amy McLean

Amy McLean holds a PhD from the Michigan State University, with a thesis on donkey (Equus asinus) nutrition, training, and management. Throughout her academic career, she has received several scholarships for research related to pain assessment in donkeys and mules, as well as mule behavior. Her research interests focus on improving equine management, behavior, and welfare with a specific interest in donkeys, mules, hinnies, and working equids in developing countries.

She is currently involved in several projects, including "Mule Behavior with two main focuses: Pain Assessment- focus on facial grimace scale and Behavior development- early foal handling", "Hematology and blood biochemistry reference values in clinically healthy mules and hinnies", "Comparing morphometric measurements in donkeys, mules, and hinnies", "Integrated approaches to improving dairy donkey welfare and donkeys being used in production by using physiological and morphometric measurements", "Comparing welfare indicators in working equids in developing countries", "Comparing heart rate variability and behavioral indicators when using restraint mechanisms in Mules", and "Participatory approaches to working with and surveying mule and donkey owners in developing countries". She has published several articles on these topics in academic journals. Currently, Dr. McLean serves as an assistant professor of equine science education at the University of California, Davis. She has been the co-chair of the Donkey Welfare Symposium (Donkey Welfare Conference) and sits on several boards helping working donkeys and mules such as the Equitarian Initiative and American Fondouk.

Preface

Recent research has emphasised the importance of understanding reproductive biology and physiology, behavior, pain management, internal medicine, pathogen prevalence, dental health, and overall wellbeing—all areas critical to the conservation and sustainable management of donkey and mule populations. Studies have also addressed anatomical and pharmacokinetic differences between horses and donkeys, highlighting their unique needs.

Assisted reproductive techniques (ARTs), such as artificial insemination (AI), sperm cryopreservation, and embryo transfer (ET), which are explored in this Special Issue, play a pivotal role in preserving genetic resources.

Animal genetic resources are priceless heritage, making it crucial to safeguard genetic diversity across species, breeds, varieties, and native ecotypes. Many donkey breeds are currently classified as endangered and face significant challenges, including inbreeding, poor reproductive management, advanced age, and low rates of reproduction.

This Special Issue brings together contributions from researchers worldwide, showcasing the latest advancements in donkey and mule research. It provides an updated overview of the primary conditions affecting the health of these species, offering valuable insights into the development of new strategies and paving the way for future studies.

Ana Martins-Bessa and Amy McLean
Guest Editors

animals

Editorial

Advances in Donkey and Mule Research

Ana Martins-Bessa [1,2,*] and Amy K. McLean [3]

[1] Animal and Veterinary Research Centre (CECAV), University of Trás-os-Montes and Alto Douro (UTAD), 5000-801 Vila Real, Portugal
[2] Associate Laboratory for Animal and Veterinary Sciences (AL4AnimalS), 5000-801 Vila Real, Portugal
[3] Department of Animal Sciences, University of California, Davis, CA 95616, USA; acmclean@ucdavis.edu
* Correspondence: abessa@utad.pt

Donkeys (*Equus asinus*) and mules represent approximately 50% of the entire domestic equine population in the world and play an essential role in the lives of thousands of people, primarily in developing countries [1]. Many donkey breeds are considered endangered and presently facing major problems such as inbreeding, poor reproductive management and old age. On the other hand, there is an increased use for donkeys as pets and production animals and mules as performance animals with superior genetics from both the donkey and horse side, which has increased the interest and value in these animals. New findings focused on improved understanding of basic physiology, behaviour, pain, internal medicine, pathogen frequency, and subjects related to their overall wellbeing (e.g., nutrition, pharmacokinetics, dentistry) are considered areas of interest for preserving and maintaining donkey and mule populations.

Assisted reproductive techniques (ART) like artificial insemination (AI), sperm cryopreservation or embryo transfer (ET) are unique tools to preserve genetic resources. Several studies of the Special Issue were conducted in donkey ART. The study of Lago-Alvarez et al. [2] analyzed the processes of cooling and freezing of epididymal donkey sperm harvested at two different moments: immediately after castration, and 24 or 48 h after shipment. The study allowed the conclusion that freshly harvest and cooled-shipped allowed satisfactory preservation of semen parameters.

Several works were devoted to male donkey reproduction physiology. Martins-Bessa et al. [3] evaluated two predictive models for donkey sperm quality by using the following independent variables: ejaculate's volume, sperm concentration, total sperm number, motility and sperm morphology. Models included combinations of age as a covariate and biometric and testicular measurements as independent factors. Results evidenced that the goodness-of-fit was similar for both models—hence, the combination of biometry and testicular factors presented improved predictive power. The application of the present models may be useful to gather relevant information that could be used hereafter for ART in donkeys.

Donkeys and mules are treated like horses in many cases, with the physiological differences between these species usually not considered [1]. One significant difference between horses and donkeys' breeding behaviour is in the additional time donkeys need to achieve erection and ejaculation. Donkeys are characterized by long courtships needed to attain excitation and erection, which can cause donkey semen collections sessions of up to 90 min. The study of Panzani et al. [4] verified the possibility of using Prostaglandin F2α analog, cloprostenol sodium, to hasten the onset of erection and ejaculation. Cloprostenol sodium significantly hastened treatment-to-erection and treatment-to-ejaculation times from 12.0 ± 1.6 to 6.0 ± 1.6 min and from 14.0 ± 1.4 to 9.6 ± 1.4 min, respectively. There were no effects of its administration on semen parameters, being suggested that cloprostenol sodium administration immediately prior to semen collection hastened time to collect semen in donkeys with no detrimental effects on semen quality and can be used by practitioners to circumvent long delays in donkey semen collection.

Citation: Martins-Bessa, A.; McLean, A.K. Advances in Donkey and Mule Research. *Animals* **2024**, *14*, 2238. https://doi.org/10.3390/ani14152238

Received: 13 July 2024
Accepted: 29 July 2024
Published: 1 August 2024

The analysis and processing of donkey's semen for assisted reproductive technologies was also the aim of several works in the Special Issue. The use of CASA (Computer Assisted Sperm Analysis) image analysis system provides an automatic, objective and reliable sperm evaluation; however systematic preparations of the samples should be made. The work of Gacem et al. [5] showed that analyzing a minimum of nine fields at 250 frames per second from the center to the edges in Spermtrack®10 chamber using a dilution of 30×10^6 sperm/mL offers the best choice for donkey computerized sperm motility analysis. Semen cryopreservation is also a very important tool for establishing germ plasm banks in order to preserve samples of endangered breeds. In the work of Hidalgo et al. [6], it was found that donkey sperm vitrification in spheres using non-permeable cryoprotectants exhibited better sperm motility and viability parameters after warming than sperm vitrification using extenders containing permeable cryoprotectants.

The use of single layer centrifugation (SLC) of fresh donkey semen was also investigated and its impact on sperm quality parameters and on the modulation of endometrial reaction following semen deposition using an in vitro model. It was concluded that SLC increases the proportions of functionally intact and motile spermatozoa and appears to remove the seminal plasma proteins that inhibit sperm-PMN binding [7].

Also, embryo technology was investigated in donkeys. The work of Dorado et al. [8] aimed to investigate the factors influencing the success of this ART. Donor jenny, donor age, successive cycle within donor, day of flushing, number of flushing, and jack affected embryo recovery rate. The identification of these key points is crucial to achieve a higher efficiency of embryo transfer and vitrification processes, before considering their application in the conservation of endangered donkey breeds.

Research was also conducted in order to find molecular markers that can be used to identify and isolate specific development stages of germ and Leydig cells in the donkey testicle. Several experiments were conducted by Choi et al. [9] with this aim, suggesting that the protein gene product (PGP) 9.5 can be used to identify and isolate spermatogonia at the seminiferous tubules.

Other articles focused on infectious diseases affecting both donkeys and mules. Realizing that many diseases could compromise donkey well-being, performance and even put the breed at risk in some parts of the world [1]. Review article of these authors provided an update on viruses that may affect donkeys and mules may endanger another species' health. Health concerns have been raised with the removal and relocation of feral donkeys to offsite adoption locations, as suggested by the work of Goodrich et al. [10]. Previous exposure of the animals to equine herpesvirus 1 (EHV-1), equine influenza (EIV), West Nile virus (WNV), and *Borrelia burgdorferi* (the causative agent of Lyme disease) was assessed. Results indicated that feral equid populations are mostly naïve and likely susceptible to these common equid pathogens upon removal from the wild. Relocation of feral donkeys and possible disease transmission between equine species and the ability to identify early stages of disease by using inflammatory markers such as serum amyloid A (SAA) the aim of the study of Jerele et al. [11]. Results suggest that donkeys do not appear to be a substantial risk for disease transmission to horses but could be if they carried strangles or other processes in which AHV-2 and *Streptococcus zooepidemicus* were involved. The study showed SAA levels may suggest disease as levels were highest in foals who had higher evidence of AHV strains compared to adults.

The study of Tirosh-Levy et al. [12] evaluated the serologic exposure of donkeys to the parasites *Toxoplasma gondii* and *Neospora* spp. in Israel. The high prevalence found in the study suggested that donkeys may have a role in the maintenance of these parasites in the area, thus serving as a source of infection for the definitive hosts.

Although belonging to the *Equus* genus, anatomical differences exist, with implications at management and welfare level [13]. This work revealed clear differences at skull morphology between horses and donkeys. The calculation of ratios of skull measurements from horses and donkeys allowed the observation that donkeys skulls have larger forehead than horses and the olfactory bulb was smaller and rotated more forward than horses.

On the other hand, pharmacological differences between mules and horses were confirmed in the work of Bazzano et al. [14], in which the efficacy and pharmacokinetics of the anthelmintic compound ivermectin administrated orally to mules at the same dosage of horses was evaluated. Results evidenced intermediate pharmacokinetic parameters between horses and donkeys, suggesting an optimization of the dosage.

Skin diseases caused by fungi and bacteria are quite common in donkeys and mules, which have been reported in several regions of the world as case reports with occasionally retrospective studies. Clinical and pathological features of skin diseases were reviewed and updated by Lima et al. [15], suggesting that geographical variations may results in significant differences in the prevalence of skin diseases. The evaluation of clinical and pathological behaviour of skin diseases directs the implementation of control strategies.

On the contrary to horses, donkeys tend to express pain and discomfort more subtlety. In consequence, a state of pain could only be recognized in an advanced degree of disease [16]. The study of these authors focused on the development of a scoring system donkey grimace scale based on body language of face and overall body posture and appearance. This system proved to be accurate and effective in identifying discomfort related to pain associated to surgical castration.

Welfare was also a relevant topic in the Special Issue. Concerning reports of welfare issues were related in the work of Farhat et al. [17], with working donkeys in Egypt showing many types of wounds associated with parts of the harness and use of excessive force or beating. The study introduced methods to measure the welfare status in working donkeys and found an association between health risks, behavioral parameters and body condition. Overall welfare of working donkeys should be enhanced in order to improve the production of the animals and the reduction of the wounds. Besides, observation of the body condition should be promoted together with the owners, in order to avoid compromising the body condition and welfare.

In countries like France, Italy and Brazil the asinine milk production has some economic relevance due to the milk qualities for human consumption. The use of donkeys for this purpose must guarantee their welfare [18]. The work of these researchers investigated whether a 2h separation period, used to achieve milking efficiency, was stressful for jennies and their foals. For that, the behavioural and physiological changes of the animals when exposed to the milking routine were assessed. Behavioral assessments and determination of cortisol salivary concentration were collected before and after separation in jennies and foals. Considering the collected variables, results of the work suggest that the milking routine did not appear to be stressful for the animals. More studies are desirable in order to measure the impact of milking routine and to guarantee the welfare of the animals in that industry.

The role of donkeys in the reduction of fuel biomass of forest was verified in the work of Bartolomé at al. [19], being suggested the combined use of donkeys and goats, as both animals feed on forest species, and could act complementary to prevent fire in the Mediterranean forests.

Some works of the Special Issue improved the knowledge about medical conditions in donkeys, which will contribute to the provision of appropriate medical care. The asinine metabolic syndrome and the pituitary pars intermedia dysfunction were reviewed by Gehlen et al. [20]. A special emphasis was given in this review to the differences evidenced in donkeys, as reference values and physiological reaction to dynamic tests.

Some health issues are still neglected in donkeys. Dental health has been neglected in donkeys [21], despite the frequency of dental problems in this specie. The work of these authors contributed to knowledge of donkey oral microbiome sequencing in association with dental care treatment. Future works should promote a more abroad understanding of the impact of disease as periodontitis in donkey dental health [21].

The study of passive transfer of immunity in donkey foals compared to horse foals has been studied very little. The work of Turini et al. [22] determined the IgG serum concentration in donkeys' foals by different analytic methods in the first 24 h. Methods revealed good or strong correlation to each other. Based on feasibility, serum total protein (TP) refractometry could be a useful method to estimate transfer of immunity to donkey foals.

The infrared thermography was applied comparatively in donkeys and horses in the work of Domino et al. [23]. Similarities were found in thermal patterns of both species; however average surface temperatures were higher in horses which was related to different thermal properties of the skin and hair coat.

Finally, two case reports described the implantation of a transvenous single-chamber pacemaker due to severe bradycardic arrythmia in five-month-old jenny with successful outcome [24] and the medical management of four donkeys with various degrees of external severe burn injuries along with clinical findings and evolution [25].

Author Contributions: Writing—original draft preparation, review and editing, A.M.-B. and A.K.M. All authors have read and agreed to the published version of the manuscript.

Funding: This work was supported by the projects UIDB/00772/2020 (Doi:10.54499/UIDB/00772/2020) funded by the Portuguese Foundation for Science and Technology (FCT).

Acknowledgments: The guest Editors would like to express their gratitude to all contributors of the Special Issue their contribution and the reviewers for their work in evaluating the submitted manuscripts.

Conflicts of Interest: The authors declare no conflict of interest.

References

1. Câmara, R.; Bueno, B.; Resende, C.; Balasuriya, U.; Sakamoto, S.; Reis, J. Viral Diseases that Affect Donkeys and Mules. *Animals* **2020**, *10*, 2203. [CrossRef] [PubMed]
2. Lago-Alvarez, Y.; Podico, G.; Segabinazzi, L.; Cunha, L.; Barbosa, L.; Arnold, C.; Lima, F.; King, L.; McLean, A.; Canisso, I. Donkey Epididymal Transport for Semen Cooling and Freezing. *Animals* **2020**, *10*, 2209. [CrossRef] [PubMed]
3. Martins-Bessa, A.; Quaresma, M.; Leiva, B.; Calado, A.; Navas González, F. Bayesian Linear Regression Modelling for Sperm Quality Parameters Using Age, Body Weight, Testicular Morphometry, and Combined Biometric Indices in Donkeys. *Animals* **2021**, *11*, 176. [CrossRef] [PubMed]
4. Panzani, D.; Quaresma, M.; Fanelli, D.; Camillo, F.; Moroni, R.; Rota, A.; Martins-Bessa, A.; Nóvoa, M.; Catalán, J.; Canisso, I.; et al. Hastening Time to Ejaculation in Donkey Jacks Treated with the PGF2α Analog, Cloprostenol Sodium. *Animals* **2020**, *10*, 2231. [CrossRef] [PubMed]
5. Gacem, S.; Catalán, J.; Valverde, A.; Soler, C.; Miró, J. Optimization of CASA-Mot Analysis of Donkey Sperm: Optimum Frame Rate and Values of Kinematic Variables for Different Counting Chamber and Fields. *Animals* **2020**, *10*, 1993. [CrossRef] [PubMed]
6. Hidalgo, M.; Diaz-Jimenez, M.; Consuegra, C.; Pereira, B.; Dorado, J. Vitrification of Donkey Sperm: Is It Better Using Permeable Cryoprotectants? *Animals* **2020**, *10*, 1462. [CrossRef] [PubMed]
7. Papas, M.; Catalán, J.; Recuero, S.; Morrell, J.; Yeste, M.; Miró, J. Single Layer Centrifugation Improves the Quality of Fresh Donkey Semen and Modifies the Sperm Ability to Interact with Polymorphonuclear Neutrophils. *Animals* **2020**, *10*, 2128. [CrossRef]
8. Dorado, J.; Bottrel, M.; Ortiz, I.; Díaz-Jiménez, M.; Pereira, B.; Consuegra, C.; Carrasco, J.; Gómez-Arrones, V.; Domingo, A.; Hidalgo, M. Factors Affecting Embryo Recovery Rate, Quality, and Diameter in Andalusian Donkey Jennies. *Animals* **2020**, *10*, 1967. [CrossRef]
9. Choi, Y.; Jung, Y.; Kim, S.; Kim, J.; Jung, H.; Yoon, M. Stage-Dependent Expression of Protein Gene Product 9.5 in Donkey Testes. *Animals* **2020**, *10*, 2169. [CrossRef]
10. Goodrich, E.; McLean, A.; Guarino, C. A Pilot Serosurvey for Selected Pathogens in Feral Donkeys (*Equus asinus*). *Animals* **2020**, *10*, 1796. [CrossRef]
11. Jerele, S.; Davis, E.; Mapes, S.; Pusterla, N.; Navas González, F.; Iglesias Pastrana, C.; Abdelfattah, E.; McLean, A. Survey of Serum Amyloid A and Bacterial and Viral Frequency Using qPCR Levels in Recently Captured Feral Donkeys from Death Valley National Park (California). *Animals* **2020**, *10*, 1086. [CrossRef] [PubMed]
12. Tirosh-Levy, S.; Steinman, A.; Minderigiu, A.; Arieli, O.; Savitski, I.; Fleiderovitz, L.; Edery, N.; Schvartz, G.; Leszkowicz Mazuz, M. High Exposure to *Toxoplasma gondii* and *Neospora* spp. in Donkeys in Israel: Serological Survey and Case Reports. *Animals* **2020**, *10*, 1921. [CrossRef] [PubMed]
13. Merkies, K.; Paraschou, G.; McGreevy, P. Morphometric Characteristics of the Skull in Horses and Donkeys—A Pilot Study. *Animals* **2020**, *10*, 1002. [CrossRef] [PubMed]
14. Bazzano, M.; Di Salvo, A.; Diaferia, M.; Veronesi, F.; Galarini, R.; Paoletti, F.; Tesei, B.; McLean, A.; Veneziano, V.; Laus, F. Anthelmintic Efficacy and Pharmacokinetics of Ivermectin Paste after Oral Administration in Mules Infected by Cyathostomins. *Animals* **2020**, *10*, 934. [CrossRef] [PubMed]

15. Lima, T.; Silva, R.; Pereira, R.; Soares, K.; Santos, N.; Sousa, M.; Mendonça, F.; Lucena, R. Skin Diseases in Donkeys and Mules—An Update. *Animals* **2021**, *11*, 65. [CrossRef] [PubMed]
16. Orth, E.; Navas González, F.; Iglesias Pastrana, C.; Berger, J.; Jeune, S.; Davis, E.; McLean, A. Development of a Donkey Grimace Scale to Recognize Pain in Donkeys (*Equus asinus*) Post Castration. *Animals* **2020**, *10*, 1411. [CrossRef] [PubMed]
17. Farhat, S.; McLean, A.; Mahmoud, H. Welfare Assessment and Identification of the Associated Risk Factors Compromising the Welfare of Working Donkeys (*Equus asinus*) in Egyptian Brick Kilns. *Animals* **2020**, *10*, 1611. [CrossRef] [PubMed]
18. de Souza Farias, S.; Montechese, A.; Bernardino, T.; Rodrigues, P.; de Araujo Oliveira, C.; Zanella, A. Two Hours of Separation Prior to Milking: Is This Strategy Stressful for Jennies and Their Foals? *Animals* **2021**, *11*, 178. [CrossRef] [PubMed]
19. Bartolomé, J.; Miró, J.; Panadès, X.; Broncano, M.; Plaixats, J.; Rigau, T.; Milán, M.; Baraza, E. Preference by Donkeys and Goats among Five Mediterranean Forest Species: Implications for Reducing Fire Hazard. *Animals* **2020**, *10*, 1302. [CrossRef]
20. Gehlen, H.; Schwarz, B.; Bartmann, C.; Gernhardt, J.; Stöckle, S. Pituitary Pars Intermedia Dysfunction and Metabolic Syndrome in Donkeys. *Animals* **2020**, *10*, 2335. [CrossRef]
21. Zhu, Y.; Jiang, W.; Holyoak, R.; Liu, B.; Li, J. Investigation of Oral Microbiome in Donkeys and the Effect of Dental Care on Oral Microbial Composition. *Animals* **2020**, *10*, 2245. [CrossRef] [PubMed]
22. Turini, L.; Bonelli, F.; Nocera, I.; Meucci, V.; Conte, G.; Sgorbini, M. Evaluation of Different Methods to Estimate the Transfer of Immunity in Donkey Foals Fed with Colostrum of Good IgG Quality: A Preliminary Study. *Animals* **2021**, *11*, 507. [CrossRef] [PubMed]
23. Domino, M.; Romaszewski, M.; Jasiński, T.; Maśko, M. Comparison of the Surface Thermal Patterns of Horses and Donkeys in Infrared Thermography Images. *Animals* **2020**, *10*, 2201. [CrossRef] [PubMed]
24. Sedlinská, M.; Kabeš, R.; Novák, M.; Kološ, F.; Melková, P. Single-Chamber Cardiac Pacemaker Implantation in a Donkey with Complete AV Block: A Long-Term Follow-Up. *Animals* **2021**, *11*, 746. [CrossRef]
25. Lohse, J.; Pietrantoni, P.; Tummers, C. Management of Thermal Injuries in Donkeys: A Case Report. *Animals* **2020**, *10*, 2131. [CrossRef]

animals
MDPI

Case Report

Single-Chamber Cardiac Pacemaker Implantation in a Donkey with Complete AV Block: A Long-Term Follow-Up

Markéta Sedlinská [1], Radovan Kabeš [1], Miroslav Novák [2], Filip Kološ [1] and Pavlína Melková [1,*]

1 Equine Clinic, Faculty of Veterinary Medicine, University of Veterinary and Pharmaceutical Sciences Brno, Palackého tř. 1946/1, 612 42 Brno, Czech Republic; sedlinskam@vfu.cz (M.S.); kabesr@vfu.cz (R.K.); kolosf@vfu.cz (F.K.)
2 St. Anne's University Hospital, Pekařská 53, 659 91 Brno, Czech Republic; 2608@mail.muni.cz
* Correspondence: melkovap@vfu.cz

Simple Summary: Cardiac pacing is widely used in human and small-animal medicine for the treatment of many symptomatic bradyarrhythmias. Although the implantation of a pacemaker has previously been described in a horse and donkey, few data are available about the long-term follow-up. This article describes a pacemaker implantation in an African donkey with a complete AV block, and reports complications associated with this procedure, which included lead dislodgement and pacemaker pocket infection. After implantation, the donkey showed an improved quality of life and was used for the animal-assisted therapy of disabled children. The function of the pacemaker was checked regularly and, to date, the pulse generator has been replaced twice. Cardiac examination eighteen years after pacemaker implantation revealed that the pacemaker is working appropriately and no morphological changes were observed in the heart.

Citation: Sedlinská, M.; Kabeš, R.; Novák, M.; Kološ, F.; Melková, P. Single-Chamber Cardiac Pacemaker Implantation in a Donkey with Complete AV Block: A Long-Term Follow-Up. *Animals* **2021**, *11*, 746. https://doi.org/10.3390/ani11030746

Academic Editors: Ana Martins-Bessa, Amy McLean and Enrica Zucca

Received: 2 February 2021
Accepted: 4 March 2021
Published: 9 March 2021

Abstract: A five-month-old African jenny was presented with a history of exercise intolerance and syncopal episodes. Severe bradycardic arrhythmia due to a high-grade second-degree atrioventricular (AV) block with progression to complete AV block was diagnosed. The jenny underwent a transvenous single-chamber pacemaker implantation. The implantation procedure was performed in a lateral recumbency and the ventricular lead was inserted through the jugular vein. Positioning of the lead was guided by echocardiography. The pacemaker was programmed to VVI mode with a minimal ventricular rate of 40 pulses per minute, a pulse amplitude of 2.4 V, a pulse width of 0.5 ms and sensing amplitude of 2.5 mV. Short-term complications associated with the procedure included lead dislodgement and pacemaker pocket infection. The long-term outcome was satisfactory; the jenny showed improvement in heart function and quality of life after pacemaker implantation. The pulse generator replacement was performed twice (at nine-year intervals) and the intervention was always associated with a local inflammatory reaction around the pacing device. Cardiac examination 18 years after pacemaker implantation revealed no morphological changes in the heart; the electrode lead was still in the correct position and successful pacing and sensing of the ventricle were obtained. Regular follow-up checks are important to evaluate pacemaker function.

Keywords: arrhythmia; cardiac pacing; pacemaker pocket infection; donkey

1. Introduction

Cardiac pacing is an effective therapy for various arrhythmias in humans and a therapeutic implantation of a cardiac pacemaker has also been described in a horse [1–4] and a donkey [5,6]. Indications for pacemaker implantation are symptomatic bradycardic arrhythmias such as third-degree AV block, high-grade second-degree AV block, sick sinus syndrome or persistent atrial standstill [1,3,5–10]. Transvenous pacemaker implantation is a relatively safe and simple procedure. Complications resulting from permanent pacemaker implantation are well-known in human and small-animal medicine and include lead malposition or displacement, pneumothorax, haemothorax, myocardial perforation, infection of pacemaker pocket, bacterial endocarditis and venous thrombosis [7,11–13]. In equids, pacemakers are not commonly used and little information is known about the long-term outcome. Hamir and Reef [14] described the complications and postmortal findings associated with a permanent transvenous pacing device in a horse thirty-four months after implantation. This article reports a pacemaker implantation in a five-month-old jenny, the short-term complications after intervention and an eighteen-year follow-up.

2. Materials and Methods

2.1. History and Clinical Findings

A five-month-old African jenny was presented to the Equine Clinic of the Veterinary University Brno with a history of exercise intolerance and frequent syncopal episodes. The syncopal episodes had started one month before and their frequency gradually increased. At the time of admission, short syncopal events (5–10 s) were observed approximately ten times per day.

The jenny was in good body condition (80 kg), bright and alert. Heart rate varied between 30 and 60 beats per minutes (bpm) and heart rhythm was irregularly irregular. No heart murmur was heard on auscultation. In both nostrils, a small amount of serous discharge was noticed. Respiratory rate was 24 breaths per minutes and lung sounds were normal. Neurological examination revealed no abnormalities. The complete blood count and biochemistry were unremarkable.

2.2. Electrocardiography and Cardiac Ultrasound

An electrocardiogram (ECG) showed a resting heart rate of 24 bpm with a high-grade second-degree AV block (Figure 1a). Impulse transmission from atria to ventricles was variable, with a conduction ratio of up to 13:1. The heart rate increased with physical activity or during manipulation with the animal. If the sinus rate increased over 80 bpm, a third-degree AV block occurred and complete atrial and ventricular dissociation was present. A junctional rhythm of a frequency around 50 bpm was associated with polymorphic ventricular complexes (Figure 1b). When the frequency of the ventricles was too low (pauses longer than 10 s), the syncope had occurred.

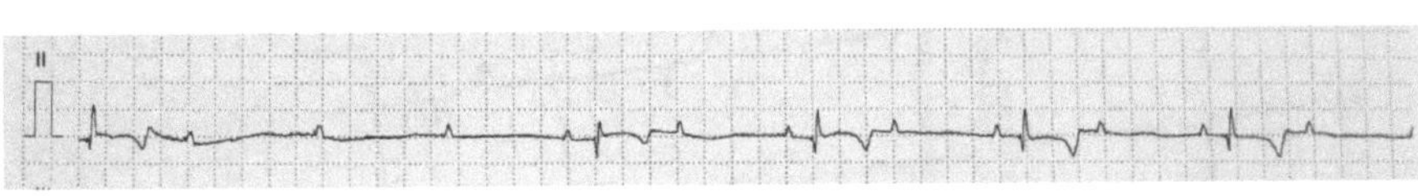

(a)

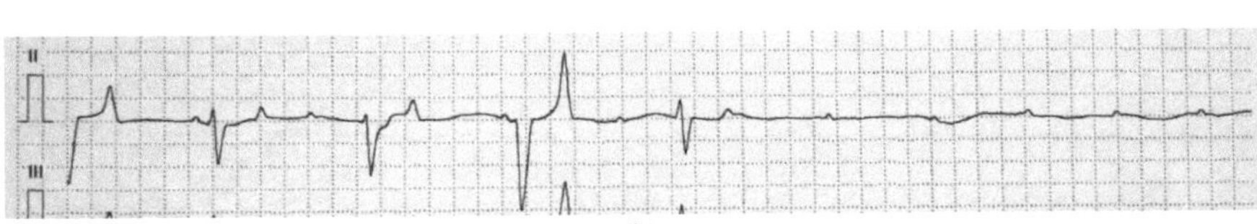

(b)

Figure 1. Base apex electrocardiogram of the jenny (**a**) A high-grade, second-degree atrioventricular (AV) block. Three consecutive P-waves are not followed by a QRS complex (**b**) A complete AV block with an independent atrial and ventricular rate. The tracing shows polymorphic QRS complexes and the interruption of the ventricular escape rhythm (25 mm/s, 1 cm/mV).

A transthoracic echocardiogram was performed using a 1.5–3.6 MHz transducer (Eub-6500, Hitachi, Tokyo, Japan). Two-dimensional and M-mode echocardiography did not reveal gross morphological or functional abnormalities. No valvular regurgitation was found; echogenicity of the endocardium and myocardium was normal. On the basis of these findings, a diagnosis of atrioventricular block of unknown aetiology was established.

3. Treatment

3.1. Medical Treatment

The jenny was treated with phenylbutazone (2.2 mg/kg p.o. q12 h) and box-rested for 14 days. The incidence of syncopes during the treatment gradually decreased until they disappeared completely. However, on the ECG, the third-degree AV block was still present. The medication was discontinued but, on the fourth day after this, the syncopes reappeared. In view of this fact, the need for a pacemaker implantation was indicated.

3.2. Surgery

The jenny was premedicated with penicillin (30,000 IU/kg i.v.) and gentamicin (6.6 mg/kg i.v.), along with atropine (0.05 mg/kg i.m.), and placed under general anesthesia in a lateral recumbency position. Anesthesia was induced with diazepam (0.1 mg/kg i.v.), butorphanol (0.015 mg/kg i.v.), ketamine (2.2 mg/kg i.v.) and guaiphenezine (25 mg/kg i.v.). The jenny was intubated using a 14 mm endotracheal tube and anesthesia was maintained with isoflurane (ET concentration 1.4 vol. % with oxygen as a carrying gas). On the left side of the neck, the hair was shaven and the skin prepared aseptically. After the skin incision, which was slightly proximal from the thoracic aperture, *vena jugularis sinistra* was prepared. The tip of a unipolar electrode was inserted under ultrasonographic control through the *vena jugularis* to the right ventricular apex. Although many attempts were made, a sufficient stimulation threshold was not found. During the procedure, periods of asystoly occurred, which were repeatedly resolved by indirect cardiac massage until the spontaneous heart rhythm was restored. The unipolar electrode was changed to bipolar with passive fixation (TIR 60 BP, Biotronik, Berlin, Germany). The length of the lead was 60 cm. The tip of the electrode was properly positioned and fastened in trabeculae of the ventricular myocardium. The proximal part of the lead was ligated to the surrounding tissue and connected to the pulse generator. A stimulation threshold was found at 0.7 V and 0.5 ms, with R waves of 7 mV. The pacemaker Kairos S (Biotronik, Berlin, Germany) was inserted into a subcutaneous pocket, which was created in the prescapular region. The subcutaneous tissue and skin were closed in a routine manner. The pacemaker was programmed to VVI mode with a minimal ventricular rate of 40 ppm (pulses per minute), a pulse amplitude of 4.8 V, a pulse width of 1.0 ms and sensing amplitude 2.5 mV. Ventricular stimulation 1:1 was confirmed by electrocardiography. Recovery from the general anaesthesia was uneventful.

The next day a mild swelling of the pacemaker surrounding tissue had developed. ECG confirmed regular heart rate and effective stimulation. Antibiotic and anti-inflammatory treatment was continued for three days (penicillin 30,000 IU/kg i.v. q8 h, gentamicin 6.6 mg/kg i.v. q24 h, flunixin 1.1 mg/kg i.v. q24 h). The jenny appeared alert and no syncopal episodes were observed.

Six days after the surgery, an ECG check was performed. It was found that the pacemaker stimulation was without response and a complete AV block was present. The heart had intrinsic ventricular activity with a frequency of 53 bpm. Lead dislodgement was confirmed by measurement of sensing parameters.

Nine days after the first surgery, the jenny was subjected to a second procedure under general anaesthesia to correct the position of the electrode using the same anaesthetic protocol. The pacemaker pocket was reopened; the lead was disconnected and its reposition was guided by echocardiography. The stimulation threshold was found at 0.6 V at a pulse width of 0.5 ms and R waves of 11 mV. The electrode was connected back to the pulse generator and the pacemaker returned to the pocket. The pocket was then sutured in layers and the surgical wound was covered with a sterile dressing. The pacemaker was set to the same values as written above. Recovery from surgery was without any complications. The next day, ECG confirmed a full response to the pacemaker stimulation.

Mild swelling around the pacemaker pocket developed again and, therefore, antibiotic and anti-inflammatory medication was prolonged. Eight days after the second surgery, the jenny exhibited lethargy and fever. A complete blood count revealed leukocytosis (23×10^9/L, reference range 5–14.5 $\times$ 10^9/L). Infection of the pacemaker pocket was diagnosed. A mild exudation appeared from the wound and a sample was taken for microbial culture. *Staphylococcus aureus* resistant to penicillin and gentamicin was isolated, and the antibiotics were changed to cefquinome (2 mg/kg i.m. q12 h). Wound-cleaning was provided on a daily basis with povidone iodine solution. The wound secretion became purulent and the area around the wound was painful; surgical revision was indicated. The surgery (12 days after pacemaker reposition) was performed under analgosedation (diazepam 0.14 mg/kg i.v., butorphanol 0.01 mg/kg i.v., romifidine 0.02 mg/kg i.v.). After opening the pacemaker pocket, the pulse generator was disconnected and removed and the cavity was thoroughly flushed with povidone iodine solution. During this procedure, unintentional displacement of the electrode occurred and the whole pacemaker stimulation system had to be removed. The wound was closed with a one position stitch and left to heal by secondary intention.

Weakness and syncopal episodes occurred again. The jenny was weak and lethargic, with a heart rate around 50–60 bpm and irregular rhythm. The wound was cleaned and lavaged with povidon iodine solution daily. Cefquinom and phenylbutazone were administrated for another 8 days. Wound-healing was satisfactory; the heart rhythm continued to be irregular.

Sixteen days after removal of the stimulation system, the jenny was subjected to the implantation of a new pacemaker, which was performed on the contralateral site of the neck. Surgical procedure was performed in the same way as described above. The pacemaker pocket was flushed with antibiotic solution (neomycin, bacitracin), and then closed routinely. Pacemaker Kairos S (Biotronik, Berlin, Germany) was set to VVI mode, a basal rate of 40 ppm, a pulse amplitude of 4.8 V, a pulse width of 1.0 ms and sensing amplitude of 2.5 mV. The stimulation threshold was 0.7 V at 0.5 ms and with R waves of 10 mV.

Over the following days, the clinical status of the jenny was very good; only mild swelling at the site of the surgical wound was noticed. Therefore, antibiotic and nonsteroidal medication was continued for 2 weeks. After that, the pacemaker function was analysed. Considering the high energy requirement for ventricular pacing, the pacemaker was reprogrammed to 2.4 V, 0.5 ms and 2.5 mV. The jenny was discharged 3 weeks after successful reimplantation.

3.3. Follow Up

The jenny was checked every six months to ensure the device was functional and performing appropriately. After three years, a regular check-up was done every 12 months. The jenny was used for animal-assisted therapy, prospered well and no side effects were observed.

Eight years later, during the regular check of the pacemaker, it was found that the pacemaker battery was low and a pacemaker replacement would be required. The jenny was admitted to the clinic. Its body condition was very good (BCS 7/9), with a heart rate around 40 bpm, and examination of the blood showed no abnormalities. On echocardiography, the lead tip was visible in the right ventricular apex, the lead surface was smooth and no irregularities or valvular lesions were detected on the tricuspid valve.

A surgical procedure was performed under general anaesthesia (premedication: penicillin 20,000 IU/kg i.v., gentamicin 6.6 mg/kg i.v., flunixin meglumine 1.1 mg/kg i.v., induction: xylazine 1.1 mg/kg i.v., ketamine 2.2 mg/kg i.v., diazepam 0.04 mg/kg i.v., anaesthesia maintained with isoflurane ET 1.4 vol. %). An incision was made in the skin above the pocket containing the pulse generator; the pacemaker pocket was opened and the device exposed. The lead was disconnected from the pulse generator. The old pacemaker was removed and replaced by the new one (Verity™ ADx XL SC 5056, St. Jude Medical, Saint Paul, MN, USA). The cavity was closed in layers. Transient bradycardia had occurred during pacemaker disconnection (for approximately 2 min). The jenny recovered uneventfully. Medication after the procedure included flunixin meglumine (1.1 mg/kg i.v. q24 h) for 4 days and antibiotics for 7 days (penicillin 20,000 IU/kg i.v. q8 h, gentamicin 6.6 mg/kg i.v. q24 h). Because a phlebitis developed at the site of the intravenous catheter insertion, the catheter was removed and antibiotic treatment changed to enrofloxacin (5 mg/kg p.o. q24 h). The jenny was discharged after 3 weeks. Regular follow-up visits were done at one-year intervals.

During clinical examination before a routine vaccination 18 years after the initial implantation, an irregular heart rhythm was noticed. On that account, the jenny was admitted to the clinic. Physical examination revealed lethargic behavior, bradycardia and irregular heart rate; other organ systems were normal. Hematologic parameters were within reference ranges. ECG record confirmed a complete AV block with resting ventricular frequency 24 bpm and atrial rate around 50 bpm. Echocardiographic examination did not reveal any abnormalities (Figure 2), except mild tricuspid regurgitation visible in the color-flow doppler mode.

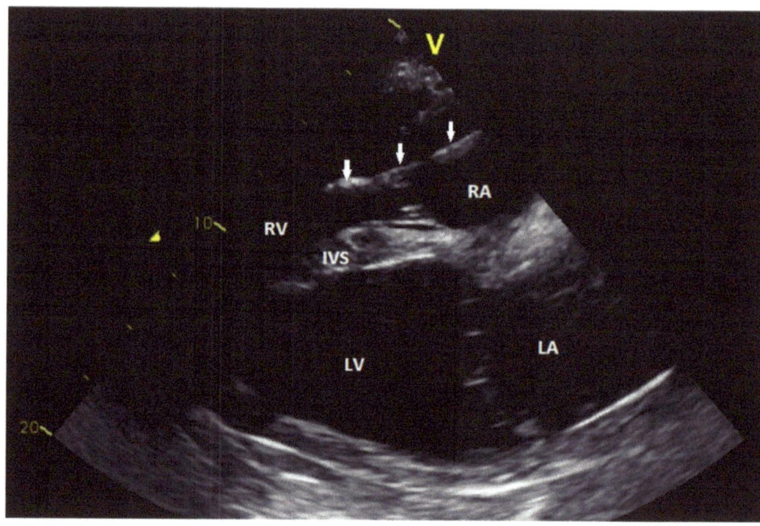

Figure 2. Echocardiographic image showing the right parasternal long-axis view. The lead (arrows) is visible in the right atrium and the right ventricle. The intraventricular part of the lead body shows slightly irregular surface (RA—rig ht atrium; RV—right ventricle; IVS—interventricular septum; LA—left atrium; LV—left ventricle).

The pacemaker function was analysed and it was found that a pulse generator exchange was needed. The pulse generator was replaced under general anaesthesia, as described above. A firm fibrous capsule had developed around the device, so it had to be opened to expose the old pulse generator. It was replaced with pacemaker Etrinsa 8 DR-T (Biotronik, Berlin, Germany). A blind connector was inserted to the atrial channel of this pacemaker. The stimulation threshold was found at 0.9 V and 0.4 ms, with R waves of 8 mV. The pacemaker was programmed to VVI mode, basic rate 45 ppm, pulse amplitude 3.0 V, pulse width 0.4 ms and sensing amplitude 2.5 mV. Before closing the pacemaker pocket, the cavity was flushed with amikacin solution. Antibiotics (enrofloxacin 5 mg/kg p.o. q24 h) were administered for 5 days and ketoprofen (2 mg/kg i.v. q24 h) for 2 days.

Eight days after the procedure, a non-painful swelling around the pacemaker pocket developed. Ultrasonographic examination revealed a seroma formation at the site of the surgical wound. Non-steroidal anti-inflammatory gel was applied topically and enrofloxacin was given orally for 2 weeks. The stitches were removed on the ninth day after surgery. The wound was clean, without secretion, and the swelling gradually subsided. Before discharging the jenny, a cardiologic examination was performed (Figure 3). The pacemaker worked properly; the jenny was reliant on pacemaker stimulation from 86%. Battery lifespan was estimated to be 9 years.

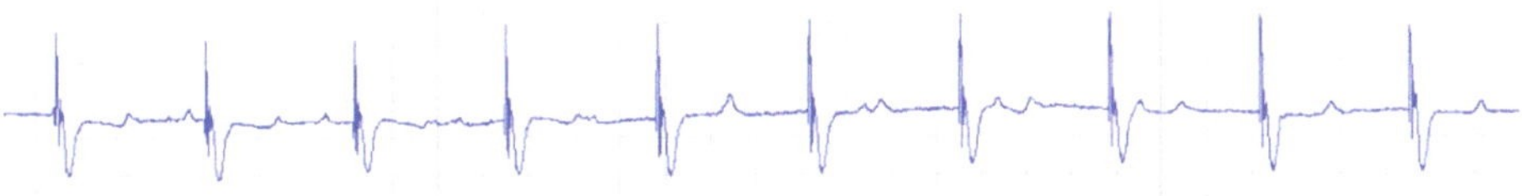

Figure 3. Base apex electrocardiogram with ventricular pacing shows regular rhythm with ventricular rate of 45 bpm. Ventricular pacing spikes precede each QRS complex (25 mm/s, 1 cm/mV).

4. Discussion

In horses, a complete AV block is uncommon and has been associated with inflammatory or degenerative diseases of the myocardial conducting tissues [10,15]. Pibarot et al. [5] reported a complete AV block in an 8-month-old female Jerusalem donkey, which started to have syncopal episodes at the age of 5 months. They assumed congenital aetiology in this particular case, because of the early appearance of the symptoms and the absence of clinical, haematological, biochemical and echocardiographic signs of chronic or acute cardiac disease. A familiar history of syncope was reported by Decloedt [16] in one donkey with several siblings suffering from episodes of collapse, suggesting a hereditary syndrome. Congenital aetiology could also be considered in our patient, when the onset of clinical signs started at the age of 4 months. However, partial improvement after phenylbutazone treatment may be indicative of inflammatory aetiology, although no clinical signs of any infectious disease were present. Serum troponin I measurement and other diagnostic tests were not performed due to strict financial constraints.

Symptomatic bradydysrhythmias, such as third-degree AV block, are the major indications for permanent pacemaker implantation. A transvenous implantation technique in horses was described in detail by van Loon at al. [2]. Although the standing position is preferably used in horses [2,4], we decided to perform the surgery under general anaesthesia in lateral recumbency to ensure a better surgical approach and to avoid a collapse of the patient during the procedure. The jugular vein was used for lead insertion. Positioning of the lead was guided by echocardiography and the correct position was confirmed by measuring the electrical characteristics of the lead.

Temporary transvenous pacing is used to avoid a syncope prior to the permanent implantation of a pacemaker [5,17]. Unfortunately, a temporary pacing system was not available in our case, and the lack of this equipment resulted in episodes of asystole during the first surgical intervention.

Although the transvenous implantation of a cardiac pacemaker is a relatively safe and simple procedure, there could be complications associated with it. Lead displacement is one of the most common major complications in both dogs and humans [7,11,13,18–20]. In our patient, lead dislodgement occurred in the early postoperative period and required correction of the lead position because of loss of capture. Thoracic radiography or fluoroscopy is used in small animals to verify a change in lead position [17]. Capture failure in the jenny was confirmed by surface ECG and the measurement of sensing parameters. The reintervention was performed nine days after the initial pacemaker implantation.

Another major complication in our patient was pacemaker pocket infection. Swelling and secretion in the surgical wound occurred a few days after lead reposition. The jenny developed a fever and became lethargic. Despite prolonged antibiotic treatment, the pacemaker infection worsened. *Staphylococcus aureus* was isolated from the wound secretion.

Cardiac implantable device infection is a serious problem even in human medicine [13,21]. Early reintervention is reported as a strong risk factor for later development of device infection [13]. Bacterial inoculation often occurs as a result of bacterial colonisation of the operative site at the time of pacemaker implantation. The *Staphylococcus* species from the skin may contaminate the wound, most likely during pocket formation, which later causes pocket infection [13]. The reported rate of infection in dogs varies from 1% to 5% [7,18–20,22], and *Staphylococcus* spp. is also the most commonly cultured isolate [20,22]. Treatment of the pacemaker pocket infection in our patient included complete system removal and prolonged antibiotic therapy. The new implantation was performed contralaterally, after complete wound-healing.

The jenny was used for animal-assisted therapy and was only ridden by children. There was no problem with this low-intensity exercise, although the VVI mode does not allow rate-adaptive pacing. A single-chamber pacemaker was sufficient for this type of mild exercise and prevented syncopes in our patient. The jenny was never used for breeding.

Follow-up visits to check the pacemaker function should be performed regularly. Pacemaker batteries are designed to have a predictable lifespan, which can be monitored by their cell voltage and cell impedance. The lifespan of a pulse generator is largely dependent on percent pacing, programmed voltage and pulse width and electrical pacing impedances. Average clinical longevity in human VVI pacemakers is reported to be around 7 years [23,24]. In our patient, the pacemaker checks were performed once a year. Surface ECG and pacing and sensing parameters of the pacemaker were measured. However, before the second pacemaker exchange, we missed the pulse generator end-of-life and the battery was completely discharged. Therefore, it would be advisable to shorten the intervals of the follow-up visits towards the end of the battery life.

The pulse generator exchange was performed twice at nine-year intervals. The procedure was associated with a localised reaction around the pacemaker device, which was resolved within a few days. Seroma formation is reported as a common minor complication in dogs and horses, and usually requires no treatment [2,7,18,19].

Long-term follow-up revealed no abnormalities on echocardiographic examination and successful pacing and sensing of the ventricle were obtained.

5. Conclusions

This case report describes a successful transvenous pacemaker implantation in a jenny with a complete AV block. Major complications associated with the procedure were lead dislodgement and pacemaker pocket infection. However, the long-term outcome was satisfactory. The jenny showed improvement in heart function and quality of life after pacemaker implantation.

Regular pacemaker checks should be undertaken to evaluate pacemaker function and battery condition. A more frequent follow up is necessary as the device approaches the end of its battery life.

Author Contributions: Conceptualization, M.S.; methodology, M.N.; validation, M.N.; investigation, R.K., F.K. and P.M.; resources, R.K. and F.K.; writing—original draft preparation, P.M.; writing—review and editing, M.S. and P.M.; supervision, M.S. and M.N. All authors have read and agreed to the published version of the manuscript.

Funding: This research received no external funding.

Institutional Review Board Statement: Ethical review and approval were waived for this study as the animal in the study was patient of the clinic and described methods were part of the therapy.

Data Availability Statement: Not applicable.

Conflicts of Interest: The authors declare no conflict of interest.

References

1. Reef, V.B.; Clark, E.S.; Oliver, J.A.; Donawick, W.J. Implantation of a permanent transvenous pacing catheter in a horse with complete heart block and syncope. *J. Am. Vet. Med. Assoc.* **1986**, *189*, 449–452.
2. Van Loon, G.; Fonteyne, W.; Rottiers, H.; Tavernier, R.; Jordaens, L.; D'Hont, L.; Copaert, R.; De Clercq, T.; Deprez, P. Dual-chamber pacemaker implantation via the cephalic vein in healthy equids. *J. Vet. Intern. Med.* **2001**, *15*, 564–571. [CrossRef]
3. Van Loon, G.; Fonteyne, W.; Rottiers, H.; Tavernier, R.; Deprez, P. Implantation of a dual-chamber, rate-adaptive pacemaker in a horse with suspected sick sinus syndrome. *Vet. Rec.* **2002**, *151*, 541–545. [CrossRef]

4. Petchdee, S.; Chanda, M.; Cherdchutham, W. Pacemaker implantation in horse with bradycardia-tachycardia syndrome. *Asian J. Anim. Vet. Adv.* **2018**. [CrossRef]
5. Pibarot, P.; Vrins, A.; Salmon, Y.; Difruscia, R. Implantation of a programmable atrioventricular pacemaker in a donkey with complete atrioventricular block and syncope. *Equine Vet. J.* **1993**, *25*, 248–251. [CrossRef]
6. De Lange, L.; Van Steenkiste, G.; Vera, L.; De Clercq, D.; Devloedt, A.; Cromheeke, K.M.C.; Van Loon, G. First successful applications of closed loop stimulation pacemakers with remote monitoring in two syncopal miniature donkey. In Proceedings of the 12th European College of Equine Internal Medicine congress (ECEIM 2019), Valencia, Spain, 20–23 November 2019; Wiley: Hoboken, NJ, USA, 2019.
7. Oyama, M.A.; Sisson, D.D.; Lehmkuhl, L.B. Practices and outcome of artificial cardiac pacing in 154 dogs. *J. Vet. Intern. Med.* **2001**, *15*, 229–239. [CrossRef]
8. Van Loon, G. *Dysrhythmias: Cardiac pacing and electrical cardioversion In Cardiology of the Horse*, 2nd ed.; Marr, C.M., Bowen, M., Eds.; Elsevier: Amsterdam, The Netherlands, 2010; pp. 179–192.
9. Santilli, R.A.; Giacomazzi, F.; Porteiro Vázquez, D.M.; Perego, M. Indications for permanent pacing in dogs and cats. *J. Vet. Card.* **2019**, *22*, 20–39. [CrossRef] [PubMed]
10. Keen, J.A. Pathological bradyarrhythmia in horses. *Vet. J.* **2020**, *259–260*, 105463. [CrossRef]
11. Pakarinen, S.; Oikarinen, L.; Toivonen, L. Short-term implantation-related complications of cardiac rhythm management device therapy: A retrospective single-centre 1-year survey. *EP Eur.* **2010**, *12*, 103–108. [CrossRef] [PubMed]
12. Genovese, D.W.; Estrada, A.H.; Maisenbacher, H.W.; Heatwole, B.A.; Powell, M.A. Procedure times, complication rates and survival times associated with single-chamber versus dual-chamber pacemaker implantation in dogs with clinical signs of bradyarrhythmia: 54 cases (1004–2009). *J. Am. Vet. Med. Assoc.* **2013**, *242*, 230–236. [CrossRef]
13. Klug, D.; Balde, M.; Pavin, D.; Hidden-Lucet, F.; Clementy, J.; Sadoul, N.; Rey, J.L.; Lance, G.; Lazarus, S.; Victor, J.; et al. PEOPLE study Group. Risk factors related to infections of implanted pacemakers and cardioverter-defibrillators: Results of a large prospective study. *Circulation* **2007**, *116*, 1349–1355. [CrossRef]
14. Hamir, A.N.; Reef, V.B. Complications of a Permanent Transvenous Pacing Catheter in a Horse. *J. Comp. Pathol.* **1989**, *101*, 317–326. [CrossRef]
15. Reef, V.B.; Marr, C.M. Dysrhythmias: Assessment and medical management. In *Cardiology of the Horse*, 2nd ed.; Marr, C.M., Bowen, M., Eds.; Elsevier: Amsterdam, The Netherlands, 2010; pp. 159–178.
16. Decloedt, A. Prevalence of cardiovascular disease in donkeys. *Vet. Rec.* **2016**, *15*, 382–383. [CrossRef]
17. Petrie, J.P. Permanent transvenous cardiac pacing. *Clin. Tech. Small Anim. Pract.* **2005**, *20*, 164–172. [CrossRef] [PubMed]
18. Wess, G.; Thomas, W.P.; Berger, D.M.; Kittleson, M.D. Applications, complications, and outcomes of transvenous pacemaker implantation in 105 dogs (1997-2002). *J. Vet. Intern. Med.* **2006**, *20*, 877–884. [CrossRef]
19. Hildebrandt, N.; Stertmann, W.A.; Wehner, M.; Schneider, I.; Neu, H.; Schneider, M. Dual chamber pacemaker implantation in dogs with atrioventricular block. *J. Vet. Intern. Med.* **2009**, *23*, 31–38. [CrossRef]
20. Swanson, L.E.; Huibregtse, B.A.; Scansen, B.A. A retrospective review of 146 active and passive fixation bradycardia lead implantations in 74 dogs undergoing pacemaker implantation in a research setting of short term duration. *BMC Vet. Res.* **2018**, *14*, 112. [CrossRef] [PubMed]
21. Kirkfeldt, R.E.; Johansen, J.B.; Nielsen, J.C. Management of cardiac electronic device infections: Challenges and outcomes. *Arrhythm. Electrophysiol. Rev.* **2016**, *5*, 183–187. [CrossRef]
22. Fine, D.M.; Tobias, A.H. Cardiovascular device infections in dogs: Report of 8 cases and review of the literature. *J. Vet. Intern. Med.* **2007**, *21*, 1265–1271. [CrossRef]
23. Katz, D.; Akiyama, T. Pacemaker longevity: The world's longest-lasting VVI pacemaker. *Ann. Noninvasive Electrocardiol.* **2007**, *12*, 223–226. [CrossRef] [PubMed]
24. Hauser, R.G.; Hayes, D.L.; Kallinen, L.M.; Cannom, D.S.; Epstein, A.E.; Almquist, A.K.; Song, L.S.; Tyers, G.F.; Vlay, S.C.; Irwin, M. Clinical experience with pacemaker pulse generators and transvenous leads: An 8-year prospective multicenter study. *Heart Rhythm* **2007**, *4*, 154–160. [CrossRef] [PubMed]

animals

MDPI

Article

Evaluation of Different Methods to Estimate the Transfer of Immunity in Donkey Foals Fed with Colostrum of Good IgG Quality: A Preliminary Study

Luca Turini [1,2,3,*], Francesca Bonelli [1,3], Irene Nocera [1], Valentina Meucci [1], Giuseppe Conte [3] and Micaela Sgorbini [1,3]

1 Department of Veterinary Sciences, University of Pisa, 56124 Pisa, Italy; francesca.bonelli@unipi.it (F.B.); irene.nocera@vet.unipi.it (I.N.); valentina.meucci@unipi.it (V.M.); micaela.sgorbini@unipi.it (M.S.)
2 Istituto Zooprofilattico Sperimentale del Lazio e della Toscana 'M. Aleandri', 00178 Rome, Italy
3 Centro di Ricerche Agro-ambientali "E. Avanzi", University of Pisa, 56122 Pisa, Italy; giuseppe.conte@unipi.it
* Correspondence: luca.turini@phd.unipi.it; Tel.: +39-347-102-2307

Simple Summary: Little is known about the passive transfer of immunity in donkey foals and about the different types of analysis that can be performed to assess it. The aims of the present study were to evaluate the correlation between IgG Serum Radial Immunodiffusion, Electrophoresis Gamma Globulins, Electrophoresis Total Protein and the Serum Total Protein analyzed by Refractometry and by Dry Chemistry Analyzer (Biuret) and to estimate serum IgG concentrations using Serum TP. IgG Serum Radial Immunodiffusion showed a good correlation with Electrophoresis Gamma Globulins and a high correlation with Total Protein Electrophoresis, Biuret and Refractometry. All the tests performed may be a useful to estimate the serum IgG in donkey foals' blood in the first day of life using a specific equation.

Abstract: The aims of the present study were to evaluate the correlation between IgG Serum Radial Immunodiffusion (SRID), Electrophoresis Gamma Globulins (EGG), Electrophoresis Total Protein (ETP) and the serum total protein (TP) analyzed by refractometry and by a dry chemistry analyzer (Biuret) and to estimate serum IgG concentrations using serum TP. A total of 36 samples collected at four different times (birth, 6, 12, 24 h after birth) from nine Amiata donkey foals were evaluated with SRID, EGG, ETP, serum TP Biuret and refractometry. SRID IgG concentration increased significantly over time until T12. Serum TP analyzed with refractometry, electrophoresis and Biuret showed a statistically significant difference between T0 and T6 vs. T12 and T24. A good or strong correlation was found between different tests performed. Equations to quantify serum IgG were created and can be used for estimating the donkey foals' serum IgG in the first day of life. Serum TP refractometry showed a high correlation with SRID IgG (0.91) which may be a particularly useful and economic instrument to estimate the transfer of immunity in donkey foals during the first day of life. Further studies evaluating a high number of animals are needed in order to set specific cut-off values.

Keywords: donkey foals; serum IgG; serum total protein; radial immunodiffusion; refractometer; electrophoresis

Citation: Turini, L.; Bonelli, F.; Nocera, I.; Meucci, V.; Conte, G.; Sgorbini, M. Evaluation of Different Methods to Estimate the Transfer of Immunity in Donkey Foals Fed with Colostrum of Good IgG Quality: A Preliminary Study. *Animals* **2021**, *11*, 507. https://doi.org/10.3390/ani11020507

Academic Editors: Marina Von Keyserlingk and Amy McLean

Received: 6 January 2021
Accepted: 13 February 2021
Published: 15 February 2021

1. Introduction

Despite donkey foals being born with a higher level of circulating antibodies compared to the equine foals, they still required a good amount of high quality colostrum due to the type of placentation [1–5]. Donkey placenta is diffuse and epitheliochorial with numerous microplacentomes consisting of a fetal microcotyledonary and a maternal microcaruncular part [1]. A well-managed passive transfer of immunity allows the donkey foal to achieve an adequate serum IgG concentration at 24h of life. The most common consequences connected with low quality colostrum in neonatal foals is sepsis which can manifest in several different ways, such as bacteremia, pneumonia, enterocolitis, omphalitis/omphalophlebitis, meningoencephalitis and septic arthritis [6]. Despite the relevance of these topics, very few studies evaluated the peripartum field in donkey and no cut-off values have been reported for the quantification of the transfer of immunity in donkey foals. Based on data about equine foals, transfer of immunity is considered to be adequate when serum IgG are $\geq$ 8 g/l, while partial failure of transfer is considered with serum IgG levels between 4 and 8 g/L and total failure of transfer with < 4 g/L [7–10]. An early detection of partial or total failure of transfer of immunity is crucial for the health of the donkey foals [7–9].

In equine foals, several tests were used to assess the transfer of immunity and the IgG concentrations evaluated by the single radial immunodiffusion (SRID) is considered the gold standard test. However, the SRID test is expensive, needing 18–24 h to obtain results and it is susceptible to human error during the measurement of the ring diameter. Results can also vary between the different commercial kits used [11–13]. Electrophoresis can be an appropriate test for the evaluation of the transfer of immunity, but it is rarely used. Serum electrophoresis results need only few hours for being ready and they do not differ between different kits as SRID. Moreover, a recent study showed a good correlation between SRID and electrophoretic gamma globulins [14]. Other tests used which provide quick results are the zinc sulfate turbidity, the glutaraldehyde coagulation, and the ELISA-based tests, such as the Snap Foal IgG test [13,15]. Refractometry represents an inexpensive, rapid, and accurate test for the failure of transfer of passive immunity (FTPI) evaluation [16–19].

The use of refractometry for serum total protein (TP) evaluation is mainly used in dairy farms for calves' immunity transfer assessment [16] while it is rarely used in foals as an indicator of transfer of immunity [17,18]. Only a few studies have been done to evaluate the use of the refractometry as indicator of FTPI in foals because the value of serum TP can be overestimated by high levels of glucose, urea and plasma lipids [19]. Novel, more accurate and less influenced methods to quantify serum TP has been found recently; serum IgG could be estimated with the use of serum protein concentrations measured with an automated chemistry analyzer in foals [17]. Compared to serum TP evaluated with a refractometer the automatic analyzer presented the advantage of not being influenced by the concentrations of plasma lipids (cholesterol, lipoproteins), glucose, urea, or excessive ethylenediaminetetracetic acid (EDTA). The disadvantages are the cost and the low feasibility in the field [17].

Despite the fact that the literature is wide for equine foals, there is a lack of knowledge about the assessment of transfer of immunity in donkey foals. Studies in different fields of donkey foal's neonatology [5,20,21] showed that differences between these two species exist and are crucial. These considerations make the investigation about other methods, such as electrophoresis, refractometry and Biuret analysis of some potential interest for the rapid assessment of passive transfer of maternal antibodies in neonatal donkey foals. Thus, the aims of the present study were to assess IgG and serum TP at different sampling times, to evaluate the correlation between IgG SRID, Electrophoresis Gamma Globulins (EGG), Electrophoresis Total Protein (ETP) and the serum TP analyzed by refractometry and by dry chemistry analyzer (Biuret) and to estimate serum IgG concentrations using serum TP.

2. Materials and Methods

2.1. Animals

Nine Amiata donkey foals were included in this study. All the donkey foals belong to the Regional studfarm "Le Bandite di Scarlino" (Grosseto, Italy). The Ethical Committee of the University of Pisa (Organismo Preposto Benessere Animale, OPBA) approved the study with protocol number 22/19. An owner's written consent was also obtained. All the donkey foals were born at the Veterinary Teaching Hospital "Mario Modenato", Department of Veterinary Sciences, University of Pisa during the foaling season 2019–2020. Immediately after birth all the donkey foals were evaluated with a dedicated APGAR score [22]. All the donkey foals were fed by their own dam with a good quality of colostrum within the first 3 h of life [23,24]. Colostrum has been evaluated as reported in Turini et al. 2020 [3]. The mean ± standard deviation of the colostrum IgG concentration of the 9 Jennies before the foal's nursing was 89.32 ± 26.43 g/L. A complete physical examination as reported in [25] was performed at each sampling time in order to assess the health status of the donkey foals. Special attention was made to those parameters evaluating the absence of infection.

2.2. Sampling Procedures

Ten mL of blood were collected immediately after birth, before the first colostrum feeding, (T0), and at 6 (T6), 12 (T12) and 24 (T24) h after birth from jugular vein of each donkey foal in order to assess IgG and TP concentrations. Samples were harvested in red-top Vacutainer tubes (10-mL BD Vacutainer glass serum tube, silicone-coated; Becton Dickinson and Co., Franklin Lakes, NJ, USA) and immediately centrifuged (Legend RT, Sorvall; ThermoFisher Scientific Inc., Waltham, MA, USA) at 1.565× *g* for 15 min in order to collect the serum. The serum samples were then stored at −20 °C until the evaluation. The day of the analysis, the serum samples were defrosted at room temperature and vortexed for 10 s just before the dilution.

2.3. Determination of Serum IgG and Total Protein

Studies have shown a strong homologue between donkey IgG and horse IgG [26,27]. Because of the lack of a specific SRID for the donkey, and because of the close similarities of the immune systems of horses and donkeys, a SRID assay specific for horse was used to analyze IgG colostrum concentrations [26,27]. All the serum samples were analyzed in a single batch. SRID was performed using a Horse IgG IDRing (R) Test (IDBiotech, Issoire, France). Test results were determined by comparison with a standard curve prepared using equine immunoglobulin standards (25, 50, 100 and 200 μg/mL) supplied with the Kit. Serum samples were diluted 1/150 before IgG determination, as suggested by the manufactory instruction, because of their high IgG concentrations.

Serum TP was measured using a temperature-compensating digital refractometer (AR200; Reichert Analytical Instruments, Reichert Inc., Depew, NY, USA). Electrophoresis gamma globulin (EGG) and electrophoresis total protein (ETP) have been analyzed with a fully automated electrophoresis instrument (Pretty Interlab, Sebia Company, Rome, Italy). Concentrations of TP were also measured using a dry chemistry analyser (SAT450, Assel, Aprilia, Italy) by biuret assay according to the manufacturer's instructions.

2.4. Statistic Analysis

A G-power analysis was performed in order to calculate the minimum number of animals that should be included. G-power analysis showed that the minimum number of animals needed was 4, considering an effect size (Cohen coefficient) of 0.8, an alpha value of 0.05 and a power of 0.8.

Data for normal distribution were evaluated by the Shapiro-Wilk test. Data on the IgG SRID, EGG, TP Biuret, TP Refractometry and ETP were analyzed by the following linear model, using JMP software (SAS Institute Inc., Cary, NC, USA):

$$y_{ijz} = \mu + T_i + A_z + \varepsilon_{ijz}$$

where y_{ijz} = dependent variables; Ti = fixed effect of the ith time of sampling (T0, T6, T12, T24); A_z = random effect of the zth animal (9 levels); ε_{ijz} = random residual.

Least-square means with their standard errors were reported, and treatment effects were declared significant at $p < 0.05$.

Correlations between IgG SRID, EGG, TP Biuret, TP Refractometry and ETP were evaluated calculating Pearson's coefficient; only correlations with a *p*-value of the linear model below 0.05 were considered as significant.

The linear contrasts were tested in the first model by the t-test with Tukey's adjustment within each parity level.

3. Results

A total of 36 samples (four sampling times × nine donkey foals) were analyzed. None of the included donkey foals showed signs of disease or discomfort during the time of the study. The temperature of the donkey foals evaluated of each sampling time was always within the normal range [25]. Data were normally distributed.

The mean ± standard deviation for each method used at different collection time were reported in Table 1.

Table 1. Results concerning serum IgG concentrations or total protein concentration evaluated by different methods, expressed in mean and standard deviation, in a population of nine donkey foals (total of 36 samples) at different collection time. Legend: IgG—immunoglobulin type G; SRID—serum radial immunodiffusion; EGG—electrophoresis gamma globulin; TP—total protein; ETP—electrophoresis total protein. Legend: A ≠ B ≠ C $p < 0.05$ between values on the same column; *** = p-value < 0.001.

Hours Post-Partum.	*n*	IgG SRID (g/dL)	EGG (g/dL)	TP Biuret (g/dL)	TP Refractometry (g/dL)	ETP (g/dL)
0	9	0.9 ± 0.12 ^{C}	0.28 ± 0.15 ^{A}	4.44 ± 0.31 ^{B}	4.29 ± 0.33 ^{B}	4.43 ± 0.32 ^{B}
6	9	1.18 ± 0.08 ^{B}	0.41 ± 0.25 ^{A}	4.80 ± 0.24 ^{B}	4.67 ± 0.26 ^{B}	4.94 ± 0.44 ^{B}
12	9	1.49 ± 0.16 ^{A}	0.48 ± 0.31 ^{A}	5.58 ± 0.39 ^{A}	5.44 ± 0.36 ^{A}	5.59 ± 0.38 ^{A}
24	9	1.49 ± 0.20 ^{A}	0.57 ± 0.25^A	5.57 ± 0.44 ^{A}	5.46 ± 0.42 ^{A}	5.59 ± 0.45 ^{A}
All time points	36	1.26 ± 0.29	0.43 ± 0.26	5.1 ± 0.6	4.96 ± 0.61	5.14 ± 0.62
SE		0.049	0.082	0.118	0.116	0.134
p-value		***	Ns	***	***	***

Correlation analysis results and p-value between the different methods used for the evaluation of donkey foals' serum immunoglobulins and TP were reported in Table 2.

Table 2. Values concerning the coefficient of correlation between different methods for the evaluation of immunoglobulins type G and serum total protein in a population of nine donkeys. Legend: IgG—immunoglobulin type G; SRID—serum radial immunodiffusion; EGG—electrophoresis gamma globulin; TP—total protein; ETP—electrophoresis total protein; ** = p-value between 0.01 and 0.001; *** = p-value < 0.001.

Methods	IgG SRID (g/dL)	EGG (g/dL)	TP Biuret (g/dL)	TP Refractometry (g/dL)	ETP (g/dL)
IgG SRID (g/dL)	1.00	** 0.56	*** 0.89	*** 0.91	*** 0.87
EGG (g/dL)		1.00	** 0.50	** 0.52	** 0.59
TP Biuret (g/dL)			1.00	*** 0.99	*** 0.94
TP Refractometry (g/dL)				1.00	*** 0.95
ETP (g/dL)					1.00

A strong significant relation between serum IgG SRID and serum TP Biuret ($p < 0.0001$, $R^2 = 0.80$) has been found. The value of serum IgG can be calculated using a serum TP Biuret with the following formula: SRID IgG (g/dL) = −0.90 + 0.42 × TP Biuret value (g/dL). Additionally, the relation between serum IgG SRID and serum TP Refractometry was highly significant ($p < 0.0001$, $R^2 = 0.82$). The value of serum IgG can be calculated using a serum TP Refractometry with the following formula: SRID IgG (g/dL) = −0.85 + 0.43 × TP Refractometry value (g/dL). A strong significant relation between serum IgG SRID and serum ETP ($p < 0.0001$, $R^2 = 0.76$) has been found. Serum IgG can be calculated using a serum TP Biuret with the following formula: SRID IgG (g/dL) = −0.80 + 0.4 × ETP value (g/dL). The confidence interval of each methods investigated was reported in Figure 1.

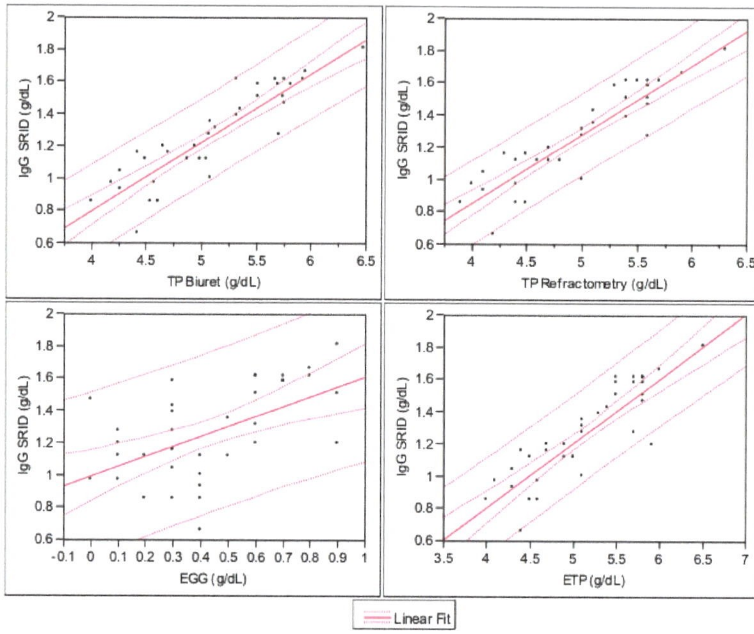

Figure 1. Confidence interval of each methods investigated. Internal dashed lines = Prediction lines. External dashed lines = Confidence lines. Legend: IgG SRID—immunoglobulin type G evaluated by the single radial immunodiffusion; TP Refractometry– total protein evaluated by refractometry; EGG—electrophoresis gamma globulin; ETP—electrophoresis total protein.

4. Discussion

In foals, SRID is considered the gold standard method to quantify serum IgG concentrations [11,28,29]. However, the SRID method could have a high individual error based on different interpretations of the precipitin ring. Additionally, results can be influenced by time of incubation and temperature which might be different based on the SRID assays from varies manufacturers [11,13,29]. Moreover, due to the lack of studies about validation of SRID analysis in donkey foals, this technique could not represent the gold standard one for this species. The aims of the present study were to assess IgG and serum TP at different sampling times, to evaluate the correlations between IgG SRID, EGG, ETP and the serum TP analyzed by refractometry and by dry chemistry analyzer and to estimate serum IgG concentration using serum TP.

SRID IgG concentration increased significantly over time until T12, remaining stable between T12 and T24. Serum TP analyzed with refractometry, electrophoresis and dry chemistry analyzer showed a statistically significant difference between T0 and T6 vs. T12 and T24, while no differences were found between T0 vs. T6 and T12 vs. T24. Studies performed in equine foals recommended determining serum IgG levels at around 18 h of life [30] or even earlier [31]. The small intestine remains permeable to macromolecules, most importantly IgG, during the first 18–24 h after birth allowing the absorption of ingested colostrum antibodies [32]. The most important risk factors for FPTI in the foal include the delay in suckling time, feeding foal with a low quality of colostrum or the lack of absorption of the IgG [32]. Thus, knowing as early as possible the level of passive transfer of immunity in donkey foals leads to a prompt veterinary intervention in case of need and to a better prognosis [32]. From the results of the present study, we can assume that an early assessment of transfer of immunity made by total protein evaluation may be done at around 12 h of life even in donkey foals [30,31].

Gamma globulins evaluated by electrophoresis showed an increasing pattern over time, however no statistically significant differences were found between different sampling times. These results were in line with equine foals in which EGG concentration increased during the first hour of life, especially between 4 and 8 h after birth [31]. However, the lack of a statistically significant increase over time is surprising and seems in contrast with other findings [15]. Despite the number of animals included in the present study being in line with what was requested by the G-power analysis, increasing the population for further studies may be indicated. Additionally, EGG performed in donkeys showed a separation of IgGa from IgGb, while horses presented IgGab and IgGc subtypes [27]. This characteristic related to the species may have influenced our results.

IgG SRID showed a good correlation with EGG and a high correlation with TP Biuret, TP Refractometry and ETP. Very few studies reported methods comparison for the evaluation of transfer of immunity in equine foals, and the differences between methods and statistical analysis approach used made it difficult to compare results between papers. The coefficient of correlation between IgG SRID and TP Refractometry was slightly higher in donkey foals (0.91) compared to equine foals (0.85 and 0.73, respectively) [11,18]. The correlation between EGG and TP Biuret was lower (0.50) in our donkey foals population compared to a study performed in equine foals (0.93) [14]. This difference may be due to the evaluation of TP Biuret made in the present study instead of the Biuret gamma globulins performed in the research of the Swiss colleagues. This difference in methods may explain the discrepancy between results. The same study also compared IgG SRID to total globulins made by Biuret analysis which found a correlation of 0.79. Our results showed a correlation of 0.89 between IgG SRID and TP Biuret leading to preferring the evaluation of total protein compared to total globulins for immunity transfer assessment in donkey foals [14].

Results from the present study showed that all the methods used for the serum TP assessment presented a high correlation between each other and could be used to estimate serum IgG with the formula found. However, due to the high dispersion around the regression curve the prediction of plasma IgG concentration can only be made within a large prediction interval. The predominant immunoglobulin in equids colostrum is IgG, and FTPI is assessed by the evaluation of IgG in foals' serum during the first hours of life [33–35]. As previously said, this is possible only by using laboratory techniques which may not be feasible for the owners under field conditions. The evaluation of serum TP levels is usually considered more practical because is cheap, ready to use and not influenced by operator skills, or environmental conditions (e.g., temperature, humidity, etc.) [19]. Due to these characteristics, TP evaluated by refractometer is a technique used by farmers for the evaluation of ruminant neonates [16]. Different equations have been calculated to estimate serum IgG from total proteins value obtained by different methods. These equations could be very suitable for owners or clinicians especially when there is no possibility to evaluate IgG levels by using the gold standard technique.

5. Conclusions

This study suggests a good correlation between IgG SRID, EGG, Biuret TP, Refractometry TP and ETP. Based on our results, serum TP refractometry showed the highest correlation with SRID IgG, along with TP Biuret and ETP. Due to its feasibility in the field, TP refractometry may be a particularly useful and economic instrument to estimate the transfer of immunity in donkey foals' during the first day of life. Further studies evaluating a higher number of animals would be needed in order to set specific cut-off values.

Author Contributions: Conceptualization, methodology, writing—original draft preparation L.T. and F.B. Methodology I.N. and V.M. Statistical analysis G.C. Conceptualization, methodology and writing—review and editing M.S. All authors have read and agreed to the published version of the manuscript.

Funding: This research received no external funding.

Institutional Review Board Statement: The study was conducted according to the guidelines of the Declaration of Helsinki, and approved by the Ethics Committee of the University of Pisa (Organismo Preposto Benessere Animale, OPBA) (protocol code 22/19, 31/05/2019).

Informed Consent Statement: Not applicable.

Data Availability Statement: The data presented in this study are available in article.

Conflicts of Interest: The authors declare no conflict of interest.

References

1. Saber, A.; Abd-Elnaeim, M.; Hembes, T.; Pfarrer, C.; Salim, A.; Leiser, R. Light and Scanning Electron Microscopic Study on the Blood Vascular System of the Donkey Placenta. *Anat. Histol. Embryol.* **2008**, *37*, 86–94. [CrossRef] [PubMed]
2. Carluccio, A.; Panzani, S.; Tosi, U.; Riccaboni, P.; Contri, A.; Veronesi, M.C. Morphological features of the placenta at term in the Martina Franca donkey. *Theriogenology* **2008**, *69*, 918–924. [CrossRef] [PubMed]
3. Turini, L.; Bonelli, F.; Nocera, I.; Battaglia, F.; Meucci, V.; Panzani, D.; Mele, M.; Sgorbini, M. Evaluation of jennies' colostrum: IgG concentrations and absorption in the donkey foals. A preliminary study. *Heliyon* **2020**, *6*, e04598. [CrossRef]
4. Veronesi, M.C.; Dall'Ara, P.; Gloria, A.; Servida, F.; Sala, E.; Robbe, D. IgG, IgA, and lysozyme in Martina Franca donkey jennies and their foals. *Theriogenology* **2014**, *81*, 825–831. [CrossRef]
5. Girardi, A.M.; Toledo, C.Z.P.; Sabes, A.F.; Oliveira, J.A.; Marques, L.C. Serum protein fractions, hematological and biochemical variables of the Pêga donkey (Equus asinus) breed in the first year of age. *Vet. Arhiv.* **2016**, *86*, 335–349.
6. Taylor, S. A review of equine sepsis. *Equine Vet. Educ.* **2015**, *27*, 99–109. [CrossRef]
7. Hines, M.T. Immunodeficiencies of foals. In *Current Therapy in Equine Medicine*, 5th ed.; Robinson, N.E., Ed.; WB Saunders: Philadelphia, PA, USA, 2003; pp. 692–697.
8. Koterba, A.M.; Brewer, B.; Drummond, W.H. Prevention and control of infection. *Vet. Clin. N. Am. Equine Pract.* **1985**, *1*, 41–50. [CrossRef]
9. Morris, D.D.; Meirs, D.A.; Merryman, G.S. Passive transfer failure in horses: Incidence and causative factors on a breeding farm. *Am. J. Vet. Res.* **1985**, *46*, 2294–2299. [PubMed]
10. Tyler-McGowan, C.M.; Hodgson, J.L.; Hodgson, D.R. Failure of passive transfer in foals: Incidence and outcome on four studs in New South Wales. *Aust. Vet. J.* **1997**, *75*, 56–59. [CrossRef]
11. Davis, D.G.; Schaefer, D.M.W.; Hinchcliff, K.W.; Wellman, M.L.; Willet, V.E.; Fletcher, J.M. Measurement of Serum IgG in Foals by Radial Immunodiffusion and Automated Turbidimetric Immunoassay. *J. Vet. Intern. Med.* **2005**, *19*, 93–96. [CrossRef] [PubMed]
12. Clabough, D.L.; Conboy, H.S.; Roberts, M.C. Comparison of four screening techniques for the diagnosis of equine neonatal hypogammaglobulinemia. *J. Am. Vet. Med. Ass.* **1989**, *194*, 1717–1720.
13. Davis, R.; Giguere, S. Evaluation of five commercially available assays and measurement of serum total protein concentration via refractometry for the diagnosis of failure of passive transfer of immunity in foals. *J. Am. Vet. Med. Ass.* **2005**, *227*, 1640–1645. [CrossRef]
14. Tscheschlok, L.; Venner, M.; Howard, J. Comparison of IgG concentrations by radial immunodiffusion, electrophoretic gamma globulin concentrations and total globulins in neonatal foals. *Equine Vet. J.* **2017**, *49*, 149–154. [CrossRef]
15. Ujvari, S.; Schwarzwald, C.; Fouché, N.; Howard, J.; Schoster, A. Validation of a Point-of-Care Quantitative Equine IgG Turbidimetric Immunoassay and Comparison of IgG Concentrations Measured with Radial Immunodiffusion and a Point-of-Care IgG ELISA. *J. Vet. Intern. Med.* **2017**, *31*, 1170–1177. [CrossRef] [PubMed]
16. Thornhill, J.B.; Krebs, G.L.; Petzel, C.E. Evaluation of the Brix refractometer as an on-farm tool for the detection of passive transfer of immunity in dairy calves. *Aust. Vet. J.* **2015**, *93*, 26–30. [CrossRef]
17. Hurcombe, S.D.; Matthews, A.L.; Scott, V.H.; Williams, J.M.; Kohn, C.W.; Toribio, R.E. Serum protein concentrations as predictors of serum immunoglobulin G concentration in neonatal foals. *J. Vet. Emerg Crit Care* **2012**, *22*, 573–579. [CrossRef] [PubMed]
18. Elsohaby, I.; Riley, C.B.; McClure, J.T. Usefulness of digital and optical refractometers for the diagnosis of failure of transfer of passive immunity in neonatal foals. *Equine Vet. J.* **2019**, *51*, 451–457. [CrossRef] [PubMed]
19. George, J.W. The usefulness and limitations of hand-held refractometers in veterinary laboratory medicine: An historical and technical review. *Vet. Clin. Pathol.* **2001**, *30*, 201–210. [CrossRef]
20. Bonelli, F.; Rota, A.; Corazza, M.; Serio, D.; Sgorbini, M. Hematological and biochemical findings in pregnant, postfoaling, and lactating jennies. *Theriogenology* **2016**, *85*, 1233–1238. [CrossRef]
21. Sgorbini, M.; Bonelli, F.; Rota, A.; Baragli, P.; Marchetti, V.; Corazza, M. Hematology and Clinical Chemistry in Amiata Donkey Foals from Birth to 2 Months of Age. *J. Equine Vet. Sci.* **2013**, *33*, 35–39. [CrossRef]
22. Bonelli, F.; Nocera, I.; Conte, G.; Panzani, D.; Sgorbini, M. Relation between Apgar scoring and physical parameters in 44 newborn Amiata donkey foals at birth. *Theriogenology* **2019**, *142*, 310–314. [CrossRef] [PubMed]
23. McCue, P.M. Colostrum banking. In *Equine Reproductive Procedures*, 1st ed.; Dascanio, J., McCue, P.M., Eds.; Wiley and Sons: Hoboken, NJ, USA, 2014; pp. 299–303.
24. Turini, L.; Nocera, I.; Bonelli, F.; Mele, M.; Sgorbini, M. Evaluation of Brix Refractometry for the Estimation of Colostrum Quality in Jennies. *J. Equine Vet. Sci.* **2020**, *92*, 103172. [CrossRef] [PubMed]

25. Nógrádi, N.; Magdesian, N.G. Physical Examination of the Neonatal Foal. In *Manual of Clinical Procedures in the Horse*, 1st ed.; Costa, L.R.R., Paradis, M.R., Eds.; John Wiley & Sons: Hoboken, NJ, USA, 2017; pp. 416–426.
26. Allen, P.Z.; Dalton, E.J. Studies on equine immunoglobulins IV. Immunoglobulins of the donkey. *Immunology* **1975**, *28*, 187–197. [PubMed]
27. Pescovitz, M.D. Immunology of horses and donkeys. In *Handbook of Veterinary Immunology*; Pastoret, P., Griebel, P., Bazin, H., Govaerts, A., Eds.; Elsevier: Amsterdam, The Netherlands, 1998; pp. 343–371.
28. Baird, A.N.; Pugh, D.G.; Rupp, G.P.; Shull, J.W.; Field, R.W. Detection of immunoglobulin G in the neonate. *J. Equine Vet. Sci* **1987**, *7*, 124–129. [CrossRef]
29. Wong, D.M.; Giguére, S.; Wendel, M.A. Evaluation of a point-of-care portable analyzer for measurement of plasma immunoglobulin G, total protein, and albumin concentrations in ill neonatal foals. *J. Am. Vet. Med. Ass.* **2013**, *242*, 812–819. [CrossRef]
30. LeBlanc, M.M. Update on passive transfer of immunglobulins in the foal. *Pferdeheilkunde* **2001**, *17*, 662–665. [CrossRef]
31. Kummer, L.L.; Govaere, J.; Egri, B. Comparison of the reliability of snap foal Ig test, Gamma-Check E test, refractometry and electrophoresis for determining the immune status of newborn foals in the first hours of life. *Acta Vet. Hung.* **2018**, *66*, 573–586. [CrossRef]
32. McKenzie, H.C., III. Disorders of Foals. In *Equine Internal Medicine*, 4th ed.; Reed, S.M., Bayly, W.M., Sellon, D.C., Eds.; Elsevier: Amsterdam, The Netherlands, 2018; pp. 1365–1459.
33. Sheoran, A.S.; Timoney, J.F.; Holmes, M.A.; Karzenski, S.S.; Crisman, M.V. Immunoglobulin isotypes in sera and nasal mucosal secretions and their neonatal transfer and distribution in horses. *Am. J. Vet. Res.* **2000**, *61*, 1099–1105. [CrossRef]
34. Aoki, T.; Chiba, A.; Itoh, M.; Nambo, Y.; Yamagishi, N.; Shibano, K.I.; Cheong, S.H. Colostral and foal serum immunoglobulin G levels and associations with perinatal abnormalities in heavy draft horses in Japan. *J. Equine Sci.* **2020**, *31*, 29–34. [CrossRef]
35. Pusterla, N.; Pusterla, J.B.; Spier, S.J.; Watson, J.L.; Puget, B. Evaluation of the SNAP foal IgG test for the semiquantitative measurement of immunoglobulin G in foals. *Vet. Rec.* **2002**, *151*, 258–260. [CrossRef]

animals

MDPI

Article

Two Hours of Separation Prior to Milking: Is This Strategy Stressful for Jennies and Their Foals?

Sharacely de Souza Farias [1,*], Ana Carolina Dierings Montechese [1], Thiago Bernardino [1], Paulo Henrique Mazza Rodrigues [2], Chiara Albano de Araujo Oliveira [3] and Adroaldo José Zanella [1]

1 Department of Preventive Veterinary Medicine and Animal Health, School of Veterinary Medicine and Animal Science, Campus "Fernando Costa", University of São Paulo, Pirassununga 13635-900, Brazil; ana.montechese@usp.br (A.C.D.M.); thiagobernardino@usp.br (T.B.); adroaldo.zanella@usp.br (A.J.Z.)
2 Department of Animal Nutrition and Production, School of Veterinary Medicine and Animal Science, Campus "Fernando Costa", University of São Paulo, Pirassununga 13635-900, Brazil; pmazza@usp.br
3 Department of Preventive Veterinary Medicine and Animal Production, School of Veterinary Medicine and Animal Science, Federal University of Bahia, Salvador 40170-110, Brazil; oliveirachiara@gmail.com
* Correspondence: shrfarias@usp.br

Simple Summary: The economic importance of donkeys has decreased in Brazil, which has led to their mass abandonment. Asinine milk production is a potential solution to the reintroduction of donkeys into the Brazilian social and economic scenario. The milk has nutraceutical properties that make it valuable for human consumption, and thus a donkey dairy industry is likely to help stop their abandonment. That said, in any such industry, the welfare of jennies maintained for milk production must be guaranteed. Few studies have been published measuring the impact of milking management on the welfare of jennies and foals, and the potential behavioural and physiological challenges it may cause. It is also unknown whether these animals adapt to the milking routine. The goal of this study was to assess the impact of separating Pêga jennies from their foals for 2 h on indicators of welfare. Animal welfare was analysed through behavioural and hormonal assessments, their potential adaptive responses and effects on milk yield. Few significant alterations were found in behaviour, salivary cortisol concentrations, or milk yield as a result of the 2-h separation, which could indicate that the welfare of the animals was not compromised; however, the adaptation of jennies and foals to separation stress remains to be fully verified. The 2-h separation period, based on the reported data, is possibly not a stressful experience for the assessed group of Pêga jennies and foals. The reported protocol, which included frequent positive interactions with the animals, may be useful to assure acceptable animal welfare levels for donkeys in small-scale dairy production settings.

Citation: de Souza Farias, S.; Montechese, A.C.D.; Bernardino, T.; Rodrigues, P.H.M.; de Araujo Oliveira, C.A.; Zanella, A.J. Two Hours of Separation Prior to Milking: Is This Strategy Stressful for Jennies and Their Foals?. *Animals* **2021**, *11*, 178. https://doi.org/10.3390/ani11010178

Received: 10 November 2020
Accepted: 7 January 2021
Published: 14 January 2021

Abstract: The goal of this study was to assess whether or not a separation period of 2 h is stressful for jennies and foals, as measured by changes in behaviour, salivary cortisol, and milk production. This study was reviewed and approved by the Committee for the Use and Care of Animals in Research (CEUA) of the School of Veterinary Medicine and Animal Science of the University of São Paulo. Fourteen multiparous Pêga jennies (245 kg average body weight) and their foals were assessed from day 45 to 135 of lactation. Dams and foals were separated for 2 h prior to milking. Behavioural assessments and saliva samples were collected before and after separation, every 15 days, resulting in 14 samples per individual animal. Behavioural states (affiliative and inactivity) and events (agonistic, abnormal, eliminative and vocalisations) of the jennies were observed during 6 min in both periods. Moreover, milk yield was measured. Few significant behavioural and salivary cortisol changes were observed, and milk yield was not affected by cortisol levels in response to the separation. The 2-h separation period, on the basis of the collected variables, did not appear to be stressful for the assessed group of Pêga jennies or foals; however, their ability to adapt to milking routine stress remains to be investigated.

Keywords: animal welfare; behaviour; donkey; cortisol; milking management

1. Introduction

Donkeys have been losing their relevance in Brazilian social and economic scenarios, having been less used in recent decades in their traditional role as animals of draft and burden. This trend can be ascribed, essentially, to the diffusion of mechanisation in agriculture [1], and the subsequent decrease in the number of donkeys used on rural properties. As a result, they have been omitted from official data and statistics collected by the Brazilian government [2], and their abandonment has been intensified, leading to increasing numbers of wandering animals with compromised welfare [3].

There are three registered Brazilians donkey breeds: Nordestino, Paulista, and Pêga [4]. Out of these breeds, the Pêga has been most developed for its genetic potential. It was developed in Brazil in 1810 [5], and is composed of medium sized donkeys, primarily bred to supply the market with mules [6]. Pêga donkeys are most commonly found in the Southeast region, and although they have a clear economic importance, productive donkey farms in Brazil are scarce and these animals are still generally left out of the social and economic scenario in the country.

In other countries, such as France and Italy, donkeys are still valued because of their milk which is used for human nutrition [7,8]. Asinine milk possesses similar chemical and organoleptic qualities to human milk [7,9], representing one of the best nourishment options, besides maternal milk, for human babies [10] that cannot be breast-fed and for consumers suffering from cow milk protein allergies [11,12]. The milk of Brazilian Pêga donkeys may have similar nutritional potential to that of Italian and French breeds, and thus their use for sustainable, high welfare, donkey dairy production is a possible means of reintroducing donkeys as an important species in the Brazilian socioeconomic scenario. The use of animals for milk production must guarantee their welfare. There is a limited number of studies regarding stressors that trigger physiological and behavioural changes, generated by the milking management in donkeys [13].

The milk storage capacity in this species is low (less than 2.5 L) [14], thus milk production is dependent on its removal from the mammary gland, generally by milking or suckling. In the latter, milk ejection is triggered by a foal's sucking, which triggers the release of oxytocin that in turn induces the contraction of myoepithelial cells [15]. Milking of jennies by humans, in terms of both human and animal safety and for optimal milk extraction, is more manageable when foals are not physically present [16].

In order to achieve efficient milking, jennies must be milked after 2 to 3 h of physical separation from their foals [17]. Long intervals between milking events may cause a rise in intra-udder pressure, inducing early cessation of glandular activity [18], due to the udder size and its low storage capacity. Therefore, donkeys may need to be milked multiple times a day [19].

The social structure of donkeys is composed of a territorial-based system [20–22], with complex hierarchies within groups [23]. The only permanent bond among donkey social structures is between jennies and their foals [24].

In precocious animals such as donkeys, the neonatal period is characterized by intense interactions between mothers and newborns, which are important for bonding and allow for the development of autonomy in the offspring, including motor, sensorial and cognitive processes [24]. In natural conditions, jennies begin approaching their foals less frequently after the first day post-partum, grazing at further distances while the foals rest and allowing them to interact with other animals [25]. During the first five days of life, donkey foals suckle every 3 to 10 min, and every 20 to 30 min by the 10th day [26]. In mule foals, the suckling frequency between 4 and 17 weeks varies between two and three bouts per hour [27].

Behavioural and social impairments have been reported in ungulates separated from their mothers for 2.5 h after birth [28,29]. The separation between jennies and foals may be stressful [30,31], and these animals could respond via behavioural and physiological changes [32]. When the restoration of homeostasis in response to a stressor is difficult, such as when animals cannot move to a more favourable environment, they may express behavioural signals such as vocalisations [33], stereotypic behaviour [34], increased inactivity [35–37] and altered social interactions [38]. Behavioural responses facilitate physiological adaptations, which may manifest via the activation of the autonomous nervous system and neuroendocrine system [39]. The stimulation of central circuits involving the amygdala, hypothalamus and periaqueductal gray (PAC) result in an increased frequency of eliminative behaviour [40], and increased releases of corticotropin-releasing factor (CRF), from the hypothalamus, and adrenocorticotropic hormone (ACTH), from the pituitary gland culminate in an increase in the secretion of glucocorticoids, such as cortisol [41], from the adrenal glands. Alveolar milk ejection could also be altered by fear or stress, due to the influence of these endocrine factors on oxytocin concentration and myoepithelial contraction [42,43].

To our knowledge, there are no scientific studies concerning the adaptive responses of jennies and foals to repeated separation and milking procedures, and so it is crucial to determine their behavioural and physiological changes when exposed to this routine. Therefore, further investigations assessing the welfare effects, if any, of milking procedures on dairy jennies and their foals are required.

This study aimed to investigate whether a separation period of 2 h in a manual milking system is stressful for Pêga jennies and foals, i.e., whether it generated changes in behaviour, or caused changes in salivary cortisol concentration and milk production, and thus, if the use of donkeys for sustainable, high welfare, donkey dairy production is a model for their economic reintroduction in Brazil. For this purpose, a manual milking protocol was proposed and implemented.

2. Animals, Materials and Methods

This study was reviewed and approved by the Committee for the Use and Care of Animals in Research (CEUA) of the School of Veterinary Medicine and Animal Science of the University of São Paulo, under the protocol number CEUA 8696141117 (ID 007216).

2.1. Animals, Housing and Management

The study was conducted in Criatório Ximbó, a donkey farm in the city of Laranjal Paulista in the state of São Paulo, Brazil. The city of Laranjal Paulista is located at an altitude of 536 m, at the coordinates 23°02′59″ latitude South and 47°50′12″ longitude West. The local climate is humid subtropical (Köeppen–Geiger classification), the yearly average temperature ranges from 13 ± 4.9 °C to 31 ± 4.7 °C, and the yearly average pluviosity is around 1177 mm.

On the farm, donkeys are kept in a semi-intensive system, and receive nutrition composed of native foliage and *Brachiaria decumbens*, as well as alfalfa hay supplementation.

The donkeys on the farm are kept in pastures during the day and are moved to stalls of 24 m^2 stalls (6 m × 4 m) overnight, in stable groups of 3 (Stall 1, 8 m^2 allowance per jenny) to 4 jennies (Stall 2, 6 m^2 allowance per jenny) and their foals in each stall. In the morning, they are released by simply opening the stall door. All animals assessed in this study were habituated to this daily routine.

The farm had no dairy production activity, and the 60 Pêga jennies on the farm were used for reproduction. Fourteen multiparous Pêga jennies (245 kg average body weight) were studied from day 45 to 135 of lactation. Of the fourteen jennies assessed, seven foaled in February 2018 and were assessed until June, and seven foaled in May and were assessed until September of the same year. No milk was collected during the first month of each foal's life as it was used exclusively for their nutrition in the month of June, around day 135 of lactation for the first group and day 45 of lactation for the second group, all fourteen jennies and their foals were assessed at once, resulting in some animals being separated for up to 3 h.

For data collection, the established groups of jennies were maintained and minimal changes to the already established farm routine were made.

2.2. Experimental Design

On each data collection day, while still inside of the pens, together with their group, the behaviour of jennies in Stall 1 was assessed and saliva was sampled from both jennies and foals. The animals were then separated for 2 h. The same procedures were repeated in Stall 2. In order to separate them, two people stood at the stall door and allowed the jennies to pass through while impeding the foals from following. The jennies were stimulated to leave the stall using visual and sound cues, such as raising hands and clapping, and the foals were stopped by standing in their path. During the separation period, the jennies were loose on the farm, while the foals were kept inside of the group stalls with no visual contact with their dams. Information regarding the stall of each animal, as well as the exact time of release from the stalls, was recorded on each data collection day in order to keep the groups and separation times constant throughout the study.

After 2 h of separation, the jennies were led, one by one, to the milking parlour. This was the first instance they were restrained, utilising a halter and loose lead rope to keep them from leaving the parlour. At this time, their respective foals were brought to the milking parlour from the stalls, in less than one minute, marking the end of the separation period. The jennies' behaviour was then assessed, and saliva samples were taken from both jennies and foals. The foals were not restrained, but were stopped from suckling by placing a hand between their mouth and the dam's teat.

It was observed that, from the third collection day onwards, for both the February and May groups, the jennies tended to wait at the milking parlour by the end of the separation time, and did not need to be brought back from elsewhere on the farm.

During milking, an additional safety measure was taken by firmly tying a lead rope to their hind limb and securing it to a fence. The milk yield was noted.

Separation of jennies and foals took place at 10:00 am. Saliva samplings were conducted, before separation, between 8:50 am and 9:50 am, and after separation, between 12:00 pm and 2:00 pm.

Behavioural assessments and saliva samplings were performed from day 45 to 135 of lactation, totalling 14 assessments per animal. All data were collected every 15 days, to assess the possible adaptation of these animals to the stress generated by the milking management routine.

2.3. Behaviour Assessments

For the behavioural assessments, jennies were identified with ribbons of different colours attached to their necks. The protocol used for behaviour assessment was focal sampling with continuous recording, performed directly by two trained assessors utilising a check sheet.

The occurrence of behavioural states (long-duration behaviours such as prolonged activities, measured in time intervals between the beginning and end of each episode) and events (instantaneous or short-duration behaviours) were observed in the jennies in the pre- and post-separation periods and were later evaluated.

The observed behavioural states were affiliative behaviour and inactivity, and the events were agonistic, abnormal and eliminative behaviours, as well as vocalisations. These behavioural categories were chosen as they could be affected by the presence of a stressor [33–38], and various aspects of behaviour were assessed in order to paint a complete picture of any alterations the jennies exhibited between the pre- and post-separation assessments. It was expected that, if these animals were stressed by the separation, the duration of these behavioural states would be altered with potential increases in inactivity [35–37] and decreases in affiliative behaviour [38] post-separation, and the frequency of occurrence of these behavioural events would increase post-separation [33,34]. For these assessments, recording sheets based on an experimental ethogram, developed in this study, were used (Table 1). All observations yielded focal observation data from each animal, with a 6-min duration for each jenny in each assessment [44].

Table 1. Experimental ethogram utilised to assess jennies' behaviour.

	Affiliative
Mutual grooming	Behaviour in which two donkeys use their teeth to simultaneously nibble any of each other's body parts.
Licking	Licking any part of the body of another donkey.
Body sniffing	Sniffing the neck, withers, flank or tail of another donkey which may or may not reciprocate.
Approaching	Moving to within 1 m of another donkey that does not immediately move away and staying there for at least 10 s without initiating physical contact with it.
Touching	Touching another donkey at the neck or head, which may or may not reciprocate.
	Agonistic
Kicking	Rapid lifting of one or both hind limbs off the ground, directed towards another donkey or the observer, in an attempt to hit them, with the ears laid back.
Pushing	Pressing head, neck, chest or shoulder against another donkey, making them move away.
Chasing	Rapid movement toward another donkey and pursuit for a distance of over three body lengths, with the ears laid back, head raised and mouth closed.
Biting	Extension of head and neck towards another donkey, with the ears laid back, head raised and mouth open, closing teeth on its body.
Fighting	Pursuing another donkey for a distance of over three body lengths, with ears laid back, head raised and mouth open, attempting to close teeth on its body.
	Abnormal
Biting the stalls or structures	Grasping of structures with incisive teeth, which may be followed by simultaneous arching of the neck and sucking of air (cribbing).
False licking	Behaviour in which the animal slowly places its tongue on the borders of the stall or trough while keeping it still and stiff, so the action does not represent true licking.
Pawing	Vigorous and persistent stomping of limbs on the ground.
	Eliminative
Urinating	Elimination of urine.
Defecating	Elimination of faeces.
	Vocalisations
Vocalisations	Expression of vocal communication, such as whinnies, snores, snorts, groans or screams.
	Inactivity
Inactivity	Absence of movement or other actions.

Behaviours considered abnormal were biting the stalls or structures, false licking and pawing. Eliminative actions were urinating and defecating [45].

The recorded vocalisations included various types of vocal communication sounds, such as whinnies, snores, snorts, groans and screams [33].

Social interactions were divided between affiliative and agonistic according to the performed action and response of the receiving animal. In the absence of signs of aggression [46], interactions were considered affiliative, and behaviours linked to aggression were considered agonistic. The observed affiliative interactions were grooming, licking, sniffing, approaching and touching [46,47]. Agonistic interactions were kicking, pushing, chasing, biting and fighting [36]. All social interactions were performed between jennies and foals or other jennies.

After the behavioural assessment of the jennies, saliva was sampled from all animals in the pre- and post-separation periods.

2.4. Saliva Sampling

Saliva samples were collected from each animal using an individual sampler developed for this study, which did not require the animals to be restrained. For the jennies, the collector was made of ground *rapadura* (sugarcane candy) wrapped in gauzes and a cotton string. The inclusion of *rapadura* was necessary to stimulate saliva production in the jennies. For the foals, only gauzes and cotton strings were used.

The samplers were presented to all animals by holding them stretched, and both jennies and foals voluntarily approached the assessors to chew on the samplers. Sometimes, the samplers were secured to the animals' necks with the cotton strings, while the procedure was carried out on the rest of the animals.

Jennies and foals chewed the collectors for 2 min, after which the strings were cut and discarded, and the gauzes were placed in 15 mL Falcon tubes with their respective identifications. The tubes were stored in sealed styrofoam boxes lined with reusable gel ice packs.

2.5. Milking

Milking procedures began with the cleaning of the jennies' udders and teats with soap and water and drying with paper towels. They were milked manually.

The milking stopped once the udders were fully emptied, after which they were cleaned and dried again. After these procedures, each teat was submerged in a post-dipping solution (Dermasoft 2.5%, composed of Povidone-iodine (2.5 g) and purified water (100 mL)) for at least 15 s.

Milk yield was noted for each jenny on every assessment day.

2.6. Salivary Cortisol Analysis

The 15 mL Falcon tubes containing the saliva from jennies and foals were stored at –20 °C until the salivary cortisol analysis, which occurred between June 2018 and May 2019. For storage, the samples were thawed in the fridge and extracted from the gauze via centrifugation. The gauzes containing the samples were centrifuged for 15 min at 1000× *g*, and the extracted fluids were placed in 1.5 mL microtubes. These were then frozen again until analysis. The analysis was performed by trained professionals following EIA protocols, developed and validated by previously reported publications [48,49].

Additionally, *rapadura* was added to a standard curve, and no effect was observed in the performance of the assay.

2.7. Data Analysis

The data for the affiliative and inactivity behaviours were studied through the Poisson distribution, according to the PROC GLIMMIX of SAS, utilising a randomised block design with repeating measurements for the duration of occurrence of the observation in question, over time. Blocks were defined by the days of lactation. The model includes the effect of observation time in two different periods (before and after separation).

Data from the events of the behavioural categories agonistic, abnormal, eliminative and vocalisation were studied through the Poisson distribution, according to the PROC GLIMMIX of SAS, utilising a randomised block design with repeating measurements over time. Blocks were defined by days of lactation. The model includes the effect of observation time in two different periods (before and after separation).

For the salivary cortisol data, the Shapiro–Wilk test was conducted to analyse the normality of the residues, and the fixed effects were analysed by PROC GLIMMIX. The studied model included the effects of observation in two different periods (pre- and post-separation).

The milk yield data were analysed in a randomised block design. The statistical model considered the day of lactation to be a fixed factor and the animal (block) effect to be a random factor, defined by the RANDOM command. Fisher's Least Significant Difference was used when the fixed factors were significant for both analyses. The PROC CORR procedure was used for determining the Pearson correlation between milk yield and salivary cortisol concentration for the jennies.

All analyses were done in the Statistical Analysis Software 9.4 (SAS) [50]; the adopted significance level was set at $p < 0.05$.

3. Results

3.1. Behaviour

The frequency of occurrence of abnormal behaviour, vocalisation, eliminative behaviour and agonistic behaviour of jennies on day 45, 60, 75, 90, 105, 120 and 135 of lactation, before and after 2 h of separation from their foals for manual milking, are presented in Figures 1–4.

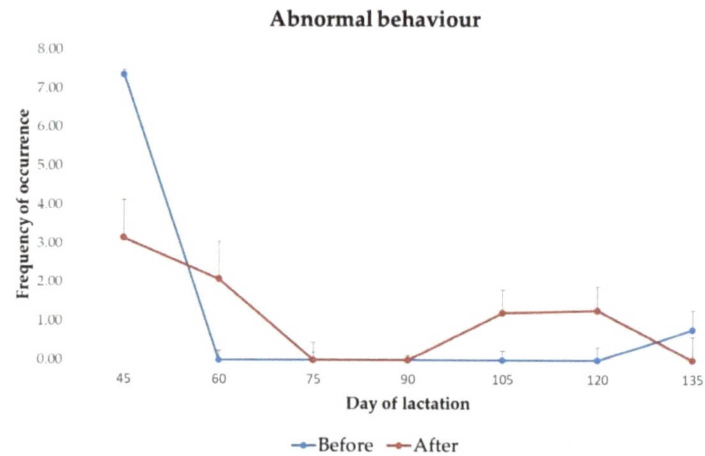

Figure 1. Mean frequency and standard deviation of occurrence of abnormal behaviour from jennies, before and after 2 h of separation from their foals for milking, on day 45 ($p = 0.99$), 60 ($p = 0.99$), 75 ($p = 0.99$), 90 ($p = 1.00$), 105 ($p = 0.99$), 120 ($p = 0.99$) and 135 ($p = 1.00$) of lactation.

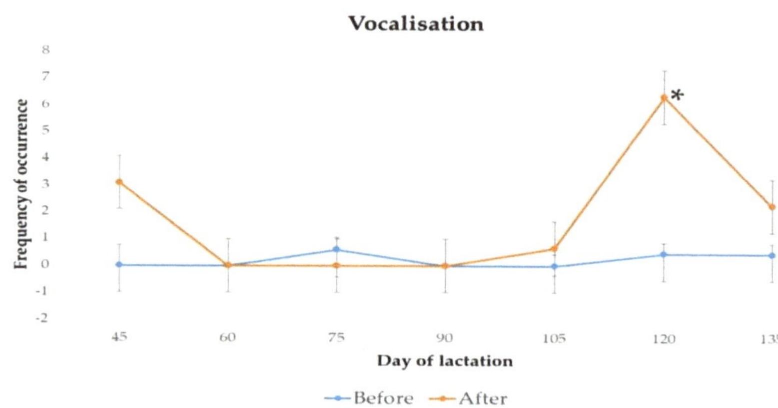

Figure 2. Mean frequency and standard deviation of occurrence of abnormal behaviour from jennies, before and after 2 h of separation from their foals for milking, on day 45 ($p = 0.99$), 60 ($p = 0.99$), 75 ($p = 0.99$), 90 ($p = 1.00$), 105 ($p = 0.99$), 120 ($p = 0.03$) and 135 ($p = 0.09$) of lactation. * indicates a statistically significant difference between frequencies.

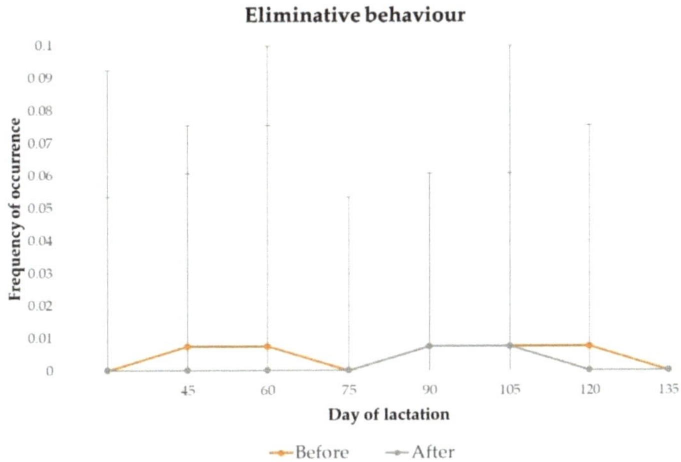

Figure 3. Mean frequency and standard deviation of occurrence of eliminative behaviour from jennies, before and after 2 h of separation from their foals for milking, on day 45 (p = 0.99), 60 (p = 0.99), 75 (p = 0.99), 90 (p = 0.99), 105 (p = 0.99), 120 (p = 0.99) and 135 (p = 0.99) of lactation.

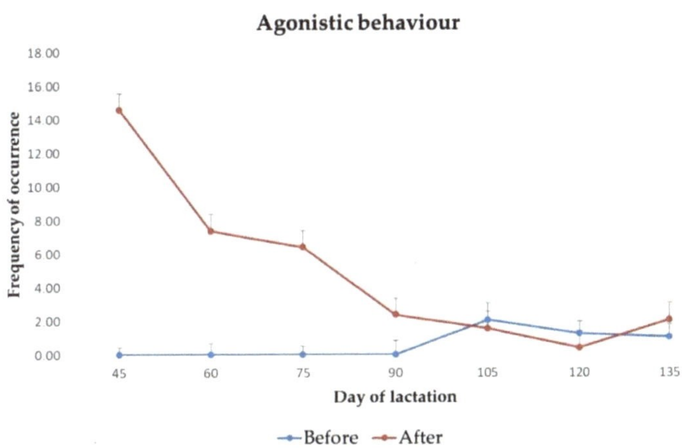

Figure 4. Mean frequency and standard deviation of occurrence of agonistic behaviour from jennies, before and after 2 h of separation from their foals for milking, on day 45 (p = 0.99), 60 (p = 0.99), 75 (p = 0.99), 90 (p = 0.99), 105 (p = 0.69), 120 (p = 0.40) and 135 (p = 0.40) of lactation.

Statistically significant differences were found for the frequency of vocalisations (p = 0.03) from jennies, on day 120 of lactation. No significant differences were found for the frequency of abnormal, eliminative or agonistic behaviours throughout lactation.

The duration of occurrence of affiliative behaviour and inactivity of jennies on day 45, 60, 75, 90, 105, 120 and 135 of lactation, before and after 2 h of separation from their foals for manual milking, are presented in Figures 5 and 6.

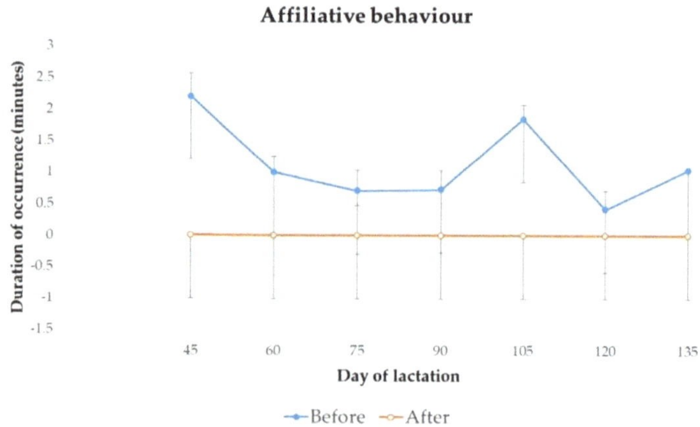

Figure 5. Mean duration and standard deviation of occurrence of affiliative behaviour from jennies, before and after 2 h of separation from their foals for milking, on day 45 (p = 0.99), 60 (p = 0.99), 75 (p = 0.99), 90 (p = 0.99), 105 (p = 0.99), 120 (p = 0.99) and 135 (p = 0.99) of lactation.

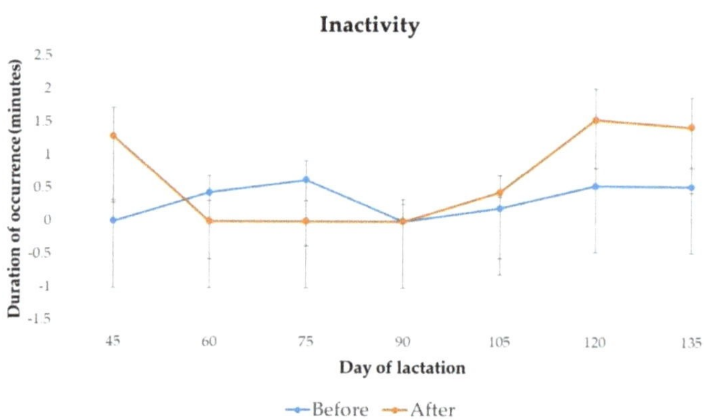

Figure 6. Mean duration and standard deviation of occurrence of inactivity from jennies, before and after 2 h of separation from their foals for milking, on day 45 (p = 0.99), 60 (p = 0.99), 75 (p = 0.99), 90 (p = 0.99), 105 (p = 0.44), 120 (p = 0.07) and 135 (p = 0.11) of lactation.

No significant differences were found for the duration of affiliative behaviour and inactivity from jennies throughout lactation.

3.2. *Salivary Cortisol Concentration*

The salivary cortisol concentration of jennies on day 45, 60, 75, 90, 105, 120 and 135 of lactation, before and after 2 h of separation from their foals for manual milking, are presented in Figure 7.

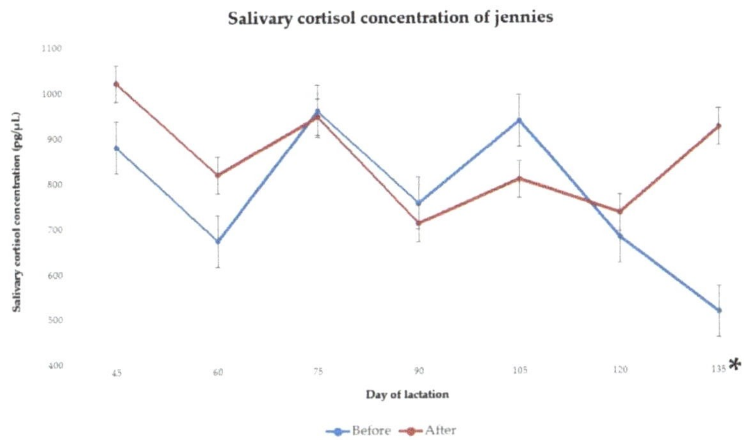

Figure 7. Mean and standard error of salivary cortisol concentration of jennies before and after 2 h of separation from their foals for milking, on day 45 ($p = 0.41$), 60 ($p = 0.40$), 75 ($p = 0.94$), 90 ($p = 0.79$), 105 ($p = 0.44$), 120 ($p = 0.77$) and 135 ($p = 0.02$) of lactation. * indicates a statistically significant difference.

A statistically significant difference between salivary cortisol concentrations of jennies before and after separation was found on day 135 of lactation ($p = 0.02$), but not on day 45, 60, 75, 90, 105 or 120 of lactation ($p > 0.05$).

The salivary cortisol concentrations of foals on day 45, 60, 75, 90, 105, 120 and 135 of lactation, before and after 2 h of separation from their dams for manual milking, are presented in Figure 8.

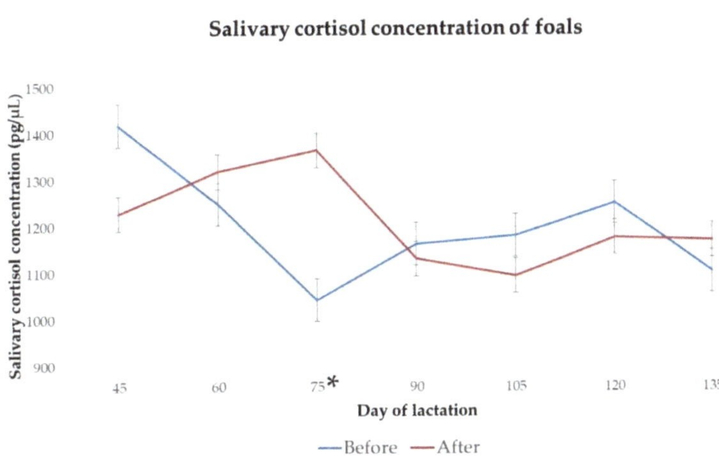

Figure 8. Mean and standard error of cortisol concentration of foals before and after 2 h of separation from their dams for milking, on day 45 ($p = 0.19$), 60 ($p = 0.63$), 75 ($p = 0.03$), 90 ($p = 0.82$), 105 ($p = 0.62$), 120 ($p = 0.61$) and 135 ($p = 0.64$) of lactation. * indicates a statistically significant difference.

A statistically significant difference between salivary cortisol concentrations of foals before and after separation was found on day 75 of lactation ($p = 0.03$), but not on day 45, 60, 90, 105, 120 and 135 or lactation ($p > 0.05$).

3.3. Milk Yield

The average milk yield of the Pêga jennies was 566.4 ± 205.2 mL/animal/milking.

There was no correlation found between the milk yield and salivary cortisol concentration pre- or post-separation (Table 2).

Table 2. Correlation between milk yield and salivary cortisol concentrations of jennies pre- and post-2 h of separation from their foals.

	Milk Yield (mL/day)	Cortisol before 2 h Separation (nmol/L)	Cortisol after 2 h Separation (nmol/pL)
Milk yield (mL/day)	1.00	−0.131	−0.044
Cortisol before 2 h separation (nmol/L)		1.00	0.432
Cortisol after 2 h separation (nmol/pL)			1.00

4. Discussion

4.1. Behaviour

Behavioural observation is considered the most reliable and immediate way to assess the perception and interaction of an animal with its environment [51]. However, the social behaviour of donkeys has not been sufficiently studied [35].

In this study, foals were separated from the jennies for milking management starting at 45 days of age. It is likely that the age-dependent reduction in the proximity between jennies and their foals partially explains the small behavioural responses reported in this study [25].

Additionally, jennies and foals were allowed to remain in physical proximity and maintained vocal communication, as the jennies were aware of the location of their foals during the separation period. Donkey foals begin drinking water and graze by themselves at four weeks of age [26], and mule foals have been observed at distances of 50 to 100 m from their dams starting from the 3rd week of life, and distances of over 100 m after the 11th week [27].

The increase in the frequency of vocalisations post-separation on day 120 of lactation, in comparison to the pre-separation period, may have been generated by various factors. Vocalisations are important to maintain the interactions between jennies and their foals, e.g., to signal the start of nursing bouts or direct the activities of the foal [25], and the increase might represent the fact that the animals were not in visual contact. It is known that equines utilise vocal communication to express many emotional states, ranging from curiosity, playfulness, and anticipation to distress signals, discomfort, frustration, and stress [23]. As this increase was only observed on one assessment day, it is not possible to determine if the increased frequency of vocalisations were a response to potential stress from the 2-h separation period, or an attempt to communicate in the absence of visual contact. The relevance of vocalisations as indicators of emotionality in animals must be analysed together with the other parameters.

The absence of significant differences in behavioural measures may indicate that both jennies and foals coped with the 2-h separation period with biologically acceptable responses that maintained good levels of animal welfare. When faced with routine changes, external stressors, or poor welfare conditions, animals tend to demonstrate behavioural signs such as a rise in inactivity [25–27], elevated frequency of urination and defecation [43,44], altered social interactions [28] and a rise in abnormal behaviours [26].

Animals also tend to perform greater amounts of abnormal and agonistic behaviour when responding to adverse situations, which may relate to stressors caused by housing problems and/or improper handling [26]. Changes in the environment and activities performed by the animals may generate alterations in the social environment [45,46]. Such changes were not observed in the present study, in which management alterations and separation between dams and foals for 2 h did not significantly impact social behaviour. The study population of purebred Pêga donkeys is unique as they encounter a wealth of human–animal interactions throughout all developmental stages, which may have mitigated their responses.

It is important to mention that the absence of alterations in social behaviour of the jennies in this study may be explained by the fact their social groups were not changed, minimising potential conflicts related to hierarchy, and might also indicate that the 2-h separation from their foals did not challenge social stability.

Affiliative behaviours among equids provide several social benefits [47]. Some known affiliative behaviours described for equines are mutual grooming, touching between the muzzle and body, playing, approaching, and following [36,49], though the occurrence of grooming and greeting are considered rare in wild jennies [48]. The occurrence of these actions is influenced by age, reproductive stage, hormones, social structures, and ecological conditions [47,49,50], and their quality and quantity may also be altered according to the quality of their habitat. Animals may display an increased frequency of affiliative behaviours to ease tensions or in situations of low perceived risk; contrarily, they may decrease their frequency to avoid imminent conflicts or in risky situations [47].

The results presented here may indicate an absence of stress in jennies when separated from their foals for 2 h, but further investigations are needed in regard to the normal social behaviours of donkeys and how they vary in response to adverse situations.

Even though no significant behavioural alterations were observed in the post-separation periods, further investigations are required in respect to the affiliative, agonistic, abnormal, and eliminative behaviours and inactivity in order to determine if the 2-h separation is a stressor for these animals. The study is unique in that it monitored the responses of purebred Pêga jennies and their foals. The animals were handled on a routine basis for other purposes, and this could have mitigated the response to the separation.

4.2. Salivary Cortisol Concentration

Cortisol was measured from saliva. This collection method is non-invasive [51–54] and reflects the biologically active portion of the total circulating concentration [52,53,55]. It is thus less likely to induce increases in cortisol concentration when compared to plasma cortisol sampling [54,56,57]. The aversive stimuli of drawing blood in dairy jennies may cause more intense stress than milking [31]. Significant differences between pre- and post-separation samples were only observed from jennies on day 135 of lactation, and from foals on day 75 of lactation. On all other assessment days, no significant differences between pre- and post-separation samples were observed. The significant rise in cortisol concentration in jennies after separation on day 135 of lactation, when compared to before separation, might have been the result of changes in management, which caused some animals to remain separated for longer than 2 h.

Few studies regarding the response and adaptation of jennies to milking have been performed [13,31], and no significant variance has been found in salivary cortisol concentrations before and after milking, even though donkeys can show great reactivity to milking procedures [13].

The average concentration of salivary cortisol from jennies, before milking was 790 pg/μL (217.93 nmol/L), taken between 9:00 am and 10:00 am, and 840 pg/μL (231.72 nmol/L), taken between 12:00 pm and 2 pm, before and after separation, respectively. The sampling period can alter salivary cortisol concentrations, which can reach values of 531.72 nmol/L when taken after milking [13]. In non-pregnant mares, basal salivary cortisol concentrations vary between 110.34 nmol/L and 331.03 nmol/L [58]. Significant differences have been reported between salivary cortisol concentrations of donkey stallions and equine mares or geldings [59,60], which have been ascribed to species variation [56].

The time of day in which samples are taken also affects results, due to circadian rhythms. Cortisol concentrations follow a clear diurnal pattern in horses, with the highest concentrations in the morning and the lowest in the late afternoon and evening [58,61–63]. This trend has also been observed in donkeys, with high plasma cortisol levels found in jennies milked at 8:00 am and lower values in groups milked at 4 pm [31].

As salivary cortisol levels were not measured throughout the day without routine changes, in the present study, and the circadian rhythm may influence basal cortisol levels, it is uncertain whether the absence of significant difference between cortisol levels before and after separation and milking is due to the absence of stress for the animals or lower basal levels at later times of the day [31].

It has been stated that inherent diurnal rhythms can be easily disturbed by minor challenges [59,61] and factors such as weather and ambient temperature, and interactions within groups may cause transient alterations [59]. Experimentally induced increases in salivary cortisol are often relatively small, and hardly exceed the range of physiological variations [54]. More research is needed regarding the variations in cortisol level in jennies according time of day and seasons of the year [13,31,56].

In horses, stressful events like separation from conspecifics acutely stimulates cortisol release [62,64], but they quickly habituate to these situations, and there are no lasting effects on diurnal rhythm. Furthermore, repeated stressful events also result in subsequent decreased cortisol levels [60].

The average concentration of salivary cortisol from foals before the separation from their dams was numerically higher than those of jennies, in agreement with studies performed in horses, which reported higher cortisol levels in suckling foals compared to their dams [59]. The elevated levels of cortisol in foals may be due to immaturity, following the same pattern found in gilts and humans [65–67], in which cortisol levels are initially high and gradually lower while forming a circadian rhythm.

The fact that the 2-h separation period, in all but two assessments, did not generate a significant difference between pre- and post-separation salivary cortisol concentrations in jennies or foals may indicate that this interval was not a stressful factor capable of altering the HPA axis, and may not compromise the welfare of the animals involved. Therefore, the ability of jennies and foals to adapt over time in response to stress remains unclear.

4.3. Milk Yield

The milk yield data differ from other studies, which, working with jennies of the Pêga breed in an extensive farming system in the drought season and without nutritional supplementation, reported an average milk yield in two daily milkings of 0.614 kg/day [68].

Ragusana donkeys receiving hay ad libitum and 3.5 kg feed/day have higher yields than those reported in this study, ranging from 0.56 to 0.59 kg/milking, from two and eight milkings, respectively [18]. This suggests that the difference in milk production of the different breeds could be linked to their diet [69].

The ejection of alveolar milk may be altered by stress, due to oxytocin concentration changes and myoepithelial contraction [42]. Jennies submitted to milking without previous training show lower milk yield when compared to jennies previously habituated to milking management, possibly due to reduced oxytocin supply via vasoconstriction or blocking of its receptors in the myoepithelial cells of the udder alveoli [13]. If we consider the density of Pêga asinine milk to be 1.03 g/mL [69], the milk yield was 0.6 kg/animal/milking, which represents approximately 0.25% of the jennies' average body weight. Since these animals may be milked two [10] to eight times a day [18], the total milk yield from these jennies may be up to 4.8 kg/animal/day.

Milk yield remained constant for all animals after the 2-h separation period on all days of lactation, and there was no correlation between the volume of milk produced and cortisol concentration pre- and post-separation, thus separation was not a stressor that impaired milk ejection. However, many other factors can alter milk ejection, and must be considered before concluding that there was no stressor present during the 2 h separation.

It is important to consider the fact that the animals had interactions with humans on a regular basis and some of the management practices involved short-term separation of jennies and foals.

5. Conclusions

The results presented in this study show that the behavioural categories assessed for the jennies were only mildly altered by the 2-h separation from their foals. The only behavioural variable that showed significant changes was vocalisation frequency, which may express social signalling in the absence of visual contact, and this was only observed on one assessment day.

The 2-h separation period also failed to generate significant changes in the majority of salivary cortisol concentration levels of jennies or foals, with the exception of one assessment in each animal category. Therefore, it does not appear to be a stressor capable of altering the HPA axis. Additionally, there was no apparent relationship between milk ejection and salivary cortisol concentrations.

We acknowledge several limitations to this study, such as the absence of a control group, the lack of true measures of basal salivary cortisol, and the low quantity of milk yield measurements to support more robust conclusions. Behavioural observations of foals will enhance our understanding of the impact of separation on their welfare. Further research is needed to determine whether the separation of these animals is indeed a stressor that could result in severe welfare problems. Additional studies are also required to determine the long-term consequences of the separation event, as well as the results of more frequent separation periods, on the lifelong trajectory of these animals.

Considering these results, it is important to emphasise that they are limited to one group of purebred Pêga jennies and foals in Brazil. It is possible that their responses were confounded by the fact that these animals were handled non-aversively on a regular basis. We hope that this study is useful to people interested in milking donkeys.

Author Contributions: Conceptualization, S.d.S.F. and A.J.Z.; methodology, S.d.S.F., A.J.Z. and P.H.M.R.; software, S.d.S.F., T.B. and P.H.M.R.; validation, S.d.S.F. and A.J.Z.; formal analysis, S.d.S.F.; investigation, S.d.S.F. and A.J.Z.; resources, S.d.S.F. and A.J.Z.; data curation, S.d.S.F., T.B. and C.A.d.A.O.; writing—original draft preparation, S.d.S.F. and A.C.D.M.; writing—review and editing, S.d.S.F., A.C.D.M., T.B., P.H.M.R., C.A.d.A.O. and A.J.Z.; visualization, S.d.S.F., A.C.D.M., T.B.; supervision, P.H.M.R., C.A.d.A.O. and A.J.Z.; project administration, A.J.Z.; funding acquisition, A.J.Z. All authors have read and agreed to the published version of the manuscript.

Funding: This research was funded by Conselho Nacional de Desenvolvimento Científico e Tecnológico—CNPq (Proc. Proc. 306464/2016-0 and 423833/2018-9).

Institutional Review Board Statement: The study was approved by the Committee for the Use and Care of Animals in Research (CEUA) of the School of Veterinary Medicine and Animal Science of the University of São Paulo, under the protocol code CEUA 8696141117 (ID 007216), approved on 17/11/2019.

Informed Consent Statement: Not applicable.

Data Availability Statement: Publicly available datasets were analyzed in this study. This data can be found here: https://data.mendeley.com/datasets/7n9t5fn99b/2.

Acknowledgments: The authors acknowledge the contribution of the members of the Center for Comparative Studies in Health, Sustainability and Welfare (CECSBE), with special thanks to João Augusto Metzner for providing training and assistance in all laboratory procedures; the Department of Preventive Veterinary Medicine and Animal Health (VPS) of the School of Veterinary Medicine and Animal Science of the University of São Paulo; the Conselho Nacional de Desenvolvimento Científico e Tecnológico (CNPq) for the funding provided; Rupert Palme of University of Veterinary Medicine, Vienna, for supplying the reagents utilized for laboratory analyses; The Donkey Sanctuary for help fostering our network with the Northeast of Brazil to gather information on the issue of abandonment of donkeys; and Criatório Ximbó for providing the animals and facilities for this study.

Conflicts of Interest: The authors declare no conflict of interest.

References

1. Bordonaro, S.; Dimauro, C.; Criscione, A.; Marletta, D.; Macciotta, N.P.P. The mathematical modeling of the lactation curve for dairy traits of the donkey (equus asinus). *J. Dairy Sci.* **2013**, *96*, 4005–4014. [CrossRef] [PubMed]
2. IBGE. Produção da pecuária municipal 2012 (In portuguese). *Inst. Bras. Geogr. Estatística IBGE* **2012**, *40*, 71.
3. Ochieng, F.; Alemayahu, M.; Smith, D. Improving the productivity of donkeys in Ethiopia. In *Responding to the Increasing Global Demand for Animal Products: Programme and Summaries 42*; Department for International Development: Addis Ababa, Ethiopia, 2004; pp. 93–94.
4. Mariante, A.D.S.; Albuquerque, M.D.S.M.; Egito, A.A.; McManus, C.; Lopez, M.A.; Paiva, S.R. Present status of the conservation of livestock genetic resources in Brazil. *Livest. Sci.* **2009**, *120*, 204–212. [CrossRef]
5. McManus, C.; Paiva, S.; Louvandini, H.; Melo, C.; Seixas, L. Jumentos no Brasil. In *INCT Informaçãos Genético-Sanitária da Pecuária*; 2010; pp. 1–10.
6. Girardi, A.M.; Marquez, L.C.; de Toledo, C.Z.P.; de Campos Filho, E. Hematological variables of the Pêga donkey (Equus asinus) breed: Influence of age and sex. *Comp. Clin. Pathol.* **2015**, *24*, 335–342. [CrossRef]
7. Chiofalo, B.; Salimei, E.; Chiofalo, L. Ass's milk: Exploitation of an alimentary resource. *Riv. Folium* **2001**, *1*, 235–241.
8. Veneziano, V.; Di Loria, A.; Masucci, R.; Di Palo, R.; Brianti, E.; Gokbulut, C. Efficacy of eprinomectin pour-on against Dictyocaulus arnfieldi infection in donkeys (Equus asinus). *Vet. J.* **2011**, *190*, 414–415. [CrossRef]
9. Souroullas, K.; Aspri, M.; Papademas, P. Donkey milk as a supplement in infant formula: Benefits and technological challenges. *Food Res. Int.* **2018**, *109*, 416–425. [CrossRef]
10. D'Alessandro, A.G.; Martemucci, G. Lactation curve and effects of milking regimen on milk yield and quality, and udder health in Martina Franca jennies (*Equus asinus*). *J. Anim. Sci.* **2012**, *90*, 669–681. [CrossRef]
11. Tesse, R.; Paglialunga, C.; Braccio, S.; Armenio, L. Adequacy and tolerance to ass's milk in an Italian cohort of children with cow's milk allergy. *Ital. J. Pediatr.* **2009**, *35*, 4–7. [CrossRef]
12. Carroccio, A.; Cavataio, F.; Montalto, G.; D'Amico, D.; Alabrese, L.; Iacono, G. Intolerance to hydrolysed cow's milk proteins in infants: Clinical characteristics and dietary treatment. *Clin. Exp. Allergy* **2000**, *30*, 1598–1603. [CrossRef]
13. De Palo, P.; Maggiolino, A.; Albenzio, M.; Caroprese, M.; Centoducati, P.; Tateo, A. Evaluation of different habituation protocols for training dairy jennies to the milking parlor: Effect on milk yield, behavior, heart rate and salivary cortisol. *Appl. Anim. Behav. Sci.* **2018**, *204*, 72–80. [CrossRef]
14. D'Alessandro, A.G.; Mariano, M.; Martemucci, G. Udder characteristics and effects of pulsation rate on milking machine efficiency in donkeys. *J. Dairy Res.* **2015**, *82*, 121–128. [CrossRef] [PubMed]
15. Brito, M.D.F.; Seppa, G.S.; Teixeira, L.G.; Rocha, T.G.; França, T.D.N.; Hess, T.M.; Peixoto, P.V. Mammary adenocarcinoma in a mare. *Cienc. Rural* **2008**, *38*, 556–560. [CrossRef]
16. Simoni, A.; Salimei, E.; Varisco, G. *Struttura e Routine di Mungitura e Caratteristiche Della Produzione di Latte di Asina, Alimento Ipoallergenico per l'infanzia*; Atti VI convegno società Italiana di ippologia: Campobasso, Italy, 2004; pp. 85–98.
17. Burden, F.; Thiemann, A. Donkeys are different. *J. Equine Vet. Sci.* **2015**, *35*, 376–382. [CrossRef]
18. Alabiso, M.; Giosuè, C.; Alicata, M.L.; Mazza, F.; Iannolino, G. The effects of different milking intervals and milking times per day in jennet milk production. *Animal* **2009**, *3*, 543–547. [CrossRef]
19. Doreau, M.; Boulot, S. Recent knowledge on mare milk production: A review. *Livest. Prod. Sci.* **1989**, *22*, 213–235. [CrossRef]
20. Henry, M.; McDonnell, S.M.; Lodi, L.D.; Gastal, E.L. Pasture mating behaviour of donkeys (*Equus asinus*) at natural and induced oestrus. *J. Reprod. Fertil. Suppl.* **1991**, *44*, 77–86.
21. Klingel, H. Observations on Social Organization and Behaviour of African and Asiatic Wild Asses (*Equus africanus* and *E. hemionus*). *Z. Tierpsychol.* **1977**, *44*, 323–331. [CrossRef]
22. Woodward, S.L. The Social System of Feral Asses (*Equus asinus*). *Z. Tierpsychol.* **1979**, *49*, 304–316. [CrossRef]
23. McGreevy, P. *Equine Behavior: A Guide for Veterinarians and Equine Scientists*, 2nd ed.; W. B. Saunders: Edinburgh, UK, 2012.
24. Mazzatenta, A.; Veronesi, M.C.; Vignola, G.; Ponzio, P.; Carluccio, A.; De Amicis, I. Behavior of Martina Franca donkey breed jenny-and-foal dyad in the neonatal period. *J. Vet. Behav.* **2019**, *33*, 81–89. [CrossRef]
25. French, J.M. Mother-offspring relationships in donkeys. *Appl. Anim. Behav. Sci.* **1998**, *60*, 253–258. [CrossRef]
26. Rashek, V.A. Details of feeding and feeding behaviour in young wild asses on Barsa-Kelmes Island (Aral Sea). *Zoologicheskiĭ Zhurnal* **1976**, *55*, 784–786.
27. Smith-Funk, E.D.; Crowell-Davis, S.L. Maternal behavior of draft mares (*Equus caballus*) with mule foals (*Equus asinus × Equus caballus*). *Appl. Anim. Behav. Sci.* **1992**, *33*, 93–119. [CrossRef]
28. Rossdale, P.D. Perinatal behavior in the throughbred horse. In *Abnormal Behavior in Animals*; Fox, M.W., Ed.; W. B. Saunders: Philadelphia, PA, USA, 1968; pp. 227–237.
29. Scott, J.P. Social behavior, organisation, and leadership in a small flock of domestic sheep. *Comp. Psychol. Monogr.* **1945**, *18*, 1–29.
30. Fazio, E.; Medica, P.; Grasso, L.; Messineo, C.; Ferlazzo, A. Changes of circulating β-endorphin, adrenocorticotrophin and cortisol concentrations during growth and rearing in Thoroughbred foals. *Livest. Sci.* **2009**, *125*, 31–36. [CrossRef]
31. Fazio, E.; Medica, P.; Cravana, C.; Ferlazzo, A. Adrenocortical response of jennies to milking stress. *Livest. Sci.* **2011**, *137*, 278–281. [CrossRef]

32. Moberg, G.P. Biological Response to Stress: Key to Assessment of Animal Well-Being? *Anim. Stress* **1985**, 27–49. [CrossRef]
33. Yeon, S.C. Acoustic communication in the domestic horse (*Equus caballus*). *J. Vet. Behav. Clin. Appl. Res.* **2012**, *7*, 179–185. [CrossRef]
34. Wickens, C.L.; Heleski, C.R. Crib-biting behavior in horses: A review. *Appl. Anim. Behav. Sci.* **2010**, *128*, 1–9. [CrossRef]
35. Fureix, C.; Meagher, R.K. What can inactivity (in its various forms) reveal about affective states in non-human animals? A review. *Appl. Anim. Behav. Sci.* **2015**, *171*, 8–24. [CrossRef]
36. Zanella, A.J.; Broom, D.M.; Hunter, J.C.; Mendl, M.T. Brain opioid receptors in relation to stereotypies, inactivity, and housing in sows. *Physiol. Behav.* **1996**, *59*, 769–775. [CrossRef]
37. McPhee, M.E.; Carlstead, K. The importance of maintaining natural behaviors in captive mammals. In *Wild Mammals in Captivity: Principles and Techniques for Zoo Management*, 2nd ed.; University of Chicago Press: Chicago, IL, USA, 2010; pp. 303–313.
38. Beery, A.K.; Kaufer, D. Stress, social behavior, and resilience: Insights from rodents. *Neurobiol. Stress* **2015**, *1*, 116–127. [CrossRef]
39. Andrade, A.; Pinto, S.C.; de Oliveira, R.S. *Animais de Laboratório: Criação e Experimentação*; FIOCRUZ: Rio de Janeiro, Brazil, 2006; ISBN 8575410156.
40. Steimer, T. The biology of fear- and anxiety-related behaviors. *Dialogues Clin. Neurosci.* **2002**, *4*, 231–249. [PubMed]
41. Reis, L.S.L.S.; Pardo, P.E.; Frazatti-Gallina, N.M.; Paoli, R.L.; Oba, E.; Kronka, S.N.; Camargos, A.S. Effects of primovaccination and booster vaccination on serum cortisol and humoral immune response in cattle. *Adv. Biosci. Biotechnol.* **2013**, *04*, 607–611. [CrossRef]
42. Bruckmaier, R.M.; Wellnitz, O. Induction of milk ejection and milk removal in different production systems. *J. Anim. Sci.* **2008**, *86*, 15–20. [CrossRef] [PubMed]
43. Scheidegger, M.D.; Gerber, V.; Ramseyer, A.; Schüpbach-Regula, G.; Bruckmaier, R.M.; van der Kolk, J.H. Repeatability of the ACTH stimulation test as reflected by salivary cortisol response in healthy horses. *Domest. Anim. Endocrinol.* **2016**, *57*, 43–47. [CrossRef]
44. Martin, P.; Bateson, P. *Measuring Behaviour: An Introductory Guide*; Cambridge University Press: Cambridge, UK, 2007.
45. Regan, F.H.; Hockenhull, J.; Pritchard, J.C.; Waterman-Pearson, A.E.; Whay, H.R. Behavioural repertoire of working donkeys and consistency of behaviour over time, as a preliminary step towards identifying pain-related behaviours. *PLoS ONE* **2014**, *9*. [CrossRef]
46. Cozzi, A.; Sighieri, C.; Gazzano, A.; Nicol, C.J.; Baragli, P. Post-conflict friendly reunion in a permanent group of horses (Equus caballus). *Behav. Processes* **2010**, *85*, 185–190. [CrossRef]
47. Heitor, F.; do Mar Oom, M.; Vicente, L. Social relationships in a herd of Sorraia horses. Part I. Correlates of social dominance and contexts of aggression. *Behav. Processes* **2006**, *73*, 170–177. [CrossRef]
48. Schmidt, E.; Zillikens, D. Modern diagnosis of autoimmune blistering skin diseases. *Autoimmun. Rev.* **2010**, *10*, 84–89. [CrossRef]
49. Schmidt, E.; Dähnrich, C.; Rosemann, A.; Probst, C.; Komorowski, L.; Saschenbrecker, S.; Schlumberger, W.; Stöcker, W.; Hashimoto, T.; Bröcker, E.B.; et al. Novel ELISA systems for antibodies to desmoglein 1 and 3: Correlation of disease activity with serum autoantibody levels in individual pemphigus patients. *Exp. Dermatol.* **2010**, *19*, 458–463. [CrossRef] [PubMed]
50. SAS Institute Inc. *SAS/SHARE®9.4: User's Guide*, 2nd ed.; SAS Institute Inc.: Cary, NC, USA, 2016.
51. Möstl, E.; Palme, R. Hormones as indicators of stress. *Domest. Anim. Endocrinol.* **2002**, *23*, 67–74. [CrossRef]
52. Lebelt, D.; Schönreiter, S.; Zanella, A.J. Salivary cortisol in stallions: The relationship with plasma levels, daytime profile and changes in response to semen collection. *Pferdeheilkunde* **1996**, *12*, 411–414. [CrossRef]
53. Kirschbaum, C.; Hellhammer, D.H. Salivary cortisol. In *Encyclopedia of Stress*; Fink, G., Ed.; Academic Press: San Diego, CA, USA, 2000; Volume 3, pp. 379–384.
54. Schmidt, A.; Möstl, E.; Wehnert, C.; Aurich, J.; Müller, J.; Aurich, C. Cortisol release and heart rate variability in horses during road transport. *Horm. Behav.* **2010**, *57*, 209–215. [CrossRef]
55. Riad, F.D.; Read, G.F.; Walker, R.F. Variation in Endocrine Activity. *J. Steroid Biochem.* **1983**, *19*, 265–272. [CrossRef]
56. Bonelli, F.; Rota, A.; Aurich, C.; Ille, N.; Camillo, F.; Panzani, D.; Sgorbini, M. Journal of Equine Veterinary Science Determination of Salivary Cortisol in Donkey Stallions. *J. Equine Vet. Sci.* **2019**, *77*, 68–71. [CrossRef]
57. Hopster, H.; Van Der Werf, J.T.N.; Erkens, J.H.F.; Blokhuis, H.J. Effects of Repeated Jugular Puncture on Plasma Cortisol Concentrations in Loose-Housed Dairy Cows. *J. Anim. Sci.* **1999**, *77*, 708–714. [CrossRef]
58. Aurich, J.; Wulf, M.; Ille, N.; Erber, R.; Von Lewinski, M.; Palme, R.; Aurich, C. Effects of season, age, sex and housing on salivary cortisol concentrations in horses. *Domest. Anim. Endocrinol.* **2015**. [CrossRef]
59. Schmidt, A.; Biau, S.; Möstl, E.; Becker-Birck, M.; Morillon, B. Changes in cortisol release and heart rate variability in sport horses during long-distance road transport. *Domest. Anim. Endocrinol.* **2010**, *38*, 179–189. [CrossRef]
60. Schmidt, A.; Hödl, S.; Möstl, E.; Aurich, J.; Müller, J.; Aurich, C. Cortisol release, heart rate, and heart rate variability in transport-naive horses during repeated road transport. *DAE* **2010**, *39*, 205–213. [CrossRef]
61. Irvine, C.H.G.; Alexander, S.L. Factors affecting the circadian rhythm in plasma cortisol concentrations in the horse. *Domest. Anim. Endocrinol.* **1994**, *11*, 227–238. [CrossRef]
62. Nagel, C.; Erber, R.; Bergmaier, C.; Wulf, M.; Aurich, J.; Möstl, E.; Aurich, C. Cortisol and progestin release, heart rate and heart rate variability in the pregnant and postpartum mare, fetus and newborn foal. *THE* **2012**, *78*, 759–767. [CrossRef] [PubMed]

63. Bohák, Z.; Szabó, F.; Beckers, J.; De Sousa, N.M.; Kutasi, O.; Nagy, K.; Szenci, O. Domestic Animal Endocrinology Monitoring the circadian rhythm of serum and salivary cortisol concentrations in the horse. *Domest. Anim. Endocrinol.* **2013**, *45*, 38–42. [CrossRef] [PubMed]
64. Alexander, S.L.; Irvine, C.H.G. The effect of social stress on adrenal axis activity in horses: The importance of monitoring corticosteroid-binding globulin capacity. *J. Endocrinol.* **1998**, *157*, 425–432. [CrossRef] [PubMed]
65. Vermes, I.; Dohanics, J.; Tóth, G.; Pongrácz, J. Maturation of the circadian rhythm of the adrenocortical functions in human neonates and infants. *Horm. Res.* **1980**, *12*, 237–244. [CrossRef] [PubMed]
66. Franks, R.C. Diurnal Variation of Plasma 17-Hydroxycorticosteroids in Children. *J. Clin. Endocrinol. Metab.* **1967**, *27*, 75–78. [CrossRef]
67. Evans, F.D.; Christopherson, R.J.; Aherne, F.X. Development of the Circadian Rhythm of Cortisol in the Gilt from Weaning Until Puberty. *Can. J. Anim. Sci.* **1988**, *68*, 1105–1111. [CrossRef]
68. Santos, I.C.B. Produção e Composição do Leite de Jumentas da Raça Pêga. Master's thesis, Federal University of Bahia, Salvador, Brazil, 2017.
69. Martini, M.; Altomonte, I.; Licitra, R.; Salari, F. Nutritional and Nutraceutical Quality of Donkey Milk. *J. Equine Vet. Sci.* **2018**, *65*, 33–37. [CrossRef]

animals

MDPI

Article

Bayesian Linear Regression Modelling for Sperm Quality Parameters Using Age, Body Weight, Testicular Morphometry, and Combined Biometric Indices in Donkeys

Ana Martins-Bessa [1,2,3,*], Miguel Quaresma [1,2,3], Belén Leiva [4], Ana Calado [1,2] and Francisco Javier Navas González [5]

1 Department of Veterinary Sciences, School of Agrarian and Veterinary Sciences, University of Trás-os-Montes and Alto Douro, 5000-801 Vila Real, Portugal; miguelq@utad.pt (M.Q.); anacalad@utad.pt (A.C.)
2 CECAV, Animal and Veterinary Research Center, University of Trás-os-Montes and Alto Douro, 5000-801 Vila Real, Portugal
3 Veterinary Teaching Hospital, University of Trás-os-Montes and Alto Douro, Quinta de Prados, 5000-801 Vila Real, Portugal
4 AEPGA-Association for the Study and Protection of Donkeys, Atenor, 5225-011 Miranda do Douro, Portugal; belenleiva.aepga@gmail.com
5 Genetics Department, Veterinary Sciences, Rabanales University Campus, University of Córdoba, Madrid-Cádiz Km. 396, 14014 Cordoba, Spain; fjng87@hotmail.com
* Correspondence: abessa@utad.pt; Tel.: +351-2593-50634

Simple Summary: The prediction of sperm output and other reproductive traits based on testicular biometry is an important tool in the reproductive management of stallions. Nevertheless, corresponding research in donkeys remains scarce. Several donkey breeds in Europe face a compromising threat of extinction, which has been accelerated by the low renovation of populations and their inbreeding levels. Although research on female reproductive physiology has made crucial advances, much less is known about the physiology of the male. In the present work, two Bayesian models were built to predict for sperm output and quality parameters in donkeys. Models included combinations of age as a covariate and biometric and testicular measurements as independent factors. Results evidenced that the goodness-of-fit was similar for both models—hence, the combination of biometry and testicular factors presented improved predictive power. The application of these models may assist in the process of making decisions in respect to the reproductive/biological, clinical, and selection handling of the animals.

Abstract: The aim of the present study is to define and compare the predictive power of two different Bayesian models for donkey sperm quality after the evaluation of linear and combined testicular biometry indices and their relationship with age and body weight (BW). Testicular morphometry was ultrasonographically obtained from 23 donkeys (six juveniles and 17 adults), while 40 ejaculates from eight mature donkeys were analyzed for sperm output and quality assessment. Bayesian linear regression analyses were considered to build two statistical models using gel-free volume, concentration, total sperm number, motility, total motile sperm, and morphology as dependent variables. Predictive model 1 comprised the covariate of age and the independent factors testicular measurements (length, height and width), while model 2 included the covariate of age and the factors of BW, testicular volume, and gonadosomatic ratio. Although goodness-of-fit was similar, the combination of predictors in model 1 evidenced higher likelihood to predict gel-free volume (mL), concentration ($\times 10^6$/mL), and motility (%). Alternatively, the combination of predictors in model 2 evidenced higher predictive power for total sperm number ($\times 10^9$), morphologically normal spermatozoa (%), and total motile sperm count ($\times 10^9$). The application of the present models may be useful to gather relevant information that could be used hereafter for assisted reproductive technologies.

Keywords: Bayesian models; testicular; sperm output; biometry; donkey

check for updates

Citation: Martins-Bessa, A.; Quaresma, M.; Leiva, B.; Calado, A.; Navas González, F.J. Bayesian Linear Regression Modelling for Sperm Quality Parameters Using Age, Body Weight, Testicular Morphometry, and Combined Biometric Indices in Donkeys. *Animals* **2021**, *11*, 176. https://doi.org/10.3390/ani11010176

Received: 9 October 2020
Accepted: 14 November 2020
Published: 13 January 2021

1. Introduction

Measuring testicular size reports an approximate measurement of the amount of testicular parenchyma present in a certain individual, which in turn determines the potential for sperm production [1,2]. Studies on equine testicular biometry are fairly common, and the correlation between testicular dimensions and the capacity for sperm production has frequently been addressed in the literature, allowing the establishment of a predictive formula for daily sperm output (DSO) [2–4]. Still, contextually, corresponding publications on male donkey remain scarce. Besides biometric testicular data such as length, width, and volume, other factors might be investigated as independent variables (covariates) in DSO prediction, like age, body weight (BW), or the gonadosomatic ratio (GSI), which is the gonadal weight/BW ratio [5]. Previous studies evidenced that knowledge from stallions could not be directly assumed for donkeys, as their reproductive physiology may present some specific particularities. For instance, donkeys present spermatogenesis cycles that last for 47.2 days, with each spermatogenic stage lasting for 10.5 days [6].

The numbers of the population of the Burro de Miranda's (*Equus asinus*), as it occurs in other donkey breeds in southern Europe countries [7], dramatically fall, with a mature male population of around 40 jackasses and 300 breeding females. Besides, reproduction rates in such populations are low, with an excessive overuse of the same males in the past, which derived in the occurrence of genetic bottlenecks. Although the reproductive physiology of Miranda donkey females has already been studied [8], no research focusing on the biometry or reproductive physiology of Miranda donkey males has been reported to date. Considering the actual population structure, the interest in applying assisted reproductive technologies (ARTs) have raised. However, the implementation of these techniques requires consistent knowledge of reproductive biology in order to select the best males for ARTs programs to be successful. Testicular biometry, due to its correlation with sperm production, is a valuable tool to estimate male fertility and is also an essential element of the breeding soundness evaluation (BSE). In light of the aforementioned points, the accurate prediction of sperm output and quality parameters obtained from biometrical and testicular morphometric parameters, will improve the effectiveness of reproductive management in male donkeys.

In this context and bearing in mind the small but diverse mature male population, some issues in regards the statistical approach to follow may arise. For these reasons, the statistical tools used in this study were chosen to fit the characteristics of the data to be analyzed. According to Oravecz and Muth [9], the popularity of growth curve modelling (GCM) lies in its flexibility to simultaneously analyze within-individual changes (e.g., changes with age, change due to intervention, due to natural changes occurring along the life of the individuals, etc.) and between-individual effects (i.e., individual differences). In other words, GCM may be useful to model inter-individual differences and intra-individual variation. GCM has been successfully used to model the evolution of semen parameters in males from other species, such as boars [10].

An individual's specific growth trajectory, specified as a mathematical function that describes how variables reciprocally relate over time, captures how an individual uniquely changes. GCM covers situations that range from those for which the change function is linear to other occasions when curvilinear polynomial functions are fitted (for instance, quadratic, cubic, etc.), which means that modelling is not limited to consider straight-line functional growth. Beyond handling varying growth functions, GCM can flexibly handle unbalanced designs, meaning study individuals may be measured at different occasions and need not be excluded from the analysis, even if some of their measurements are missing [8].

In these regards, Bayesian inference potentiates the flexibility of GCM, given Bayesian analyses do not assume large samples, as it would happen in maximum likelihood estimation (either it is nonparametric or parametric inference). Besides, smaller data sets can be evaluated preventing power loss and retaining precision, as suggested by Hox, et al. [11] and Lee and Song [12]. In small sample size conditions, the probability of finding significant results decreases [13]. Given power issues, this limitation often translates into an increased hardness to obtain meaningful results [14].

According to Stoltzfus [15], the basic assumptions that must be met for the outputs of regression analyses to be valid include independence of errors, linearity in continuous variables, the absence of multicollinearity, and a lack of strongly influential outliers. Additionally, there should be an adequate number of events per independent variable (covariate) to avoid an overfit model. Commonly recommended minimum "rules of thumb" range from 10 to 20 events per covariate. This would be supported by Chen, et al. [16], who suggested that the usual minimum number of observations for running a linear regression to be 30 to obtain statistically significant estimates. The same authors would even state that sometimes this requirement cannot be met, for instance when the number of individuals in the sample is limited, which is common to all donkey breeds [17]. Consequently, the general rule of thumb explains that, to succeed when conducting a linear regression analysis, the number of observations must not be smaller than 30 or $3\times$ (k + 1), where k represents the number of independent variables (covariates); hence, the sample size used in the present study fulfils all the assumptions to be used in linear regression analyses.

Contextually, Bayesian estimation methods have been reported to require a much smaller ratio of parameters to observations (1:3 instead of 1:5); that is, Bayesian inference maximizes the ability to determine significant effects for relatively limited sample sizes. These sample limitations are reflected in the broadening of confidence intervals, which must be accompanied by an acceptable Bayes factor value.

To the best of our knowledge, no previous study has reported an estimation of donkey sperm output and quality traits using a Bayesian approach. In this context, the aims of the present study are to define and compare the predictive power of two Bayesian predictive models for sperm quality parameters using linear testicular measurements, combined biometrical indices, and their relationship with age and BW as predictive factors.

2. Materials and Methods

2.1. Animals

The study was carried out in the Veterinary Teaching Hospital of University of Trás-os-Montes and Alto Douro (VTH-UTAD, Vila Real, Portugal). Animals have been evaluated with approval and in collaboration with the Association for Study and Protection of the Donkey Breed Burro de Miranda (AEPGA), in the behalf of a scientific protocol of caooperation signed between both institutions. All animal procedures were conducted in accordance with national laws for animal welfare and experimentation as with the EU Directive 2010/63/EU for animal experiments and the approval of the Directive Hospital Committee (Approval Ref. 408/VTH-UTAD).

Animals were clinically examined, and the genital tract was palpated previous to ultrasonographic (US) evaluation. Body weight (BW) (kg) was assessed using an equine digital floor weighting scale. For the testicular morphometric evaluation, 23 Miranda donkeys were considered. Animals were allocated to two groups by age; juvenile to prepubertal ($n = 6$) ($\leq$14 months) and mature ($n = 17$) ($\geq$24 months) (Table 1). Only clinically healthy animals with normal size and consistency symmetrical testis and epididymis showing no echogenic changes in the testicular parenchyma were included. Epididymis and spermatic cords were included. Either in the juvenile or adult group, only animals with both testicles at scrotal position were considered. For the assessment of sperm, a sub-group of eight Miranda mature donkeys was selected from the mature group for further sperm collection and evaluation. Exams were performed during the autumn–winter season in 2018–2019 (US evaluations of the juveniles); and spring–summer in 2019 (semen collections and US examination of adult males).

Table 1. Donkeys enrolled in the study.

N	Age Range	Age Percentile	Median Weight Evolution Per Percentile
6	7 to 14 months	14 months (P25)	200 kg
11	15 to 95 months	40 months (P50/Median)	248 kg
6	≥96 months	96 months (P75)	302 kg

P25: value at 25% of observations; P50: value at 50% of observations and P75: value at 75% of observations.

2.2. Testicular Morphometry Evaluation

US testicular measurements were obtained from 23 donkeys aged from 7 to 259 months old and with 120 to 400 kg of weight (juvenile donkeys means: 11.17 ± 2.77 months old and 160.33 ± 24.66 kg; mature donkeys means: 79.94 ± 50.25 months old and 279.47 ± 54.43 kg, respectively). US measurements were performed with Philips® CX30 Portable Ultrasound (Philips®, Amsterdam, Holland) with a sectorial 3.0–7.0 MHz transducer, following the previously described technique for stallion measurements [4]. Longitudinal and transversal plans were performed in each testicle, being the epididymis excluded from testicular US measurements. The electronic cursors were placed at the limit of tunica albuginea, and after three consecutive scans, the following parameters were obtained considering the largest measurement (cm): right and left length (L), height (H), and width (W) (cm). Right and left testicular volume (TV) were calculated using the Lambert formula, TV = L × W × H × 0.5233, used to measure the volume of an ellipsoid [4]. Total testicular volume (TTV), which represents the sum of the right and left TV, was obtained for each donkey. To compute gonadosomatic ratio (GSI) (%), i.e., testicular weight/BW, TTV (cm^3) was directly converted into grams, based on the fact that testis volume density in mammals is very close to one [18]. After US measurements, routine orchiectomy was performed on five juvenile and two adult donkeys. After surgery, the extirpated testis (n = 14) were measured—the same measurements as in vivo—using precision sliding calipers.

2.3. Semen Collection and Evaluation

A sub-group of eight jackasses, ranging between 34 and 259 months of age (214–400 kg), was selected from the mature group for semen collection and further evaluation. A total of 40 ejaculates (five ejaculates per jackass) was collected. Donkeys had been successfully used in previous natural services. Before starting the experiment, sperm collections were performed for three consecutive days to minimize the number of sperm from extra-gonadal reserves, as it has been previously reported for donkeys [19]. Collections were performed at two-day intervals and using an artificial vagina (AV) (Hannover model—Minitub Iberica S.L., Tarragona, Spain) lubricated with non-spermicidal gel (ReproJelly—Minitub Iberica S.L., Tarragona, Spain), using a jenny in heat as a mount. The AV was filled with warm water to reach and maintain an inner temperature of 50–55 °C. A sterile semen collection bottle was used in each collection. The gel fraction was removed by filtering the whole ejaculate with a nylon filter (Minitub Iberica S.L., Tarragona, Spain). Gel-free ejaculate was immediately evaluated for volume (mL), motility (%), concentration ($\times 10^6$/mL), and percentage of morphologically normal (%). Volume was measured in a graduated semen collection bottle. Then, each collected ejaculate was evaluated for sperm motility and concentration. For sperm motility evaluation, an aliquot of gel-free ejaculate was immediately extended 1:1 (vol/vol) with INRA 96 extender at 37 °C. Sperm motility was blind and subjectively estimated by the same experienced operator after the evaluation of motile spermatozoa (%) considering five different fields under light microscopy (×200), placing a semen droplet in a prewarmed (37 °C) slide covered by a cover slip. Concentration was determined using an improved Neubauer hemocytometer. Total sperm number (TSN, $\times 10^9$) was computed considering the product between the volume of gel-free ejaculates and sperm concentration, whereas total motile sperm count (TMS, $\times 10^9$) was obtained by computing the product between motility and TSN. Sperm morphology defects (head, intermediary piece, tail) were evaluated in eosin-nigrosin stained smears using light microscopy in oil immersion objective lens (×1000), counting a total of 200 sperm cells [20].

2.4. Statistical Analysis

2.4.1. Parametric Assumptions Testing and Approach Decision

Since sample size was a limitation in this study, parametric assumptions were tested to decide on the most appropriate statistical approach to follow to analyse the present data. The Shapiro–Francia W′ test (for $50 < n < 2500$ samples), Shapiro–Wilk test (for $n < 50$ samples), and Levene's test were used to discard gross violations of parametric assumptions (normality and homoscedasticity). The Shapiro–Francia W′ test was performed using the Shapiro–Francia normality routine of the test and distribution graphics package of the Stata Version 15.0 software (StataCorp [21]. Supplementary Tables S1 and S2 report a gross violation of normality assumption occurred in all variables of testicular biometry and sperm parameters ($p < 0.01$), respectively, except for gel-free volume (mL) and sperm concentration ($\times 10^6$/mL). Homoscedasticity was violated as well ($p < 0.01$); hence, a nonparametric approach was suggested.

All statistical tests, including all Bayesian procedures, were performed using the explore procedure of the descriptive statistics package in SPSS Statistics (Version 25.0, IBM Corp., Armonk, NY, USA) [22].

2.4.2. Comparative Analysis of US and Caliper Testicular Morphometry between Juvenile and Mature Jacks

Bayesian one-way ANOVA procedure was used to detect differences in the means for testicular measurements between juvenile and mature jackstocks using the Bayesian ANOVA task from the Bayesian statistics procedure of SPSS Statistics, Version 25.0, IBM Corp. [22].

2.4.3. Analysis of US Testicular Morphometry, Age and BW

Bayesian inference of Pearson's correlation was used to characterize the posterior distribution of the linear correlation between age and BW, US testicular measurements, and composite indices using the Pearson correlation task from the Bayesian statistics procedure of SPSS Statistics, Version 25.0, IBM Corp. [22]. The correlation methods used and discussed in this paper can be validly used even if we work with repeated measures as we tested independent data [23]. Furthermore, in case variable pairs tested held a perfect linear correlation $r_{xy} = 1$, the integral equation to perform Bayesian inference for Pearson's correlation would not have converged [24].

2.4.4. Analysis of US and Caliper Testicular Morphometry

Bayesian inference of Pearson's correlation was used to characterize the posterior distribution of the linear correlation between caliper testicular biometry variables ($n = 14$ testis) and US testicular biometry variables ($n = 46$ testis). The Pearson's correlation coefficient measures the pairwise linear relation between the dependent variable y and the independent variable x. When $r_{xy} = |1|$, the dependent variable y is perfectly linearly correlated with the independent variable x. Then, following a decreasing order, a coefficient of $|0.8| < r_{xy} < |1|$ suggests a strong linear correlation; a coefficient of $|0.3| < r_{xy} < |0.6|$ suggests a moderate correlation; and a coefficient of $0 < r_{xy} < |0.3|$ suggests a weak correlation, respectivelyProfillidis and Botzoris [25].

The two methods were compared to decide on whether to use US or real biometric parameters or a combination of both to build Bayesian regression models. Bayesian inference for Pearson correlation was performed using the Pearson correlation task from the Bayesian statistics procedure of SPSS Statistics, Version 25.0, IBM Corp. [22]. The aforementioned test evidenced that US and caliper measuring methods were significantly correlated.

Supplementary Table S3 summarizes the estimated Pearson's correlation pairwise coefficients and respective Bayes factors. For all measurement pairs, the estimated Pearson's correlation coefficient was always higher than 0.938, with corresponding Bayes factor of <0.001. As a result, the use of US measurements was exclusively selected to integrate the models for predicting sperm output, provided measurements were taken in vivo, hence they had a higher clinical applicability.

According to Doğan [26], although correlation analyses may erroneously detect the occurrence of incidental relationships instead of meaningful clinical/biological association, these may be a preferable choice under certain contexts. For instance, the same authors reported that one of the critical problems in other presumably more robust techniques such as the Bland–Altman analysis relies on the need for the data to meet the assumption of a normal distribution. Contrastingly, when testing Pearson's correlations, the pairs of continuous variables need not be normally distributed, although their differences should. To determine the violation of this assumption, data may be tested against the normal distribution using classical methods such as the Shapiro–Wilk test or the Kolmogorov–Smirnov test.

Additionally, the same authors reported the fact that the Bland–Altman analysis is not an appropriate method to compare items for which repeated measurements were considered, as in the present study. In these regards, Batterham [27] would suggest that in a spreadsheet-based simulation of calibration and validity studies, a Bland–Altman plot of difference versus mean values for the instrument and criterion may show a systematic proportional bias in the instrument's readings, even though none is present. This artifactual bias arises in a Bland–Altman plot of any measures with substantial random error. In contrast, a regression analysis of the criterion versus the instrument shows no bias. In this context, a regression analysis also provides complete statistics for recalibrating the instrument, if bias develops, or if random error changes since the last calibration. Consequently, the Bland–Altman analysis of validity should therefore be abandoned in favor of regression, as was performed in our study.

2.4.5. Bayesian Linear Regression Modelling for Sperm Quality and Output Predictions

Gel-free volume (mL), concentration ($\times 10^6$/mL), TSN ($\times 10^9$), motility (%), morphologically normal (%), morphologically abnormal (%), gonadosomatic ratio (GSI) (%), and TMS ($\times 10^9$) were considered the dependent variables in our study. Two separate statistical models were built, in which the predictive power of combinations of certain independent factors was evaluated.

Each of the regression models used in this study followed the general equation $y_i = X_1\beta_1 + \ldots X_i\beta_i + \varepsilon_i$, where i = 1,2, ... i is the ith number of factors; y_i is the vector of records for the aforementioned dependent variables with dimension n (217 records belonging to 31 jacks); X_i is the appropriate incidence matrix for factors; and β_i are the standardized regression coefficients for the ith number of factors and covariates considered, respectively. The general regression equation for model 1 was Y = Intercept + $\beta_{\text{age (months)}}$·age (months) + $\beta_{\text{Length LT (cm)}}$·length LT (cm) + $\beta_{\text{Length RT (cm)}}$ ·length RT (cm) + $\beta_{\text{Height LT (cm)}}$· height LT (cm) + $\beta_{\text{Height RT (cm)}}$· height RT (cm) + $\beta_{\text{Width LT (cm)}}$· width LT (cm) + $\beta_{\text{Width RT (cm)}}$·width RT (cm). Oppositely, the general regression equation for model 2 was Y = Intercept + $\beta_{\text{Age (months)}}$·Age (months) + $\beta_{\text{BW (kg)}}$· BW (kg) + $\beta_{\text{TTV (cm3)}}$·TTV (cm^3) + β_{GSI}·GSI, except for gonadosomatic ratio (GSI) (%), for which the last term in the equation was not included, provided this term refers to gonadosomatic ratio (GSI) (%) itself (β_{GSI}·GSI).

According to Carlin [28], Bayesian inferences are sensitive to the dependence of variables on time (conditional on θ and x). If such dependence is large, it needs to be modeled, or the inferences will not be appropriate. For this reason, age was considered in the models. Under this design, time (age) plays a similar role to a blocking variable or covariable. For example, suppose that $E(y \mid x, \theta)$ has a linear trend in time (age) but that this dependence is not modeled (that is, suppose that a model is fit ignoring time (age)). Then, posterior means of factors or covariables in the model will tend to be reasonable, but posterior standard deviations will be too large, because this design yields treatment assignments that, compared to complete randomization, tend to be more balanced for time (age).

As Brewer [29] suggested, in our case, the use of an intercept was necessary as an empirical need for it was detected (for instance when unstandardized coefficients are used as in the present study). In these regards, confidence intervals for the estimated intercept were used as empirical indicators for the need of the intercept. Residual effects (ε_i) were assumed to follow a normal distribution as $\varepsilon_i|X_i N(0, \sigma^2_{\varepsilon i})$, where $X_{\varepsilon i}$ is an identity matrix and $\sigma^2_{\varepsilon i}$ is residual variance, respectively. For a continuous predictor variable, such as those in the present study, unstandardized coefficients are produced by the linear regression model using the independent variables measured in their original scales.

Unstandardized coefficients β_i can be interpreted considering what was stated by Hayes, et al. [30]—that all other variables being held constant, an increase of one unit in X_i is associated with an average increase of β_i units in Y. In the sections below, a detailed summary of the priors and posterior distributions used in this study is reported. A full description of the algorithms used by SPSS to perform Bayesian Inference on Multiple Linear Regression Models in this study can be found in the public document IBM SPSS Statistics Algorithms v. 25.0. by IBM Corp. [24].

When large number of parameters are being considered in a model, quadratic approximation has been reported to be computationally faster in terms of discretization and computing the likelihood over all possible parameter combinations compared to other approximations such as the Markov Chain Monte Carlo (MCMC) methods used in this study. However, the use of this quadratic approach was not feasible given it assumes the posterior distribution follows a normal distribution. In the context of our data, this assumption cannot be presumed provided the gross violation reported for the distribution properties reported at previous assumption testing stage.

After Bayesian Pearson's correlation coefficients across variables had been performed, two distinct combinations of factors were evaluated. First, model 1 comprised the covariate of age (months) and the independent factors of LLT (length of left testicle) (cm), LRT (length of right testicle) (cm), HLT (length of left testicle) (cm), HRT (height of right testicle) (cm), WLT (cm) (width of left testicle), and WRT (width of right testicle) (cm). Second, model 2 comprised the covariate of age (months) and the factors BW (kg), TTV (total testicular volume) (cm^3) and GSI (%). Lowest correlations were found for age and any of the rest of variables, hence, the covariate was retained in both models. The value of almost 1 for the correlation found between VLT (cm^3) (volume of left testicle) and VRT (volume of right testicle) (cm^3) and TTV (cm^3) was the basis to decide on using composite TTV (cm^3), given the reduced number of variables in model 2. BW was only considered in model 2, given the high generalized close to or above 0.9 correlations that it held with biometric caliper measurements. Bayesian linear regression analyses were performed using the linear regression package from the Bayesian statistics task of SPSS Statistics, Version 25.0, IBM Corp. [22]. The Bayesian Linear Regression test routine of the linear regression and related package of the Stata Version 16.0 software process was used to compute posterior distribution statistics for each factor included in each model to predict for each dependent variable. Once the analysis had been performed, we interpreted the estimated effect of the factors considered in the resulting predictive models, their confidence intervals, and the posterior distribution statistics.

2.4.6. Jeffrey–Zellner–Siow (JZS) Mixture of g-Priors

For the present analyses, the Jeffrey–Zellner–Siow mixture of *g*-priors [31] was used. Jeffrey–Zellner–Siow's prior somehow appears as a data-dependent prior through its dependence on X_i, but this has been reported not to be a drawback since regression models are conditional on X_i. As suggested by Heck [32], JZS prior could be an alternative that may satisfy several theoretical requirements such as the equality constraint on the test-relevant parameters, for instance of β, which leads to the null hypothesis $H_0 = \beta = \beta_0$ [33]. The benefits of JSZ prior distribution had also been reported by Rouder, et al. [34] and Liang, et al. [31]. Contextually, conditional on the residual variance ($\sigma^2_{\varepsilon i}$), the JZS prior defines a multivariate Cauchy distribution for the slope parameters of the full model, as follows

$(\beta_i|\sigma^2_{\varepsilon i}) \sim MVC(0_P, \gamma_i{}^2\sigma^2_{\varepsilon i}C_i{}^{-1})$, which is defined by a P-dimensional zero vector (location vector) and a scale matrix. The constant γ_i determines the amount of scaling, which is chosen by the user a priori, the residual variance $\sigma^2_{\varepsilon i}$, and the matrix $C_i = X_i'X_i/N_i$, which is the covariance matrix of the centred design matrix X_i.

There are qualities of the JZS prior [34] that make it especially appropriate when performing linear regression analyses. Among these, the prior is symmetric and centered at zero in line with the predictive matching criterion as reported by Bayarri, et al. [35]. As a result, positive and negative values of the slope parameters have a priori the same probability to occur. Furthermore, JZS prior is scale invariant, thus the resulting Bayes factor does not depend on the scale of both the dependent variable and factors or covariates, hence results do not change when different unit variables are evaluated together, which is common in field conditions studies.

This independence from the measurements of model elements is achieved by scaling the multivariate Cauchy distribution by the residual variance $\sigma^2_{\varepsilon i}$ (a priori, a larger residual variance implies larger slopes) and by the inverse of the covariance matrix C_i (a priori, a covariate with a larger variance implies smaller slopes). It may be worth considering that the procedure of defining a scaled prior for unstandardized coefficients (β_i) equals the process of defining a prior for standardized coefficients (β_i^*) [34].

Third, the scale parameter γ is fixed to a constant by the user, which allows prior beliefs to be specified about the expected effect size. IBM Corp. [24] algorithm guide reports that the algorithm of JZS prior for linear regression analyses, to compute the Bayes Factor uses the default value of $\gamma = 2\sqrt{\pi} = 3.5$, which reflects a prior belief of a medium effect size. For a single covariate x, this choice implies that the standardized regression slope $\beta_i^* = \beta_i \cdot SD(x_i)/\sigma_i$ has an a priori probability of 53.2% of being in the range [−0.50, +0.50].

Authors such as Rouder and Morey [36] also reported additional theoretical advantages of the JZS prior, such as its consistency in model selection (the fact that the Bayes factor, goes to infinity as the number of observations N increases without bound-favoring the data-generating model) or consistency in information (the Bayes factor for a certain effect goes to infinity as the proportion of explained variance or R Squared (R^2) increases to 1). Additionally, Bayes factors for JZS prior can be relatively easily and highly precisely computed [37] and has been adapted for the default *t*-test [38], ANOVA [34], and linear regression [32].

2.4.7. Factor and Covariate Effects Bayesian Modelling (FCEBM)

Being y_i, any of the effects of any of the independent variables (covariates) considered in this study, the posterior distribution of y_i in the context of the data D is

$$p(y_i/D) = \sum_{i=20}^{i} p(y_i|M_i, D)\, p(M_i|D)$$

This is an average of the posterior distributions of each model, weighted by the corresponding posterior model probabilities. In the previous equation, the posterior predictive distribution of y_i given a particular model M_i is

$$p(y_i|M_iD) = \int p(y_i|\beta_i, M_i, D)p(\beta_i|M_iD)d\beta_i$$

and the posterior probability of model M_i is given by

$$p(M_i|D) = \frac{p(D|M_i)p(M_i)}{\sum_{i=20}^{i} p(D|M_i)p(M_i)}$$

where

$$p(M_i|D) = \int p(D|\beta_i, M_i)p(\beta_i|M_i)d\beta_i$$

is the integrated likelihood of model M_i, β_i is the vector of parameters of model M_i, $p(\beta_i|M_i)$ is prior density of β_i under model M_i, $p(D|\beta_i|M_i)$ is the likelihood, and $p(M_i)$ is the prior probability that M_i is the true model.

For a problem with P potential covariates, the number of models, K, can be enormous ($K = 2^P$ in the absence of other constraints). Only a small number of these models will have much support from the data, thus be selected by SPSS for each covariate. Marginal posterior distributions of all unknowns were estimated using the Gibbs sampling algorithm.

2.4.8. Factors and Covariate Effect Bayesian Interpretation (CEBI)

The checklist proposed by Depaoli and Van de Schoot [39] was used to detect issues to check before estimating the model, (b) issues to check after estimating the model but before interpreting results, (c) understanding the influence of priors, and (d) actions to take after interpreting results.

Interpreting the effect of each particular covariate (independent variables used in this study) can be made as follows.

First, the posterior probability $p[\beta_i^* \neq 0/D]$ expresses the likelihood that the factor or covariate has an effect on each particular variable. Standard rules of thumb [40] for interpreting this posterior probability are as follows: <50% evidence against the effect; 50–75% weak evidence for the effect; 75–95% positive evidence; 95–99% strong evidence; >99% very strong evidence, whose results are comparable to commonly used thresholds to define significance of evidence through Bayes factor (BF) as reported in Supplementary Table S4.

Second, posterior distribution estimates (means) are used to measure the magnitude of the effect of a particular factor and covariate. For continuous predictor variables (metric covariates), such as the numeric variables used in this study, the regression coefficient represents the difference in the predicted value of the response variable for each one-unit change in the predictor variable, assuming all other predictor variables are held constant. When response variables are metric and can readily be interpreted in terms of impact, such as the ones in our study, β regression coefficients effect sizes by themselves.

Third, the 95% credibility interval shows that there is a 95% probability that these regression coefficients (posterior distribution mean value for each covariate and factor) in the population lie within the corresponding credibility intervals. When 0 is not contained in the credibility interval, a significant effect for such factor is detected.

Supplementary Table S5 report a summary of posterior distribution statistics from Bayesian unstandardized linear (β) regression coefficients for each of the aforementioned variables considered in the analyses and a summary of Bayesian ANOVA outputs to test for differences in the means for US and caliper testicular measurements between juvenile ($n = 6$) and mature donkeys ($n = 17$).

2.4.9. Convergence Criterion

The rounds of iteration continued until a tolerance convergence criterion of 10^{-8} was reached as suggested in literature [41]. Once the convergence criterion was reached, initial parameters were set, and model fitting properties were evaluated. The maximum number of iteration rounds used for each analysis was 2000 as suggested in IBM SPSS Statistics Algorithms version 25.0 by IBM Corp. [24]. This convergence criterion was chosen provided it has been used in Bayesian ANOVA and linear regression analyses in research contexts of limited sample sizes [42].

2.4.10. Model Validity, Explanatory Power of Present Data, and Predictive Power of Future Data

The process of validation and comparison of Bayesian model is fully mathematically described in Geweke [43]. In this context, some authors [44] have suggested a correct proof for model validation should be based on the mean square error (MSE) of the models being evaluated. Additionally, although mean square residual or error (MSE) and minimum mean-square residual or error (MMSE) have been used and widely reported to measure how close a regression line is to a set of points (how good a certain model fits the data being observed), mean square prediction error or MSPE (= RSS/no. of observations) was chosen to measure error variation given MSE has been reported to be influenced by the number of predictors [26] in cases of reduced sample sizes [45,46].

Residual sum of squares (RSS) measures the amount of variance in a data set that is not explained by a regression model. That is, if we consider a regression to be a measure of the strength of the relationship between a dependent variable and an independent variable from a set of independent variables, then the RSS measures the amount of error remaining between the regression function and the data set—hence, it essentially determines how well a regression model explains or represents the data in the model. A smaller RSS figure represents a better suitability of the regression function to model for the data that it is intended to model.

In Bayesian inference, Monte Carlo Standard Error (MCSE) is another measure of accuracy of the chains. It is defined as the standard deviation of the chains divided by their effective sample size. MCSE has been reported to be the nonparametric or Bayesian counterpart of MSPE, and has been suggested to be used as the validation criteria in Bayesian Linear Regression model comparison studies [47].

Bayes factor (BF) provides an indirect measure of the explanatory power of the model to describe presently observed data (in our study). Larger BFs imply higher likelihoods for the combination of factors considered to explain the response variables being modelled. Commonly used thresholds to define significance of evidence following the premises by Jeffreys [48] and Lee and Wagenmakers [49] are reported in Table S4. Intrinsically related to BF, Bayesian R^2 can be considered as a data-based estimate of the proportion of variance explained for data. Additionally, acceptance rate, efficiency, and Monte Carlo standard error (MCSE) were used to determine the validity of the Bayesian methods implemented. Supplementary Table S4 reports a summary of the description and interpretation of each model validity parameter used. Bayesian statistics predictive accuracy of the model [50] can be estimated through posterior predictive checking [51] (Supplementary Table S6).

BIC was then calculated, as it explains how well the model will predict on new data. Bayesian information criterion (BIC) or Schwarz information criterion (also SIC, SBC, SBIC) was computed as follows:

$$\text{BIC} = \text{N} * \text{N}\ \ln(\text{MSPE}) + \text{K} * \ln(\text{N}) \tag{1}$$

where MSPE is the mean squared prediction error, N is the number of observations or records, and K is the number of independent parameters of the model.

BIC was evaluated to compare predictive power across models. To summarize, BIC considers both the statistical goodness of fit and the number of parameters that have to be estimated to achieve this particular degree of fit, by imposing a penalty the number of parameters is increased [52,53]. BIC measures the trade-off between model fit and complexity of the model to determine [54]. Lower BIC values suggest that a particular model should have improved prediction properties in comparison to models for which higher values have been reported. In these regards, Bayesian R^2 answers a different question as Bayesian R^2 estimates the explanatory power of observed data, when the model is regression and non-adjusted R^2 is used.

Frequently, when more variables are added, model predictive accuracy decreases. Consequently, a model with higher R^2 will have higher-hence, worse-BIC values. The addition of "noise" variables to the fit (for which a relationship has not been suggested) will increase R^2 values, but it will also decrease predictive power of the model. Hence, the model with more "noise" variables will have higher R^2 and higher BIC.

3. Results

3.1. Descriptive Analysis for US Testicular Morphometry, Combined Biometric Indices and Sperm Output

Table 2 reports a summary of the descriptive statistics for age, BW, ultrasonographic (US) and caliper testicular morphometry (cm); L, H, W, TTV, GSI for juvenile and mature donkeys (n = 46 testis). Descriptive statistics of US and caliper measurements were computed for each group to perform a comparative analysis (Supplementary Table S7a,b). In the juvenile group, mean TTV (cm^3) was 17.74 $\pm$ 9.89 (n = 12 testis), while in mature group, TTV (cm^3) was 271.69 $\pm$ 133.21 (n = 34 testis).

In the juvenile group, there was a progressive increase in all US testicular measurements, namely in TTV, from seven to 24 months, which was especially noticeable after 11 months of age. At 12–14 months, mean TTV was 21.05 $\pm$ 9.30 cm^3 (n = 10 testis), and at 24–26 months, TTV was 85.27 $\pm$ 18.66 cm^3 (n = 4 testis) ($P < 0.01$). Additionally, an increase in TTV was described after 150 kg of BW had been attained, which was verified in all donkeys after 12 months. On the other hand, after 168 months of age, a gradual decrease in TTV was noted. No difference between left and right testicle was found. Gonadosomatic ratio (GSI) (%) means in juveniles was 0.11 $\pm$ 0.06 and in matures 0.95 $\pm$ 0.39. Significant differences ($p < 0.001$) were found between juvenile and mature groups for all testicular biometrical parameters (Table S5).

Results of sperm output and quality parameters (n = 40 ejaculates, observational unit); gel-free volume (mL), motility (%), concentration ($\times 10^6$/mL), TSN ($\times 10^9$), TMS ($\times 10^9$), normal and abnormal sperm morphology (%) are presented in Table 1. TSN and TMS means was 18.453 $\pm$ 1.936 $\times 10^9$ sperm and 13.555 $\pm$ 1.479 $\times 10^9$ motile sperm, respectively. Sperm morphological abnormalities description can be consulted in Table S8.

3.2. Statistical Analyses

3.2.1. Bayesian Pearson's Correlation Coefficients Preliminary Testing

Following a probabilistic view of regression, it can be assumed that any dependent variable (Y) has a certain associated variance σ^2. Linear regression bases on identifying the weight vector from observed data of a dependent variable to then use it to make predictions. For the model to be stable enough, the variance of the weight vector (W_{ls}) should be low. If weight vectors variance is high, it means that the model is very sensitive to data. The weights differ largely with observed data if the variance is high. This means that the model might not perform well with observed data. When highly correlated covariables are used in regression models, the variance of the weight vector will be large. This occurs because when highly correlated features (covariates or factors) are considered, the values in the Singular Value Decomposition "S" matrix will be small. Hence inverse square of "S" matrix (S^{-2}) will be large which makes the variance of W_{ls} large. For these reasons, Pearson's correlation coefficients must be tested prior to performing regression analyses.

Table 3 summarizes the estimated sample Pearson's correlation coefficient and the Bayes factors for BW (kg), age (months), US testicular biometric parameters and composite indices. For all variable pairs, the estimated Pearson's correlation coefficient was always higher than 0.461, with a corresponding Bayes factor of <0.001, in all cases. Besides, moderate to high Bayesian inference Pearson's correlation coefficients were found between age, BW, and testicular biometric variables. Pearson's correlation coefficients between testicular biometry and BW were always >0.778, whereas Pearson's correlation coefficients between testicular biometry and age were >0.467.

Table 2. Descriptive statistics for testicular US measurements in 23 donkeys (n = 46 testis, observational sample) and precision caliper after orchiectomy in seven of these donkeys (n = 14 testis, observational sample) and sperm quality parameters (n = 40 ejaculates, observational sample).

Items	N	Mean	SEM	SD	Skewness	Kurtosis	Minimum	Percentile 25	Median	Percentile 75	Maximum
Body Weight (kg)	161	248.39	5.63	71.38	0.23	−0.56	120.00	200.00	248.00	302.00	400.00
Age (months)	161	62.00	4.66	59.07	1.77	3.28	7.00	14.00	40.00	96.00	259.00
US Length LT (cm)	161	6.94	0.19	2.42	−0.51	−1.07	2.80	3.87	7.50	8.76	10.60
US Length RT (cm)	161	6.81	0.20	2.51	−0.50	−1.07	2.36	4.10	7.57	8.87	10.10
US Height LT (cm)	161	4.16	0.12	1.57	−0.11	−1.07	1.50	2.63	4.51	5.50	6.93
US Height RT (cm)	161	3.96	0.12	1.48	0.15	−0.01	1.40	2.56	4.26	4.94	7.61
US Width LT (cm)	161	5.18	0.16	1.98	−0.44	−1.11	1.50	3.32	5.42	6.86	7.88
US Width RT (cm)	161	5.17	0.16	1.97	−0.37	−1.13	1.60	3.07	5.69	6.69	8.40
US Volume LT (cm^3)	161	106.60	6.56	83.22	0.45	−0.82	3.30	17.95	95.58	175.20	283.91
US Volume RT (cm^3)	161	98.88	6.13	77.80	0.65	−0.10	2.93	19.28	91.39	136.63	297.64
US TTV (cm^3)	161	205.44	12.62	160.08	0.52	−0.53	6.23	37.23	185.30	329.08	581.54
US GSI (%)	161	0.73	0.04	0.50	0.37	−0.60	0.04	0.22	0.72	1.09	1.86
Caliper Length LT (cm)	49	4.70	0.27	1.86	1.24	−0.06	3.30	3.50	3.70	6.40	8.50
Caliper Length RT (cm)	49	4.83	0.29	2.06	1.13	−0.13	3.00	3.20	4.00	6.60	9.00
Caliper Height LT (cm)	49	3.13	0.13	0.94	0.61	−1.37	2.20	2.30	2.70	4.50	4.50
Caliper Height RT (cm)	49	3.06	0.15	1.05	0.51	−1.46	2.00	2.00	2.50	4.50	4.60
Caliper Width LT (cm)	49	3.23	0.20	1.41	1.00	−0.37	1.90	2.00	2.50	4.50	6.00
Caliper Width RT (cm)	49	3.26	0.23	1.63	1.09	−0.31	1.90	2.00	2.50	4.80	6.50
Caliper Volume LT (cm^3)	49	35.55	5.73	40.10	1.34	0.25	7.94	8.35	14.13	67.81	120.10
Caliper Volume RT (cm^3)	49	38.85	6.70	46.93	1.33	0.25	6.59	7.33	10.46	76.25	137.76
Caliper TTV (cm^3)	49	74.40	12.43	86.99	1.34	0.26	14.53	17.89	21.46	144.06	257.86
Caliper GSI (%)	49	0.35	0.05	0.35	1.16	−0.25	0.10	0.10	0.14	0.69	1.05
Gel-free volume (mL)	40	75.09	6.49	41.07	0.51	0.19	12.00	38.25	75.25	103.50	189.00
Concentration (× 10^6/mL)	40	281.00	21.03	133.00	0.02	−0.41	45.00	213.75	282.50	363.75	540.00
TSN (× 10^9) sperm	40	18.45	1936.98	12,250.54	1.27	2.05	4560.00	8482.50	15,750.00	25,653.75	59,360.00
Motility (%)	40	72.13	2.60	16.44	−1.35	1.80	20.00	60.00	77.50	85.00	90.00
Morphologically normal sperm (%)	40	87.35	1.58	9.97	−1.46	1.95	58.00	83.00	90.00	94.00	99.00
Morphologically abnormal sperm (%)	40	12.43	1.52	9.61	1.52	2.45	1.00	6.00	10.00	17.00	42.00
TMS (× 10^9) sperm	40	13,555.38	1479.60	9357.81	0.91	0.76	1650.00	5607.75	11,264.50	20,300.00	42,642.00

LT—left testicle; RT—right testicle; TTV—total testicular volume; BW—body weight; TSN—total sperm number; GSI—gonadosomatic ratio; TMS—total motile sperm count.

Table 3. Bayesian inference Pearson's correlation output summary for BW (kg), age (months), US parameters, and composite indices.

	Body Weight (kg)	Age (months)	Length LT (cm)	Length RT (cm)	Height LT (cm)	Height RT (cm)	Width LT (cm)	Width RT (cm)	Volume LT (cm^3)	Volume RT (cm^3)	TTV (cm^3)	GSI (%)
Body Weight (kg)	1.000	0.552	0.845	0.876	0.825	0.778	0.826	0.824	0.797	0.805	0.806	0.680
Age (months)	0.552	1.000	0.511	0.600	0.479	0.523	0.522	0.556	0.467	0.539	0.505	0.461
Length LT (cm)	0.845	0.511	1.000	0.977	0.944	0.924	0.942	0.925	0.916	0.908	0.918	0.917
Length RT (cm)	0.876	0.600	0.977	1.000	0.948	0.897	0.957	0.946	0.917	0.911	0.920	0.903
Height LT (cm)	0.825	0.479	0.944	0.948	1.000	0.908	0.940	0.926	0.962	0.922	0.949	0.939
Height RT (cm)	0.778	0.523	0.924	0.897	0.908	1.000	0.901	0.858	0.897	0.930	0.919	0.914
Width LT (cm)	0.826	0.522	0.942	0.957	0.940	0.901	1.000	0.962	0.922	0.901	0.917	0.923
Width RT (cm)	0.824	0.556	0.925	0.946	0.926	0.858	0.962	1.000	0.903	0.900	0.908	0.909
Volume LT (cm^3)	0.797	0.467	0.916	0.917	0.962	0.897	0.922	0.903	1.000	0.976	0.994	0.964
Volume RT (cm^3)	0.805	0.539	0.908	0.911	0.922	0.930	0.901	0.900	0.976	1.000	0.993	0.951
TTV (cm^3)	0.806	0.505	0.918	0.920	0.949	0.919	0.917	0.908	0.994	0.993	1.000	0.964
GSI (%)	0.680	0.461	0.917	0.903	0.939	0.914	0.923	0.909	0.964	0.951	0.964	1.000

BF < 0.0001; GSI—gonadosomatic ratio (%).

3.2.2. Bayesian Linear Regression Modelling for Sperm Quality and Output Predictions

Model Explicative Power

Bayesian determination coefficients (R^2) or percentages of variance captured for each of the two models and their respective Bayes factors are provided in Table 4. Both models were considerably more likely than others comprising just the intercept.

Bayesian estimates of linear regression coefficients for predictive models 1 and 2 for gel-free volume, concentration, morphologically normal or abnormal, TSN, GSI, motility, and TMS are presented in Tables 5–7. The intercept term in the regression evidences the average expected value for the response variable when all of the predictor variables are equal to zero.

Predictive Power and Model Validity

Posterior predictive *P* values for models 1 and 2 were around 0.331. The combination of predictors in model 1 evidenced a higher likelihood to predict for gel-free volume (mL), concentration ($\times 10^6$/mL), and motility (%) (BIC: 387.587 to 534.480).

Table 4. Bayes Factor Model Summary for model 1 (comprising age and testicular morphometric parameters) and model 2 (comprising age, BW, TTV, and GSI) to predict for sperm output and quality in Miranda donkey breed.

Model 1	**Bayes Factor**	**R**	**R Squared**	**Adjusted R Squared**
Gel-free volume (mL)	406,756.54	0.855	0.731	0.682
Concentration ($\times 10^6$/mL)	1554.89	0.788	0.621	0.553
TSN ($\times 10^9$)	1308.11	0.786	0.617	0.548
Motility (%)	180.53	0.754	0.568	0.490
Morphologically normal (%)	47,305.85	0.832	0.693	0.637
Morphologically abnormal (%)	8839.07	0.812	0.660	0.598
GSI	1.38×10^{19}	0.980	0.961	0.954
TMS ($\times 10^9$)	52,401.57	0.833	0.695	0.639
Model 2	**Bayes Factor**	**R**	**R Squared**	**Adjusted R Squared**
Gel-free volume (mL)	252,538.00	0.794	0.630	0.599
Concentration ($\times 10^6$/mL)	1169.55	0.706	0.498	0.457
TSN ($\times 10^9$)	370.37	0.682	0.465	0.420
Motility (%)	5160.20	0.734	0.539	0.500
Morphologically normal (%)	1,907,536.17	0.819	0.670	0.643
Morphologically abnormal (%)	259,329.18	0.794	0.631	0.600
GSI	4.89×10^{44}	0.999	0.998	0.998
TMS ($\times 10^9$)	7080.51	0.740	0.547	0.509

TSN—total sperm number; TMS—total motile sperm count; GSI—gonadosomatic ratio.

Yet, the combination of predictors in M=model 2 evidenced a higher likelihood to predict for TSN ($\times 10^9$), morphologically normal and abnormal spermatozoa (%), TMS ($\times 10^9$) and gonadosomatic ratio (GSI) (%), (BIC: −40.559 to 34,635.240). Age-related effects were verified on the following parameters: gel-free volume, morphologically abnormal spermatozoa (%), and TSN (Tables 5 and 6). The summary of the results for the parameters of validity of both models is reported in Table 8.

Table 5. Bayesian Estimates of Unstandardized Linear Regression Coefficients for predictive model 1 for gel-free volume (mL), concentration ($\times 10^6$/mL), morphologically normal (%), and morphologically abnormal (%) sperm output in Miranda donkey breed.

Parameter	Posterior			95% Credible Interval		Parameter	Posterior			95% Credible Interval	
Gel-free volume (mL)	Mean	SD	MCSE	Lower Bound	Upper Bound	Morphologically normal (%)	Mean	SD	MCSE	Lower Bound	Upper Bound
(Intercept)	6.750	85.650	7.882	11.940	−159.039	(Intercept)	38.820	61.999	17.656	35.535	−69.634
Age (months)	0.554	0.186	0.020	0.551	0.190	Age (months)	0.009	0.120	0.033	0.016	−0.273
Length LT (cm)	−7.618	23.395	2.210	−8.748	−52.776	Length LT (cm)	7.176	15.183	4.290	8.154	−29.472
Length RT (cm)	−8.047	19.952	1.350	−7.762	−45.424	Length RT (cm)	−7.758	6.349	0.756	−7.367	−21.025
Height LT (cm)	39.055	10.577	0.513	39.257	16.789	Height LT (cm)	4.253	2.752	0.142	4.226	−1.076
Height RT (cm)	−0.142	15.488	1.598	−0.446	−30.291	Height RT (cm)	−2.344	10.261	2.868	−3.001	−20.264
Width LT (cm)	−8.293	13.214	1.203	−8.355	−35.057	Width LT (cm)	2.925	5.712	1.243	3.167	−7.979
Width RT (cm)	−0.867	16.024	1.877	−0.780	−32.031	Width RT (cm)	3.021	6.681	1.586	2.709	−10.081
Concentration ($\times 10^6$/mL)	Mean	SD	MCSE	Lower Bound	Upper Bound	Morphologically abnormal (%)	Mean	SD	MCSE	Lower Bound	Upper Bound
(Intercept)	119.325	95.837	6.792	120.692	−67.802	(Intercept)	−92.566	4.923	1.290	−92.168	−102.543
Age (months)	−1.675	0.438	0.024	−1.672	−2.510	Age (months)	0.268	0.036	0.002	0.267	0.199
Length LT (cm)	−179.851	51.735	14.252	−170.219	−289.282	Length LT (cm)	28.736	4.098	0.370	28.543	21.311
Length RT (cm)	112.455	63.065	7.805	114.114	−13.492	Length RT (cm)	1.078	5.696	0.306	1.182	−10.506
Height LT (cm)	−82.447	38.657	3.239	−84.282	−154.495	Height LT (cm)	−5.022	2.742	0.145	−5.009	−10.399
Height RT (cm)	60.562	33.694	8.149	57.459	2.457	Height RT (cm)	−22.037	2.535	0.272	−21.978	−27.089
Width LT (cm)	143.483	47.524	9.710	138.746	58.199	Width LT (cm)	8.980	3.366	0.145	8.888	2.389
Width RT (cm)	12.984	51.340	8.872	9.472	−80.658	Width RT (cm)	−15.696	3.963	0.276	−15.608	−23.572

LT—left testicle; RT—right testicle.

Table 6. Bayesian Estimates of Unstandardized Linear Regression Coefficients for predictive model 1 for TSN, gonadosomatic ratio (GSI) (%), motility (%), and TMS ($\times 10^9$) sperm output in Miranda donkey breed.

Parameter	Posterior			95% Credible Interval		Parameter	Posterior			95% Credible Interval	
TSN	Mean	SD	MCSE	Lower Bound	Upper Bound	Gonadosomatic ratio (GSI)	Mean	SD	MCSE	Lower Bound	Upper Bound
(Intercept)	23,753.340	80.646	4.744	23,753.840	23,598.750	(Intercept)	−0.600	0.035	0.003	−0.602	−0.670
Age (months)	−1416.025	443.100	133.800	−1451.571	−2085.682	Age (months)	0.000	0.000	0.000	0.000	−0.001
Length LT (cm)	8883.943	111.914	30.604	8885.097	8667.620	Length LT (cm)	0.048	0.031	0.003	0.046	−0.014
Length RT (cm)	13,450.930	78.239	21.894	13,444.170	13,319.860	Length RT (cm)	−0.086	0.036	0.004	−0.086	−0.156
Height LT (cm)	−33,418.180	131.843	37.637	−33,411.060	−33,681.770	Height LT (cm)	0.154	0.029	0.003	0.153	0.099
Height RT (cm)	3060.495	62.605	4.230	3059.644	2940.727	Height RT (cm)	0.097	0.027	0.002	0.097	0.046
Width LT (cm)	19,157.550	69.930	5.174	19,157.660	19,020.890	Width LT (cm)	0.036	0.030	0.003	0.037	−0.034
Width RT (cm)	−17,716.000	24.807	1.749	−17,716.020	−17,763.690	Width RT (cm)	0.072	0.025	0.002	0.071	0.026
Motility (%)	Mean	SD	MCSE	Lower Bound	Upper Bound	TMS ($\times 10^9$)	Mean	SD	MCSE	Lower Bound	Upper Bound
(Intercept)	−12.707	90.778	13.158	−11.443	−185.472	(Intercept)	−782.074	1094.008	328.918	−1260.849	−1857.071
Age (months)	−0.069	0.175	0.022	−0.070	−0.390	Age (months)	−612.420	250.955	74.637	−643.819	−958.823
Length LT (cm)	16.644	21.823	3.203	15.668	−24.687	Length LT (cm)	−3816.467	87.995	16.210	−3819.298	−3982.185
Length RT (cm)	12.235	11.157	0.668	11.796	−9.577	Length RT (cm)	4859.292	206.757	60.216	4806.093	4601.664
Height LT (cm)	−5.247	5.292	0.369	−5.314	−15.183	Height LT (cm)	−2809.112	2318.070	702.958	−1870.947	−7896.015
Height RT (cm)	−11.939	14.762	2.118	−11.310	−40.592	Height RT (cm)	536.574	140.014	15.828	524.424	295.207
Width LT (cm)	2.605	9.550	0.603	2.700	−16.521	Width LT (cm)	1678.912	2687.695	816.107	484.096	−742.483
Width RT (cm)	−13.494	10.278	0.916	−13.619	−33.146	Width RT (cm)	−7924.881	837.695	252.844	−7632.309	−9824.660

Table 7. Bayesian Estimates of Unstandardized Linear Regression Coefficients for predictive model 2 for gel-free volume (mL), concentration ($\times 10^6$/mL), morphologically normal (%) and morphologically abnormal (%), TSN, gonadosomatic ratio (GSI) (%), motility (%), and TMS ($\times 10^9$) sperm output in Miranda donkey breed.

Parameter		Posterior		95% Credible Interval		Parameter		Posterior		95% Credible Interval	
Gel-free volume (mL)	Mean	SD	MCSE	Lower Bound	Upper Bound	Morphologically normal (%)	Mean	SD	MCSE	Lower Bound	Upper Bound
(Intercept)	18.027	74.948	4.027	17.645	−131.021	(Intercept)	101.165	48.993	2.230	102.862	6.593
Age (months)	0.380	0.075	0.003	0.386	0.228	Age (months)	−0.088	0.028	0.001	−0.089	−0.141
BW (kg)	0.059	0.241	0.013	0.057	−0.408	BW (kg)	−0.048	0.153	0.007	−0.051	−0.353
TTV (cm^3)	0.229	0.206	0.011	0.237	−0.183	TTV (cm^3)	0.082	0.130	0.006	0.086	−0.178
GSI	−61.537	63.042	3.287	−61.751	−178.796	GSI	−16.479	39.912	1.736	−17.993	−95.823
Concentration ($\times 10^6$/mL)	Mean	SD	MCSE	Lower Bound	Upper Bound	Morphologically abnormal (%)	Mean	SD	MCSE	Lower Bound	Upper Bound
(Intercept)	34.599	87.218	4.216	29.855	−135.220	(Intercept)	−34.631	48.161	2.471	−33.664	−130.104
Age (months)	−1.281	0.246	0.012	−1.287	−1.751	Age (months)	0.098	0.027	0.002	0.098	0.045
BW (kg)	1.577	0.397	0.018	1.568	0.802	BW (kg)	0.151	0.151	0.008	0.147	−0.133
TTV (cm^3)	−0.434	0.282	0.013	−0.443	−1.002	TTV (cm^3)	−0.167	0.128	0.006	−0.168	−0.424
GSI	71.418	77.349	4.329	72.597	−84.034	GSI	43.228	39.364	1.902	43.112	−30.970
TSN	Mean	SD	MCSE	Lower Bound	Upper Bound	Gonadosomatic ratio (GSI) (%)	Mean	SD	MCSE	Lower Bound	Upper Bound
(Intercept)	11,264.220	1978.212	598.275	10,786.800	8794.520	(Intercept)	1.213	0.036	0.001	1.214	1.140
Age (months)	4577.180	417.008	125.933	4671.496	3705.060	Age (months)	−0.001	0.000	0.000	−0.001	−0.001
BW (kg)	−5091.494	439.681	133.092	−5203.422	−5634.810	BW (kg)	−0.004	0.000	0.000	−0.004	−0.004
TTV (cm^3)	2606.332	220.725	66.742	2662.469	2131.406	TTV (cm^3)	0.003	0.000	0.000	0.003	0.003
GSI	−4379.107	756.935	227.540	−4209.127	−5956.150						
Motility (%)	Mean	SD	MCSE	Lower Bound	Upper Bound	TMS ($\times 10^9$)	Mean	SD	MCSE	Lower Bound	Upper Bound
(Intercept)	10.199	65.301	3.019	8.928	−114.356	(Intercept)	−4076.090	195.502	54.547	−4028.317	−4586.811
Age (months)	−0.116	0.043	0.002	−0.116	−0.202	Age (months)	−1422.795	311.451	88.829	−1524.509	−1776.786
BW (kg)	0.223	0.206	0.010	0.221	−0.173	BW (kg)	569.508	82.412	14.404	572.642	397.312
TTV (cm^3)	−0.152	0.175	0.008	−0.156	−0.484	TTV (cm^3)	−172.885	46.034	7.844	−172.536	−261.502
GSI	51.775	53.600	2.428	54.210	−54.645	GSI	2298.200	133.412	29.547	2281.863	2070.603

TTV—total testicular volume; TSN—total sperm number; TMS—total motile sperm count; GSI—gonadosomatic ratio.

Table 8. Model validity parameters.

Model 1	**Gel-Free Volume (mL)**	**Concentration (× 10^6/mL)**	**TSN**	**Motility (%)**	**Morphologically Normal (%)**	**Morphologically Abnormal (%)**	**Gonadosomatic Ratio (GSI)**	**TMS (× 10^9)**
MCMC iterations	12.500	12.500	12.500	12.500	12.500	12.500	12.500	12.500
Burn-in	2.500	2.500	2.500	2.500	2.500	2.500	2.500	2.500
MCMC sample size	10.000	10.000	10.000	10.000	10.000	10.000	10.000	10.000
Number of obs	40.000	40.000	40.000	40.000	40.000	40.000	40.000	40.000
Acceptance rate	0.299	0.225	0.800	0.314	0.325	0.292	0.345	0.539
Min efficiency	0.007	0.001	0.001	0.005	0.001	0.001	0.015	0.001
Avg efficiency	0.023	0.010	0.011	0.019	0.008	0.029	0.037	0.002
Max efficiency	0.083	0.033	0.029	0.065	0.037	0.059	0.071	0.008
Log marginal likelihood	−209.558	−259.862	−132,346.360	−186.416	−164.579	−166.624	−31.747	−5993.399
BIC	433.872	534.480	264,707.476	387.587	343.913	348.003	78.249	12,001.553
Model 2	**Gel−Free Volume (mL)**	**Concentration (× 10^6/mL)**	**TSN**	**Motility (%)**	**Morphologically Normal (%)**	**Morphologically Abnormal (%)**	**Gonadosomatic Ratio (GSI)**	**TMS (× 10^9)**
MCMC iterations	12.500	12.500	12.500	12.500	12.500	12.500	12.500	12.500
Burn-in	2.500	2.500	2.500	2.500	2.500	2.500	2.500	2.500
MCMC sample size	10.000	10.000	10.000	10.000	10.000	10.000	10.000	10.000
Number of obs	40.000	40.000	40.000	40.000	40.000	40.000	40.000	40.000
Acceptance rate	0.302	0.410	0.467	0.348	0.353	0.402	0.300	0.705
Min efficiency	0.032	0.032	0.001	0.036	0.045	0.031	0.049	0.001
Avg efficiency	0.049	0.056	0.001	0.051	0.063	0.049	0.064	0.002
Max efficiency	0.108	0.125	0.001	0.088	0.129	0.099	0.095	0.003
Log marginal likelihood	−215.503	−263.929	−17,310.242	−186.604	−163.466	−163.531	27.657	−1746.014
BIC	445.762	542.613	34,635.240	387.964	341.687	341.817	−40.559	3506.783

4. Discussion

4.1. Testicular Morphometry (Juveniles and Matures) and Sperm Quality Parameters in Miranda Donkey Breed

The indication that testis biometry could provide a quantitative indication of sperm production has been previously reported in bulls [55,56], bucks [57,58], and dogs [59,60]. In the horse, morphometric, ultrasonographic-echotextural, and histomorphometric studies have been carried out [3,59,60], which evidenced the relation between testicular dimensions and sperm outputs [2] and the contribution of the ultrasonographic (US) evaluation in the accurate evaluation of the testicular functional status [61].

Albeit less than in horses, some studies on testicular morphometry have been conducted in donkey breeds such as Brazilian Pêga [6,62]; Ethiopian [63]; Egyptian [64,65]; and in the Italian breeds, Ragusano [19] and Martina Franca [66,67]. These studies have addressed the considerable existing variation among donkey breeds, which led to the need of investigating testicular dimensions in Miranda donkey breed. Besides, no previous works on Bayesian approaches to predict for sperm output and quality in donkeys has been conducted to the knowledge of the authors.

Mean US values of TTV of 271.69 cm^3 ($\pm$133.21) obtained in mature Miranda donkeys were higher than those found in Egyptian donkeys [64], similar to those in Brazilian Pêga donkeys [62], and lower than those reported for Ethiopian donkeys [63]. In comparison to other morphologically similar breeds to Miranda donkey, our values were similar to slightly lower than TTV values found in Ragusano and Martina Franca donkeys [19,66]. Even if all the aforementioned breeds were medium to large-sized, differences of TTV could still be attributed to BW, age and management conditions of the males selected for the studies.

In the juveniles, studies are still scarce, but the values in our study (17.74 $\pm$ 9.89 cm^3) were similar to those described for the prepubertal Egyptian [68] and Amiata donkeys [67]. Donkeys between 10 to 14 months are still in their pubertal transition period and, which reaches its end at 19–20 months of age, when testis have presumably completed their descent into the scrotum [37]. In the present work, a rapid increment of TV was verified after 11–12 months, which, besides, was simultaneous to the increase of BW. According to the work by Rota, et al. [67], a progressive increase of testicular width was noted after 10 months, and notably after 16 months of age; however, puberty—defined by the first presence in the ejaculate of 50 $\times 10^6$ sperm with at least 10% of motility-, was not attained in donkeys before 19–20 months. A previous histological work by our group evidenced that although a rapid increase of TV could be observed after 12–14 months, spermatogenesis was still incipient at that age [69]. Still, further studies should be carried out to precisely determine the age of Miranda donkey at puberty.

The comparative analysis of US measurements with those obtained using a precision caliper after orchiectomy evidenced that the former were very accurate. The precise position and orientation of the probe during US examination and the correct handling of the testicle, avoiding excessive tension during the exam, may have additionally contributed to the obtention of reliable US measurements.

The quantitative and qualitative sperm parameters obtained; total sperm number (TSN) per ejaculate, volume, concentration and morphology were within the range found for other European Donkey breeds such as Zamorano–Leonese [70], Catalonian [71], Andalusian [72–74], and Amiata donkey [67]. The values of GSI obtained for mature donkeys (0.9494) were higher than those reported in other domestic species [5]. This finding is of great interest and application when implementing ARTs' strategies and is consistent with previous studies that observed the comparatively greater efficiency for sperm production of donkeys among mammals, characterizing by a high spermatogenic efficiency and short length of spermatogenesis [6,75].

4.2. Bayesian Approach and Predictive Models

Comparative observations were taken at different time points, from a population whose membership changes over time, but retains some constant members. This sample condition is known as partially overlapping samples and for this study, it implies the fact that not all the animals which were measured for morphometric parameters were evaluated for semen parameters. As reported by Kay, et al. [76] in studies working with partially overlapping samples, when there has been a gross violation of normality, as in our study, samples should be considered independent. In a nonparametric context, the strong subdivision of samples across the different experiments may condition results, hence, a Bayesian approach was followed given smaller data sets that can be evaluated avoiding power loss and retaining precision.

Posterior predictive p values (total model probability) for models 1 and 2 of around 0.331. Indicated moderately plausible good-fitting models. Similarly, the difference of more than 3 log likelihood units can be considered as strong evidence against models 1 or 2 depending on the parameters considered. The higher value reported for this parameter may suggest the acceptance of a more parameter-rich or simpler model accordingly. BIC explains how well the model will fit for new data (instead of explaining the existing data, which is measured by Adjusted R^2). Models presenting lower BIC values evidence improved predictions for the dependent variable or variables that they model for. Frequently, adding more variables decreases predictive accuracy, and in that case, the model with even higher Adjusted R^2 will display higher BIC, decreasing its predictive power [52,76,77]. However, considering the higher Adjusted R^2 and the lower BIC, model 1 performs better at explaining and predicting than model 2 for gel-free volume (mL) and concentration ($\times 10^6$/mL). For motility (%), model 1 was more precise to predict for future data while slightly worse at explaining present data (0.01 lower R^2). The opposite situation was reported for TSN ($\times 10^9$) and TMS ($\times 10^9$), for which model 2 was more precise to predict for future data, although it may be slightly worse at explaining present data (0.13 lower R^2). For morphologically normal and abnormal spermatozoa (%) and gonadosomatic ratio (GSI) (%), model 2 suggested a higher ability to explain for present and predict for future data.

In the present research, when comparatively analyzing testis' biometry predictive power on spermatic parameters, some differences were found. The left testicle seems to exert a higher influence on gel-free volume, while, on the other hand, the biometry of the testicles seems to affect TSN differently. Concretely, the length and width of the left testicle and the height of the right testicle seem to increase in parallel with sperm quantitative parameters. Oppositely, as length and width of right testicle and height of the left testicle increase, sperms output parameters seem to decrease. Hence, the negative/positive balance between linear regression coefficients of morphometry variables (length, width and height) suggest that testis may reciprocally react to changes in the contralateral testicle, which affects almost all sperm outputs variables.

A previous work purposes the "compensation hypothesis" in birds, that states that one of the testis could serve as a "back-up" for any reduced function of the other and provides a mechanism to explain intraspecific variation in degree and direction of gonad asymmetry [77]. Another work relates that the degree of testicular asymmetry was positively correlated with inbreeding coefficient and negatively correlated with the proportion of normal sperm [78]. However, in the present work, testicular asymmetry was not found in both clinical and morphometric evaluation, as both features do not meet the inclusion criterion.

Mahmoud Ali Omar, et al. [79] reported a similar compensatory effect in the right testicle after the removal of the contralateral testicle in donkeys. Other authors have ascribed this compensation to the increase in serum LH and FSH concentrations and, potentially higher intratesticular testosterone [80]. Unilateral orchiectomy has been reported to increase the mean diameter of seminiferous tubules by 21% and of their lumina by 51% [81]. Additionally, two events in line with our results were described. A weight compensation was reported for the remaining testis, which has been already described [82]. Also, the histological examination of the testis of donkeys after unilateral orchiectomy with scrotum suture revealed hyperplasia of Leydig and Sertoli cells [79]. This had also been reported by Putra and Blackshaw [83], who suggested an increase in the number of Sertoli cells and germ cells occupying the seminiferous epithelium after unilateral orchiectomy. Our results may evidence that compensation may occur physiologically without these events, as it has also been reported in other species [78]. Still, future works are necessary in order to confirm these findings in donkeys.

In the present study, the age covariate, included in both predictive models, was significantly and positively correlated with several parameters, namely with gel-free volume and sperm output (TSN). The significant age-related positive effects on gel-free volume and TSN agreed those in previous works in stallions [84,85]. For instance, the influence of age in testicular dimensions of juvenile and peripubertal donkeys was verified by Rota, et al. [67], who suggested that age markedly influenced testicular width.

On the other hand, age shows a linear association with abnormal sperm morphology in model 1. Morphologically abnormal spermatozoa percentage slightly increases with age; while sperm concentration and morphologically normal spermatozoa linearly decrease. The negative impact of advanced age on morphology has been already described in stallions and has been ascribed to testicular degeneration, abnormal epididymal function [86] or to age-related testicular dysfunction associated with deterioration in DNA sperm motility [87]. A study in Egyptian donkeys reports that from six years onward, histological features were indicative of spermatogenic efficiency starting to decrease [65]; however, more studies should be performed before concluding that the same occurs in Miranda donkey breed.

In general, stronger correlations between BW and testicular biometry than between age and testicular biometry were verified in the present study. This agrees with the findings in a previous study conducted in stallions which emphasized the influence of body size in testicular measurements and sperm output [2]. However, the analysis of regression coefficients evidenced that the association of motility and total motile sperm (TMS) with TTV was not always constant. On the contrary, sperm motility, as well as TMS and concentration, were positively and linearly associated with gonodasomatic ratio (GSI). Overall, this supports the fact that even if the measurements of the testicular parameters could provide useful information about the potential sperm production, when it comes to predict motility, these parameters should be adjusted for the BW of the donkey, as reported by [Woodall and Johnstone [88]] when predicting for fertility in dogs. Contextually, further investigations should allow to determine and confirm the relationship between BW and TV in donkeys.

5. Conclusions

The results of the present work evidence the reliability of ultrasonographic measurements of testis, which emphasizes its importance and value to obtain reference values of donkey testicular volumes. Values of testicular volume and sperm output in the Miranda donkey breed are similar to those in other affine European donkey breeds. Gonadomatic index (GSI) is higher in the donkey than in other domestic species as previously described, which confirms the great reproductive potential of male donkeys.

Combinations of biometrical and testicular morphometric factors (age, body weight, testicular volume and GSI) will likely improve the predictive accuracy of Models than using factors separately. Besides biometry, considering data such as BW and age, testicular volume, and GSI may be systematically taken into consideration and integrated on BSE of donkeys. The present study provides new insights into donkey reproductive biology, which may be transferred to ARS strategies. Appropriate use of both models may be useful to further improve knowledge on the reproductive characteristics of donkey breeds, which may reinforce clinical purposes and maximize the outcomes from direct conservation or selection strategies.

Supplementary Materials: The following Tables are available online at https://www.mdpi.com/2076-2615/11/1/176/s1, Table S1. Testing for normality using Shapiro–Francia W′ test (for 50 < n < 2500 samples) for testicular biometry.; Table S2. Testing for normality using Shapiro–Wilk test (for n < 50 samples) for testicular biometry (n = 16 testis) and spermatic data (n = 40 ejaculates) of eight mature donkeys.; Table S3. Bayesian Inference Pearson's Correlation Coefficient function output summary for US and caliper testicle biometry.; Table S4. Commonly used thresholds to define significance of evidence through Bayes factor (BF).; Table S5. Model validity and accuracy parameters definition and interpretation.; Table S6. Descriptive statistic for US testicular in six juveniles and 17 matures donkeys (n = 46 testis) and caliper meas-urements after orchiectomy in seven of these donkeys (n = 14 testis).; Table S7a. Posterior distribution statistics for US and caliper testicular measurements in juvenile (n = 6) and mature donkeys (n = 17).; Table S7b. Summary of Bayesian ANOVA outputs to test for differences in the mean for US and caliper testicular measurements between juvenile (n = 6) and mature donkeys (n = 17).

Author Contributions: Conceptualization, A.M.-B. and F.J.N.G.; methodology, A.M.-B. and F.J.N.G.; software, F.J.N.G.; validation, A.M.-B., M.Q., and A.C.; formal analysis, F.J.N.G.; investigation, A.M.-B. and M.Q.; resources, M.Q., B.L., A.C., and A.M.-B.; data curation, F.J.N.G.; writing—original draft preparation, A.M.-B., M.Q., and A.C.; writing—review and editing, A.M.-B. and F.J.N.G.; visualization, A.M.-B. and F.J.N.G.; supervision, A.M.-B.; project administration, A.M.-B.; funding acquisition, A.M.-B. and M.Q. All authors have read and agreed to the published version of the manuscript.

Funding: This work was supported by the project UIDB/CVT/00772/2020 funded by the supported by the Portuguese Science and Technology Foundation (FCT).

Institutional Review Board Statement: The study was conducted in accordance with the Declaration of Helsinki. Animals have been evaluated with approval and in collaboration with the Association for Study and Protection of the Donkey Breed Burro de Miranda (AEPGA), in the behalf of a scientific protocol of caooperation signed between both institutions. All animal procedures were conducted in accordance with national laws for animal welfare and experimentation as with the EU Directive 2010/63/EU for animal experiments and the approval of the Directive Hospital Committee (Approval Ref. 408/VTH-UTAD).

Informed Consent Statement: Not applicable.

Data Availability Statement: Data available on request due to restrictions eg privacy or ethical.

Acknowledgments: The authors thank the Veterinary Degree' students for their willingness to participate in the present study. Also, the authors would like to thank the support of AEPGA-Association for the Study and Protection of Donkeys (Miranda do Douro, Portugal) and the Veterinary Teaching Hospital of UTAD (Vila Real, Portugal) to carry out the present work. Critical comments from A. Rocha (DACT) and T. Guimarães (DECAR) (ICBAS and CRAV, University of Oporto, Portugal) are also highly appreciated.

Conflicts of Interest: The authors declare no conflict of interest.

References

1. Thompson, D.L., Jr.; Pickett, B.W.; Squires, E.L.; Amann, R.P. Testicular measurements and reproductive characteristics in stallions. *J. Reprod. Fertil. Suppl.* **1979**, *27*, 13–17.
2. Kavak, A.; Lundeheim, N.; Aidnik, M.; Einarsson, S. Testicular measurements and daily sperm output of Tori and Estonian breed stallions. *Reprod. Domest. Anim.* **2003**, *38*, 167–169.
3. Pricking, S.; Bollwein, H.; Spilker, K.; Martinsson, G.; Schweizer, A.; Thomas, S.; Oldenhof, H.; Sieme, H. Testicular volumetry and prediction of daily sperm output in stallions by orchidometry and two-and three-dimensional sonography. *Theriogenology* **2017**, *104*, 149–155. [PubMed]
4. Love, C. How to measure testes size and evaluate scrotal contents in the stallion. In Proceedings of the 60th AAEP Annual Convention, Salt Lake City, UT, USA, 6–10 December 2014; Volume 60, pp. 302–308.
5. Lara, N.L.; Costa, G.M.; Avelar, G.F.; Lacerda, S.; Hess, R.A.; França, L.R. Testis physiology—overview and histology. In *Encyclopedia of Reproduction*; Skinner, M.K., Ed.; Academic Press: New York, NY, USA, 2018; pp. 105–116.
6. Neves, E.S.; Chiarini-Garcia, H.; França, L.R. Comparative testis morphometry and seminiferous epithelium cycle length in donkeys and mules. *Biol. Reprod.* **2002**, *67*, 247–255.
7. Kugler, W.; Grunenfelder, H.P.; Broxham, E. *Donkey Breeds in Europe: Inventory, Description, Need for Action, Conservation*; Report 2007/2008; Monitoring Institute for Rare Breeds and Seeds in Europe: St. Gallen, Switzerland, 2008; p. 26.

8. Quaresma, M.; Payan-Carreira, R. Characterization of the estrous cycle of Asinina de Miranda jennies (*Equus asinus*). *Theriogenology* **2015**, *83*, 616–624. [PubMed]
9. Oravecz, Z.; Muth, C. Fitting growth curve models in the Bayesian framework. *Psychon. Bull. Rev.* **2018**, *25*, 235–255. [CrossRef] [PubMed]
10. Knecht, D.; Jankowska-Mąkosa, A.; Duziński, K. Boar genotype as a factor shaping age-related changes in semen parameters and reproduction longevity simulations. *Theriogenology* **2017**, *98*, 50–56. [PubMed]
11. Hox, J.; Moerbeek, M.; Kluytmans, A.; Van De Schoot, R. Analyzing indirect effects in cluster randomized trials. The effect of estimation method, number of groups and group sizes on accuracy and power. *Front. Psychol.* **2014**, *5*, 78. [PubMed]
12. Lee, S.-Y.; Song, X.-Y. Evaluation of the Bayesian and maximum likelihood approaches in analyzing structural equation models with small sample sizes. *Multivar. Behav. Res.* **2004**, *39*, 653–686.
13. Peto, R.; Pike, M.; Armitage, P.; Breslow, N.; Cox, D.; Howard, S.V.; Mantel, N.; McPherson, K.; Peto, J.; Smith, P. Design and analysis of randomized clinical trials requiring prolonged observation of each patient. I. Introduction and design. *Br. J. Cancer* **1976**, *34*, 585–612.
14. Button, K.S.; Ioannidis, J.P.; Mokrysz, C.; Nosek, B.A.; Flint, J.; Robinson, E.S.; Munafò, M.R. Power failure: Why small sample size undermines the reliability of neuroscience. *Nat. Rev. Neurosci.* **2013**, *14*, 365–376. [PubMed]
15. Stoltzfus, J.C. Logistic Regression: A Brief Primer. *Acad. Emerg. Med.* **2011**, *18*, 1099–1104. [CrossRef] [PubMed]
16. Chen, Y.; Li, Y.; Wu, H.; Liang, L. Data envelopment analysis with missing data: A multiple linear regression analysis approach. *Int. J. Inf. Tech. Decis.* **2014**, *13*, 137–153. [CrossRef]
17. Navas, F.J.; Jordana, J.; León, J.M.; Barba, C.; Delgado, J.V. A model to infer the demographic structure evolution of endangered donkey populations. *Animal* **2017**, *11*, 2129–2138. [CrossRef]
18. Johnson, L.; Neaves, W.B. Age-related changes in the Leydig cell population, seminiferous tubules, and sperm production in stallions. *Biol. Reprod.* **1981**, *24*, 703–712.
19. Quartuccio, M.; Marino, G.; Zanghì, A.; Garufi, G.; Cristarella, S. Testicular volume and daily sperm output in Ragusano donkeys. *J. Equine Vet. Sci.* **2011**, *31*, 143–146. [CrossRef]
20. Kenney, R.M. *Society for Theriogenology Manual for Clinical Evaluation of the Stallion*; The Society for Theriogenology: Montgomery, AL, USA, 1983.
21. StataCorp. *Stata Statistical Software, 15*; StataCorp: College Station, TX, USA, 2017.
22. IBM Corp. *IBM SPSS Statistics for Windows*, 25th ed.; IBM Corp: Armonk, NY, USA, 2017.
23. Bakdash, J.Z.; Marusich, L.R. Repeated Measures Correlation. *Front. Psychol.* **2017**, *8*, 456. [CrossRef] [PubMed]
24. IBM Corp. *IBM SPSS Statistics Algorithms*, 25th ed.; IBM Corp: Armonk, NY, USA, 2017; p. 110.
25. Profillidis, V.A.; Botzoris, G.N. Chapter 5—Statistical Methods for Transport Demand Modeling. In *Modeling of Transport Demand*; Profillidis, V.A., Botzoris, G.N., Eds.; Elsevier: Amsterdam, The Netherlands, 2019.
26. Doğan, N.Ö. Bland-Altman analysis: A paradigm to understand correlation and agreement. *Turk. J. Emerg. Med.* **2018**, *18*, 139–141. [CrossRef]
27. Batterham, A.M. Bias in Bland-Altman but not Regression Validity Analyses. *Sportscience* **2004**, *8*, 42–47.
28. Gelman, A.; Carlin, J.; Stern, H.; Rubin, D.; Dunson, D.; Vehtari, A. Solutions to some exercises from Bayesian Data Analysis, by Gelman, Carlin, Stern, and Rubin. In *Bayesian Data Analysis*; Columbia University: New York, NY, USA, 2020.
29. Brewer, K.R. *Combined Survey Sampling Inference: Weighing Basu's Elephants*; Oxford University Press: Oxford, UK, 2002.
30. Hayes, A.F.; Glynn, C.J.; Huge, M.E. Cautions Regarding the Interpretation of Regression Coefficients and Hypothesis Tests in Linear Models with Interactions. *Commun. Methods Meas.* **2012**, *6*, 1–11.
31. Liang, F.; Paulo, R.; Molina, G.; Clyde, M.A.; Berger, J.O. Mixtures of g priors for Bayesian variable selection. *J. Am. Stat. Assoc.* **2008**, *103*, 410–423. [CrossRef]
32. Heck, D. A Caveat on the Savage-Dickey Density Ratio: The Case of Computing Bayes Factors for Regression Parameters. *Br. J. Math. Stat. Psychol.* **2019**, *72*, 316–333. [CrossRef] [PubMed]
33. Zellner, A.; Siow, A. Posterior odds ratios for selected regression hypotheses. *Trab. Estadística Y Investig. Oper.* **1980**, *31*, 585–603. [CrossRef]
34. Rouder, J.N.; Morey, R.D.; Speckman, P.L.; Province, J.M. Default Bayes factors for ANOVA designs. *J. Math. Psychol.* **2012**, *56*, 356–374. [CrossRef]
35. Bayarri, M.J.; Berger, J.O.; Forte, A.; García-Donato, G. Criteria for Bayesian model choice with application to variable selection. *Ann. Stat.* **2012**, *40*, 1550–1577. [CrossRef]
36. Rouder, J.N.; Morey, R.D. Default Bayes factors for model selection in regression. *Multivar. Behav. Res.* **2012**, *47*, 877–903. [CrossRef] [PubMed]
37. Morey, R.; Rouder, J. *BayesFactor 0.9. 12-2*; Comprehensive R Archive Network: Vienna, Austria, 2015.
38. Rouder, J.N.; Speckman, P.L.; Sun, D.; Morey, R.D.; Iverson, G. Bayesian t tests for accepting and rejecting the null hypothesis. *Psychon. Bull. Rev.* **2009**, *16*, 225–237. [CrossRef] [PubMed]
39. Depaoli, S.; Van de Schoot, R. Improving transparency and replication in Bayesian statistics: The WAMBS-Checklist. *Psychol. Methods* **2017**, *22*, 240. [CrossRef]
40. Kass, R.E.; Raftery, A.E. Bayes factors. *J. Am. Stat. Assoc.* **1995**, *90*, 773–795. [CrossRef]

41. Arora, J.S. Chapter 14—Practical Applications of Optimization. In *Introduction to Optimum Design*, 4th ed.; Arora, J.S., Ed.; Academic Press: Boston, MA, USA, 2017; pp. 601–680.
42. Pizarro Inostroza, M.G.; Navas González, F.J.; Landi, V.; León Jurado, J.M.; Delgado Bermejo, J.V.; Fernández Álvarez, J.; Martínez Martínez, M.D.A. Bayesian Analysis of the Association between Casein Complex Haplotype Variants and Milk Yield, Composition, and Curve Shape Parameters in Murciano-Granadina Goats. *Animals* **2020**, *10*, 1845. [CrossRef]
43. Geweke, J. Variable selection and model comparison in regression. In Proceedings of the Fifth Valencia International Meeting, Valencia, Spain, 5–9 June 1994.
44. Analla, M. Model validation through the linear regression fit to actual versus predicted values. *Agric. Syst.* **1998**, *57*, 115–119. [CrossRef]
45. Pizarro Inostroza, M.G.; Navas González, F.J.; Landi, V.; León Jurado, J.M.; Delgado Bermejo, J.V.; Fernández Álvarez, J.; Martínez Martínez, M.D.A. Software-Automatized Individual Lactation Model Fitting, Peak and Persistence and Bayesian Criteria Comparison for Milk Yield Genetic Studies in Murciano-Granadina Goats. *Mathematics* **2020**, *8*, 1505. [CrossRef]
46. Pizarro Inostroza, M.G.; Navas González, F.J.; Landi, V.; León Jurado, J.M.; Delgado Bermejo, J.V.; Fernández Álvarez, J.; Martínez, M.D.A.M. Goat Milk Nutritional Quality Software-Automatized Individual Curve Model Fitting, Shape Parameters Calculation and Bayesian Flexibility Criteria Comparison. *Animals* **2020**, *10*, 1693. [CrossRef] [PubMed]
47. Hall, P.; Maiti, T. Nonparametric estimation of mean-squared prediction error in nested-error regression models. *Ann. Stat.* **2006**, *34*, 1733–1750. [CrossRef]
48. Jeffreys, H. *Theory of Probability*, 3rd ed.; Oxford University Press: Oxford, UK, 1961.
49. Lee, M.; Wagenmakers, E. *Bayesian Data Analysis for Cognitive Science: A Practical Course*; Cambridge University Press: New York, NY, USA, 2013.
50. Kaplan, D.; Depaoli, S. Bayesian structural equation modeling. In *Handbook of Structural Equation Modeling*; Hoyle, H., Ed.; The Guilford Press: New York, NY, USA; pp. 650–673.
51. Gelman, A.; Carlin, J.B.; Stern, H.S.; Dunson, D.B.; Vehtari, A.; Rubin, D.B. *Bayesian Data Analysis*; CRC Press: Boca Ratón, FL, USA, 2013.
52. Drton, M.; Plummer, M. A Bayesian information criterion for singular models. *J. R. Stat. Soc. B* **2017**, *79*, 323–380. [CrossRef]
53. Clyde, M.; Cetinkaya-Rundel, M.; Rundel, C.; Banks, D.; Chai, C.; Huang, L. Bayesian Model Choice. In *An Introduction to Bayesian Thinking: A Companion to the Statistics with R Course*; Clyde, M., Cetinkaya-Rundel, M., Rundel, C., Banks, D., Chai, C., Huang, L., Eds.; Coursera: Mountain View, CA, USA, 2019; Chapter 7.
54. Gelman, A.; Goodrich, B.; Gabry, J.; Vehtari, A. R-squared for Bayesian regression models. *Am. Stat.* **2019**, *73*, 307–309. [CrossRef]
55. Kumar, S.; Srivastava, S. Testicular biometry and its correlation with body weight and semen output in murrah bull. *Buffalo Bull.* **2017**, *36*, 105–114.
56. Brito, L.F.; Silva, A.E.; Barbosa, R.T.; Kastelic, J.P. Testicular thermoregulation in Bos indicus, crossbred and Bos taurus bulls: Relationship with scrotal, testicular vascular cone and testicular morphology, and effects on semen quality and sperm production. *Theriogenology* **2004**, *61*, 511–528. [CrossRef]
57. Gemeda, A.E.; Workalemahu, K. Body weight and scrotal-testicular biometry in three indigenous breeds of bucks in arid and semiarid agroecologies, Ethiopia. *J. Vet. Med.* **2017**, *2017*. [CrossRef]
58. Kerketta, S.; Singh, M.; Patel, B.; Dutt, T.; Upadhyay, D.; Bharti, P.; Sahu, S.; Kamal, R. Relationships between age, body measurements, testicular measurements and total ejaculation of semen in local goat of Rohilkhand region. *Small Rum. Res.* **2015**, *130*, 193–196. [CrossRef]
59. Olar, T.; Amann, R.; Pickett, B. Relationships among testicular size, daily production and output of spermatozoa, and extragonadal spermatozoal reserves of the dog. *Biol. Reprod.* **1983**, *29*, 1114–1120. [CrossRef] [PubMed]
60. Mialot, J.; Guerin, C.; Begon, D. Growth, testicular development and sperm output in the dog from birth to post pubertal period. *Andrologia* **1985**, *17*, 450–460. [CrossRef] [PubMed]
61. Pozor, M.; Morrissey, H.; Albanese, V.; Khouzam, N.; Deriberprey, A.; Macpherson, M.L.; Kelleman, A.A. Relationship between echotextural and histomorphometric characteristics of stallion testes. *Theriogenology* **2017**, *99*, 134–145. [CrossRef] [PubMed]
62. Canisso, I.F.; Morel, M.D.; McDonnell, S. Strategies for the management of donkey jacks in intensive breeding systems. *Equine Vet. Educ.* **2009**, *21*, 652–659. [CrossRef]
63. Lemma, A.; Deressa, B. Study on reproductive activity and evaluation of breeding soundness of jacks (Equus asinus) in and around Debre Zeit, Ethiopia. *Livest. Res. Rural. Dev.* **2009**, *21*, 126.
64. Moustafa, M.; Sayed, R.; Zayed, A.; El-Hafeez, A.H. Morphological and morphometric study of the development of seminiferous epithelium of donkey (Equus asinus) from birth to maturity. *J. Cytol. Histol.* **2015**, *6*, 1.
65. Nipken, C.; Wrobel, K. A quantitative morphological study of age-related changes in the donkey testis in the period between puberty and senium. *Andrologia* **1997**, *29*, 149–161. [CrossRef]
66. Carluccio, A.; Panzani, S.; Contri, A.; Bronzo, V.; Robbe, D.; Veronesi, M. Influence of season on testicular morphometry and semen characteristics in Martina Franca jackasses. *Theriogenology* **2013**, *79*, 502–507. [CrossRef]
67. Rota, A.; Puddu, B.; Sabatini, C.; Panzani, D.; Lainé, A.-L.; Camillo, F. Reproductive parameters of donkey jacks undergoing puberty. *Anim. Reprod. Sci.* **2018**, *192*, 119–125. [CrossRef]
68. Abdelhafeez, H.; Moustafa, M.; Zayed, A.; Sayed, R. Morphological and Morphometric Study of the Development of Leydig Cell population of Donkey (*Equus asinus*) Testis from Birth to Maturity. *Cell Biol.* **2017**, *6*, 1. [CrossRef]

69. Calado, A.; Lemos, H.; Leiva, B.; Quaresma, M.; Martins-Bessa, A. Comparative testicular histology from adult *Equus Asinus* and *Equus Caballus*. In Proceedings of the Book INCOMAM 18—52° International Congress on Microscopy and Microanalysis, Coimbra, Portugal, 12–13 October 2018; p. 35.
70. Serres, C. Evaluación y Conservación del Semen en el Asno Zamorano-Leonés. Ph.D. Thesis, Complutense University of Madrid, Madrid, Spain, 2003.
71. Miró, J.; Lobo, V.; Quintero-Moreno, A.; Medrano, A.; Peña, A.; Rigau, T. Sperm motility patterns and metabolism in Catalonian donkey semen. *Theriogenology* **2005**, *63*, 1706–1716. [CrossRef] [PubMed]
72. Ortiz, I.; Dorado, J.; Ramírez, L.; Morrell, J.M.; Acha, D.; Urbano, M.; Gálvez, M.J.; Carrasco, J.J.; Gómez-Arrones, V.; Calero-Carretero, R.; et al. Effect of single layer centrifugation using Androcoll-E-Large on the sperm quality parameters of cooled-stored donkey semen doses. *Animal* **2013**, *8*, 308–315. [CrossRef] [PubMed]
73. Dorado, J.; Acha, D.; Ortiz, I.; Gálvez, M.; Carrasco, J.; Díaz, B.; Gómez-Arrones, V.; Calero-Carretero, R.; Hidalgo, M. Relationship between conventional semen characteristics, sperm motility patterns and fertility of Andalusian donkeys (*Equus asinus*). *Anim. Reprod. Sci.* **2013**, *143*, 64–71. [CrossRef] [PubMed]
74. Dorado, J.; Acha, D.; Ortiz, I.; Gálvez, M.; Carrasco, J.; Gómez-Arrones, V.; Calero-Carretero, R.; Hidalgo, M. Effect of extender and amino acid supplementation on sperm quality of cooled-preserved Andalusian donkey (*Equus asinus*) spermatozoa. *Anim. Reprod. Sci.* **2014**, *146*, 79–88. [CrossRef]
75. Canisso, I.F.; Panzani, D.; Miró, J.; Ellerbrock, R.E. Key Aspects of Donkey and Mule Reproduction. *Vet. Clin. Equine* **2019**, *35*, 607–642. [CrossRef]
76. Kay, G.; Tligui, N.; Semmate, N.; Azrib, R.; González, F.J.N.; Brizgys, L.; McLean, A. Determining factors and interspecific modeling for serum amyloid a concentrations in working horses, donkeys, and mules. *Res. Vet. Sci.* **2019**, *125*, 256–265. [CrossRef]
77. Calhim, S.; Birkhead, T.R. Intraspecific variation in testis asymmetry in birds: Evidence for naturally occurring compensation. *Proc. Biol. Sci.* **2009**, *276*, 2279–2284. [CrossRef]
78. Cassinello, J.; Abaigar, T.; Gomendio, M.; Roldan, E. Characteristics of the semen of three endangered species of gazelles (Gazella dama mhorr, G. dorcas neglecta and G. cuvieri). *Reproduction* **1998**, *113*, 35–45. [CrossRef]
79. Omar, M.M.A.; Hassanein, K.M.A.; Abdel-Razek, A.-R.K.; Hussein, H.A.Y. Unilateral orchidectomy in donkey (Equus asinus): Evaluation of different surgical techniques, histological and morphological changes on remaining testis. *Vet. Res. Forum* **2013**, *4*, 1–6.
80. Hoagland, T.; Ott, K.; Dinger, J.; Mannen, K.; Woody, C.; Riesen, J.; Daniels, W. Effects of unilateral castration on morphologic characteristics of the testis in one-, two-, and three-year-old stallions. *Theriogenology* **1986**, *26*, 397–405. [CrossRef]
81. Barnes, M.; Longnecker, J.; Charter, R.; Riesen, J.; Woody, C. Influence of unilateral castration and increased plane of nutrition on sexual development of Holstein bulls. I. Growth and sperm production. *Theriogenology* **1980**, *14*, 49–58. [CrossRef]
82. Barnes, M.; Longnecker, J.; Riesen, J.; Woody, C. Influence of unilateral castration and increased plane of nutrition on sexual development of Holstein bulls. III. Endocrine responses. *Theriogenology* **1980**, *14*, 67–81. [CrossRef]
83. Putra, D.H.; Blackshaw, A. Quantitative studies of compensatory testicular hypertrophy following unilateral castration in the boar. *Aust. J. Biol. Sci.* **1985**, *38*, 429–434. [CrossRef]
84. Kandiel, M.; El Khawagah, A. Evaluation of semen characteristics, oxidative stress, and biochemical indices in Arabian horses of different ages during the hot summer season. *Iran. J. Vet. Res.* **2018**, *19*, 270. [PubMed]
85. Gamal, A.; El-Maaty, A.M.A.; Rawash, Z.M. Comparative blood and seminal plasma oxidant/antioxidant status of Arab stallions with different ages and their relation to semen quality. *Asian Pac. J. Reprod.* **2016**, *5*, 428–433.
86. Dowsett, K.; Knott, L. The influence of age and breed on stallion semen. *Theriogenology* **1996**, *46*, 397–412. [CrossRef]
87. Blanchard, T.L.; Brinsko, S.P.; Varner, D.D.; Love, C.C.; Morehead, J.P. Progression of reproductive changes accompanying testicular dysfunction in aging Thoroughbred stallions: Case studies. In Proceedings of the AAEP Annual Convention, Nashville, TN, USA, 11 December 2013; Volume 59, pp. 532–536.
88. Woodall, P.; Johnstone, I. Dimensions and allometry of testes, epididymides and spermatozoa in the domestic dog (*Canis familiaris*). *Reproduction* **1988**, *82*, 603–609. [CrossRef] [PubMed]

animals

MDPI

Review

Skin Diseases in Donkeys and Mules—An Update

Telma S. Lima [1], Raquel A. F. Silva [2], Raquel M. F. Pereira [1], Karoline L. Soares [1], Nayadjala T. A. Santos [1], Mônica S. Sousa [2], Fábio S. Mendonça [1] and Ricardo B. Lucena [2,3,*]

[1] Graduate Program in Veterinary Medicine, Federal Rural University of Pernambuco (UFRPE), Rua Dom Manuel de Medeiros, S/N Dois Irmãos, 52171-900 Recife-PE, Brazil; telma.lima@ufrpe.br (T.S.L.); raquel.mota@ufrpe.br (R.M.F.P.); karoline.lacerda@academico.ufpb.br (K.L.S.); nayadjala.santos@ufrpe.br (N.T.A.S.); fabio.mendonca@ufrpe.br (F.S.M.)

[2] Graduate Program in Animal Science and Health, Rural Health and Technology Center (CSTR), Federal University of Campina Grande (UFCG), Avenida Universitária, S/N Jatobá, 58708-110 Patos-PB, Brazil; raquel.annes@estudante.ufcg.edu.br (R.A.F.S.); monica.shinneider@estudante.ufcg.edu.br (M.S.S.)

[3] Graduate Program in Animal Science, Department of Agrarian Sciences (CCA), Federal University of Paraíba (UFPB), 12, Rod. PB-079, 58397-000 Areia-PB, Brazil

* Correspondence: ricardolucena@cca.ufpb.br

Simple Summary: Equids are part of the history of many countries, including Brazil, where they were used in trade routes and expansion of the current states. Several skin diseases affect these animals; however, visibility is higher on horses than on donkeys and mules, which is linked to regional cultural and socioeconomic factors, even resulting in a decline of the world population of these animals. In this context, the objective of this study was to review which skin diseases have been reported in the scientific literature with emphasis on skin pathologies.

Abstract: The skin of donkeys and mules represents a promising source of income; however, cultural, productive, and infectious factors can directly interfere with the quality of the integumentary tissue and well-being of these species. The objective of this study is to present a literature review on equine dermatopathies. This literature review included scientific articles related to equine medicine and breeding according to pre-established search terms and expressions published in recently articles. The evaluation of the clinical and pathological behavior of dermatopathies implies the use of control strategies and the recognition of pathological patterns that may be particular to the species.

Keywords: skin; integument; donkey; mule

Citation: Lima, T.S.; Silva, R.A.F.; Pereira, R.M.F.; Soares, K.L.; Santos, N.T.A.; Sousa, M.S.; Mendonça, F.S.; Lucena, R.B. Skin Diseases in Donkeys and Mules—An Update. *Animals* **2021**, *11*, 65. https://doi.org/10.3390/ani11010065

Received: 10 November 2020
Accepted: 3 December 2020
Published: 31 December 2020

1. Introduction

Donkeys and mules represent an important share of the world's equids, totaling about 44 million in recent years, of which approximately 1 million are in Brazil [1,2]. The breeding of these animals is important in developing countries, and their introduction in the territories was quite variable, being associated with the colonial period in South America [3]. The work of donkeys and mules ranges from field work to recreation and entertainment [4], safeguarding socioeconomic importance in many states of Brazil, especially in eastern Brazil.

Many of the phenotypic and physiological characteristics of these animals favor sustainable breeding to increase regional economy. Donkey breeding, however, still presents major challenges in Brazil, which sees their population gradually decreasing over the years [5]. On the other hand, there is a trend toward expansion of the consumer market for meat, milk, and especially donkey skin, given the demand for these products in Asian countries such as China [6].

In this context, the investigation of diseases affecting donkeys and mules becomes important and gives rise to discussions on animal welfare and public health. This is demonstrated particularly in serological studies that show the participation of donkeys and mules in the epidemiological chain of reportable diseases [7,8]. It is furthermore important to highlight that cultural factors may influence the frequency of these diseases, affecting both animal and human health and welfare. Integrating these concepts within the population of an increasingly technological world is perhaps the greatest challenge of sustainable breeding of donkeys and mules today.

Regarding skin diseases in productive animals, the literature lacks studies on dermatopathies and their risk factors, especially in Brazil. Until recently, retrospective studies in ruminants and equids [9,10] have been highlighted in the country, whereas studies on donkeys are still incipient [11,12]. In this context, the objective of this study was to present a literature review on skin diseases in donkeys and mules.

To identify the main dermatopathies diagnosed in equids today, this study consists of a literature review containing the main studies on equine medicine and breeding, in which four major themes were initially proposed to direct the bibliographic research: (a) the use of donkeys and mules in Brazil and in the world; (b) the main diseases of donkeys and mules; (c) skin diseases diagnosed in donkeys and mules; and (d) equid dermatopathies in northeastern Brazil.

The search terms or expressions were chosen based on the major topics listed above and then typed into the major digital veterinary medicine libraries. Where necessary, Boolean operators such as and or not were used. The eligibility criteria were the terms and/or expressions, timeliness, status, and relevance of the publication according to the major themes. For this purpose, studies published or in press preferably from 2016 to 2020 were considered. The search was expanded to the last ten years when the terms and/or expressions were not covered in the period previously stipulated. In this case, studies from 2010 onwards were evaluated. Studies published in English were considered; however, those written in other languages, such as Portuguese, with an abstract in English, were not removed from the archives.

2. Relationship between Disease Occurrence and Type of Donkey and Mule Farming

2.1. Overview of Donkey and Mule Farming in Brazil

Donkeys and mules represent a growing source of income, especially in Brazil, which has approximately 1 million donkeys [2,4]. However, although there is a potential consumer market, the Brazilian market is still not very expressive, which may be linked to the gradually decreasing number of these animals over the years, a decline that reached 37.08% in 2016 [5].

Another factor to consider is the local culture, which associates the use of equids mostly for traction and other field activities. Despite the economic importance, the exact number of donkeys and mules is uncertain, partially due to extensive farming and absence of technical guidelines, and cultural aspects that associate these animals to poverty, generating abandonment resulting in traffic accidents in the region [5,13].

Despite similarities, donkeys and mules are distinct animals that originated from the cross-breeding of *Equus asinus* and *Equus caballus* species [14]. These animals have intermediate characteristics between the progenitor species making their rearing relatively easy due to their rusticity, easy breeding, longevity, less selective eating habits, docile temperament, and dexterity in performing agricultural activities. In Brazil, three types of donkeys which originated from Europe and Africa are recognized; the northeastern donkey is well adapted to the semiarid climate [5]. Donkeys and mules are part of the herd in all Brazilian states, which may be linked to their use in the colonial period while expanding routes in the interior of the country.

The identification of these characteristics favored the investment in research directed to the biotechnology of reproduction, serological research of microorganisms, as well as in the food sector, raising the topic of the welfare of these animals. In the state of Bahia, Brazil, a private company producing donkey meat and donkey-hide gelatin became an important exporter to China [4,5]. The significant increase in the export of donkey hide to China [4].

Similarly to ruminants, the erroneous notion that donkeys and mules are resistant to the environment and adverse conditions is disseminated among farmers, generating carelessness with nutritional and sanitary management, it is not uncommon to observe working animals with chronic disease and skin lacerations due to whipping [13,15]. The visibility that the articles achieve is essential to raise awareness and disseminate technical information. In this context, Non-Governmental Organizations and other entities commonly promote care in the farming of these animals, although government agencies take time to follow their recommendations [16].

There is further a growing number of articles published in journals relevant to the scientific community addressing the health of these animals. A previous study tracking publications on donkeys [13] showed that until 2018, there were only 114 publications from 56 different countries discussing varying topics. In this study, the themes of reproduction followed by studies on anatomy, pathology, surgery, and equine medicine were highlighted. This shows, among other things, that there is still a vast field of research to be explored with donkeys and mules.

2.2. Disease Profile of Donkeys and Mules

Undoubtedly, diseases are the main obstacle in animal husbandry, so much so that occurrence of disease outbreaks can generate significant losses. The role of diagnostic laboratories is essential, considering that infectious diseases stand out among all other diseases that affect herbivores, and are generally due to digestive disorders [17–19].

Even considering the economic relevance, the diagnosis of equine diseases is still precarious in the literature. Diseases in donkeys and mules are largely unknown [20]. Generally, these animals develop diseases similar to horses, however, certain irregularities must be considered before a presumptive diagnosis. The clinical differences tend to be quite subtle, such as demonstration of pain, which is generally not as expressive as in horses, justifying differentiated physical examination [14].

Therefore, studies aimed at diagnosing these diseases should be stimulated. The importance of these studies is even more relevant when considering zoonotic agents, such as the West Nile virus, a micro-organism that has re-emerged in recent years and that leads to severe neurological lesions in horses and humans [21]. A serological investigation was performed on donkeys and mules in southern Spain after the detection of West Nile virus infection in the region. It was observed that nine of the 90 herds evaluated contained at least one seropositive animal and antibodies against the virus were further detected in 1/4 of the donkeys coming from farms where the cases were confirmed in horses, demonstrating that serological surveillance of sentinel donkeys and mules is necessary for the epidemiological monitoring of these diseases [8].

In Brazil, a recent study demonstrated the frequency of antibodies to some equine diseases in donkeys in the state of São Paulo. Of the 85 serums evaluated, it was estimated that 50.6% exhibited antibodies against the H3N8 subtype of the influenza virus, 47% against the equine Herpesvirus, and 20% against the equine arteritis virus, demonstrating that these agents circulate among donkeys in the region and reiterating the importance of epidemiological monitoring of equids in Brazil [7].

It is important to highlight, however, that a significant portion of diseases diagnosed in domestic animals is performed through necropsy examination. Many retrospective studies include pathological results and estimate the frequency of the diseases in a certain geographical region, thus improving local clinical diagnosis, as it is possible to draw an epidemiological and clinical-pathological profile in most of these studies. In Brazil, only some studies approach the main equine diseases [17,22,23]. However, the literature is limited regarding equine diseases, with only two studies [11,12] conducted in the Brazilian semiarid region being found.

A North American study on causes of death and reason for euthanasia in equids highlighted digestive disorders and lesions in the pituitary gland and locomotor system. This study focused especially on geriatric diseases, represented by neurological, urinary, and neoplastic disorders, which alone accounted for more than 18% of the causes of death in all animals evaluated [24]. These results are important when considering that donkeys and mules have great longevity and often need appropriate veterinary medical attention.

2.3. Skin Diseases Diagnosed in Donkeys and Mules

Skin lesions have been reported in horses in several regions of the world, mainly as case reports with the occasional retrospective study. Regarding skin diseases, there is variation in the origin of the lesions, suggesting individual differences of the species or factors related to the environment, for example traumatic injuries and lesions. Infectious and parasitic dermatopathies impact the quality of life and skin byproducts of these animals. Thus, certain studies report the occurrence or frequency of integument pathologies that may interfere with donkey and mule skin health.

As for infectious and parasitic diseases, the first report of skin besnoitiosis by *Besnoitia besnnetti* in donkeys in the United Kingdom was recently published, referring to the analysis of 20 tissue samples characterized by nodular lesions on skin, mucous membranes, and muco-cutaneous transitions [25]. A very important issue in this study was the fact that the lesions were frequently attributed to sarcoid when macroscopically evaluated, the histopathological findings essential for the definitive diagnosis of a disease that had never before been diagnosed in the country, showing the need to pay more attention to routine cases for skin and mucous membrane lesions in donkeys and mules.

These studies on skin diseases further the understanding of the epidemiology and clinical presentation of diseases and infer the prognosis of the patients. A study conducted in Egypt to evaluate filarial infection in donkeys showed a significant number of filarial lesions in a period of two years, which were not restricted only to the integument. Of the 188 animals studied, 163 were parasitized by *Onchocerca cervicalis*, followed by *Setaria equina*, *Parafilaria multipapillosa*, and *Onchocerca reticulata* being the infection most common in adult males aged five to fifteen years [26].

A retrospective study on skin diseases in donkeys from European countries and the USA has recently been published, emphasizing the geographical distribution of dermatopathies on donkeys in the region. In this study, the following diseases were highlighted: hypersensitivity to insect bites, sarcoid, habronemiasis, superficial pyoderma, and dermatophytosis, among other less expressive diseases. Dermatopathies were common pathologies in equids and, as was already expected, factors such as age, sex, and diagnosis varied by geographical location, showing the importance of a thorough dermatological examination, regardless of the reason and clinical presentation of these diseases [27].

The following tables (Tables 1 and 2) show some of the main skin disorders caused by fungi, bacteria, parasites and oomycetes described in donkeys and mules in the veterinary literature.

Table 1. Clinical and pathological characteristics of skin diseases caused by fungi and bacteria in donkeys and mules.

Disease	Gross Pathology	Histopathology	Clinical	References
Dermatophilosis (*Dermatophilus congonlensis*)	Lesions are usually crusted, alopecic, circumscribed and with marked agglutination of hair, or spread diffusely.	These areas characterize an exudative dermatitis with the formation of crusts interspersed with layers of exudate.	In general, muzzle, face, eyes, limbs and back are the main affected areas, however they can manifest in a widespread way.	[12,28,29]
Dermatophytosis	Unique, multifocal areas slightly elevated and with regular edges, accompanied by alopecia, flaking and grayish crusts.	This lesion represents hyperplastic dermatitis with suppurative folliculitis, hyperkeratosis, epidermal acanthosis and microabscesses.	The lesions are generally not itchy and start in areas of abrasions with loins, rump and head, but which can expand to the back and flank.	[12,28,29]
Epizootic lymphangitis (*Histoplasma farciminosum*)	Single or multiple nodular areas, of slow growth, which ulcerate and drain purulent content. There is usually granulation tissue surrounding these lesions.	Generally granulomatous lesions with adjacent granulation tissue and intrahistiocytic and extracellular yeasts, stained positively with Grocott's methenamine silver and Periodic acid-Schiff stains; that can even show budding.	It can occur on lymphatic lines of the legs, neck region, on the skin or in the nasolacrimal region, limbs (in particular after localized trauma).	[27,28,30,31]
Glanders (*Burkholderia mallei*)	Lesions range from nodular swelling in lymph vessels (rosary beads) to abscesses, alopecia, ulcerations and edema.	These lesions are irregular and characterized by a necrosis center surrounded by a granulomatous to pyogranulomatous infiltrate and adjacent fibrous connective tissue.	The nodules usually follow the distribution of the lymphatic vessels, but are observed particularly in the limbs and flank, head and neck, and can "float" on palpation.	[32–34]

Table 2. Clinical and pathological characteristics of skin diseases caused by parasites and oomycetes in donkeys and mules.

Disease	Gross Pathology	Histopathology	Clinical	References
Habronemiasis (*Habronema* sp.)	Small, crusted nodular lesions, which progressively increase in volume and acquire a spongy and reddish appearance.	These nodulations represent severe eosinophilic dermatitis and panniculitis, with fibroplasia and the presence of intralesional larvae.	It can occur on the skin of the limbs, withers, penis or in the ocular conjunctiva. In conjunctival form it accompanies ocular discharge.	[12,27,29]
Filariosis (*Onchocerca* sp.; *Setaria equina*; *Parafilaria multipapillosa*)	Nodular swelling, of variable size, which may ulcerate.	These lesions correspond to a granulomatous inflammation that usually forms in response to the larvae.	Skin infections are often associated with *Oncocerca cervicalis*, in the nuchal ligament but can affect tendons and ligaments of the limbs.	[26,28,35]
Pythiosis (*Pythium insidiosum*)	Ulcerated nodules and drain sero-bloody secretion. Accompanied by fibrous, whitish and shiny fabric, interspersed by kunkers.	Areas of necrosis and eosinophilic infiltrate, surrounded by granulation tissue and fibrosis, with negative images of intralesional hyphae; Grocott-Gomori methenamine silver stain positive.	Lesions are seen in the limbs, ventral abdominal region, chest, neck, face, lips, breast and genitals, and can be itchy, predisposing to self-mutilation.	[12,28,36–38]
Besnoitiosis (*Besnoitia* spp.)	They may appear as small, multiple, round, yellowish-white, punctate lesions, with thickening, peeling, formation of wrinkles/folds and lichenification.	These lesions represent a mixed inflammatory infiltrate involving the cysts of *Besnoitia* spp. In addition, hypeceratosis, scales and crusts can be observed accompanying dermatitis.	Lesions can be seen in the neck, head, limbs and perineum are particularly affected. The lesions can also appear in areas that have suffered previous trauma or in self-inflicted trauma.	[25,27]

Ulcerative lymphangitis is caused by bacteria belonging to the genera *Corynebacterium pseudotuberculosis*, *Staphylococcus* spp., *Streptococcus* spp., *Pseudomonas aeruginosa* and *Rhodococcus equi*. They initially occur as swelling that fistulates and drains purulent content, progressing to cutaneous or subcutaneous abscesses accompanied by edema. Lesions can be seen particularly in the limbs, such as hocks and fetlock, usually accompanying the lymphatic chain [39–41]. As for the ectoparasites that affect the skin of donkeys and mules, there are different agents involved such as diptera causing myiasis, lice and scabies.

Dermatobia hominis, *Gasterophilus nasalis*, and *Oestrus ovis* are examples of myiasis-causing flies. Infestation by small numbers of larvae may have little or no clinical effect on the host. In general, these lesions do not require a biopsy to confirm the diagnosis. Infestations are greater in rainy periods when the proliferation of flies and mosquitoes, however the occurrence of myiasis does not follow seasonality, considering that traumatic injuries and untreated injuries that determine their occurrence. The injuries are variable and depend on the source of the injury [42].

Lice and scabies are frequently observed and their visualization already gives a diagnosis, however the skin may exhibit flaking and hyperemia resulting from the blood meal (sucking lice particularly). The parasites can be observed dispersed in the animal's coat, usually in those most weakened immunosuppressed. Secondary infections are common, particularly in traumatized skin. The distribution in the animal organism varies according to the species of mite, being *Sarcoptes scabiei* var. *equi* but common in the head and neck; *Psoroptes equi* at the base of the mane and tail hair and *Chorioptes equi* below the hock and knees. *Chorioptes equi* is the most common parasite reported in donkeys and mules, associated with intense itching, which can stimulate the habit of self-mutilation [12,27,28].

Neoplastic diseases are an important group of skin diseases in most animals (Table 3), whether during companionship or reproduction. A study conducted using the records of five North American institutions, concluded that 125 of the 357 donkeys evaluated were diagnosed with neoplasia. The skin tumors stood out, with sarcoid being the most common, followed by soft tissue sarcomas. A very important finding highlighted by the authors was the presence of different neoplasm behavior in donkeys when compared to horses. For example, squamous cell carcinoma and melanoma were considered unusual to rare tumors in donkeys, as well as lymphosarcoma, implying that there may be carcinogenesis differences among equids, which should be considered at the time of diagnosis and therapy, as well as at the time of informing veterinarians of the prevalence of tumors in these species [43].

Table 3. Clinical and pathological characteristics of the main primary cutaneous neoplasms diagnosed in donkeys and mules.

Disease	Gross Pathology	Histopathology	Clinical	References
Sarcoids	Nodulations are classified based on morphological patterns, being of the type, fibroblastic, verrucous and mixed.	These lesions are characterized as proliferative masses that have two histological constituents: an epithelial tissue and a dermal connective tissue with marked disorganized proliferation of connective tissue.	Commonly affected sites include ears, labial commissures, ventral trunk and feet.	[43,44]
Squamous cell carcinoma	The lesions are nodular, expansive, usually firm and sessile, which gradually increase in volume, ulcerating, with easy bleeding and crusting.	These nodulations correspond to infiltrative and irregular masses with cells arranged in cords or nests whose center may contain aggregates of keratin (corneal pearls) according to the degree of differentiation.	Its occurrence is often attributed to sun exposure in anatomical regions unprotected from pigmentation or hair, particularly in eyelids, ears, snout, perineum and udder.	[12]
Papilloma	The lesions are often arborescent or filiform, with a dry surface and that detach from the skin under traction.	This proliferation is benign and corresponds to an epithelial hyperplasia, forming papillary projections, with moderate supporting connective tissue.	The anatomical location is variable but can be seen particularly in the head, neck, belly, limbs, face, penis and base of the tail. They are usually benign and self-limiting.	[12,43]
Melanoma	Melanocytic tumors occur as single or multiple, shiny masses, usually blackened and multilobulated, with a high rate of metastasis.	Neoplastic melanocytes can exhibit intense pleomorphism with high mitotic activity, and are arranged in nests.	They are commonly seen in old, light-haired equines, particularly in the perineum, base of the tail and external genitalia.	[12,44]

Skin wounds are a frequent problem in equids, especially those of traumatic origin linked to traction activities.

A study evaluating 148 donkeys subjected to workloads three to five hours a day for three to five days a week in Tanzania showed the presence of wounds (one or more) in 56.1% of the animals. The lesions were mostly on the back and neck, and resulted from contact with harnesses, breast collars, and the cart. According to the authors, the lesions were variable in size and extension, most of them affecting the skin and subcutaneous tissues, some deepening to the musculature, while another portion was associated to granulation tissue, exudation, besides hemorrhages and necrosis [45]. An important finding of this study was the report of owners using substances such as motor oil for the treatment of injuries, a common practice in the northeast of Brazil, attributing to the oil protection and lubrication of the injured skin.

In Ethiopia, of a total of 997 horses, the most common skin diseases included skin wounds, followed by ectoparasites, dermatophilosis, sarcoid, and dermatophytes. A very important finding of this study refers to risk factors for the occurrence of these lesions, with the body score being very important for the occurrence of injuries, as well as what kind of work these animals performed [15]. The use of donkeys and mules for a multitude of tasks in Ethiopia is part of the culture of the country; however, sanitary and nutritional conditions do not always follow the degree of effort to which these animals are subjected to, thus generating the aforementioned injuries. The approach in these cases should be more holistic to consider cultural and socioeconomic issues and, at the same time, to improve the health and welfare of the animals.

The following table (Table 4) show some of the main skin disorders caused by factors related to the environment and traumatic injuries lesions.

Table 4. Environmental disease described in donkeys and mules.

Disease	Gross Pathology	Histopathology	Clinical	References
Photosensitization	The lesions are characterized by erosions and crusts accompanied by hyperemia, serous exudate, and, subsequently, to cracks and cutaneous detachment.	Microscopically, there are hyperkeratosis, ulcers in the epidermis, crusting and infiltration in the dermis that varies from polymorphonuclear to mononuclear cells.	It particularly affects depigmented areas such as the snout, udder, back and vulva. This condition can occur primary or secondary. The animal exhibits intense itching, contributing to self-mutilation.	[46,47]
Wounds, exuberant granulation tissue	Initial wounds can be of varied causes, leading to ulcerative and crusted lesions. These lesions can evolve into the exuberant granulation tissue becoming spongy, irregular, with no evident exudation.	In these lesions, there is a marked proliferation of fibrous connective tissue, neovascularization, fibroplasia, in addition to a chronic active inflammatory infiltrate, depending on whether the pathogenic stimulus persists or not.	Varied location, usually associated with the use of ropes, saddles and whips for containment, being commonly observed in the neck, limbs, back and tail.	[12,28,45]

Hypersensitivity to mosquito bites in general are progressive lesions starting with papules and crusts, accompanied by alopecia and erythema and may progress to ulceration. Mixed and perivascular dermatitis occurs with hyperplasia, orthokeratosis, crusts, edema, hyperkeratosis in follicles and hyperplasia of adnexal glands [27,28,48]. Typically, the disease manifests itself as chronic and itchy in character, particularly in the periocular region and external auditory canal, extending to the neck, back, abdomen, tail and limbs. Secondary mutilation and infections can occur secondarily and itching is variable, and the exposure of the injured areas to the sun's rays can cause serious complications [49,50].

Among the autoimmune diseases, pemphigus foliaceus stands out as being progressive and may contain epidermal collarette, papules and crusts, progressing to a scaly and alopecic lesion with or without exudation and easy hair removal. Microscopically these lesions correspond to a dermatitis with intraepidermal pustules with acantholysis and infiltration of intralesional and dermal neutrophils and eosinophils. Affected areas are multifocal that can coalesce and become generalized. They usually start in the face and limbs, but can affect mucocutaneous junctions [27,28,51].

3. Equine Dermatopathies in Northeast Brazil

Much of the attention given to donkey skin is due to the commercial value in the international market. In this context, similar to bovine leather, skin lesions may depreciate the final product and interfere with quality. In Brazil however, reports of integument disorders in donkeys and mules are little reported when compared to North American, European, and Asian countries. The low value attributed to these animals in the northeastern rural communities, associated with particularities of the species regarding clinical presentation of some diseases, probably impact the owner's decision not to take these animals to the veterinarian, resulting in many diseases being underdiagnosed in the region.

A study in the semiarid region of Paraíba State reported that 258 donkeys and mules seen in 10 years presented integumental diseases as the main causes of consultation (88 cases), mainly traumatic wounds, sarcoids, and abscesses in donkeys and traumatic wounds, squamous cell carcinomas, and habronemiasis in mules. A very important finding raised in this study was that most diseases diagnosed could be linked to mistreatment or lack of attention to these animals [11], a factor associated to the cultural habits of many owners.

In the same year, another retrospective study exclusively on equid skin diseases in the semiarid northeast was published [11]. According to this study, only one case of pythiosis was reported in donkeys in the northeast, which was subsequently approached regarding clinical, epidemiological, pathological, immunohistochemical, and molecular aspects [36]. According to the authors, the pattern of the lesion was similar to cases of pythiosis in cattle, and the origin of the disease was attributed to grazing in flooded areas, similarly to what is described in equines.

Other diseases that deserve to be highlighted, despite the less expressive diagnosis, are photosensitizing and allergic diseases, recently diagnosed in donkeys and mules in the northeast of Brazil. A case of allergic dermatitis to a *Culicoides* bite was reported in the state of Pernambuco, characterized by crusted papule and pruritic skin lesions with a clinical progression of two years. A very important finding highlighted in this study was the difficult clinical diagnosis due to similarities with other equid dermatopathies [48].

Primary photosensitization caused by *Froelichia humboldtiana* in equids have been reported in the semiarid region of Brazil, affecting donkeys and mules. This condition was described as exuberant, ulcerated lesions, with abundant serous exudation and crusts, alopecia, erythema, edema and areas of necrosis (Figure 1), especially in face, croup, and withers, accompanied by intense itching. The lesions were also associated with myiasis and secondary infections and, in many cases, due to the impossibility of treating the wounds, there was high mortality in the herd. The disease occurs at the end of the rainy season in pastures highly invaded by *F. humboldtiana*. Animals usually recover after their removal from areas invaded by this plant [46].

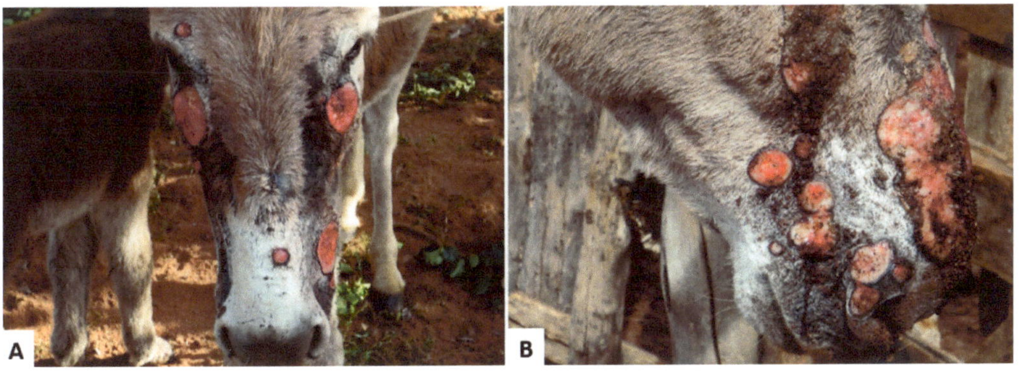

Figure 1. Donkeys naturally poisoned by *Froelichia humboldtiana*. (**A**) Multiple skin ulcers caused by secondary self-mutilation to intense itching; (**B**) Multiple and extensive ulcerated wounds, that drained serous exudate.

4. Conclusions

The study of skin diseases in equines is essential, especially in Brazil, which has an impressive number of donkeys and mules. The evaluation of the clinical and pathological behavior of these diseases implies the implementation of control strategies and the recognition of pathological patterns may be specific to the species.

Geographical variations may further result in significant differences in the prevalence of skin diseases, identifying the need for regional discussions on the emergence of common diseases. In addition, cultural factors may interfere with the frequency of these diseases, which implies losses in animal health and welfare. Incorporating these concepts in an increasingly technological world is perhaps the greatest challenge for the sustainable farming of donkeys and mules.

Author Contributions: Conceptualization, T.S.L.; F.S.M. and R.B.L.; methodology, T.S.L., R.A.F.S. and R.B.L.; investigation, T.S.L., R.B.L.; resources, R.B.L.; T.S.L. and M.S.S.; writing—original draft preparation, T.S.L., R.M.F.P., K.L.S., N.T.A.S. and R.A.F.S.; writing—review and editing, T.S.L., R.A.F.S.

and R.B.L.; supervision, R.B.L. and F.S.M. All authors have read and agreed to the published version of the manuscript.

Funding: This research was funded by Coordination for the Improvement of Higher Education Personnel grant number Code 001 And The APC was funded by Federal University of Paraíba.

Conflicts of Interest: The authors declare no conflict of interest.

References

1. The Food and Agriculture Organization-FAO. 2016. Available online: http://www.fao.org/faostat/en/#home (accessed on 23 April 2020).
2. The Food and Agriculture Organization-FAO. 2018. Available online: http://www.fao.org/faostat/en/#data/QA (accessed on 23 April 2020).
3. Luís, C.; Bastos-Silveira, C.; Cothran, E.G.; Oom, M.D.M. Iberian origins of New World horse breeds. *J. Hered.* **2006**, *97*, 107–113. [CrossRef] [PubMed]
4. Baker, M. *Sob a Pele-O Comércio Emergente de Pele de Asno e Suas Implicações Para o Bem-Estar e os Meios de Subsistência dos Asnos*, 1st ed.; The Donkey Sanctuary: Sidmouth, UK, 2017; Available online: https://www.thedonkeysanctuary.org.uk/sites/uk/files/201711/under_the_skin_report_portuguese.pdf (accessed on 10 July 2020). (In Portuguese)
5. Carneiro, G.F.; Cavalcante Lucena, J.E.; de Oliveira Barros, L. The current situation and trend of the donkey industry in South America. *J. Equine Vet. Sci.* **2018**, *65*, 106–110. [CrossRef]
6. Köhle, N. Feasting on Donkey Skin. In *Conspicuous Consumption*, 1st ed.; Smith, C.A., Kohle, N., Jaivin, L., Eds.; ANU Press: Canberra, Australia, 2018; ISBN (ebook): 9781760462031.
7. Lara, M.C.C.S.H.; Villalobos, E.M.C.; Cunha, S.E.M.; Oliveira, J.V.; Castro, C.; Nassar, A.F.C.; Silva, L.M.P.; Okuda, L.H.; Romaldini, A.H.C.N.; Cunha, M.S.; et al. Occurrence of viral diseases in donkeys (*Equus asinus*) in São Paulo State, Brazil. *Braz. J. Vet. Res. An. Sci.* **2017**, *54*, 154–158. [CrossRef]
8. García-Bocanegra, I.; Arenas-Montes, A.; Jaén-Téllez, J.A.; Napp, S.; Fernández-Morente, M.; Arenas, A. Use of sentinel serosurveillance of mules and donkeys in the monitoring of West Nile virus infection. *Vet. J.* **2012**, *194*, 262–264. [CrossRef]
9. Lima, T.S. Caracterização Clínico-patológica e Epidemiológica das Dermatopatias de Ruminantes no Agreste da Paraíba. Master's Dissertation, Universidade Federal da Paraíba, Areia, Paraíba, 2019. Available online: https://repositorio.ufpb.br/jspui/handle/123456789/14355 (accessed on 10 July 2020).
10. Assis-Brasil, N.D.; Marcolongo-Pereira, C.; Stigger, A.L.; Fiss, L.; Santos, B.L.; Coelho, A.C.B.; Sallis, E.S.V.; Fernandes, C.G.; Schild, A.L. Dermatopatias em equinos na região sul do Rio Grande do Sul: Estudo de 710 casos. *Ciên. Rural.* **2015**, *45*, 519–524. (In Portuguese) [CrossRef]
11. Pessoa, A.F.A.; Pessoa, C.R.M.; Miranda Neto, E.G.; Riet-Correa, F. Doenças de asininos e muares no semiárido brasileiro. *Pesq. Vet. Bras.* **2014**, *34*, 1210–1214. (In Portuguese) [CrossRef]
12. Pessoa, A.F.A.; Pessoa, C.R.M.; Miranda Neto, E.G.; Dantas, A.F.M.; Riet-Correa, F. Doenças de pele em equídeos no semiárido brasileiro. *Pesq. Vet. Bras.* **2014**, *34*, 743–748. (In Portuguese) [CrossRef]
13. McLean, A.K.; Navas Gonzalez, F.J. Can scientists influence donkey welfare? Historical perspective and a contemporary view. *J. Equine Vet. Sci.* **2018**, *65*, 25–32. [CrossRef]
14. Miranda, A.L.S.; Palhares, M.S. Muares: Características, origem e particularidades clínico-laboratoriais. *Rev. Cient. Med. Vet.* **2017**, 1–8. Available online: http://faef.revista.inf.br/imagens_arquivos/arquivos_destaque/qqT3b4TLANsOU7k_2018-6-30-10-43-55.pdf (accessed on 10 July 2020). (In Portuguese).
15. Tesfaye, A.; Tekle, Y.; Taddele, H.; Gezahagn, K.; Yihdego, H. Survey of Common Skin Problem of Working Equines in and Around Mekelle, North Ethiopia. *Acad. J. Anim. Dis.* **2015**, *4*, 30–38. [CrossRef]
16. Davis, E. Donkey and Mule Welfare. *Vet. Clin. N. Am. Equine Pract.* **2019**, *35*, 481–491. [CrossRef] [PubMed]
17. Pierezan, F.; Rissi, D.R.; Rech, R.R.; Fighera, R.A.; Brum, J.S.; Barros, C.S.L. Achados de necropsia relacionados com a morte de 335 eqüinos: 1968–2007. *Pesq. Vet. Bras.* **2009**, *29*, 275–280. (In Portuguese) [CrossRef]
18. Lucena, R.B.; Pierezan, F.; Kommers, G.D.; Irigoyen, L.F.; Fighera, R.A.; Barros, C.S.L. Doenças de bovinos no Sul do Brasil: 6.706 casos. *Pesq. Vet. Bras.* **2010**, *30*, 428–434. (In Portuguese) [CrossRef]
19. Marques, A.L.A.; Aguiar, G.M.N.; Lira, M.A.A.; Miranda Neto, E.G.; Azevedo, S.S.; Simões, S.V.D. Enfermidades do sistema digestório de bovinos da região semiárida do Brasil. *Pesq. Vet. Bras.* **2018**, *38*, 407–416. [CrossRef]
20. Barrandeguy, M.E.; Carossino, M. Infectious diseases in donkeys and mules: An overview and update. *J. Equine Vet. Sci.* **2018**, *65*, 98–105. [CrossRef]
21. Mavrouli, M.; Vrioni, G.; Kapsimali, V.; Tsiamis, C.; Mavroulis, S.; Pervanidou, D.; Billinis, C.; Hadjichristodoulou, C.; Tsakris, A. Reemergence of West Nile Virus Infections in Southern Greece, 2017. *Am. J. Trop. Med. Hyg.* **2019**, *100*, 420–426. [CrossRef]
22. Souza, T.M.; Brum, J.S.; Fighera, R.A.; Brass, K.E.; Barros, C.S.L. Prevalência dos tumores cutâneos de equinos diagnosticados no Laboratório de Patologia Veterinária da Universidade Federal de Santa Maria, Rio Grande do Sul. *Pesq. Vet. Bras.* **2011**, *31*, 379–382. (In Portuguese) [CrossRef]
23. Marcolongo-Pereira, C.; Estima-Silva, P.; Soares, M.P.; Sallis, E.S.V.; Grecco, F.B.; Fernandes, C.G.; Raffi, M.B.; Schild, A.L. Doenças de equinos na região Sul do Rio Grande do Sul. *Pesq. Vet. Bras.* **2014**, *34*, 205–210. (In Portuguese) [CrossRef]

24. Miller, M.A.; Moore, G.E.; Bertin, F.R.; Kritchevsky, J.E. What's New in Old Horses? Postmortem Diagnoses in Mature and Aged Equids. *Vet. Pathol.* **2015**, *53*, 390–398. [CrossRef]
25. Elsheikha, H.M.; Schares, G.; Paraschou, G.; Sullivan, R.; Fox, R. First record of besnoitiosis caused by *Besnoitia bennetti* in donkeys from the UK. *Parasites Vectors* **2020**, *13*, 1–10. [CrossRef]
26. Radwan, A.M.; Ahmed, N.E.; Elakabawy, L.M.; Ramadan, M.Y.; Elmadawy, R.S. Prevalence and pathogenesis of some filarial nematodes infecting donkeys in Egypt. *Vet. World* **2016**, *9*, 888–892. [CrossRef] [PubMed]
27. White, S.D.; Bourdeau, P.J.; Brement, T.; Vandenabeele, S.I.; Haspeslagh, M.; Bruet, V.; Van Oldruitenborgh-Oosterbaan, M.M.S. Skin disease in donkeys (*Equus asinus*): A retrospective study from four veterinary schools. *Vet. Dermatol.* **2019**, *30*, 247–276. [CrossRef] [PubMed]
28. Knottenbelt, D.C. Skin Disorders of the Donkey and Mule. *Vet. Clin Equine.* **2019**, *35*, 493–514. [CrossRef] [PubMed]
29. Alvarez, J.A.C.; Socarras, T.O.; Tous, M.G. Dermopatias en burros de trabajo (*Equus asinus*) en areas rurales de Cordoba (Colombia). *Rev. Med. Vet.* **2017**, *34*, 81–91. (In Spanish) [CrossRef]
30. Meselu, D.; Abebe, R.; Mekibib, B. Prevalence of Epizootic Lymphangitis and bodily distribuition of lesions in cart-mules in Behair Dar Town, Northwest Ethiopia. *J. Vet. Sci. Technol.* **2018**, *9*, 1–4. [CrossRef]
31. Mahendra, P. Occurrence of Cutaneous Epizootic Lymphangitis in Working Donkeys in Debre Zeit, Ethiopia. *EC Microbiol.* **2019**, *15*, 382–384.
32. Mota, R.A.; Oliveira, A.A.F.; Pinheiro Junior, J.W.; Silva, L.B.G.; Brito, M.F.; Rabelo, S.S.A. Glanders in donkeys (*Equus asinus*) in the state of Pernambuco, Brazil: A case report. *Braz. J. Microbiol.* **2010**, *41*, 146–149. [CrossRef]
33. Rabelo, S.S.A.; Soares, P.C.; Silva, L.B.G.; Cunha, A.P.; Nascimento Sobrinho, E.; Pinheiro Junior, J.W.; Barbosa, M.A.G.; Mota, R.A. Indicadores Clínicos em Muares Naturalmente Infectados pela *Burkholderia mallei*. *Vet. Zootec.* **2006**, *13*, 54–62. Available online: https://docplayer.com.br/71952462-Indicadores-clinicos-em-muares-naturalmente-infectados-pela-burkholderia-mallei-resumo.html (accessed on 10 July 2020). (In Portuguese).
34. Mota, R.A.; Brito, M.F.; Castro, F.J.C.; Massa, M. Mormo em eqüídeos nos estados de Pernambuco e Alagoas. *Pesq. Vet. Bras.* **2000**, *20*, 155–159. (In Portuguese) [CrossRef]
35. Ghahvei, Y.; Mirzaei, M.; Hashemnia, S.; Golchin, M.; Kheirandish, R.; Uni, S.; Endoza-Roldan, J.A.; Otranto, D.; Sazmand, A. Scanning electron microscopy of *Onchocerca fasciata* (Filarioidea: Onchocercidae) adults, microfilariae and eggs with notes on histopathological findings in camels. *Parasit. Vectors* **2020**, *13*, 1–10. [CrossRef]
36. Maia, L.A.; Olinda, R.G.; Araújo, T.F.; Firmino, P.R.; Nakazato, L.; Miranda Neto, E.G.; Riet-Correa, F.; Dantas, A.F.M. Cutaneous pythiosis in a donkey (*Equus asinus*) in Brazil. *J. Vet. Diagn. Invest.* **2016**, *28*, 436–439. [CrossRef] [PubMed]
37. Tabosa, I.M.; Medeiros, V.T.; Dantas, A.F.M.; Azevedo, E.O.; Maia, J.C. Pitiose cutânea em equinos no semi-árido da Paraíba. *ABMVZ* **1999**, *51*, 27–30. (In Portuguese)
38. Santos, C.E.P.; Santurio, J.M.; Colodel, E.M.; Juliano, R.S.; Silva, J.A.; Marques, L.C. Contribuição ao Estudo da Pitose Cutânea em Equídeos do Pantanal Norte, Brasil. *ARS Vet.* **2011**, *27*, 134–140. Available online: http://arsveterinaria.org.br/ars/article/download/310/337 (accessed on 10 July 2020). (In Portuguese).
39. Leite, N.M.; Rocha, M.V.; Souza, K.M.; Vago, P.B. Linfangite ulcerativa em equino: Relato de caso. *PUBVET* **2019**, *13*, 1–6. (In Portuguese) [CrossRef]
40. Sureshjani, M.H.; Atyabi, N.; Tazikeh, A.; Falahatipour, S.K.; Hashemian, M. Isolation of Rhodococcus equi from a mule with cutaneous wound. *Comp. Clin. Pathol.* **2014**, 1–3. [CrossRef]
41. Prescott, J.F. Rhodococcus equi: An animal and human pathogen. *Clin. Microbiol. Rev.* **1991**, *4*, 1–20. [CrossRef]
42. Guimarães, J.H.; Papavero, N.; Prado, A.P. As miíases na região neotropital (identificação, biologia, bibliografia). *Rev. Bras. Zool.* **1982**, *1*, 239–416. [CrossRef]
43. Davis, C.R.; Valentine, B.A.; Gordon, E.; McDonough, S.P.; Schaffer, P.A.; Allen, A.L.; Pesavento, P. Neoplasia in 125 donkeys (*Equus asinus*): Literature review and a survey of five veterinary schools in the United States and Canada. *J. Vet. Diagn. Investig.* **2016**, *28*, 662–670. [CrossRef]
44. Alberti, T.S.; Zamboni, R.; Venancio, F.R.; Scheid, H.V.; Bermann, C.S.; Raffi, M.B.; Sallis, E.S.V. Melanoma anaplásico em equino de pelagem tordilha com metástase em osso e músculo. *Cien. Anim.* **2019**, *29*, 129–134. (In Portuguese)
45. Rayner, E.; Airikkala-Otter, I.; Susheelan, A.; Gibson, A.; Itaba, R.; Mayani, T.; Mellanby, R.J.; Gamble, L. Prevalence of skin wounds in working donkeys in Bukombe, Tanzania. *Vet. Rec.* **2019**, *186*, 1–3. [CrossRef]
46. Knupp, S.N.R.; Knupp, L.S.; Riet-Correa, F.; Lucena, R.B. Plants that cause photosensitivity in ruminants in Brazil. *Semin. Ciênc. Agrár.* **2016**, *37*, 2009–2020. [CrossRef]
47. Amado, G.P.; Silva, C.C.B.; Barbosa, F.M.S.; Nascimento, H.H.L.; Malta, K.C.; Azevedo, M.V.; Lacerda-Lucena, P.B.; Lucena, R.B. Surtos de fotossensibilização e dermatite alérgica em ruminantes e equídeos no Nordeste do Brasil. *Pesq. Vet. Bras.* **2018**, *38*, 889–895. (In Portuguese) [CrossRef]
48. Silva, T.I.B.; Melchior, L.A.K.; Baptista Filho, L.C.F.; Fernandes, A.C.C.; Silva, L.G.; Vasconcelos, K.F.; Revorêdo, R.G.; Silva, D.D.; Melo, L.E.H. Dermatite alérgica à picada de Culicoides em muar: Relato de caso. *Arq. Bras. Med. Vet. Zootec.* **2017**, *69*, 1407–1412. (In Portuguese) [CrossRef]
49. Schaffartzik, A.; Hamza, E.; Janda, J.; Crameri, R.; Marti, E.; Rhyner, C. Equine insect bite hypersensitivity: What do we know? *Vet Immunol. Immunopathol.* **2012**, *147*, 113–126. [CrossRef] [PubMed]

50. Corrêa, T.G.; Ferreira, J.M.; Riet-Correa, G.; Ruas, J.L.; Schild, A.L.; Riet-Correa, F.; Guimarães, A.; Felippe-Bauer, M.L. Seasonal allergic dermatitis in sheep in southern Brazil caused by *Culicoides insignis* (Diptera: Ceratopogonidae). *Vet. Parasitol.* **2007**, *10*, 181–185. [CrossRef] [PubMed]
51. Monteiro, G.A.; Souza, M.V.; Conceição, L.G.; Balbi, C.L.; Borba, R.; Moreira, M.A.S. Pênfigo foliáceo em um equino. *Ciência Rural* **2007**, *37*, 594–598. (In Portuguese) [CrossRef]

animals

MDPI

Article

Pituitary Pars Intermedia Dysfunction and Metabolic Syndrome in Donkeys

Heidrun Gehlen [1,*], Bianca Schwarz [2], Claus Bartmann [3], Jennifer Gernhardt [1] and Sabita D. Stöckle [1]

1 Equine Clinic, Veterinary Department, The Free University of Berlin, 14163 Berlin, Germany; jennifer.gernhardt@fu-berlin.de (J.G.); sabita.d.stoeckle@fu-berlin.de (S.D.S.)
2 Pferdeinternist Dr. Bianca C. Schwarz, DipECEIM, Bei der Taffingsmühle 1, 66740 Saarlouis, Germany; dr.bianca.schwarz@googlemail.com
3 Equine Clinic, Veterinary Department, University of Giessen, 35392 Giessen, Germany; cpbartmann@gmx.de
* Correspondence: heidrun.gehlen@fu-berlin.de

Received: 5 October 2020; Accepted: 3 December 2020; Published: 8 December 2020

check for updates

Simple Summary: Donkeys are one of the six species of the equid family. Even though they may look similar to horses, there are optical, behavioral, and physiological differences between the two species. The most important endocrine diseases in horses (equine metabolic syndrome and pituitary pars intermedia dysfunction: PPID) also exist in donkeys. The key symptoms of asinine metabolic syndrome (AMS), similar to horses, are obesity, insulin dysregulation, and laminitis. It can be diagnosed with either basal glucose and insulin concentration or dynamic tests. The intravenous glucose tolerance test and the combined glucose insulin tolerance test were evaluated for donkeys. The therapy of AMS is aimed at weight and exercise management. Donkeys suffering from PPID are often laminitic. Other authors have reported on hypertrichosis as a cardinal sign. Donkey-specific differences in shedding compared to horses have to be considered. The PPID can be diagnosed with donkey-specific reference values or dynamic testing. The dexamethasone suppression test, the thyrotropin releasing hormone (TRH) test, and the combined dexamethasone suppression/TRH test were evaluated for donkeys.

Abstract: Appropriate medical care for donkeys is challenging despite being important working animals in non-industrialized countries and pets in first world countries. Although the same principles of diagnosis and therapy as in horses are commonly applied, there are differences in reference values and physiologic reaction to dynamic tests. However, donkeys seem to suffer from typical equine diseases, such as metabolic syndrome and pituitary pars intermedia dysfunction (PPID). Asinine metabolic syndrome (AMS) comprises obesity, insulin dysregulation, and laminitis. The principles of diagnosis are similar to horses. Donkey-specific reference ranges for insulin and glucose have been evaluated previously. Examinations regarding dynamic testing revealed differences in the intravenous glucose tolerance test and the combined insulin tolerance test compared to horses. The therapy of AMS is based mainly on weight loss and exercise. There are conflicting data regarding the incidence of PPID in donkeys. Laminitis and hypertrichosis were described as the main clinical signs. Species-specific and seasonal reference ranges were defined to diagnose PPID in donkeys. Furthermore, the dexamethasone suppression test, the thyrotropin releasing hormone (TRH) test and the combined dexamethasone suppression/TRH test were evaluated. Pergolide is commonly recommended for treatment.

Keywords: ACTH; PPID; metabolic syndrome; insulin dysregulation

1. Introduction

The equid family comprises donkeys and zebras in addition to the horse, as well as their hybrids [1]. Donkeys are commonly used as working animals and support humans mostly in farm work and transport [2,3]. They are occasionally used for milk, leather, and meat production [3,4]. Mules and hinnies, hybrids of horses and donkeys [5,6], that are also used as working animals, are especially sure of step and, therefore, often used as pack animals in the mountains and under adverse climatic conditions [3–8]. Donkeys and their hybrids are no longer required as working animals in many European and North American countries but are kept as pets [2,9]. Consequently, the donkey population has decreased significantly [2].

Even though donkeys are important working animals in developing countries and areas that are difficult to access otherwise, and even kept as pets in first world countries [2], adequate medical care for these animals is not ensured [10,11]. Donkeys show different pharmacokinetic characteristics than the horse because of different genetic and physiological properties [11–13].

Endocrinologic diseases in the horse, such as pituitary pars intermedia dysfunction (PPID) and equine metabolic syndrome (EMS), are a common diagnosis. These diseases also occur in donkeys, but there are species-specific characteristics.

2. Asinine Metabolic Syndrome

2.1. Epidemiology

Donkeys are adapted to rough environmental conditions with extremes in temperature, low-quality diets, and a high workload [2]. Under these harsh conditions, the donkey developed energy-efficiency traits, with an efficiency to rapidly mobilize fat in situations of increased energy demands or when food is scarce. Due to this evolution, the donkey can be considered as a typical "easy-keeper" [14]. Therefore, obese donkeys are often encountered in developed countries if high-quality and/or calorie-rich food is provided. In the authors' experience, owners of donkeys are often not aware that their animal is obese.

2.2. Clinical Signs

Donkeys have a much better feed conversion ratio than horses: a donkey requires only 57–67% of the digestible energy the pony needs compared to ponies of the same size, another point leading donkeys to be considered as typical easy keepers [15–17]. As fat distribution and neck morphology in donkeys differs from that in horses, a donkey-specific body condition score (BCS), and neck scoring system were developed (see Tables 1 and 2). It is important to keep in mind that there is a possibility of calcification of fat deposits in donkeys [15]. These no longer respond to weight loss and should not be incorporated in the BCS [15]. Furthermore, donkeys may have intra-abdominal fat depots [16] that may, for example, cover the linea alba [18,19].

Similar to horses, obesity and insulin resistance may contribute to laminitis [21]. Laminitis often results in irreversible changes of the anatomy of the foot, which are caused by failure of the suspensory apparatus of the distal phalanx [22,23]. These changes lead to altered biomechanics and, therefore, also to secondary pathologies and altered hoof horn production, which, in turn, leads to changes in the hoof conformation [24]. This disease causes foot pain and lameness [24]. In a study of radiological signs for laminitis in donkeys, Collins et al. [24] based the clinical diagnosis of acute laminitis on stance and gait irregularities, increased digital pulse amplitude, increased sensitivity to hoof testers in the dorsal aspect of the foot, increased hoof temperature, the presence of supracoronary depression and behavioral changes indicative of pain. Regarding the hooves themselves, the presence of divergent growth rings, widening of the white line, extensive flattening of the sole, sole hemorrhage in the region adjacent to the dorsodistal margin of the distal phalanx, distortion to the hoof capsule, dorsal concavity of the hoof wall, and/or perioplic hyperproliferation were defined as signs consistent with laminitis [24]. Radiographic signs in laminitic donkeys include a rotation, distal displacement, and/or morphometric change of the coffin bone and increases in the integument depth [24,25]. The radiological diagnosis of laminitis in donkeys, however, cannot be based on baseline data established for the horse [24]. Chronically laminitic hooves of a donkey are shown in Figure 1.

Table 1. Donkey-specific body condition scoring provided by the Donkey Sanctuary (https://www.thedonkeysanctuary.org.uk/sites/uk/files/2018-10/body-scoring-chart.pdf).

Condition Score	Neck and Shoulders	Withers	Ribs and Belly	Back and Loins	Hindquarters
1 Poor (Very thin) 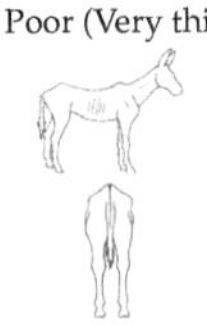	Neck thin, all bones felt easily. Neck meets shoulder abruptly, shoulder bones felt easily, angular.	Dorsal spine and withers prominent and felt easily.	Ribs can be seen from a distance and felt easily. Belly tucked up.	Backbone prominent, dorsal, and transverse processes felt easily.	Hip bones visible and felt easily (dock and pin bones). Little muscle cover. May be cavity under tail.
2 Moderate (Underweight) 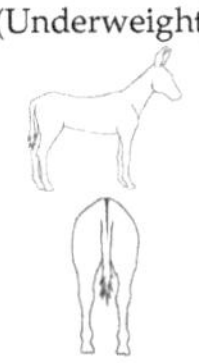	Some muscle development overlying bones. Slight step where neck meets shoulders.	Some muscle development overlying bones. Slight step where neck meets shoulders.	Ribs not visible but can be felt easily.	Dorsal and transverse processes felt with light pressure. Poor muscle development either side of midline.	Poor muscle cover on hindquarters, hip bones felt easily.
3 Ideal 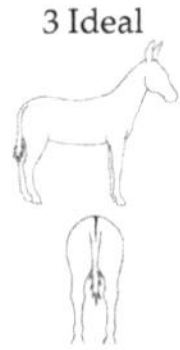	Good muscle development, bones felt under light cover of muscle/fat. Neck flows smoothly into shoulder, which is rounded.	Good cover of muscle/fat over dorsal spinous processes, withers flow smoothly into back.	Ribs just covered by light layer of fat/muscle, ribs can be felt with light pressure. Belly firm with good muscle tone and flattish outline.	Can feel individual spinous or transverse processes with pressure. Muscle development either side of midline is good.	Good muscle cover over hindquarters, hip bones rounded in appearance, can be felt with light pressure.
4 Overweight (Fat) 	Neck thick, crest hard, shoulder covered in even fat layer.	Withers broad, bones felt with pressure.	Ribs dorsally only felt with firm pressure, ventral ribs may be felt more easily. Belly overdeveloped.	Can only feel dorsal and transverse processes with firm pressure. May have slight crease along midline.	Hindquarters rounded, bones felt only with pressure. Fat deposits evenly placed.
5 Obese (Very fat) 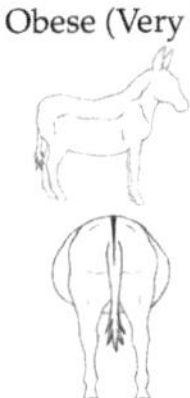	Neck thick, crest bulging with fat and may fall to one side.	Shoulder rounded and bulging with fat. Withers broad, bones felt with firm pressure.	Large, often uneven fat deposits covering dorsal and possibly ventral aspect of ribs. Ribs not palpable dorsally. Belly pendulous in depth and width.	Back broad, difficult to feel individual spinous or transverse processes. More prominent crease along midline fat pads on either side. Crease along midline, bulging fat either side.	Cannot feel hip bones, fat may overhang either side of tail head, fat often uneven and bulging.

Table 2. Donkey-specific neck scoring system provided by Mendoza et al. (2015) [20].

Score	Description
0 	Neck thin with absence of a visible and palpable crest.
1 	Neck still slightly thin. Crest not visible, but palpable. Normal appearance.
2 	Neck moderately fatty. Noticeable crest that can be palpated from withers to poll. Patchy fat deposits can be palpated. Bony prominences cannot be felt.
3 	Neck thick and rounded. Crest is enlarged, thickened and hard. It is palpable from withers to poll. Crest begins to make longitudinal fat deposit to both sides of the neck. Fat deposited from the middle of the neck to withers. Crest width is increased.
4 	Neck thick and rounded. Crest grossly enlarged and thickened. Large fat deposits from poll to withers, forming longitudinal hard bands of fat at both neck sides that can be grasped with the hand. Crest width is grossly expanded and it cannot be grasped with one hand. Occasionally in large breeds, crest may drop to one side.

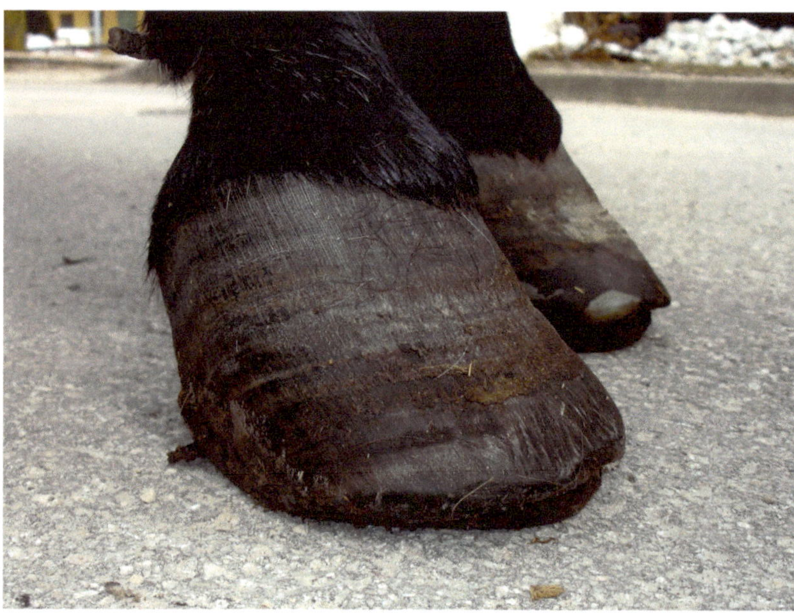

Figure 1. Hooves of a donkey with signs of chronic laminitis. Note the divergent growth rings.

Insulin concentrations in obese donkeys were shown to be higher when compared to moderate and thin donkeys [26]. Furthermore, an overall trend for an increasing BCS to lower insulin sensitivity has been suspected previously [26]. Additionally, there are changes in the lipid and lipoprotein metabolism in obese donkeys which may be a predisposition to hyperlipemia [27]. Compared to horses, the information available on AMS and insulin resistance in donkeys is scarce [28]. Figure 2 shows an obese donkey with a BCS of 5/5.

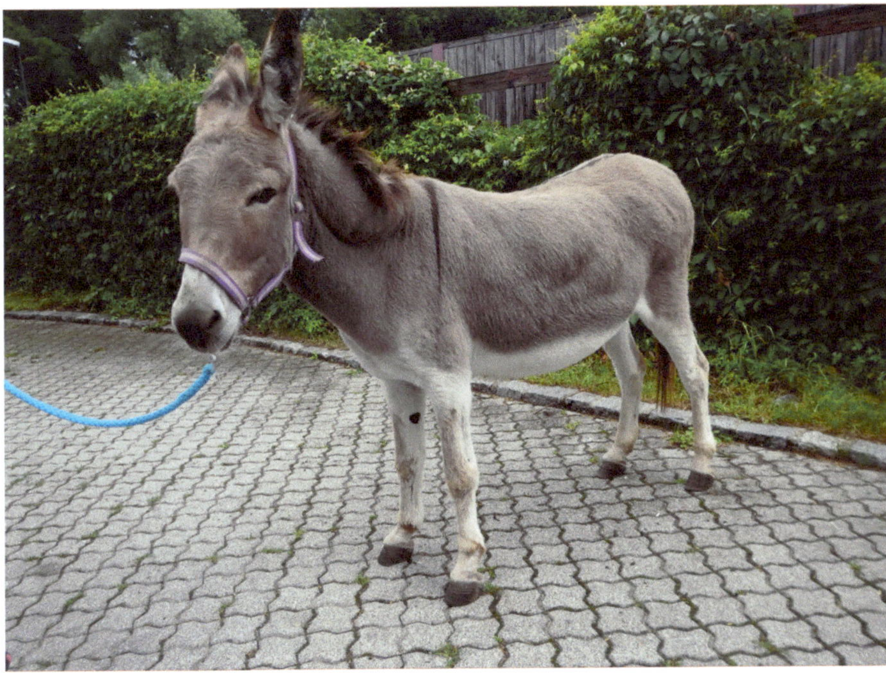

Figure 2. Obese donkey with a body condition score of 5.

2.3. Pathogenesis

The EMS is a collection of risk factors for the development of endocrinopathic laminitis [29]. This includes obesity (regional and/or generalized) and systemic insulin dysregulation/resistance [29]. The failure of the tissues to respond to insulin adequately and/or an altered insulin clearance is defined as insulin dysregulation/resistance [29]. Genetic predisposition may be important in horses for the development of obesity and insulin dysregulation [30]. Donkeys developed energy-efficiency traits, with an efficiency to rapidly mobilize fat in situations of increased energy demands or when food is scarce as they are commonly used as working animals under harsh environmental conditions. This may predispose the animals to obesity [14].

Insulin dysregulation plays a key role in EMS and hyperinsulinemia is probably the most important pathophysiologic component of insulin dysregulation in horses [31,32]. Donkeys have a decreased insulin sensitivity [16] compared to horses and donkeys with insulin resistance often have higher insulin plasma concentrations than horses with insulin dysregulation [28]. Insulin concentrations in obese donkeys are often high, suggesting insulin dysregulation. Hyperinsulinemia in horses has previously been identified as the most probable cause of endocrinopathic laminitis [32–35], but hyperinsulinemia has not yet been identified as the direct cause of asinine endocrinopathic laminitis [17]. However, one study suggested that insulin concentrations higher than 20 μIU/mL were associated with laminitis in donkeys [36]. The clinical relevance of this or higher cutoff values to diagnose a risk of endocrinopathic laminitis remains unclear [17], especially as there are age-related variations in insulin concentration. Geriatric donkeys often show lower insulin concentrations, indicating a reduced β-cell mass or less β-cell sensitivity to glucose [20]. The gender of the animal may influence the concentration of metabolic parameters significantly [37,38]. Non-pregnant jennies have lower glucose and higher triglyceride concentrations compared to male animals [20].

2.4. Diagnostic Testing for Insulin Dysregulation

The diagnostic principles of insulin dysregulation in donkeys do not differ from horses. In horses, baseline insulin and glucose concentrations are recommended as a screening test for metabolic syndrome (high specificity), but the sensitivity is not adequate for ruling out EMS [39]. Newer research suggests that the cut-off values for suspected insulin dysregulation/resistance in horses might be lower than initially anticipated [40,41]. Furthermore, glucose and insulin concentrations in donkeys may be altered by previous transportation [42]. The literature suggests not providing feed containing more than 10% nonstructural carbohydrates for 6–8 h before testing a horse for insulin dysregulation/resistance [29]. For donkeys, there is no consensus on how to prepare the animals for testing, however, Mendoza et al. recommend allowing access to a flake of hay to reduce stress and avoid hyperlipemia [28]. If results of baseline testing are variable or fasting insulin is normal despite a strong clinical suspicion of EMS, dynamic testing is recommended [28]. Only a few studies have evaluated glucose-insulin dynamics in donkeys [21,37,38,43]. The combined glucose-insulin test (CGIT) and the intravenous glucose tolerance test (IVGTT) [37,43–45] for dynamic testing were evaluated in donkeys.

2.4.1. Combined Glucose Insulin Test

The CGIT assesses insulin sensitivity by determining the time when blood glucose concentrations return to baseline values after the simultaneous intravenous administration of dextrose/glucose (150 mg/kg, 50 % dextrose/40% glucose solution) and insulin (0.1 IU/kg) [46]. The animals need to be fasted for 6 h but are allowed free access to water. An intravenous catheter is also required. The blood glucose concentration is determined at 0, 1, 5, 15, 25, 35, 45, 60, 75, 90, 105, 120, 135, and 150 min, and insulin concentrations are measured at 0 and 45 min. The glucose curve in insulin-sensitive donkeys is shifted to the right, reaching its lowest concentration at 120 min (horse: 75 min) and it takes 240 min for glucose concentrations to return to baseline (horse: 150 min) (Figure 3) [43,47]. Insulin dysregulation in horses is suspected if the blood glucose concentration is above baseline at 45 min or the insulin concentration is greater than 100 μIU/mL 45 min after dextrose/glucose administration [43,47].

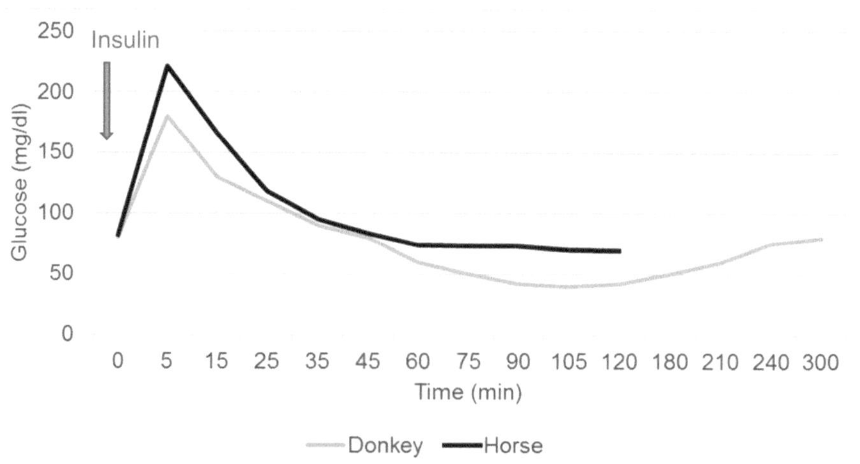

Figure 3. Typical combined glucose insulin test in donkeys compared to horses.

2.4.2. Intravenous Glucose Tolerance Test

The IVGTT is a method frequently used for glucose tolerance in horses [46]. The animals need to be fasted for 6–8 h for the IVGTT. Dextrose/glucose (150–300 mg/kg) is administered intravenously as a 50% or 40% solution, respectively, after collection of a baseline glucose and insulin sample. Afterwards, samples for blood glucose determination are collected every 30 min for the next 180 min. Compared to horses (150 min), donkeys also showed a right shift of the glucose curve in this test. In donkeys, the positive glucose phase lasted 160.9 ± 13.3 min (Figure 4) [43]. Furthermore, donkeys showed a negative phase in the IVGTT that is usually not present in the horse, which suggests a delayed biological efficiency of insulin [43].

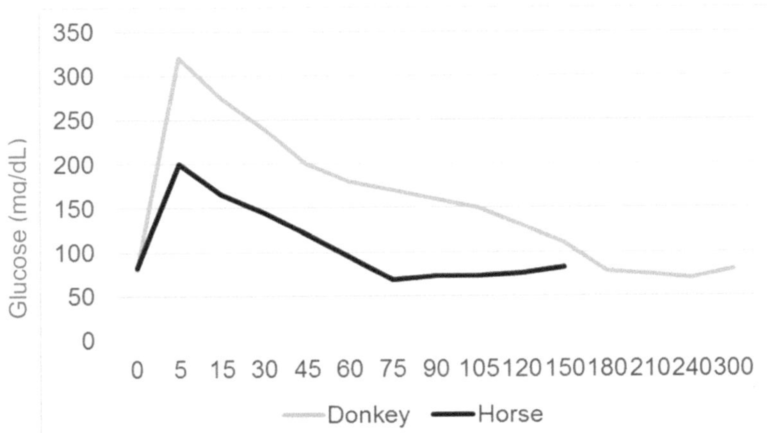

Figure 4. Typical intravenous glucose tolerance test in donkeys compared to horses.

Table 3 summarizes the concentrations of hormones and metabolites reported in donkeys that are typically used to evaluate metabolic disease in equids. However, it has to be mentioned that there were different methods used regarding both laboratory methods and fasting. Mendoza et al. examined 63 healthy donkeys which were mostly of Andalusian breed (9 geldings and 53 non-pregnant jennies) based on history and clinical examination as well as on hematology and blood biochemistry results [20]. In this study, the animals were starved for 12 h. Glucose and triglycerides were determined by spectrophotometry. Commercially available radioimmunoassay kits were used to determine the leptin, total adiponectin and active ghrelin concentrations [20]. These radioimmunoassay kits had been validated for horses and donkeys previously [48–50]. Insulin, IGF-1 and glucagon human radioimmunoassay kits were validated for donkeys by assessing the specificity, sensitivity and intra-assay precision [20], as Midgley described previously [51].

Table 3. Biochemical parameters and dynamic testing to evaluate metabolic disease in donkeys.

Parameter	Dynamic Testing	Mean ± SE	Mean ± SE	Mean ± SE	Mean ± SE	Mean ± SE
Glucose (mg/dL)		79.2 ± 3.5 [a]	84.1 ± 3.7 [a]	80.01 ± 12.25 [d]		
Triglycerides (mg/dL)		75.3 ± 10.1[a]	66.6 ± 10.5 [a]	58.9 ± 3.6 [b]	66.4 ± 34.2 [c]	75.25 ± 43.75 [d]
Insulin (μIU/mL)		8.1 ± 0.8 [a]	8.9 ± 0.7 [a]	10.1 ± 0.5 [b]	2.1 ± 2.05 [c]	4.16 ± 3.46 [d]
Glucagon (pg/mL)		144.2 ± 6.7 [b]				
Leptin (ng/mL)		2.7 ± 0.3 [b]				
Adiponectin HE (ng/mL)		458.1 ± 11.8 [b]				
Ghrelin HE (pg/mL)		45.1 ± 1.6 [b]				
IGF-1 (ng/mL)		234.9 ± 13.5 [b]				
CGIT	positive phase (min)	44.1 ± 3.01 [a]				
	negative phase (min)	255.9 ± 3.01 [a]				
IVGTT	positive phase (min)	160.9 ± 13.3 [a]				
	negative phase (min)	139.1 ± 13.3 [a]				

[a]: [43], [b]: [36], [c]: [52], [d]: [53], SE: Standard error of the mean; CGIT: Combined glucose–insulin test; IVGTT: Intravenous glucose tolerance test.

Another study evaluated the CGIT and IVGTT in ten healthy female non-pregnant Andalusian jennies. Health was based on the history, clinical examination and blood work [43]. This reference also gives two means for fasting glucose concentration. These were determined before the IVGTT and the CGIT, respectively. Glucose and triglyceride concentrations were determined by spectrophotometry. A commercial radioimmunoassay kit, that had previously been validated for donkeys [20], was used for the insulin determination. Before testing, the donkeys were housed overnight (22:00–8:00) and supplied with one flake of hay and water ad libitum.

Dugat et al. determined basal insulin and glucose concentrations in 44 mammoth donkeys and 1 miniature donkey. Animals were classified as healthy based on history and clinical examination. Of these animals, 36 were non-pregnant jennies, 5 pregnant jennies, 3 intact males, and 1 castrated male. There is neither information on feeding prior to testing nor on the laboratory methods used [52].

Gehlen et al. included data of 35–44 healthy donkeys based on history and clinical examination. The night before testing, the animals had access to grass, hay and/or straw but not to concentrates. A chemiluminescence assay was applied for insulin determination. Triglycerides and glucose were determined by photometry [53].

2.5. Treatment

No specific treatments have been described for donkeys suffering from ASM so far. The same principles for treating EMS generally apply [17,28]. The most important aspect of ASM therapy is clearly weight loss, which can be achieved by offering a controlled diet rich in crude fiber and low in starch [15,16]. Weight loss of 2% bodyweight per month until optimal weight and BCS is reached is targeted [28]. Despite the need for weight loss and food restriction, it has to be kept in mind that donkeys are prone to develop hyperlipemia in cases of caloric restriction/anorexia [16].

Pharmacologic treatments described for horses, for example, metformin, have not been critically evaluated in donkeys [17,28] are discussed controversially [16].

3. Pituitary Pars Intermedia Dysfunction

3.1. Epidemiology

According to McGowan, 20% of all horses, ponies, and donkeys over the age of 15 years suffer from PPID [54]. There is no gender or breed predisposition [54]. Other authors report that PPID rarely occurs in donkeys [55] and is less common than expected [56]. Cox et al. published a questionnaire study in the United Kingdom (1114 questionnaires were returned) in 2010, in which no cases of donkeys suffering from PPID were recorded [57]. Morrow et al. performed necropsies of 1444 donkeys, of which 1.9% (n = 27) had PPID [57]. Of these, 96.3% also had a foot disorder. This publication also reports on adrenal changes in 127/1444 donkeys (8.8%). These changes, however, are not classified further [57].

3.2. Clinical Signs

Signs of PPID appear to be similar in horses and donkeys. According to some authors, chronic laminitis is the cardinal symptom for PPID in donkeys [16,56], whereas others report hypertrichosis as the clinical sign observed most frequently [17,27]. However, donkeys have a physiologically longer and thicker haircoat and start shedding earlier in the autumn than horses [58]. Furthermore, shedding in donkeys takes more time in spring than it does in horses [58]. Remainders of the winter coat may still sporadically be seen in the summer [58].

Other signs, such as insulin resistance, abnormal fat distribution and muscle wasting (pot belly), that are common in horses [59] seem to also be common in donkeys suffering from PPID [17,28]. Polyuria has been reported in approximately 30% of horses suffering from PPID [60–69], but it has not yet been reported in donkeys [17]. Fertility problems, secondary infections, and lethargy have not been reported in donkeys suffering from PPID [28]. Lethargy may easily be misdiagnosed due to the donkey-specific behavior [28].

3.3. Pathogenesis

The pars intermedia of the pituitary gland is inhibited by dopamine provided by the hypothalamus [70]. Upon activation by dopamine, the D2 receptors mediate inhibition of proopiomelanocortin (POMC) mRNA expression and POMC-derived hormone release [71]. Degeneration of hypothalamic dopaminergic neurons in the PPID patient leads to a loss of inhibition of the melanotropes located in the pars intermedia of the pituitary gland [72]. The increased melanotrope activity results in adenoma formation and dysregulated POMC secretion [59]. The hormone precursor molecule, POMC, is synthesized in the anterior (corticotropes) and intermediate lobe (melanotropes) of the pituitary gland [73]; however, the cleavage of POMC is different between the anterior and intermediate pituitary lobes [74,75].

The POMC is split into adrenocorticotropic hormone (ACTH) and β-lipotropin in the anterior lobe. The ACTH is further cleaved in the intermediate lobe into α-melanocyte-stimulating hormone and corticotropin-like intermediate lobe peptide. Furthermore, β-lipotropin is processed to β-endorphin [74,76], of which a significantly more active form is present in the pituitary gland of PPID patients than in normal tissues [75].

The ACTH in healthy horses is only a minor (2%) cleavage product of POMC in the pars intermedia [77]. The POMC production and cleavage in the pars intermedia seems to be intact, but the post-cleavage modification of the peptides is different [73,75]. The exact consequences of these deviating concentrations remain unclear [78–80].

Cortisol may be responsible for some of the clinical signs associated with PPID, but many horses suffering from PPID have cortisol concentrations within or below the reference range. A recent study reported on similar total cortisol plasma concentrations in healthy and diseased horses [61,81] but found an increased free cortisol concentration (active form) in PPID patients compared to normal horses [82].

This might contribute to the clinical signs, even without an elevated total plasma cortisol concentration [82]. Similar to horses, hypercortisolemia is a rare finding in donkeys suffering from PPID [27].

3.4. Diagnosis

The diagnosis of PPID in donkeys relies on the same principles as in horses [17,28,59]. In horses, the TRH test is recommended for early PPID and the basal ACTH concentration for moderate to advanced PPID [83]. Plasma ACTH concentrations show seasonal changes in both horses and donkeys [53]. Furthermore, the ACTH and cortisol concentration in donkeys may be increased even after short transportation [84]. Ethylenediaminetetraacetic acid plasma is used to determine the ACTH. It is important that the sample remains cooled during overnight transportation to the laboratory [83].

The Donkey Sanctuary mentions that the same diagnostic tests can be used in the donkey for dynamic testing as those in the horse: The TRH test, the dexamethasone suppression test and the combined dexamethasone suppression/TRH test [56]. The Equine Endocrinology Group recommends firstly determining the basal ACTH concentration. If the basal ACTH concentration is within the reference range but there is a high suspicion of PPID in the patient, the TRH test is recommended for diagnostic testing [83]. At the time of writing this, performing the TRH test in autumn is not recommended because reference values for this season have not yet been established [83].

Table 4 summarizes the reference concentrations of ACTH and cortisol for donkeys provided in literature.

Table 4. Concentrations of ACTH and cortisol in donkeys reported in literature

Parameter	Mean ± SE	Mean ± SE	Mean ± SE
ACTH (pg/mL)	66.7 ± 20.7 [a] (May/June)	62.93 ± 37.4 [b] (Aug) 20.6 ± 14.06 [b] (Feb, May, Nov)	21.3–24.7 [c] 25.9–36.9 (Jul–Oct) [c] 16.5–19.5 [c]
Cortisol (μg/dL)	4.0 ± 1.2 [a]		

[a]: [52], no information on feeding status and laboratory method; [b]: [53], animals had no access to concentrate feed the night before testing but unlimited access to hay, grass, and/or straw, chemiluminescence immunoassay; [c]: [20].

3.4.1. TRH Test

The basal ACTH concentration is determined for the TRH test. Afterwards, the patient receives 0.5 (equids < 250 kg) or 1 mg (equids > 250 kg) TRH intravenously and the ACTH concentration is determined again 10 min after TRH administration. Transient side effects of TRH administration in horses include coughing, flehmen, and yawning [83]. A recent study evaluated the TRH test with ACTH determination in donkeys with the same criteria as in horses (positive if ACTH > 110 pg/mL 10 min post TRH administration). The test correctly identified 6/6 clinically suspect donkeys as PPID positive. The ACTH peaked at 10–20 min post TRH and returned to baseline at 30 min [85]. The TRH test can be performed after hay has been fed [86,87], but stimulation after a grain meal has not yet been examined. The Equine Endocrinology Group does not recommend testing after a grain meal [83]. The TRH test with cortisol determination is no longer recommended. Furthermore, the conduction of the TRH test in autumn is not recommended (no clinical data available) [83].

3.4.2. Dexamethasone Suppression Test

Dexamethasone administration does not suppress the cortisol release in the PPID patient as much as in the healthy equid. After determination of the basal cortisol concentration, 40 μg/kg dexamethasone is administered intramuscularly [88]. A second sample is collected 12–24 h later and the cortisol is determined. A significant decrease in cortisol concentration can be observed in the healthy patient. There is no decrease in the cortisol concentration in the PPID patient [59]. A recent study on a small number of donkeys applied the same criteria as those reported for horses in this test, i.e., cortisol concentration higher than 27.6 nmol/L at 19 h post dexamethasone administration as positive. Three out of six clinically suspect animals were identified as PPID positive in this test [85].

3.4.3. Combined Dexamethasone Suppression/TRH Test

The combined dexamethasone suppression/TRH test recombines the two tests mentioned previously.

A basal sample for cortisol concentration determination is collected. Then, 40 μg/kg dexamethasone is administered intramuscularly. A second sample for cortisol determination is collected 3 h after dexamethasone administration and 1 mg TRH is administered intravenously. Several blood samples (15, 31, 45, and 60 min, 21 h) are collected afterwards [89]. A recent study evaluated the combined dexamethasone suppression/TRH test in donkeys with the same criteria as those in horses (positive if cortisol concentration >66% baseline value at 195 min post dexamethasone). This test correctly identified 67% of clinically suspect donkeys tested (4/6) as PPID-positive [85].

3.5. *Treatment*

Pergolide, a dopamine agonist derived from ergot alkaloids, is most commonly used for PPID treatment in the horse as its efficacy and safety have been demonstrated by various studies [64,90,91]. Pergolide has a high affinity to dopamine D2 receptors of the pars intermedia and, therefore, inhibits the production of POMC successfully. Based on clinical results, pergolide dosing in donkeys seems to be similar to that in horses [17,28]. A starting dose of 0.002 (0.001–0.003 mg/kg) per os once daily is recommended [16,88,92].

Pergolide is licensed for the symptomatic treatment of PPID in horses (*Prascend®*, Boehringer Ingelheim Vetmedica GmbH, Ingelheim/Germany) but not in donkeys. The side effects reported in donkeys are diarrhea, colic, depression, and anorexia [16,56,88,92].

4. Conclusions

The ASM and PPID in donkeys have been insufficiently studied. Diagnostic testing has recently been examined in small animal groups. Further studies are required, especially regarding the treatment of these endocrine disorders in donkeys.

Author Contributions: Conceptualization: H.G.; Supervision: H.G.; Visualization: C.B.; writing—original draft: H.G. and S.D.S.; Writing—review & editing: H.G., B.S., C.B., J.G. and S.D.S. All authors have read and agreed to the published version of the manuscript.

Funding: This research received no external funding.

Conflicts of Interest: The authors declare no conflict of interest.

References

1. Spektrum.de. Kompaktlexikon der Biologie-Equidae. 26 July 2020. Available online: https://www.spektrum.de/lexikon/biologie-kompakt/equidae/3779 (accessed on 11 November 2020). (In German).
2. Starkey, P.; Starkey, M. Regional and world trends in donkey populations. In *Donkeys, People, and Development*; Animal Traction Network for Eastern and Southern Africa, Ed.; ATNESA: Wageningen, The Netherlands, 2004; pp. 10–21.
3. Pearson, R.A.; Sims, B.; Abound, A. Equines. In *Livestock and Wealth Creation: Improving the Husbandry of Animals Kept by Resource-Poor People in Developing Countries*; Owen, E., Kitalyi, A., Jayasuriya, N., Smith, T., Eds.; DFID Department of International Development; Nottingham University Press: Nottingham, UK, 2005; pp. 489–508.
4. Blakeway, S.J. The Welfare of Donkeys. Master's Thesis, University of Edinburgh, Edinburgh, UK, 1994.
5. Hutchins, B. The modern ass and mule, a part of equine practice. *J. Equine Vet. Sci.* **1983**, *3*, 30–31. [CrossRef]
6. McDonnell, S.M. Reproductive behavior of donkeys (*Equus asinus*). *Appl. Anim. Behav. Sci.* **1998**, *60*, 277–282. [CrossRef]

7. Von Gugelberg, H.; Bähler, C.; Bruns, U. *Alles über Maultiere*, 1st ed.; Müller Rüschlikon: Cham, Switzerland, 1994. (In German)
8. Rennenkampff, F.; Bartmann, C. Das Gebirgstragtierwesen im Neuen Heer. *DGWMP* **2008**, *31*, 53–55. (In German)
9. Trawford, A. Donkey welfare internationally—Current research. In Proceedings of the 50th British Equine Veterinary Association Congress, Liverpool, UK, 7–10 September 2011.
10. Burnham, S.L. Anatomical differences of the donkey and mule. In Proceedings of the 48th Annual Convention of the American Association of Equine Practitioner, Orlando, FL, USA, 4–8 December 2002; pp. 102–109.
11. Grosenbaugh, D.; Reinemeyer, C.; Figueiredo, M. Pharmacology and therapeutics in donkeys. *Equine Vet. Educ.* **2011**, *23*, 523–530. [CrossRef]
12. Lizarraga, I.; Sumano, H.; Brumbaugh, G. Pharmacological and pharmacokinetic differences between donkeys and horses. *Equine Vet. Educ.* **2004**, *16*, 102–112. [CrossRef]
13. Senior, J. Not small horses: Improving treatments for donkeys. *Vet. Rec.* **2013**, *173*, 292–293. [CrossRef]
14. Mendoza, F.J.; Toribio, R.E.; Perez-Ecija, A. Metabolic and endocrine disorders in donkeys. *Vet. Clin. Equine Pract.* **2019**, *35*, 399–417. [CrossRef]
15. Burden, F. Practical feeding and condition scoring for donkeys and mules. *Equine Vet. Educ.* **2012**, *24*, 589–596. [CrossRef]
16. Schwarz, B.; Anen, C. *Eselmedizin: Basiswissen*; Continuing Veterinary Education, Veterinär-Verlag: Gnarrenburg, Germany, 2014. (In German)
17. Mendoza, F.J.; Toribio, R.E.; Perez-Ecija, A. Aspects of clinical relevance in donkeys. In *Equine Internal Medicine*; Reed, S.M., Baly, W.M., Sellon, D.C., Eds.; Elsevier: St. Louis, MO, USA, 2018; pp. 1513–1520.
18. Thiemann, A. Clinical approach to the dull donkey. *In Pract.* **2013**, *35*, 470–476. [CrossRef]
19. Thiemann, A.K.; Rickards, K.J.; Getachew, M.; Paraschou, G. Colic in the donkey. In *The Equine Acute Abdomen*; Blikslager, A.T., White, N.A., Moore, J.N., Mair, T.S., Eds.; Wiley Blackwell: Hoboken, NJ, USA, 2017; pp. 471–488.
20. Mendoza, F.J.; Estepa, J.C.; Gonzalez-De Cara, C.A.; Aguilera-Aguilera, R.; Toribio, R.E.; Perez-Ecija, A. Energy-related parameters and their association with age, gender, and morphometric measurements in healthy donkeys. *Vet. J.* **2015**, *204*, 201–207. [CrossRef]
21. Du Toit, N.; Trawford, A. Determination of serum insulin and insulin resistance in clinically normal donkeys and donkeys with a history of laminitis (obese and non-obese). *J. Vet. Intern. Med.* **2010**, *24*, 779.
22. Pellmann, R.; Budras, K.; Bragulla, H. Structure and function of the suspensory apparatus of the coffin bone in the horse. *Pferdeheilkunde* **1997**, *13*, 53.
23. Collins, S.N.; Van Eps, A.W.; Pollitt, C.C.; Kuwano, A. The lamellar wedge. *Vet. Clin. N. Am. Equine Pract.* **2010**, *26*, 179–195. [CrossRef] [PubMed]
24. Collins, S.N.; Dyson, S.J.; Murray, R.C.; Burden, F.; Trawford, A. Radiological anatomy of the donkey's foot: Objective characterisation of the normal and laminitic donkey foot. *Equine Vet. J.* **2011**, *43*, 478–486. [CrossRef]
25. Walker, M.; Taylor, T.; Slater, M.; Hood, D.; Weir, V.; Elslander, J. Radiographic appearance of the feet of mammoth donkeys and the finding of subclinical laminitis. *Vet. Radiol. Ultrasound* **1995**, *36*, 32–37. [CrossRef]
26. Pritchard, A.; Nielsen, B.; McLean, A.; Robison, C.; Yokoyama, M.; Hengemuehle, S.; Bailey, S.; Harris, P. Insulin resistance as a result of body condition categorized as thin, moderate, and obese in domesticated U.S. donkeys (*Equus asinus*). *J. Equine Vet. Sci.* **2019**, *77*, 31–35. [CrossRef] [PubMed]
27. Watson, T.D.; Packard, C.J.; Shepherd, J.; Fowler, J.N. An investigation of the relationships between body condition and plasma lipid and lipoprotein concentrations in 24 donkeys. *Vet. Rec.* **1990**, *127*, 498–500.
28. Mendoza, F.J.; Toribio, R.E.; Perez-Ecija, A. Donkey internal medicine—Part I: Metabolic, endocrine, and alimentary tract disturbances. *J. Equine Vet. Sci.* **2018**, *65*, 66–74. [CrossRef]
29. Burns, T.A.; Toribio, R.E. Insulin dysregulation and equine metabolic syndrome. In *Equine Internal Medicine*; Reed, S.M., Baly, W.M., Sellon, D.C., Eds.; Elsevier: St. Louis, MO, USA, 2018; pp. 1085–1100.
30. Jeffcott, L.B.; Field, J.R.; McLean, J.G.; O'Dea, K. Glucose tolerance and insulin sensitivity in ponies and Standardbred horses. *Equine Vet. J.* **1986**, *18*, 97–101. [CrossRef]
31. De Laat, M.A.; McGree, J.M.; Sillence, M.N. Equine hyperinsulinemia: Investigation of the enteroinsular axis during insulin dysregulation. *Am. J. Physiol. Endocrinol. Metab.* **2016**, *310*, E61–E72. [CrossRef]

32. Durham, A.E.; Frank, N.; McGowan, C.M.; Menzies-Gow, N.J.; Roelfsema, E.; Vervuert, I.; Feige, K.; Fey, K. ECEIM consensus statement on equine metabolic syndrome. *J. Vet. Intern. Med.* **2019**, *33*, 335–349. [CrossRef] [PubMed]
33. Asplin, K.E.; Sillence, M.N.; Pollitt, C.C.; McGowan, C.M. Induction of laminitis by prolonged hyperinsulinaemia in clinically normal ponies. *Vet. J.* **2007**, *174*, 530–535. [CrossRef] [PubMed]
34. De Laat, M.A.; McGowan, C.M.; Sillence, M.N.; Pollitt, C.C. Equine laminitis: Induced by 48 h hyperinsulinaemia in Standardbred horses. *Equine Vet. J.* **2010**, *42*, 129–135. [CrossRef] [PubMed]
35. de Laat, M.A.; Sillence, M.N.; McGowan, C.M.; Pollitt, C.C. Continuous intravenous infusion of glucose induces endogenous hyperinsulinaemia and lamellar histopathology in Standardbred horses. *Vet. J.* **2012**, *191*, 317–322. [CrossRef] [PubMed]
36. Carter, R.A.; Geor, R.J.; Staniar, W.B.; Cubitt, T.A.; Harris, P.A. Apparent adiposity assessed by standardised scoring systems and morphometric measurements in horses and ponies. *Vet. J.* **2009**, *179*, 204–210. [CrossRef] [PubMed]
37. McLean, A.; Nielsen, B.D.; Yokoyama, M.; O'Connor-Robison, C.; Hengemuehle, S.; Wang, W.; Geor, R.J.; Harris, P. Insulin resistance in standard donkeys (Equus asinus) of three body conditions-thin, moderate, and obese. *J. Equine Vet. Sci.* **2009**, *5*, 406–407. [CrossRef]
38. Pleasant, R.S.; Suagee, J.K.; Thatcher, C.D.; Elvinger, F.; Geor, R.J. Adiposity, plasma insulin, leptin, lipids, and oxidative stress in mature light breed horses. *J. Vet. Intern. Med.* **2013**, *27*, 576–582. [CrossRef]
39. Dunbar, L.; Mielnicki, K.A.; Dembek, K.A.; Toribio, R.E.; Burns, T.A. Evaluation of four diagnostic tests for insulin dysregulation in adult light-breed horses. *J. Vet. Intern. Med.* **2016**, *30*, 885–891. [CrossRef]
40. Olley, R.B.; Carslake, H.B.; Ireland, J.L.; McGowan, C.M. Comparison of fasted basal insulin with the combined glucose-insulin test in horses and ponies with suspected insulin dysregulation. *Vet. J.* **2019**, *252*, 105351. [CrossRef]
41. Van Den Wollenberg, L.; Vandendriessche, V.; van Maanen, K.; Counotte, G.H.M. Comparison of two diagnostic methods to detect insulin dysregulation in horses under field conditions. *J. Equine Vet. Sci.* **2020**, *88*, 102954. [CrossRef]
42. Forhead, A.; Smart, D.; Smith, R.F.; Dobson, H. Transport-induced stress responses in fed and fasted donkeys. *Res. Vet. Sci.* **1995**, *58*, 144–151. [CrossRef]
43. Mendoza, F.J.; Aguilera-Aguilera, R.; Gonzalez-De Cara, C.; Toribio, R.E.; Estepa, R.E.; Perez-Ecija, A. Characterization of the intravenous glucose tolerance test and the combined glucose–insulin test in donkeys. *Vet. J.* **2015**, *206*, 371–376. [CrossRef] [PubMed]
44. June, V.; Soderholm, V.; Hintz, H.F.; Butler, W.R. Glucose tolerance in the horse, pony and donkey. *J. Equine Vet. Sci.* **1992**, *12*, 103–105. [CrossRef]
45. Forhead, A.J.; Dobson, H. Plasma glucose and cortisol responses to exogenous insulin in fasted donkeys. *Res. Vet. Sci.* **1997**, *62*, 265–269. [CrossRef]
46. Frank, N. Equine metabolic syndrome. *Vet. Clin. Equine Pract.* **2011**, *27*, 73–92. [CrossRef]
47. Eiler, H.; Frank, N.; Andrews, F.M.; Oliver, J.W.; Fecteau, K.A. Physiologic assessment of blood glucose homeostasis via combined intravenous glucose and insulin testing in horses. *Am. J. Vet. Res.* **2005**, *66*, 1598–1604. [CrossRef]
48. Barsnick, R.J.I.M.; Hurcombe, S.D.A.; Dembek, K.; Frazer, M.L.; Slovis, N.M.; Saville, W.J.A.; Toribio, R.E. Somatotropic axis resistance and ghrelin in critically ill foals. *Equine Vet. J.* **2014**, *46*, 45–49. [CrossRef]
49. Salimei, E.; Rosi, F.; Maglieri, C.; Magistrelli, D.; Fantuz, F. Leptin in milk and plasma of dairy asses. *Ital. J. Anim. Sci.* **2007**, *6*, 654–656. [CrossRef]
50. Gordon, M.E.; McKeever, K.H.; Betros, C.L.; Manso Filho, H.C. Plasma leptin, ghrelin and adiponectin concentrations in young fit racehorses versus mature unfit standardbreds. *Vet. J.* **2007**, *173*, 91–100. [CrossRef]
51. Midgley, A.R.; Niswender, G.D.; Rebar, R.W. Principles for the assessment of the reliability of radioimmunoassay methods (precision, accuracy, sensitivity, specificity). *Eur. J. Endocrinol.* **1969**, *62*, S163–S184. [CrossRef]
52. Dugat, S.L.; Taylor, T.S.; Matthews, N.S.; Gold, J.R. Values for triglycerides, insulin, cortisol, and ACTH in a herd of normal donkeys. *J. Equine Vet. Sci.* **2010**, *30*, 141–144. [CrossRef]
53. Gehlen, H.; Twickel, S.; Stöckle, S.; Weber, C.; Bartmann, C.P. Diagnostic orientation values for ACTH and other parameters for clinically healthy donkeys and mules (insulin, triglycerides, glucose, fructosamines, and γ-GT). *J. Anim. Physiol. Anim. Nutr.* **2020**, *104*, 679–689. [CrossRef] [PubMed]
54. McGowan, C.M. Hyperadrenocorticism (pituary pars intermedia dysfunction) in horses. In *Clinical Endocrinology of Companion Animals*; Rand, J., Ed.; John Wiley & Sons, Ltd.: Hoboken, NJ, USA, 2013; pp. 100–114.
55. Anonymous. Results of postmortem examinations at the Donkey Sanctuary. *Vet. Rec.* **2006**, *159*, 582. [CrossRef]

56. Sprayson, T. The care of the geriatric donkey. In *The Professional Handbook of the Donkey*; Svendsen, E.D., Duncan, J., Hadrill, D., Eds.; Whitted Books: Wiltshire, UK, 2008; pp. 10–27.
57. Morrow, L.D.; Smith, K.C.; Piercy, R.J.; du Toit, N.; Burden, F.A.; Olmos, G.; Gregory, N.G.; Verheyen, K.L.P. Retrospective analysis of post-mortem findings in 1444 aged donkeys. *J. Comp. Pathol.* **2011**, *144*, 145–156. [CrossRef]
58. Hafner, M. *Esel Halten*, 3rd ed.; Verlag Eugen Ulmer: Stuttgart, Germany, 2012.
59. McFarlane, D. Equine pituitary pars intermedia dysfunction. *Vet. Clin. N. Am. Equine Pract.* **2011**, *27*, 93–113. [CrossRef]
60. Hillyer, M.H.; Taylor, F.G.R.; Mair, T.S.; Murphy, D.; Watson, T.D.G.; Love, S. Diagnosis of hyperadrenocorticism in the horse. *Equine Vet. Educ.* **1992**, *4*, 131–134. [CrossRef]
61. Boujon, C.E.; Bestetti, G.F.; Meier, H.P.; Straub, R.; Junker, U.; Rossi, G.L. Equine pituitary adenoma: A functional and morphological study. *J. Comp. Pathol.* **1993**, *109*, 163–178. [CrossRef]
62. Couëtil, L.; Paradis, M.R.; Knoll, J. Plasma adrenocorticotropin concentration in healthy horses and in horses with clinical signs of hyperadrenocorticism. *J. Vet. Intern. Med.* **1996**, *10*, 1–6. [CrossRef]
63. Muñoz, M.C.; Doreste, F.; Ferrer, O.; González, J.; Montoya, J.A. Pergolide treatment for Cushing's syndrome in a horse. *Vet. Rec.* **1996**, *139*, 41–43. [CrossRef]
64. Donaldson, M.T.; LaMonte, B.H.; Morresey, P.; Smith, G.; Beech, J. Treatment with pergolide or cyproheptadine of pituitary pars intermedia dysfunction (equine Cushing's disease). *J. Vet. Intern. Med.* **2002**, *16*, 742–746.
65. Perkins, G.A.; Lamb, S.; Erb, H.N.; Schanbacher, B.; Nydam, D.V.; Divers, T.J. Plasma adrenocorticotropin (ACTH) concentrations and clinical response in horses treated for equine Cushing's disease with cyproheptadine or pergolide. *Equine Vet. J.* **2002**, *34*, 679–685. [CrossRef] [PubMed]
66. McGowan, C.M.; Neiger, R. Efficacy of trilostane for the treatment of equine Cushing's syndrome. *Equine Vet. J.* **2003**, *35*, 414–418. [CrossRef] [PubMed]
67. Donaldson, M.T.; Jorgensen, A.J.; Beech, J. Evaluation of suspected pituitary pars intermedia dysfunction in horses with laminitis. *J. Am. Vet. Med. Assoc.* **2004**, *224*, 1123–1127. [CrossRef] [PubMed]
68. Sgorbini, M.; Panzani, D.; Maccheroni, M.; Corazza, M. Equine cushing-like syndrome: Diagnosis and therapy in two cases. *Vet. Res. Commun.* **2004**, *28* (Suppl. 1), 377–380. [CrossRef]
69. Carter, R.A.; Treiber, K.H.; Geor, R.J.; Douglass, L.; Harris, P.A. Prediction of incipient pasture-associated laminitis from hyperinsulinaemia, hyperleptinaemia and generalised and localised obesity in a cohort of ponies. *Equine Vet. J.* **2009**, *41*, 171–178. [CrossRef]
70. Saland, L.C. The mammalian pituitary intermediate lobe: An update on innervation and regulation. *Brain Res. Bull.* **2001**, *54*, 587–593. [CrossRef]
71. Goudreau, J.L.; Lindley, S.E.; Lookingland, K.J.; Moore, K.E. Evidence that hypothalamic periventricular dopamine neurons innervate the intermediate lobe of the rat pituitary. *Neuroendocrinology* **1992**, *56*, 100–105. [CrossRef]
72. McFarlane, D. Advantages and limitations of the equine disease, pituitary pars intermedia dysfunction as a model of spontaneous dopaminergic neurodegenerative disease. *Ageing Res. Rev.* **2007**, *6*, 54–63. [CrossRef]
73. Wilson, M.G.; Nicholson, W.E.; Holscher, M.A.; Sherrell, B.J.; Mount, C.D.; Orth, D.N. Proopiolipomelanocortin peptides in normal pituitary, pituitary tumor, and plasma of normal and Cushing's horses. *Endocrinology* **1982**, *110*, 941–954. [CrossRef]
74. Orth, D.N.; Nicholson, W.E. Bioactive and immunoreactive adrenocorticotropin in normal equine pituitary and in pituitary tumors of horses with Cushing's disease. *Endocrinology* **1982**, *111*, 559–563. [CrossRef]
75. Millington, W.R.; Dybdal, N.O.; Dawson, R., Jr.; Manzini, C.; Mueller, G.P. Equine Cushing's disease: Differential regulation of β-endorphin processing in tumors of the intermediate pituitary. *Endocrinology* **1988**, *123*, 1598–1604. [CrossRef] [PubMed]
76. Castro, M.G.; Morrison, E. Post-translational processing of proopiomelanocortin in the pituitary and in the brain. *Crit. Rev. Neurobiol.* **1997**, *11*, 35–57. [PubMed]
77. Martin, K.J.; Hruska, K.A.; Freitag, J.J.; Klahr, S.; Slatopolsky, E. The peripheral metabolism of parathyroid hormone. *N. Engl. J. Med.* **1979**, *301*, 1092–1098. [CrossRef] [PubMed]
78. Haritou, S.J.; Zylstra, R.; Ralli, C.; Turner, S.; Tortonese, D.J. Seasonal changes in circadian peripheral plasma concentrations of melatonin, serotonin, dopamine and cortisol in aged horses with Cushing's disease under natural photoperiod. *J. Neuroendocrinol.* **2008**, *20*, 988–996. [CrossRef] [PubMed]

79. Beech, J.; Boston, R.C.; McFarlane, D.; Lindborg, S. Evaluation of plasma ACTH, α-melanocyte–stimulating hormone, and insulin concentrations during various photoperiods in clinically normal horses and ponies and those with pituitary pars intermedia dysfunction. *J. Am. Vet. Med. Assoc.* **2009**, *235*, 715–722. [CrossRef]
80. Copas, V.E.; Durham, A.E. Circannual variation in plasma adrenocorticotropic hormone concentrations in the UK in normal horses and ponies, and those with pituitary pars intermedia dysfunction. *Equine Vet. J.* **2012**, *44*, 440–443. [CrossRef]
81. Heinrichs, M.; Baumgärtner, W.; Capen, C.C. Immunocytochemical demonstration of proopiomelanocortin-derived peptides in pituitary adenomas of the pars intermedia in horses. *Vet. Pathol.* **1990**, *27*, 419–425. [CrossRef]
82. Hart, K.A.; Wochele, D.M.; Noron, N.A.; McFarlane, D.; Wooldridge, A.A.; Frank, A. Effect of age, season, body condition, and endocrine status on serum free cortisol fraction and insulin concentration in horses. *J. Vet. Intern. Med.* **2016**, *30*, 653–663. [CrossRef]
83. Equine Endocrinology Group, Pituitary Pars Intermedia Dysfunction (PPID). 2019. Available online: https://sites.tufts.edu/equineendogroup/files/2019/12/2019-PPID_EEGbooklet.pdf (accessed on 11 November 2020).
84. Fazio, E.; Medica, P.; Cravana, C.; Aveni, F.; Ferlazzo, A. Comparative endocrinological responses to short transportation of Equidae (*Equus asinus* and *Equus caballus*). *Anim. Sci. J.* **2013**, *84*, 258–263. [CrossRef]
85. Mejia-Pereira, S.; Perez-Ecija, A.; Buchanan, B.R.; Toribio, R.E.; Mendoza, F.J. Evaluation of dynamic testing for pituitary pars intermedia dysfunction diagnosis in donkeys. *Equine Vet. J.* **2019**, *51*, 481–488. [CrossRef]
86. Diez de Castro, E.; Lopez, I.; Cortes, B.; Pineda, C.; Garfia, B.; Aguilera-Tejero, E. Influence of feeding status, time of the day, and season on baseline adrenocorticotropic hormone and the response to thyrotropin releasing hormone-stimulation test in healthy horses. *Domest. Anim. Endocrinol.* **2014**, *48*, 77–83. [CrossRef] [PubMed]
87. Restifo, M.M.; Frank, N.; Hemida, P.; Sanchez-Londoño, A. Effects of withholding feed on thyrotropin-releasing hormone stimulation test results and effects of combined testing on oral sugar test and thyrotropin-releasing hormone stimulation test results in horses. *Am. J. Vet. Res.* **2016**, *77*, 738–748. [CrossRef]
88. Rickards, K. Treatment of hyperadrenocorticism in donkeys. *Vet. Rec.* **2010**, *166*, 152. [CrossRef]
89. Frank, N.; Andrews, F.M.; Sommardahl, C.S.; Eiler, H.; Rohrbach, B.W.; Donnell, R.L. Evaluation of the combined dexamethasone suppression/thyrotropin-releasing hormone stimulation test for detection of pars intermedia pituitary adenomas in horses. *J. Vet. Intern. Med.* **2006**, *20*, 987–993. [CrossRef] [PubMed]
90. Schott, H.C., II; Coursen, C.L.; Eberhart, S.W.; Nachreiner, R.J.; Refsal, K.R.; Ewart, S.; Marteniuk, J.V. The Michigan Cushing's Project. In Proceedings of the 47th Annuual Convention of American Association of Equine Practitioners, San Diego, CA, USA, 28 November 2001.
91. Durham, A. The pharmacologic basis for the treatment of endocrinopathic laminitis. *Vet. Clin. N. Am. Equine Pract.* **2010**, *26*, 303–314. [CrossRef] [PubMed]
92. Catling, S.; Horspool, L.; Thiemann, A. Appendix 2: Formulary (doses of commonly used medicines). In *The Professional Handbook of the Donkey*; Svendsen, E.D., Duncan, J., Hadrill, D., Eds.; Whitted Books: Wiltshire, UK, 2008; pp. 385–399.

Publisher's Note: MDPI stays neutral with regard to jurisdictional claims in published maps and institutional affiliations.

animals

MDPI

Communication

Investigation of Oral Microbiome in Donkeys and the Effect of Dental Care on Oral Microbial Composition

Yiping Zhu [1], Wuyan Jiang [1], Reed Holyoak [2], Bo Liu [1] and Jing Li [1,*]

[1] Equine Clinical Diagnostic Center, College of Veterinary Medicine, China Agricultural University, Beijing 100193, China; sy20203050871@cau.edu.cn (Y.Z.); jiangwyan@cau.edu.cn (W.J.); bliu2017@cau.edu.cn (B.L.)

[2] College of Veterinary Medicine, Oklahoma State University, Stillwater, OK 74078, USA; reed.holyoak@okstate.edu

* Correspondence: jlivet@cau.edu.cn; Tel.: +86-135-52228206

Received: 13 November 2020; Accepted: 25 November 2020; Published: 30 November 2020

check for updates

Simple Summary: Dental health in donkeys has long been neglected, even though it is quite common for them to have dental problems. Therefore, dental care, as basic as dental floating, can be a good start to improve their dental condition. Oral microbiome sequencing is a reliable way to reflect the oral health of animals. However, little is known on the effect of dental care on the oral microbiome of donkeys. Hence, a research project was undertaken to investigate the relationship between dental floating and oral microbial changes using a current sequencing technique. We found that the changes of the oral microbiome were not significant, probably due to the necessity of more specific and consistent treatment. However, the study provided an insight of the oral microbial composition and helped increase awareness of dental care in donkeys.

Abstract: The objective of this study was to investigate the oral microbial composition of the donkey and whether basic dental treatment, such as dental floating, would make a difference to the oral microbial environment in donkeys with dental diseases using high-throughput bacterial 16S rRNA gene sequencing. Oral swab samples were collected from 14 donkeys with various dental abnormalities on day 0 (before treatment) and day 20 (twenty days after treatment). It is the first report focusing on the oral microbiome in donkeys with dental diseases and the impact of common dental procedures thereon. Identified in group Day 0 and group Day 20, respectively, were 60,439.6 and 58,579.1 operational taxonomic units (OTUs). Several taxa in Day 0 differed significantly from Day 20 at the phylum and genus levels, but no statistically significant difference was observed in richness and diversity of Day 0 and Day 20. The results also indicated that a larger-scale study focusing on healthy donkey oral microbiome, as well as the correlation of dental diseases and oral microbiomes at different time frames following more specific and consistent dental treatment, are warranted.

Keywords: oral microbiome; 16S rRNA; dental treatment; donkeys; China

1. Introduction

Dental disease is a large welfare concern in donkeys [1]. The incidence of severe dental diseases such as diastema, overgrown teeth and periodontal diseases is high, [2] yet little attention has been paid to those conditions, and in fact, it is a common problem that most of donkeys do not receive dental examinations and preventative treatments regularly, either due to economic reasons or ignorance of dental care [3]. Diastema is the most common dental disease seen in donkeys [4] causing food entrapment and bacteria building up. The result is usually gingivitis and periodontitis and, in some cases, deep periodontal pockets [5] due to food mechanical irritation [6] and chemical destruction from bacterial infection [7]. In donkey farms in China, most donkeys are raised in large numbers and in high density. Additionally, donkeys are usually very stoic and good at hiding their discomfort, thereby eliminating clinical signs, even with severe dental diseases. As a result, their dental abnormalities are usually undiagnosed, causing severe pain, weight loss or even colic, until declining health leads to further examination and discovery of the dental disease [8]. Due to the high prevalence and negative impact of dental diseases in donkeys, dental care, especially the basic dental floating, is the most necessary treatment that can be provided routinely. There have been many studies conducted to investigate equine oral microbiome and what kinds of bacteria predominate or are associated with periodontal diseases in horses [8,9]; however, no report has focused on the donkey oral microbiome. Therefore, this study was intended to explore the oral microbial composition in donkeys and assess the impact of dental floating on the oral microbiome.

2. Materials and Methods

2.1. Animal Selection

The donkeys involved in this study were of the Liaoxi breed from a donkey farm in Northeast China. Donkeys in this farm were mainly female, used for milk and reproduction. No dental care had been performed on any of these animals before this study, and no antimicrobial medications had been administered in the previous 8 weeks to any of them selected for this study. Fourteen donkeys were randomly selected from a group of approximately 100 donkeys on Day 0. All were female, aged between 2 to 10 years old (mean age was 6). They were fed with mainly crop residue (corn stalk and millet stake) and supplemented with a small amount of homemade concentrate (maize, bran, soybean meal and salt) three times a day for years. None of the donkeys wore halters daily, only for special procedures—for example, the dental exam in this study. The examination started with a visual inspection of the skull symmetry and detection of ocular or nasal discharge. There was minimum-to-no obvious pain reactions such as head tossing noted while palpating and pressing the cheeks against the cheek teeth. Incisors were examined first to look for any abnormalities, such as brachygnathism, prognathism or occlusal surface abnormalities such as a slant, "frown" or extreme "smile". Whereafter, they were each sedated, and the oral cavity was flushed with tap water. An incisor speculum (Hausmann speculum) was then applied to facilitate examination. Other tools used to detect dental abnormalities of cheek teeth included a head light, a dental mirror, a dental explorer and a periodontal probe. Dental disorders were recorded, including severe sharp enamel points, periodontal pockets, hooks, diastemata, shear mouth, step mouth, wave mouth, accentuated transverse ridges and caries. Following the dental examination, sample procurement from the oral cavity was conducted in all 14 donkeys.

2.2. Sample Collection

On Day 0, after a thorough oral exam, samples were collected by using a sterile swab rubbed around the gingival margin and on top of the gingival sulcus on the buccal side bilaterally with sufficient pressure [10]. Each swab was then individually placed in a sterile tube containing 1-mL sterile Tris-EDTA buffer solution and stored in ice until arrival at the laboratory and then deep-frozen at −20 °C awaiting analysis. After sampling, the mouth was rinsed off with a chlorohexidine solution, and dental floating was performed for each of the 14 donkeys to help reduce the severity of dental abnormalities and improve occlusion. Twenty days after the dental treatment, another dental examination was performed, and a second set of oral samples was collected and processed in the same way.

2.3. Sample Processing and DNA Extraction

All samples were maintained at -20 °C before DNA extraction. Sample DNA was obtained from the oral swabs using a Hi-Swab DNA Kit (Tiangen, Beijing, China) in accordance with the manufacturer's instructions. The DNA concentration and purity were monitored by 1% agarose gel electrophoresis. According to the concentration, DNA was diluted to 1 ng/μL using sterile water.

2.4. 16S rRNA Sequencing

The V4 hyper-variable regions of the 16S rRNA genes were amplified by Novogene using the specific primers 515F (5′-3′ GTGCCAGCMGCCGCGGTAA) and 806R (5′-3′ GGACTACHV GGGTWTCTAAT), which target the V4 region of the bacterial 16S rRNA gene. Each PCR reaction was carried out in total volume of 30 μL with 15 μL of Phusion® High-Fidelity PCR Master Mix (New England Biolabs, Ipswich, MA, USA); 0.2 μm of forward and reverse primers, 10-ng template DNA and ultrapure water. Thermal cycling consisted of initial denaturation at 98 °C for 60 s, 30 cycles of denaturation at 98 °C for 10 s, 50 °C for 30 s, 72 °C for 30 s and one final cycle of 72 °C for 300 s.

2.5. Statistical Analysis

For α-diversity, the richness of samples representing the number of different bacterial taxa in each group was assessed using the number of observed operational taxonomic units (OTUs) and the Shannon index. In addition, the Simpson index was used for assessing evenness. The Illumina-sequenced paired-end reads were merged by using FLASH (V1.2.7). Sequences were processed by using the QIIME (V1.9.1, Quantitative Insights Into Microbial Ecology) software package for quality filtering and construction of the OTUs [11]. Sequences with ≥97% similarity were assigned to the same OTUs. The RDP (ribosomal database project) classifier was used to annotate the taxonomic information for each representative sequence.

The Kolmogorov-Smirnov test and the Shapiro-Wilk test were used to assess continuous data for normal distribution. *T*-test for paired samples was applied for the comparisons of the Shannon index. β-diversity analysis was performed to compare samples. Principal component analysis (PCA) was used to evaluate the general distribution of the resulting bacterial community composition. The linear discriminant analysis effect size (LEfSe) pipeline was also applied.

Paired *t*-test (normally distributed values) and Mann-Whitney U test (non-normally distributed values) were used for the statistical evaluation of continuous variables.

3. Results

3.1. Clinical Findings

On Day 0, all 14 donkeys in this study had mild-to-severe sharp enamel points and a varied degree of periodontal pockets (Table 1). According to the incisor abnormalities, only four (29%) donkeys had mild brachygnathism. No other disorders, such as smile, slant or prognathism, were noted. In this study group, 50% of the donkeys had hooks and diastema. Additionally, nearly half of the donkeys (43%) suffered from oral ulceration, and fibrously healed ulcers were observed in some of the donkeys. Four (29%) donkeys had a transverse ridge, all of which also had diastema. The occurrence of wave mouth (7%), step (14%), shear mouth (7%) and caries (14%) constituted the remainder of the dental disorders in this study. On Day 20, there were no donkeys with sharp enamel points or hooks. The diastema found in the seven donkeys on Day 0 was still there; however, less food was trapped, especially those with concurrent transverse ridges on Day 0. Most of the ulcers noted on Day 0 started healing, and some were almost healed. No significant changes were noted on periodontal pocket depths. Steps, wave mouth and shear mouth were still noted, but the evenness was improved.

Table 1. Prevalence of dental disorders (n = 14 donkeys).

Dental Disorders	Number of Donkeys	Prevalence
Periodontal pockets (2–5 mm)	14	100%
Sharp enamel points	14	100%
Hooks	7	50%
Diastemata	7	50%
Ulcers	6	43%
Disorders in the occlusal surface	5	36%
Transverse ridge	4	29%
Brachygnathism	4	29%
Caries	2	14%
Step	2	14%
Shear mouth	1	7%
Wave mouth	1	7%

3.2. Microbiome Profile Analysis

The mean values of the sequences were 60,439.6 and 58,579.1 (Mann-Whitney U test, $P = 0.401$) from samples collected on Day 0 and Day 20, respectively. The microbial richness and diversity were evaluated with α-diversity indexes, including the Chao1 index, observed OTUs, Shannon index and Simpson index. The richer bacterial microbiomes are indicated by a higher Chao1 index or observed OTUs, while more diverse bacterial microbiomes are indicated by a higher Shannon index or lower Simpson index. There was no significant difference of the mean number of OTUs (paired t-test, $P = 0.414$), Chao1 index ($P = 0.291$), Shannon index ($P = 0.779$) and observed OTUs ($P = 0.306$) (Figure 1) between the two groups, meaning the richness of Day 0 and Day 20 were not significantly different.

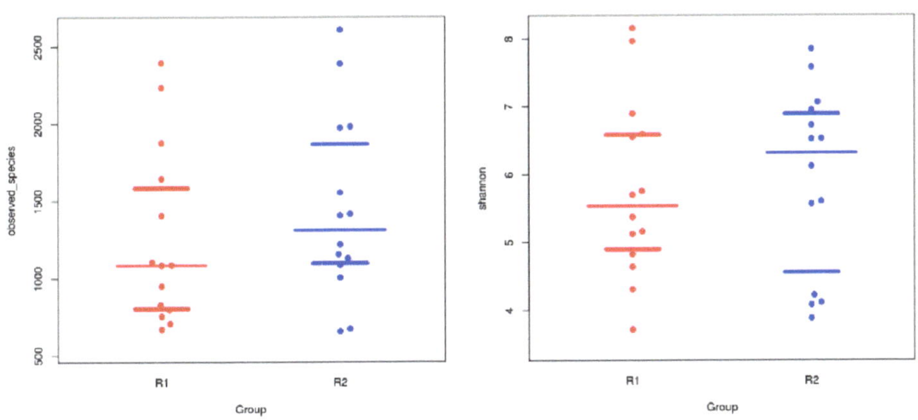

Figure 1. Observed operational taxonomic units (OTUs) (**left**) and Shannon index (**right**). Red color indicates the Day 0 group, while blue indicates the Day 20 group.

3.3. Compositional Analysis between the Groups

When the taxonomic abundance of the two groups of sequences was compared at the phylum level, the microbial community of the two groups featured a similar profile. The relative abundance of the top nine phyla in both groups are displayed in Figure 2a, which are *Firmicutes*, *Proteobacteria*, *Actinobacteria*, *Bacteriodetes*, *Fusobacteria*, *Cyanobacteria*, *Acidobacteria*, *Gemmatimonadetes*, *Chloroflexi* and *Spirochaetes*. Among these phyla, each of the first five phyla (*Firmicutes*, *Proteobacteria*, *Actinobacteria*, *Bacteriodetes* and *Fusobacteria*) accounted for >1% of the total sequences amongst both groups. *Firmicutes* and *Proteobacteria* were the most common bacterial phyla in the Day 0 group and Day 20 group, which accounted for 81.1–81.9% (Figure 2a). *Cyanobacteria* was 1.8% on Day 20, which increased significantly compared to

0.06% on Day 0 ($P = 0.004$). Actinobacteria had significantly lower relative abundance (3.3%) on Day 20 after dental treatment compared to the data (8.8%) on Day 0 ($P = 0.035$).

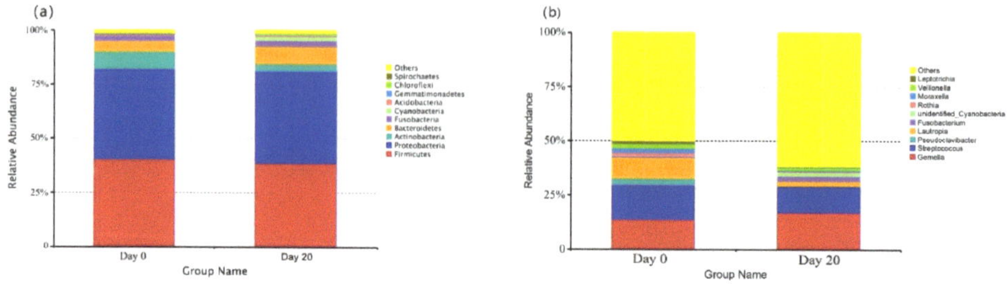

Figure 2. (**a**) Microbial composition of Day 0 and Day 20 at the phylum level. (**b**) Microbial composition of Day 0 and Day 20 at the genus level.

There were also some differences of the taxonomic abundance at the genus level (Figure 2b). Eight hundred and ninety-one operational taxonomic units (OTUs) were identified at the genus level. The differences of taxonomic abundance were more notable from the two groups, and the shifts of the bacterial taxa were more significant at this level. *Gemella* and *Streptococcus* were the most common bacterial genera in both groups. Higher levels of *Pseudoclavibacter* (2.97%), *Lautropia* (9.74%), *Moraxella* (2.17%), *Leptotrichia* (1.64%) and *Rothia* (1.35%) were found in the Day 0 group.

B-diversity was accessed by a principal coordinate analysis (PCoA), which did not show statistic difference at the phylum level, while there was a greater overall variance in the Day 20 group than in the Day 0 group (Figure S1). The linear discriminant analysis (LDA) effect size (LEfSe) was applied, and ten discriminant taxa (Figure 3) at the genus or higher levels were found significantly different between Day 0 and Day 20 (LDA score >2, $P < 0.05$). The genera found most associated with the Day 0 group were *Lautropia* and *Pseudoclavibacter*, while *Fusobacterium* was the genus most associated with the Day 20 group. *Burkholderiaceae* and *Bacteoidales* yielded the highest LDA scores among the periodontal biomarkers in the Day 0 group and Day 20 group, respectively.

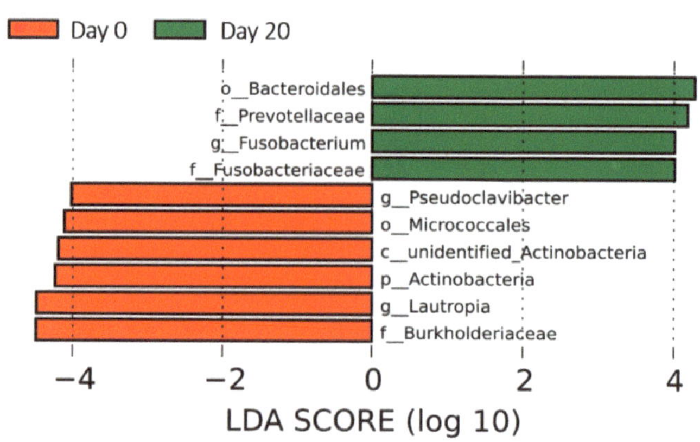

Figure 3. Linear discriminant analysis (LDA) effect size analysis indicating the bacterial taxa (genus or higher level) most associated with the Day 0 group and Day 20 group, respectively. Only 10 taxa that had LDA scores above 2 were shown. The taxa were ranked by the effect size in linear discriminant analysis effect size (LEfSe).

4. Discussion

This is the first report investigating the oral microbiome in donkeys and comparing the microbial changes before and after dental floating. All donkeys in this study were female due to the production purpose of this farm. There has been no report of the correlation of sex and oral microbial composition, while gut microbiota variance has exhibited an association with sex in other species [12], including horses, through poorly understood mechanisms [13]. Therefore, the oral cavity, a part of the digestive system, may also have different microbiomes between male and female donkeys. A study carried out in Mexico involving 203 donkeys (69% males and 31% females) showed that sex was significantly associated with dental diseases, probably due to the utility of nosebands in male donkeys causing more traumatic dental and soft tissue buccal injuries. [5]. Hence, it is reasonable to consider that the individual's sex may have some influence on the oral microbial composition. While our study was not confounded by animal sex influence, further investigation in this area is warranted.

The feeding pattern of donkeys in this farm only allowed them to chew for short periods of time during the day, which may cause the teeth wearing down inadequately and unevenly [14]. Therefore, it could lead to an increased risk of dental disorders such as sharp enamel points. There was also much evidence showing that fiber functioned as a mechanical force on dental calculus in some animal studies [15,16], which could impact the oral microbiome associated with different dental conditions. Additionally, carbohydrates, lipids and minerals, as well as vitamins, were all factors that had exhibited correlations with some bacterial taxa in human studies [17,18]. Therefore, further investigations targeting the diet composition and oral microbiome of donkeys will help to promote scientific diets for them and be beneficial to their welfare.

The environment is another popular variable input while studying a microbiome, yet no report regarding the association of environmental changes and oral microbiota composition in animals has been published. However, environmental factors such as husbandry conditions in terms of stable types and grazing periods have been proven to have a great influence on the richness and diversity of gut microbiota in horses [19]. Derived from these results, we hypothesized that oral microbiota may change in different environmental conditions. Since, in this study, all donkeys were from the same farm, a controlled study designed in different local donkey farms would be helpful to investigate this hypothesis.

Oral microbiome changes have been associated with dental disease, especially periodontal diseases in humans and many animal species, including horses [9,11] dogs [20], cats [21] and cattle [22], but not yet in donkeys suffering from dental abnormalities. Dental floating, the most common dental treatment, plays an important role in dental disease control within farm-raised donkeys. Therefore, it is meaningful to know if there is an impact of dental floating on the oral microbiome in donkeys so that dental care can be conducted in a more scientific way in the future.

The donkeys selected in our study had severe sharp enamel points prevalently, which are especially common in younger donkeys [1], and periodontal pockets, which are commonly associated with diastema in horses [23]. In equids, diastema is considered related to periodontal diseases, such as periodontitis [24], which usually further impacts the oral microbiome [23]. Diastema generally requires a series of dental treatments, similarly to other severe dental abnormalities seen in this study, including shear mouth, wave mouth and step mouth. Dental care, especially dental floating, is usually the fundamental step of treating most dental disorders and could also improve those dental disorders to some extent. For example, decreasing the amount of food trapped into the diastema by reducing the transverse ridges on the opposite site [25]; therefore, we proposed that it might subsequently lead to some microbial changes before (Day 0) and after (Day 20) our dental treatment.

In the current study, 16S rRNA sequencing revealed that the composition of the oral microbiome of the two groups (Day 0 and Day 20) were not significantly different in richness and diversity, indicating that dental floating did not render many changes at the microbial level. However, the relative differences indicated a shift in relative abundance, indicating a reassortment in dominance. Hence, more specific and consistent dental treatments on dental disorders may yield more optimal outcomes, and more studies focusing on these topics are worthwhile for the better control and improvement of donkey dental conditions. In previous research using the same 16S rRNA sequencing technique, *Actinobacteria, Bacteriodetes, Firmicutes, Proteobacterio* and *Spirochaetes*, which were found predominant in both groups at the phylum level in this study, were also commonly found in the mouths of healthy humans [26], equines [5] and canines [27], with slightly different proportions between the oral microbiomes. This suggests that the oral microbiome in the donkey share some similarities with those of humans and other animals. *Firmicutes*, the most common bacteria phyla in this study, was also a common inhabitant of the oral cavity, especially in the subgingival area. However, some studies focusing on oral microbiomes in dogs and cats with periodontitis showed that it was one of the predominant phyla in the periodontitis group. *Firmicutes* was also abundant in horse subgingival plaque, likely acting as an opportunistic pathogen and contributing to periodontitis [8]. *Proteobacteria*, the second-highest bacterial phylum in this study, was also abundant in horse subgingival plaques [9]. Further evaluations with specific sampling may provide more insight on whether they are associated with periodontal diseases or not in donkeys.

Gemella spp. and *Streptococcus* spp. were the most common bacterial genera in this study, as well as in many animals in other studies [28]. *Gemella* was slightly more predominant in the Day 20 group, while *Streptococcus* was slightly lower in the Day 20 group. *Gemella* has been shown more prominent in healthy horses [8], and *Streptococcus* has been reported to be the most common genera of *Firmicutes* in the healthy human oral microbiome [29], but indeed, some species have been associated with active caries [30]. Dental floating may have contributed to the increase of *Gemella* in the Day 20 group but not enough for notable changes.

Fusobacterium, an obligate, anaerobic bacterium, is considered as a common oral niche inhabitant in humans and some other animals, including horses [31]. In this study, it was shown to be most associated with the Day 20 group, while *Lautropia* and *Pseudoclavibacter* were the genera most associated with Day 0 in LEfSe. These changes were likely an indication of an expansion of the lesser phyla allowed by suppression of the more dominant microflora present in Day 0. Its population can increase significantly as an opportunistic pathogen in oral disease and has been cultured from a horse and a donkey with dental abscesses [32]. Some specific species of *Fusobacterium*, such as *F. nucleatum*, have been associated with periodontal diseases in many animals, including dogs, cats and horses [33–35]. However, in this study, its population did not show a dramatic increase. The sampling specifically from periodontal pockets from donkeys only with periodontitis may help reveal better associations with dental treatment. Additionally, there was a relatively rare bacterial taxon at the genus level identified on Day 0 that was *Pseudoclavibacter*, belonging to the *Actinobacteria* phylum. So far, nothing has been reported about this bacterial genus in oral microbiome studies either in humans or in animals. It was not identified in the Day 20 group and only associated with the Day 0 group according to the LDA score, which may indicate some associations between this bacteria genus and donkey dental diseases. Further studies involving more individuals will be helpful to decide whether *Pseudoclavibacter* is specific in donkeys and associated with dental diseases or merely an environmental contaminant.

5. Conclusions

In conclusion, this is the first report of the donkey oral microbiome in association with basic dental treatment. Our results did not reveal significant differences of the oral microbial composition in selected donkeys; however, there were bacterial genera showing significant differences between the Day 0 and Day 20 groups, as well as a strong association to each group. These shifts of relative diversity and dominance indicated a reassortment relative to the dental treatment. Furthermore, our study revealed that the donkeys shared many similarities with other animals at the phylum and genus levels. Additional studies to investigate the donkey oral microbiome in healthy conditions, as well as its associations with specific types of oral diseases, such as periodontitis and diastema, by selecting samples specifically will promote an understanding of the impact of these diseases on donkey dental health. It will also build up a basis for studying the influences of other potential factors, including diet, feeding pattern and farm environment on the donkeys' dental condition, so that more scientific raising methods can be established to improve their quality of life.

Supplementary Materials: The following are available online at http://www.mdpi.com/2076-2615/10/12/2245/s1, Figure S1: Principal coordinates analysis (PCoA) plot showing overlap of oral microbiomes in group Day 0 and group Day 20.

Author Contributions: Conceptualization, J.L.; methodology, W.J.; software, W.J. and B.L.; formal analysis, Y.Z. and W.J.; investigation, Y.Z. and J.L.; data curation, Y.Z. and B.L.; writing—original draft preparation, Y.Z.; writing—review and editing, R.H. and J.L.; supervision, J.L. and project administration, J.L. All authors have read and agreed to the published version of the manuscript.

Funding: This research received no external funding.

Conflicts of Interest: The authors declare no conflict of interest.

References

1. Burden, F.; Du Toit, N.; Thiemann, A. Nutrition and dental care of donkeys. *Practice* **2013**, *35*, 405–410. [CrossRef]
2. Du Toit, N.; Gallagher, J.; Burden, F.A.; Dixon, P.M. Post mortem survey of dental disorders in 349 donkeys from an ages population (2005–2006). Part 1: Prevelance of specific dental disorders. *Equine Vet. J.* **2010**, *40*, 204–208. [CrossRef] [PubMed]
3. Rodrigues, J.B.; Lilly, G. Dental disorders of Donkey. *Vet. Clin. N. Am. Equine Pract.* **2019**, *35*, 529–544. [CrossRef] [PubMed]
4. Du Toit, N.; Dixon, P.M. Common dental disorders in the donkey. *Equine Vet. Educ.* **2012**, *24*, 45–51. [CrossRef]
5. Du Toit, N.; Burden, F.A.; Dixon, P.M. Clinical dental findings in 203 working donkeys in Mexico. *Vet. J.* **2008**, *178*, 380–386. [CrossRef]
6. Du Toit, N.; Burden, F.A.; Baedt, L.G.; Shaw, D.J.; Dixon, P.M. Dimensions of Diastemata and Associated Periodontal Food Pockets in Donkey Cheek Teeth. *J. Vet. Dent.* **2009**, *26*, 10–14. [CrossRef]
7. Walker, H.; Chinn, E.; Holmes, S.; Barwisemunro, L.; Bobertson, V.; Bradley, S.; Shaw, D.J.; Dixon, P.M. Prevalence and some clinical characteristics of equine cheek teeth diastemata in 471 horses examined in a UK first-opinion equine practice (2008 to 2009). *Vet. Rec.* **2012**, *171*, 44. [CrossRef]
8. Du Toit, N.; Burden, F.A.; Dixon, P.M. Clinical dental examination of 357 donkeys in the UK. Part 1: Prevalence of dental disorders. *Equine Vet. J.* **2009**, *41*, 390–394. [CrossRef]
9. Gao, W.; Chan, Y.; You, M.; Lacap-Bugler, D.C.; Leung, W.K.; Watt, R.M. In-depth snapshot of the equine subgingival microbiome. *Microb. Pathog.* **2015**, *94*, 76–89. [CrossRef]
10. Kennedy, R.; Lappin, D.F.; Dixon, P.M.; Buijs, M.J.; Zaura, E.; Crielaard, W.; O'Donnell, L.; Bennett, D.; Brandt, B.W.; Riggio, M.P. The microbiome associated with equine periodontitis and oral health. *Vet. Res.* **2016**, *47*, 49–58. [CrossRef]
11. Caporaso, J.G.; Kuczynski, J.; Stombaugh, J.; Bittinger, K.; Bushman, F.D.; Costello, E.K.; Fierer, N.; Pena, A.G.; Goodrich, J.K.; Gordon, J.I.; et al. QIIME allows analysis of high-throughput community sequencing data. *Nat. Methods* **2010**, *7*, 335–336. [CrossRef]
12. Freire, A.C.; Basit, A.W.; Choudhary, R.; Piong, C.W.; Merchant, H.A. Does sex matter? The influence of gender on gastrointestinal physiology and drug delivery. *Int. J. Pharm.* **2011**, *415*, 15–28. [CrossRef]
13. Mshelia, E.S.; Adamu, L.; Wakil, Y.; Turaki, U.A.; Gulani, I.A.; Musa, J. The association between gut microbiome, sex, age and body condition scores of horses in Maiduguri and its environs. *Microb. Pathog.* **2018**, *118*, 81–86. [CrossRef]
14. Feeding Practices May Impact Horse Dental Health. Available online: https://ker.com/equinews/feeding-practices-may-impact-horse-dental-health/ (accessed on 4 November 2020).
15. Van, A.; Beynen, A.C.; Altena, F.V.; Visser, E.A. Beneficial effect of a cellulose-containing chew treat on canine periodontal disease in a double-blind, placebo-controlled trial. *Am. J. Anim. Vet. Sci.* **2010**, *5*, 192–195.
16. Logan, E.I. Oral cleansing by dietary means: Results of six-months studies. In Proceedings of the Companion Animal Oral Health Conference, Lawrence, KS, USA, 1–3 March 1996.
17. Murtaza, N.; Burke, L.M.; Vlahovich, N.; Charlesson, B.; O'Neill, H.M.; Ross, M.L.; Campbell, K.L.; Krause, L.; Morrison, M. Analysis of the effects of dietary pattern on the oral microbiome of elite endurance athletes. *Nutrients* **2019**, *11*, 614. [CrossRef]

18. Kato, I.; Vasquez, A.; Moyerbrailean, G.; Land, S.; Djuric, Z.; Sun, J.; Lin, H.; Ram, J.L. Nutritional correlates of human oral microbiome. *J. Am. Coll. Nutr.* **2017**, *36*, 88–98. [CrossRef]
19. Kaiser-Thom, S.; Hilty, M.; Gerber, V. Effects of hypersensitivity disorders and environmental factors on the equine intestinal microbiota. *Vet. Q.* **2020**, *40*, 97–107. [CrossRef]
20. Ruparell, A.; Iuni, T.; Staunton, R.; Wallis, C.; Deusch, O.; Holcombe, L.J. The canine oral microbiome: Variation in bacterial populations across different niches. *BMC Med.* **2020**, *20*, 42. [CrossRef]
21. Dewhirst, F.E.; Klein, E.A.; Bennett, M.L.; Croft, J.M.; Harris, S.J.; Marshall-Jones, Z.V. The feline oral microbiome: A provisional 16S rRNA gene based on taxonomy with full-length reference sequences. *Vet. Microbiol.* **2015**, *175*, 294–303. [CrossRef]
22. Borsanelli, A.C.; Lappin, D.F.; Viora, L.; Bennett, D.; Dutra, I.S.; Brandt, B.W.; Riggio, M.P. Microbiomes associated with bovine periodontitis and oral health. *Vet. Microbiol.* **2018**, *218*, 1–6. [CrossRef]
23. Chinkangsadarn, T. Clinical and Microbiological Aspects of Periodontal Disease in Horses in South-East Queensland, Australia. Ph.D. Thesis, The University of Queensland, Brisbane, Australia, 2015.
24. Kennedy, R.S.; Dixon, P.M. The aetiopathogenesis of equine periodontal disease-a fresh perspective. *Equine Vet. Educ.* **2018**, *30*, 161–168. [CrossRef]
25. Collins, N.M.; Dixon, P.M. Diagnosis and management of equine diastema. *Clin. Tech. Equine Pract.* **2005**, *4*, 148–154. [CrossRef]
26. Palmer, R.J. Composition and development of oral bacterial communities. *Periodontology* **2014**, *64*, 20–39. [CrossRef]
27. Holcombe, L.J.; Patel, N.; Colyer, A.; Deusch, O.; O' Flynn, C.; Harris, S. Early canine plaque biofilms: Characterization of key bacterial interactions involved in initial colonization of enamel. *PLoS ONE* **2014**, *9*, e113744. [CrossRef]
28. Bik, E.M.; Long, C.D.; Armitage, G.C.; Loomer, P.; Emerson, J.; Mongodin, E.F.; Nelson, K.E.; Gill, S.R.; Fraser-Liggett, C.M.; Relman, D.A. Bacterial diversity in the oral cavity of 10 healthy individuals. *ISME J.* **2010**, *4*, 962–974. [CrossRef]
29. McLean, J.S. Advancements toward a systems level understanding of the human oral microbiome. *Front. Cell. Infect. Microbiol.* **2014**, *4*, 98. [CrossRef]
30. Dewhirst, F.E.; Chen, T.; Izard, J.; Paster, B.J.; Tanner, A.C.R.; Yu, W.; Lakshmanan, A.; Wade, W.G. The human oral microbiome. *J. Bacteriol.* **2010**, *192*, 5002–5017. [CrossRef]
31. Darrene, L.N.; Cecile, B. Experimental models of oral biofilms developed on inert substrates: A review of the literature. *BioMed Res. Intern.* **2016**, *2016*, 1–8. [CrossRef]
32. Takada, K.; Hayashi, K.; Sato, Y.; Hirasawa, M. Prevotella dentasini sp. nov., a black-pigmented species isolated from the oral cavity of donkeys. *Int. J. Syst.* **2010**, *60*, 1637–1639. [CrossRef] [PubMed]
33. Mallonee, D.H.; Harvey, C.E.; Venner, M.; Hammond, B.F. Bacteriology of periodontal disease in the cat. *Arch. Oral Biol.* **1988**, *33*, 677–683. [CrossRef]
34. Sturgeon, A.; Pinder, S.L.; Costa, M.C.; Weese, J.S. Characterization of the oral microbiota of healthy cats using next-generation sequencing. *Vet. J.* **2014**, *201*, 223–229. [CrossRef]
35. Sturgeon, A.; Stull, J.W.; Costa, M.C.; Weese, J.S. Metagenomic analysis of the canine oral cavity as revealed by high-throughput pyrosequencing of the 16S rRNA gene. *Vet. Microbiol.* **2013**, *162*, 891–898. [CrossRef]

animals

Article

Hastening Time to Ejaculation in Donkey Jacks Treated with the PGF2α Analog, Cloprostenol Sodium

Duccio Panzani [1,*], Miguel Quaresma [2], Diana Fanelli [1], Francesco Camillo [1], Rebecca Moroni [1], Alessandra Rota [1], Ana Martins-Bessa [2], Miguel Nóvoa [3], Jaime Catalán [4], Igor F. Canisso [5], Giuseppe Conte [6] and Jordi Mirò [4]

1 Department of Veterinary Clinical Sciences, University of Pisa, via Livornese 1286, San Piero a Grado, 56122 Pisa (PI), Italy; diana.equirepro@gmail.com (D.F.); francesco.camillo@unipi.it (F.C.); r.moroni1@studenti.unipi.it (R.M.); alessandra.rota@unipi.it (A.R.)

2 Center of Animal and Veterinary Science (CECAV), University of Trás-os-Montes e Alto Douro (UTAD), Quinta de Prados, 5000-801 Vila Real, Portugal; miguelq@utad.pt (M.Q.); abessa@utad.pt (A.M.-B.)

3 Association for the Study and Protection of Donkeys (AEPGA), Largo da Igreja, 5225-011 Atenor, Portugal; miguelnovoa@aepga.pt

4 Equine Reproduction Service, Department of Animal Medicine and Surgery, Veterinary Faculty, Autonomous University of Barcelona, E-08193 Bellaterra, Cerdanyola del Valles, Spain; dr.jcatalan@gmail.com (J.C.); Jordi.Miro@uab.cat (J.M.)

5 Department of Veterinary Clinical Medicine, College of Veterinary Medicine, University of Illinois Urbana-Champaign, Champaign, IL 61802, USA; canisso@illinois.edu

6 Department of Agriculture, Food and Environment, University of Pisa, via del Borghetto 80, 56124 Pisa, Italy; giuseppe.conte@unipi.it

* Correspondence: duccio.panzani@unipi.it

Received: 2 November 2020; Accepted: 25 November 2020; Published: 27 November 2020

Simple Summary: Semen collection in donkey jacks can last up to 90 min due to the long courtship needed for this species' males to obtain sexual excitation and erection. In several domestic animals, ProstaglandinF2α successfully stimulated excitement in the male prior to collection and enhanced semen production. In our study, the prostaglandin analog cloprostenol sodium, administered prior to the semen collection, hastened erection and ejaculation in almost all donkey jacks. No differences have been found in semen production compared to control.

Abstract: Due to the long courtship needed to attain excitation and erection, donkey semen collection can take up to 90 min. ProstaglandinF2α (PGF2α) has been reported to hasten the onset of erection and ejaculation in domesticated mammals, presumably by inducing smooth muscle contractions in the internal genitalia. However, while it has been anecdotally used in donkeys, it has yet to be critically evaluated. This study aimed to compare behavioral and semen parameters in Catalan, Balearic, Amiata, and Miranda jacks treated with the PGF2α analogue cloprostenol sodium immediately prior to exposure to an estrus jenny. Nineteen donkeys were assigned in a crossover design to receive cloprostenol sodium (125 µg, i.m.; n = 53 collections) or saline (1 mL, i.m.; n = 53 collections). There were no differences for erection (52/53 vs. 52/53) or ejaculation (52/53 vs. 48/53) for collection attempts assigned to saline or cloprostenol sodium, respectively. Cloprostenol sodium significantly hastened treatment-to-erection and treatment-to-ejaculation times from 12.0 ± 1.6 to 6.0 ± 1.6 min and from 14.0 ± 1.4 to 9.6 ± 1.4 min, respectively. Significant effects of breed and age were observed in behavioral and parameters, but there were no effects of cloprostenol sodium administration on semen parameters. In conclusion, cloprostenol sodium administration immediately prior to semen collection hastened time to collect semen in donkeys with no detrimental effects on semen quality and can be used by practitioners to circumvent long delays in donkey semen collection.

Keywords: donkey; semen collection; cloprostenol sodium; erection; ejaculation

1. Introduction

Throughout the centuries, domestication of donkeys has had numerous functions, such as transportation, plowing plantation fields, mining, packing animals, and producing mules (hybrids with horses) [1–3]. After the mechanization of agriculture, donkeys lost their value as agricultural commodities in industrialized [4] and developing countries [5]. However, in developing countries, donkeys still playing a paramount role as a valuable agricultural commodity [5].

The decreased importance of donkeys and their hybrids, particularly in Western Europe, has resulted in a drastic reduction in donkey numbers. Most importantly, European donkey breeds are at risk of extinction or have a low minimal critical number (<500) of animals recommended by the conservation entities [4,6,7]. One of the ways to circumvent extinction is developing germplasm banks of cryopreserved semen and embryos [6]. However, donkeys present a particular challenge during intensive breeding management situations, as semen collection cannot, at times, be consistently carried out. For instance, weather conditions or minor changes in management can drastically affect time to collect semen, especially in young donkeys that take much longer to collect than older mature donkeys [8]. In the American continent, donkeys are used primarily to produce show mules or to herd beef cows [9]. As most mares are not receptive to donkeys [10], semen is typically collected, extended, and mares are artificially inseminated [6,9]. Under intensive regimen of semen collection for cryopreservation or immediate artificial insemination, procurement of semen in a consistent and reasonable timely fashion is highly desirable and challenging in donkeys [8]. Collecting semen from donkey jacks is often a time consuming and frustrating practice. The time needed to obtain an erection followed by ejaculation varies from 6 to 32 min when mounting estrus jennies [11–13] or estrus mares [14]. In a study of natural mating, the time between the introduction of the jack in the 'jennies' pen and the first mount with ejaculation varied between 25 and 93 min [15]. This study illustrated that donkeys could take a long time to collect semen and take a long time to naturally cover jennies.

The luteolytic activity was the first reproductive action described for Prostaglandin F2α (PGF2α) [16]. Prostaglandin F2α is known to induce smooth muscle contractions in several species' male internal genitalia, including humans, horses, and bulls [14,17]. This effect has been used to influence the volume and composition of the ejaculate in domesticated mammals [15,17–22]. Administration of PGF2α immediately before semen collection increases the total sperm number ejaculated in bulls, rabbits, and stallions [17–21].

Synthetic prostaglandins with more potent luteolytic activity have been developed in the last two decades. Uterine contractibility effect shows a high variability depending on the type of prostaglandin and the stage of the estrous cycle. Cloprostenol, a PGF2α analog, induces uterine contractions, nevertheless, causes no significant changes in any stage of the estrous cycle in cows [23]. Administration of PGF2α analogs immediately before semen collection enhanced libido of boars with known decreased sex drive [20,24] and reduced the time needed to train young boars for collecting semen and the number of false mounts in boars trained to the dummy mount [25]. Similarly, the administration of PGF2α to low libido bulls improved their libido [21]. It is unclear how PGF2α administration enhances sexual behavior in bulls or boars, but it is known that PGF2α can stimulate several areas of the brain in sows, which increase C-Fos mRNA expression and brooding behavior [26]. In horses, repeated treatment with PGF2α prior to semen collections initially increased the number of ejaculated sperm, and then an increase in semen volume and a reduction in sperm concentration was observed after 18 injections in nine weeks [27]. The same study reported no apparent deleterious effects on sperm production and quality associated with nine weeks and 18 injections of PGF2α.

Prostaglandin F2α administration has been used in clinical practice to circumvent problems with low libido in donkeys and to reduce time to collect semen under an intensive breeding management regimen [6,9]. However, the use of this eicosanoid has not been tested under controlled conditions. An increase of PGFM, an inactive metabolite assessed as a proxy of PGF2α, has been found in the plasma of jacks right before and after ejaculation time [13,17], suggesting that PGFα plays a role in erection and ejaculation in this species.

The objectives of this study were to assess semen and behavioral parameters of four European donkey breeds treated with cloprostenol sodium. We hypothesized that the administration of PGF2α pre-courtship reduces the time to collect semen in donkeys successfully.

2. Materials and Methods

This study was conducted after revision and authorization by the Research Ethics Committee of the Pisa University Ethics Committee under protocol# 8/2020, 31/01/2020.

2.1. Animals and Locations

A total of 19 jacks were enrolled in the study, and a total of 108 semen collections were attempted. The breeds of donkeys used in the study were as follows: Amiata (n = 4, age 6 ± 0 years, body weight 266 ± 26 kg, height: 135 ± 6 cm, Body Condition Score 3.1 ± 0.1), Balear (n = 4, age 8 ± 5 years, body weight 333 ± 20 kg, height: 142 ± 2 cm, Body Condition Score: 3.6 ± 0.5), Catalan (n = 5, age 6 ± 2 years, body weight: 406 ± 4 6 kg, height: 154 ± 4 cm, Body Condition Score: 3.5 ± 0.0), and Miranda (n = 6, age: 8 ± 6 years, weight 272 ± 34 kg, height 132 ± 5 cm, Body Condition Score 3.2 ± 0.4). The study was performed during the spring of 2020 in three research centers located in Pisa, Italy (Amiata), Barcelona, Spain (Catalan and Balearic), and Vila Real, Portugal (Miranda). Body condition score was calculated using the scale going from 1 to 4 described by Vall et al. [28].

During each semen collection, a jenny in good standing estrus was used as a mounting female for all three locations. Good standing estrus was defined as standing to be mounted and full display of "mouth clapping" [29]. One week before the beginning of the study, clean out semen collections were performed Monday, Wednesday, and Friday to standardize extra-current duct sperm reservoirs and to minimize possible effects of sexual inactivity. During the study, all jacks were housed in individual stalls far away from females to avoid audio and visual stimulatory effects. Jacks were collected with a Colorado (Amiata) or Hannover (Catalan, Balearic, and Miranda) artificial vagina models. After the cleanout week, jacks in each of the three research centers were randomly assigned in a cross over design between a treatment and control group. Immediately before exposure to an estrus jenny, the treatment group received cloprostenol sodium (125 μg cloprostenol sodium, i.m., 0.5 mL Estrumate®, MSD Animal Health SRL, Milan, Italy, PGF), whereas the control group received 1 mL of saline, i.m. (CTRL). After the cleanout week, jacks in each of the three research centers were randomly assigned in a crossover design, using the Latin square method. On consecutive collection days, the jacks would alternate between the treatment and control group. The order of collection between the different jacks would also be different in each collection session. Each Amiata, Catalan, and Miranda jack was submitted to three collections preceded by saline and three collections preceded by cloprostenol sodium treatment. Balearic donkeys instead had a total of two collections preceded by saline and two collections preceded by cloprostenol sodium treatment, following the same collection design used for the other breeds. Seventy-two hours of an interval between two collections were followed throughout the experiment across all breeds.

2.2. Behavioral Assessment

During all semen collection attempts, one observer recorded jacks' sexual behavior in the presence of an estrus jenny. Specifically, the number of mounts, number of erections, time from treatment to erection, and time from treatment to ejaculation were recorded and used for comparisons across groups and breeds. The number of erection failures, ejaculation failures, persistent erections, and the number of sweating animals were accounted for and used for comparisons across groups and breeds. It was deemed erection failure if a jack did not achieve an erection by 90 min after exposure to an estrus female. Ejaculation failure was defined as the failure to successfully ejaculate within 100 min after exposure to an estrus jenny. Immediately after collection, jacks were assessed for a second persistent erection, every 15 min for 2 h after treatment administration. An erection lasting for 15 min or more was defined as "persistent erection".

2.3. Semen Assessment

Immediately after collection, semen was assessed for total (pre-filtration) and gel-free (post-filtration) volume and concentration using a hemocytometer counting chamber. An aliquot of semen was extended (1:1) with a temperature-matched commercially available horse extender (INRA 96, IMV, l'Aigle, France). The extended semen was placed on a warm slide, covered with a coverslip, and then subjective sperm motility was assessed at 37 °C and 200× magnification under a phase-contrast microscope.

2.4. Statistical Analyses

Data analyses were carried out with JMP software (JMP 7 SAS Institute Inc., Cary, NC, USA). Fisher's exact test was used to compare breeds, treatment regarding erection failure, ejaculation failure, persistent erection, and sweating after each collection attempt. A Shapiro–Wilk test was conducted to assess the normality of all variables and all dependent variables were shown to be normally distributed. Additionally, a linear model was created to determine the effects of multiple variables as described below.

$$y_{ijzpq} = m + Breed_i + Treatment_j + Age_z + Treatment\ order_q + Breed_i \times Treatment_j + Age_z \times Treatment_j + Animal_k(Breed_i) + e_{ijzpq}$$

where:

y_{ijzpq} = number of mounts, number of erections, treatment to erection interval (minutes), treatment to ejaculation interval (minutes), pre-filtration semen volume (mL), post-filtration semen volume (mL), subjective sperm motility (%), response sperm concentration (spz/mL), total sperm number.

$$m = mean$$

Fixed effects were age, breed, treatment, order of treatment. Jacks were accounted as random effect.
$Breed_i$ = fixed effect of the $_i$th breed (Amiata, Balearic, Catalan, Miranda).
$Treatment_j$ = fixed effect of the $_j$th treatment (CTRL, PGF).
Age_z = fixed effect of the $_z$th age level (≤5, >5).
$Treatment\ order_q$ = fixed effect of the qth treatment order level (6 levels).
$Animal_k$ = random effect of the $_j$th donkey (19 levels).

Since age is known to affect semen parameters and behavioral parameters [14], all jacks were grouped together and then split into groups according to age to assess those effects. Arbitrarily jacks were classified as young (i.e., ≤5 years old, n = 6), or mature (i.e., >5 years old, n = 13). Thereafter, data analyses were carried to assess the age and body weight effects on behavioral and semen parameters using the same similar linear model. Least-square means with their standard errors were reported, and treatment effects were declared significant at $P < 0.05$. The linear contrasts were tested in the first model by the t-test with Tukey's adjustment within each parity level.

3. Results

An overall 3.7% erection failure was recorded in all 108 collections (Table 1). One Amiata donkey failed to achieve an erection while receiving saline treatment, and one Miranda jack failed to achieve an erection in the PGF group during a collection. Out of the 108 collection attempts, ejaculation failure was recorded in 5.5% of the collections (Table 1). One Amiata jack failed to ejaculate in one collection when assigned to the control group, whereas five jacks failed to ejaculate on one occasion (n = 1 Amiata,

n = 1 Balearic, n = 2 Catalan, n = 1 Miranda) when assigned to the PGF group. There were no significant differences between groups regarding failures to achieve an erection or ejaculation (Table 1).

Table 1. Erection failure, ejaculation failure, persistent erection, and sweating in jacks treated with 125 µg of cloprostenol sodium (PGF) or 1 mL of saline (CTRL) prior to semen collection mount in an estrus jenny. Three semen collections were performed in each of the assigned groups (PGF vs. Control) at 72 h intervals for Amiata, Catalan, and Miranda, whereas only two collections per group were performed in Balearic jacks at 72 h intervals.

	Amiata (n = 4)		**Balear (n = 4)**		**Catalan (n = 5)**		**Miranda (n = 6)**		**Total (n = 19)**	
	CTRL	**PGF**	**CTRL**	**PGF**	**CTRL**	**PGF**	**CTRL**	**PGF**	**CTRL**	**PGF**
Erection failure	1/12 (0%)	0/12 (0%)	0/8 (0%)	0/8 (0%)	0/15 (0%)	0/15 (0%)	0/18 (0%)	1/18 (5.5%)	1/53 (3.7%)	1/53 (3.7%)
Ejaculation failure	1/12 (8%)	1/12 (8%)	0/8 (0%)	1/8 (12.5%)	0/15 (0%)	2/15 (13.3%)	0/18 (0%)	1/18 (5.5%)	1/53 (0.2%)	5/53 (9.4%)
Persistent erection	0/12 (0%)	12/12 (100%)	0/8 (0%)	0/8 (0%)	0/15 (0%)	0/15 (0%)	0/18 (0%)	1/18 (5.5%)	0/53 (0%)	13/53 (24.5%)
Sweating	0/12 (0%)	0/12 (0%)	0/8 (0%)	0/8 (0%)	0/15 (0%)	0/15 (0%)	0/18 (0%)	6/18 (33.3%)	0/53 (0%)	6/53 (11.3%)

There were no differences between groups for any of the variables assessed. Only the Miranda jacks sweated in response to PGF.

In all semen collection attempts, jacks performed at least one false mount. All jacks attaining an erection had it after interacting with the estrus jenny. In the PGF group, a second persistent erection, lasting for a minimum of 30 min, was observed after all collections in all the Amiata jacks and after one collection in a Miranda jack, respectively. Around thirty min of sweating was observed in two Miranda jacks in all cloprostenol sodium-treated collection attempts.

There were significant effects of breed, treatment, and breed by treatment interaction for behavioral parameters (Table 2). Cloprostenol sodium administration, compared to control, reduced the number of mounts per collection attempt (P = 0.025). Catalan and Balearic jacks had the lowest number of mounts per collection attempt in comparison with the other two breeds (P < 0.001) that were not significantly different from one another. The number of erections per attempt was affected by breed (P = 0.012), but not by treatment (P = 0.11), and it tended to have an interaction between breed and treatment (P = 0.06). There were effects of the breed (P < 0.001), treatment (P < 0.001), and breed by treatment interaction (P < 0.001) for the variables time from treatment to erection and time from treatment to ejaculation (Table 2). There were breed effects (P < 0.001) but no effect of treatment (P > 0.6) or breed by treatment interaction (P > 0.4) for all semen parameter evaluation (Table 3).

Table 2. Behavioral parameters in 106 semen collections performed in 19 jacks of different breeds (Amiata, n = 4; Balearic N = 4; Catalan n = 5; Miranda n = 6). In half of the collections, jacks were treated with 125 µg of cloprostenol sodium (PGF) or 1 mL of saline (CTRL) prior to semen collection mounting an estrus jenny.

Breeds (n)	**Number of Mounts Mean ± SEM**		**Number of Erections Mean ± SEM**		**Treatment to Erection Mean ± SEM (min)**		**Treatment to Ejaculation Mean ± SEM (min)**	
	CTRL	**PGF**	**CTRL**	**PGF**	**CTRL**	**PGF**	**CTRL**	**PGF**
Amiata (n = 4)	6.8 ± 1.5 ^{a}	6.8 ± 0.9 ^{a}	1.3 ± 0.2	1.5 ± 0.3	18.7 ± 3.0 ^{A}	6.3 ± 0.4 ^{C}	21.6 ± 3.0 ^{a}	15.6 ± 5.0 ^{b}
Balearic (n = 4)	3.6 ± 0.3 ^{b}	1.1 ± 0.1 ^{c}	1.2 ± 0.2	1.0 ± 0.0	18.4 ± 1.7 ^{A}	6.7 ± 0.8 ^{C}	21.0 ± 1.3 ^{a}	9.1 ± 0.8 ^{c}
Catalan (n = 5)	3.5 ± 0.2 ^{b}	1.6 ± 0.2 ^{c}	1.7 ± 0.2	1.1 ± 0.1	14.3 ± 1.1 ^{B}	6.4 ± 0.6 ^{C}	17.7 ± 0.8 ^{b}	9.9 ± 0.8 ^{c}
Miranda (n = 6)	6.9 ± 1.5 ^{a}	4.8 ± 1.0 ^{b}	1.1 ± 0.1	1.1 ± 0.2	7.4 ± 0.6 ^{C}	6.0 ± 0.4 ^{C}	8.1 ± 0.5 ^{c}	7.2 ± 1.3 ^{c}
Mean (n = 19)	6.0 ± 0.8 ^{a}	4.0 ± 0.8 ^{b}	1.0 ± 0.2	1.1 ± 0.2	12.0 ± 1.6 ^{B}	6.0 ± 1.6 ^{C}	14.0 ± 1.4 ^{b}	9.6 ± 1.4 ^{c}

$^{a–c}$: Means within rows and columns within the same analyzed parameter, with different letters, significantly differ ($P \leq 0.05$); $^{A–C}$: Means within a column with different letters significantly differ ($P \leq 0.01$).

Table 3. Seminal parameters in 106 semen collections performed in 19 jacks of different breeds (Amiata, $n = 4$; Balearic $N = 4$; Catalan $n = 5$; Miranda $n = 6$). In half of the collections, jacks were treated with 125 μg of cloprostenol sodium (PGF) or 1 mL of saline (CTRL) prior to semen collection mounting an estrus jenny ($P > 0.05$).

	Semen Volume (mL)				Total Motility Mean ± SEM (%)		Sperm Concentration Mean ± SEM (×106/mL)		Total Sperm Ejaculated Mean ± SEM (×109)	
	Pre-Filtration Mean ± SEM		Post-Filtration Mean ± SEM							
	CTRL	PGF	CTRL	PGF	CTRL	PGF	CTRL	PGF	CTRL	PGF
Amiata (n = 4)	33.8 ± 3.1	45.4 ± 8.6	31.7 ± 3.2	43.6 ± 8.6	70.0 ± 3.2	173.6 ± 4.6	425.5 ± 93.2	555.7 ± 138.9	11.3 ± 1.9	19.4 ± 4.4
Balearic (n = 4)	79.7 ± 7.5	69.7 ± 7.2	78.1 ± 7.3	69.1 ± 7.2	87.0 ± 1.5	88.3 ± 1.5	175.3 ± 15.3	198.3 ± 17.9	13.0 ± 0.9	13.1 ± 0.9
Catalan (n = 5)	68.0 ± 6.5	64.3 ± 7.4	67.1 ± 6.3	63.9 ± 7.6	88.2 ± 1.4	85.3 ± 1.9	237.8 ± 21.7	214.6 ± 22.1	14.8 ± 1.1	12.9 ± 1.7
Miranda (n = 6)	88.2 ± 12.7	81.3 ± 10.1	74.6 ± 11.0	71.8 ± 10.2	65.7 ± 6.1	68.9 ± 6.4	266.4 ± 39.4	227.4 ± 41.6	17.9 ± 3.5	15.63 ± 3.5
Mean (n = 19)	74.0 ± 7.5	70.0 ± 7.5	66.3 ± 7.0	64.5 ± 7.0	74 ± 3.1	76.0 ± 3.1	274.7 ± 42.4	279.3 ± 42.4	14.45 ± 1.85	15.3 ± 1.85

There were no differences between groups for any of the variables assessed.

The number of mounts on the female before ejaculation was significantly higher in young compared to mature jacks ($P = 0.001$) as well as semen volume and total sperm number were higher in mature jacks compared to young ones ($P < 0.001$) (Tables 4 and 5). Semen motility was, instead, higher in young ones ($P = 0.021$). Total sperm ejaculated was not significantly higher in the mature group.

Table 4. Behavioral parameters in 106 semen collections performed in 19 jacks classified as young (≤5 years old) and mature (>5 years old). In half of the collections, jacks were treated with 125 μg of cloprostenol sodium (PGF) or 1 mL of saline (CTRL) prior to semen collection mounting an estrus jenny.

Groups	Treatment	Young ($n = 6$; Mean ± SEM)	Mature ($n = 13$; Mean ± SEM)
Number of mounts	CTRL	8.1 ± 1.7 ^{A,X}	4.8 ± 0.6 ^{B,X}
	PGF	4.8 ± 1.2 ^{A,Y}	3.6 ± 0.5 ^{B,Y}
	Total	6.5 ± 1.1	4.1 ± 0.4
Number of erections	CTRL	1.4 ± 0.2	1.2 ± 0.1
	PGF	1.1 ± 0.1	1.2 ± 0.1
	Total	1.2 ± 0.1	1.2 ± 0.1
Treatment to erection (min)	CTRL	11.5 ± 1.2	12.4 ± 1.6
	PGF	6.4 ± 0.4	6.2 ± 0.3
	Total	8.9 ± 0.9	9.32 ± 1.0
Treatment to ejaculation (min)	CTRL	13.6 ± 1.3 ^{X}	14.3 ± 1.7 ^{X}
	PGF	10.1 ± 1.3 ^{Y}	9.4 ± 2.0 ^{Y}
	Total	11.8 ± 0.9	11.8 ± 1.3

$^{A,B:}$ means within a line with different letters significantly differ ($P \le 0.01$). $^{X,Y:}$ For each parameter, means within a column with different letters significantly differ ($P \le 0.01$).

There were effects of semen collection order for the number of mounts: In the second and third collection, a reduced number of mounts per collection attempts was observed in the cloprostenol sodium treated jacks, while treatment to erection and to ejaculation remained significantly shorter in the cloprostenol sodium treated group compared to the CTRL (Figure 1). No differences between treatment groups and semen collection order were found in seminal analyzed parameters ($P > 0.05$).

Table 5. Seminal parameters in 106 semen collections performed in 19 jacks classified as young (≤5 years old) and mature (>5 years old). In half of the collections, jacks were treated with 125 µg of cloprostenol sodium (PGF) or 1 mL of saline (CTRL) prior to semen collection mounting an estrus jenny.

Groups	Treatment	Young (n = 6; Mean ± SEM)	Mature (n = 13; Mean ± SEM)
Pre-filtration semen volume (mL)	CTRL	50.4 ± 5.4	84.1 ± 7.6
	PGF	50.3 ± 5.7	78.0 ± 6.3
	Total	50.3 ± 3.6 ^{A}	81.0 ± 4.8 ^{B}
Post-filtration semen volume (mL)	CTRL	44.8 ± 5.7	75.9 ± 6.6
	PGF	46.3 ± 5.9	72.7 ± 6.3
	Total	45.5 ± 3.8 ^{A}	74.3 ± 4.4 ^{B}
Total motility (%)	CTRL	81.3 ± 2.4	71.1 ± 3.8
	PGF	80.8 ± 2.3	74.1 ± 3.7
	Total	81.1 ± 1.7 ^{A}	72.5 ± 2.5 ^{B}
Sperm concentration (×106/mL)	CTRL	271.6 ± 19.5	276.0 ± 47.5
	PGF	243.3 ± 31.1	295.2 ± 62.0
	Total	257.5 ± 12.0	285.7 ± 36.8
Total sperm ejaculated (×109)	CTRL	10.7 ± 1.0	17.7 ± 1.9
	PGF	11.2 ± 1.1	17.3 ± 2.2
	Total	10.9 ± 0.7	17.5 ± 1.4

$^{A,B:}$ means within a line with different letters significantly differ ($P \leq 0.01$).

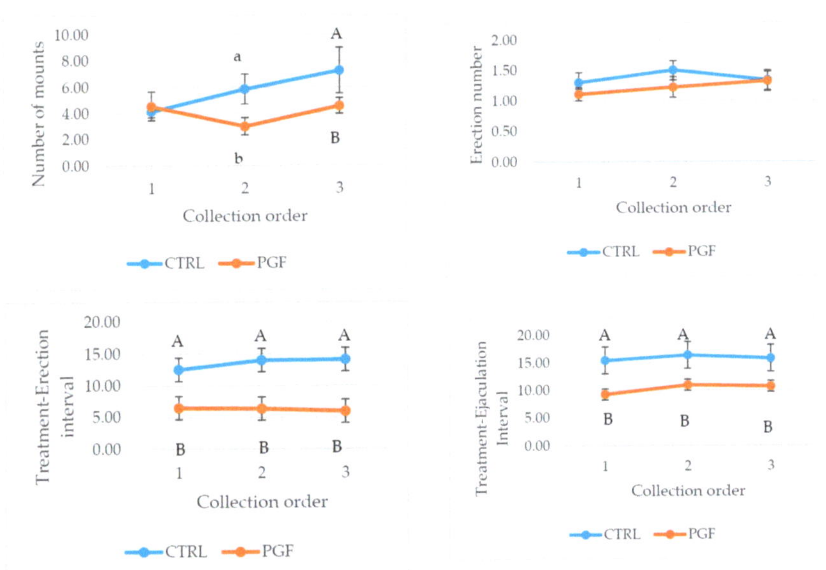

Figure 1. Effect of semen collection order (1, 2, or 3) and treatment (125 µg of cloprostenol = PGF, or 1 mL of saline = Control) interaction for behavioral features of donkeys: Jumps on teaser jenny, number of erections, treatment-erection interval (minutes), treatment-ejaculation interval (minutes). Data are expressed as least squared mean ± standard error. a,b: means ± SEM within rows and columns within the same analyzed parameter, with different letters significantly differ ($P \leq 0.05$); A,B: means ± SEM within a column with different letters significantly differ ($P \leq 0.01$).

4. Discussion

This is the first study comparing the sexual behavior and semen parameters of donkey jacks receiving cloprostenol sodium to hasten time to semen collection. While studying donkeys in multiple centers, we evaluated four donkey breeds: The Balearic donkey is considered a large donkey breed, as well as the Catalonian, which is one of the founding breeds of the American Mammoth Jack; the Miranda, a Portuguese donkey breed, which would be considered a small standard jack based on size and the dairy Italian breed Amiata, which could be viewed as a standard size jack.

While we did not fully assess courtship behavior, cloprostenol sodium did not seem to drastically affect normal behavior, other than shortening the pre-copulatory time and decreasing the number of mounts without an erection. Worth noting that mounts with or without a full erection are part of a normal donkey courtship behavior during mating [15] semen collection mounting jennies [11] or mares. In the present study, cloprostenol sodium reduced the number of mounts per collection, likely because of the time to obtain an ejaculate. The dose of cloprostenol sodium used herein was half of the typical amount used to induce luteolysis in mares and jennies. The minimal effective dose of cloprostenol sodium has not been studied in donkeys or horses. In clinical practice, cloprostenol sodium (125–250 μg) or dinoprost (2.5–5 mg) have been used as either half or full luteolytic dose. Herein, half of the dose usually administered to induce luteolysis in equids was able to hasten the erection and ejaculation more rapidly and efficiently (with fewer false mounts before ejaculation) than the control collections.

Erection was obtained in all but three collections (one after saline and two after cloprostenol sodium administration) after female contact, showing the necessity of sexual stimulation to awake sexual activity, and, probably, an exclusive action of cloprostenol sodium on the penis erectility, as demonstrated in vitro in several species [18,30]. The interval from courtship to mount and ejaculation in the donkey species is longer than in horses [12,13,15,30]. The inter-male variation regarding the minimal time necessary for semen collection or mating has been well documented. Our findings are certainly encouraging to suggest that cloprostenol sodium administration can be an alternative to hasten the time necessary to attain a semen collection in donkeys.

Although it is impossible to recreate an identical environment during semen in three breeding centers across Europe, attempts were carried out to keep the conditions as similar as possible. The authors conducted extensive discussions to ensure standard conditions across centers. Therefore, breed differences observed herein for all the variables (i.e., the number of false mounts, number of erections before ejaculation, and the interval between treatment/female contact and ejaculation, in both PGF treated and untreated jacks) are likely real.

Ejaculate volume (pre- and post-filtration) and semen concentration varied with breeds but not between cloprostenol sodium vs. control treatment within the same breed. Semen concentration was just affected by breed, while treatment, age and weight had no influence on the number of ejaculated spermatozoa. Interestingly, 10 mg of dinoprost (natural PGF2α) resulted in increased post-filtration volume and decreased sperm concentration in horses [27]. Breed differences in reproductive parameters of donkeys have already been anecdotally reported [9]. One of the limitations of the present study is that a small number of animals represented each breed, thus, it is possible that if a larger number of jacks were enrolled in the study, the differences in breeds could have been different.

Surprisingly, age did not affect the behavioral parameters evaluated, except for the number of mounts per collection. These results could be real or be due to the fact that the number of animals in each group (i.e., ≤5 years old young, $n = 5$ vs. >5 years old mature, $n = 14$) was skewed. Previously, it has been shown that younger donkeys took longer times to be collected than mature donkeys [5,14]. In the present study, mature donkeys had greater pre- and post-filtration ejaculate volume and lower total motility, regardless of treatment received. Mature donkeys had greater total sperm ejaculated compared to young donkeys. This could be because three of the donkeys in the young group were still developing and not reached the supposed peak in sperm production expected to happen around five years of age.

Differences among subsequent treatments are not evident within treated and untreated jacks for all the parameters studied (differently than from the horse stallion [27]). No negative or positive effects on studied parameters could be attributed to this dose of cloprostenol sodium in jackasses included in this study.

The sweating response found in two Miranda jacks, subsequent to cloprostenol sodium treatment, is consistent with what was previously reported in horses [19], probably due to the release of epinephrine from the adrenal medulla [31]. The jacks having this side effect were the smaller of the jacks included in the study of their breed, a consequence of the dose/weight, and of a breed-enhanced sensibility to cloprostenol sodium could be advocated.

Persistent erection in all the Amiata jacks treated with cloprostenol sodium could be due to the spastic contractions of the smooth muscle apparatus of the reproductive tract caused by PGF2α as seen in several species, such as humans [18], horses, and bulls [27,32]. The absence of this side effect in two out of four breeds could be due to the inherent breed variations, or body size and the only case in 1/4 breeds included could be due to the individual and breed differences already shown in this species [33].

5. Conclusions

In conclusion, the administration of cloprostenol sodium was able to hasten erection and ejaculation without affecting semen quality. The present multi-centric study certainly supports the previously anecdotal use of cloprostenol sodium in clinical practice, a multi-breed controlled design study described herein.

Author Contributions: Conceptualization, D.P., J.M. and I.F.C.; Methodology, D.P., M.Q., I.F.C., J.M.; Software, G.C.; Validation, G.C. and I.F.C.; Formal Analysis, G.C., F.C., A.R.; Investigation, D.P., M.Q., D.F., F.C., R.M., A.R., A.M.-B., M.N., J.C. and J.M.; Resources, F.C., M.Q., J.M., M.N. and J.C.; Data Curation, G.C. and D.P. Writing–Original Draft Preparation, D.P. and I.C.; Writing–Review & Editing, D.P., M.Q., D.F., F.C., R.M., A.R., A.M.-B., M.N., J.C., I.F.C., G.C. and J.M.; Visualization, D.P., M.Q., D.F., F.C., R.M., A.R., A.M.-B., M.N., J.C., I.F.C., G.C. and J.M.; Supervision, F.C., J.M., A.R., A.M.-B.; Project Administration, D.P.; Funding Acquisition, F.C., M.Q., J.M., M.N. and J.C. All authors have read and agreed to the published version of the manuscript.

Funding: This research was funded by: Department of Veterinary Clinical Sciences, University of Pisa, via Livornese 1286, San Piero a Grado, Pisa, 56122 (PI), Italy; Center of Animal and Veterinary Science (CECAV) University of Trás-os-Montes e Alto Douro (UTAD), Quinta de Prados, 5000-801 Vila Real, Portugal; Association for the Study and Protection of Donkeys (AEPGA), Largo da Igreja, 5225-011 Atenor, Portugal; Equine Reproduction Service, Department of Animal Medicine and Surgery, Veterinary Faculty, Autonomous University of Barcelona, 08193 Bellaterra, Cerdanyola del Valles, Spain.

Conflicts of Interest: The authors declare no conflict of interest.

References

1. Lu, C.J.; Xie, W.M.; Su, R.; Ge, Q.L.; Chen, H.; Shen, S.Y.; Lei, C.Z. African origin of Chinese domestic donkeys. *Yi Chuan Hered.* **2008**, *30*, 324–328. [CrossRef]
2. Beja-Pereira, A.; England, P.R.; Ferrand, N.; Jordan, S.; Bakhiet, A.O.; Abdalla, M.A.; Mashkour, M.; Jordana, J.; Taberlet, P.; Luikart, G. African origins of the domestic donkey. *Science* **2004**, *304*, 1781. [CrossRef] [PubMed]
3. Kimura, B.; Marshall, F.; Beja-Pereira, A.; Mulligan, C. Donkey Domestication. *Afr. Archaeol. Rev.* **2013**, *30*, 83–95. [CrossRef]
4. Camillo, F.; Rota, A.; Biagini, L.; Tesi, M.; Fanelli, D.; Panzani, D. The Current Situation and Trend of Donkey Industry in Europe. *J. Equine Vet. Sci.* **2018**, *65*, 44–49. [CrossRef]
5. Carneiro, G.F.; Cavalcante Lucena, J.E.; de Oliveira Barros, L. The Current Situation and Trend of the Donkey Industry in South America. *J. Equine Vet. Sci.* **2018**, *65*, 106–110. [CrossRef]
6. Kugler, W.; Grünenfelder, H.P.; Broxham, E. Donkey Breeds in Europe. 2008. Available online: http://www.agrobiodiversity.net/topic_network/pdf/donkey.pdf (accessed on 1 September 2020).
7. Lesté-Lasserre, C. Donkeys face worldwide existential threat. *Science* **2019**, *366*, 1294–1295. [CrossRef]
8. Canisso, I.F.; Davies Morel, M.C.G.; McDonnell, S. Strategies for the management of donkey jacks in intensive breeding systems. *Equine Vet. Educ.* **2009**, *21*, 652–659. [CrossRef]

9. Canisso, I.F.; Panzani, D.; Miró, J.; Ellerbrock, R.E. Key Aspects of Donkey and Mule Reproduction. *Vet. Clin. Equine Pract.* **2019**, *35*, 607–642. [CrossRef]
10. Lodi, L.D.; Henry, M.; Paranhos da Costa, M.J.R. Behavior of Donkey Jacks (Equus asinus) Breeding Horse Mares (Equus caballus) at Pasture. *Biol. Reprod.* **1995**, *52*, 591–598. [CrossRef]
11. Gastal, M.O.; Henry, M.; Beker, A.R.; Gastal, E.L.; Goncalves, A. Sexual behavior of donkey jacks: Influence of ejaculatory frequency and season. *Theriogenology* **1996**, *46*, 593–603. [CrossRef]
12. Rota, A.; Sgorbini, M.; Panzani, D.; Bonelli, F.; Baragli, P.; Ille, N.; Gatta, D.; Sighieri, C.; Casini, L.; Maggiorelli, M.M.; et al. Effect of housing system on reproductive behaviour and on some endocrinological and seminal parameters of donkey stallions. *Reprod. Domest. Anim.* **2018**, *53*, 40–47. [CrossRef] [PubMed]
13. Rota, A.; Puddu, B.; Sabatini, C.; Panzani, D.; Lainé, A.L.; Camillo, F. Reproductive parameters of donkey jacks undergoing puberty. *Anim. Reprod. Sci.* **2018**, *192*, 119–125. [CrossRef] [PubMed]
14. Canisso, I.F.; Carvalho, G.R.; Morel, M.C.G.D.; Guimarães, J.D.; McDonnell, S.M. Sexual behavior and ejaculate characteristics in Pêga donkeys (Equus asinus) mounting estrous horse mares (Equus caballus). *Theriogenology* **2010**, *73*, 56–63. [CrossRef] [PubMed]
15. Henry, M.; McDonnell, S.M.; Lodi, L.D.; Gastal, E.L. Pasture mating behaviour of donkeys (Equus asinus) at natural and induced oestrus. *J. Reprod. Fertil.* **1991**, *44*, 77–86.
16. Pharriss, B.B.; Wyngarden, L.J. The Effect of Prostaglandin F2α on the Progestogen Content of Ovaries from Pseudopregnant Rats. *Proc. Soc. Exp. Biol. Med.* **2016**, *130*, 92–94. [CrossRef]
17. Jeremy, J.Y.; Mikhailidis, D.P. Prostaglandins and the penis: Possible role in the pathogenesis and treatment of impotence. *Sex. Marital. Ther.* **2007**, *5*, 155–165. [CrossRef]
18. Veronesi, M.C.; De Amicis, I.; Panzani, S.; Kindahl, H.; Govoni, N.; Probo, M.; Carluccio, A. PGF(2α), LH, testosterone, oestrone sulphate, and cortisol plasma concentrations around sexual stimulation in jackass. *Theriogenology* **2011**, *75*, 1489–1498. [CrossRef]
19. Cornwell, J.C.; Koonce, K.L.; Kreider, J.L. Effect of prostaglandin F2 alpha on seminal characteristics of the stallion. *J. Anim. Sci.* **1974**, *38*, 226.
20. Hafs, H.D.; Louis, T.M.; Stellflug, J.N. Increased sperm numbers in the deferent duct after prostaglandin F2alpha in rabbits. *Proc. Soc. Exp. Biol. Med.* **1974**, *145*, 1120–1124. [CrossRef]
21. Hafs, H.D.; Louis, T.M.; Waters, R.J.; Stellflug, J.N.; Haynes, N.B. Increased sperm output of rabbits and bulls treated with prostaglandin F2α. *Prostaglandins* **1974**, *8*, 417–422. [CrossRef]
22. Hemsworth, P.H.; Donnelly, J.; Findlay, J.K.; Galloway, D.B. The effects of prostaglandin F2α on sperm output in boars. *Prostaglandins* **1977**, *13*, 933–941. [CrossRef]
23. Stolla, R.; Schmid, G. Effects of natural and synthetic PGF2 alpha preparations on the uterine contractility of cattle. *Berl. Munch. Tierarztl. Wochenschr.* **1990**, *103*, 198–202.
24. Szurop, I.; Nagy, A.; Jöchle, W. Stimulation of Libido in Pubertal and Mature Boars with Prostaglandin F2α Analogs: Clinical Observations. *Reprod. Domest. Anim.* **1986**, *21*, 83–86. [CrossRef]
25. Estienne, M.J. A review of the effects of prostaglandins on sexual behavior in boars. *Appl. Anim. Behav. Sci.* **2014**, *154*, 1–7. [CrossRef]
26. Burne, T.H.J.; Murfitt, P.J.E.; Gilbert, C.L. c-fos mRNA expression associated with PGF2α-induced nest-building behaviour in female pigs. *Mol. Brain Res.* **2002**, *104*, 31–37. [CrossRef]
27. Kreider, J.L.; Ogg, W.L.; Turner, J.W. Influence of prostaglandin F2α on sperm production and seminal characteriticss of the stallion. *Prostaglandins* **1981**, *22*, 903–913. [CrossRef]
28. Vall, E.; Ebangi, A.L.; Abakar, O. A method for estimating body conditions score (BCS) in donkeys. In *Working Animals in Agriculture and Transport: A Collection of Some Current Research and Development Observations, Proceedings of the EAAP Annual Conference 53, Caire, Égypte, 1–4 September 2002*; Anne, P.R., Philippe, L., Markku, S., William, M.R., Eds.; EAAP, University of Edinburgh, CIRAD, MTT Agrifood Research Finland, INRA; Wageningen Academic Publishers: Wageningen, The Netherlands, 2003; pp. 93–102.
29. Canisso, I.F.; da Silva Coutinho, M.A.; Davies Morel, M.C.G.; McDonnel, S. How to manage jacks to breed mares. In Proceedings of the Annual Convention of the AAEP, Phoenix, AZ, USA, 9 December 2009; pp. 342–348.
30. Gastal, M.O.; Henry, M.; Beker, A.R.; Gastal, E.L. Effect of ejaculation frequency and season on donkey jack semen. *Theriogenology* **1997**, *47*, 627–638. [CrossRef]
31. Lauderdale, J.W. Catecholamines and rectal temperature after PGF2a. *J. Anim. Sci.* **1975**, *41*, 365.

32. Klinge, E.; Sjöstrand, N.O. Comparative Study of Some Isolated Mammalian Smooth Muscle Effectors of Penile Erection. *Acta Physiol. Scand.* **1977**, *100*, 354–367. [CrossRef]
33. Camillo, F.; Vannozzi, I.; Tesi, M.; Sabatini, C.; Rota, A.; Paciolla, E.; Dang-Nguyen, I.; Panzani, D. Induction of ovulation with buserelin in jennies: In search of the minimum effective dose. *Anim. Reprod. Sci.* **2014**, *151*, 56–60. [CrossRef]

Publisher's Note: MDPI stays neutral with regard to jurisdictional claims in published maps and institutional affiliations.

animals

MDPI

Review

Viral Diseases that Affect Donkeys and Mules

Rebeca Jéssica Falcão Câmara [1], Bruna Lopes Bueno [1], Cláudia Fideles Resende [1], Udeni B. R. Balasuriya [2], Sidnei Miyoshi Sakamoto [3] and Jenner Karlisson Pimenta dos Reis [1,*]

1 Laboratório de Retroviroses, Departamento de Medicina Veterinária Preventiva, Escola de Veterinária, Universidade Federal de Minas Gerais, Belo Horizonte 31270-901, Brazil; rebeca.jfc@gmail.com (R.J.F.C.); brulbueno18@gmail.com (B.L.B.); c.fideles91@gmail.com (C.F.R.)

2 Louisiana Animal Disease Diagnostic Laboratory and Department of Pathobiological Sciences, School of Veterinary Medicine, Louisiana State University, River Rd, Room 1043, Baton Rouge, LA 70803, USA; balasuriya1@lsu.edu

3 Laboratório Multidisciplinar do Centro de Ciências Biológicas e da Saúde, Departamento de Ciências da Saúde (DCS), Universidade Federal Rural do Semi-Árido, Rio Grande do Norte 59625-900, Brazil; sakamoto@ufersa.edu.br

* Correspondence: jenner@ufmg.br; Tel.: +55-31-3409-2100

Received: 19 October 2020; Accepted: 19 November 2020; Published: 25 November 2020

check for updates

Simple Summary: Donkeys have been neglected and threatened by abandonment, indiscriminate slaughter, and a lack of proper sanitary management. They are often treated as "small horses." However, donkeys and horses have significant genetic, physiological, and behavioral differences. Specific knowledge about viral infectious diseases that affect donkeys and mules is important to mitigate disease outbreaks. Thus, the purpose of this review is to provide a brief update on viral diseases of donkeys and mules and ways to prevent their spread.

Abstract: Donkeys (*Equus asinus*) and mules represent approximately 50% of the entire domestic equine herd in the world and play an essential role in the lives of thousands of people, primarily in developing countries. Despite their importance, donkeys are currently a neglected and threatened species due to abandonment, indiscriminate slaughter, and a lack of proper sanitary management. Specific knowledge about infectious viral diseases that affect this group of *Equidae* is still limited. In many cases, donkeys and mules are treated like horses, with the physiological differences between these species usually not taken into account. Most infectious diseases that affect the *Equidae* family are exclusive to the family, and they have a tremendous economic impact on the equine industry. However, some viruses may cross the species barrier and affect humans, representing an imminent risk to public health. Nevertheless, even with such importance, most studies are conducted on horses (*Equus caballus*), and there is little comparative information on infection in donkeys and mules. Therefore, the objective of this article is to provide a brief update on viruses that affect donkeys and mules, thereby compromising their performance and well-being. These diseases may put them at risk of extinction in some parts of the world due to neglect and the precarious conditions they live in and may ultimately endanger other species' health and humans.

Keywords: donkeys; *Equus asinus*; mules; viral diseases; health; infectious diseases

1. Introduction

Horses (*Equus caballus*) and donkeys (*Equus asinus*) are mammals belonging to genus, *Equus*, which emerged approximately 4.5 million years ago [1]. Although there is an inheritance shared between them, the horses and the donkeys are notably different in their physical and behavioral characteristics [2]. The crossing of these two species results in a sterile hybrid, named mule, which has the characteristics of the two progenitor species, generating a great diversity of body types and temperaments, however, their physiological characteristics resemble those of the horse [3,4].

Currently, the world population of Equidae is estimated at 116.7 million animals, comprising 57.7 million horses, 50.4 million donkeys and 8.5 million mules [5]. Most donkey and mule populations are concentrated in the Asian continent, as well as in some countries of Central America, South America and Africa, ensuring the livelihood of 500 million people in poor communities in developing countries around the world. These donkeys and mules are used mainly as work animals (transporting cargo and traction) and for transporting people [6]. However, the industrialization and mechanization of agriculture has also led to an increasing and global abandonment of animals, which start to live in an "errant" way, roaming the roads and causing traffic accidents, in addition these stray animals representing a potential source of transmission of infectious diseases for other species, due to the lack of sanitary control [7]. Analysis of the complete genome of horses and donkeys showed divergences in genes directly involved in the inflammatory response to trauma (ITIH4) and in the regulation of cholesterol synthesis (HMGCR). These genes were positively selected in donkeys [1,8] and, they produce significant differences in physiological and biochemical parameters when compared to horses (Table 1). However, many benchmarks are still being established for specific breeds of donkeys [2,9,10].

Table 1. Comparison of the physiological parameters between donkeys and horses. Source [11,12].

Physiological Parameters	Donkeys	Horses
Temperature (°C)	36.5–37.8	37.5–38.5
Pulse (Beats/Minute)	36–52	30–40
Respiration (Breaths/Minute)	12–28	18–20
Hematology and Biochemistry Parameters		
RBC Count ($10^6/\mu L$)	4.4–7.1	6.0–10
Hemoglobin (g/dL)	8.9–14.7	12–17
MCV (fl)	53–67	42–58
Triglyceride (mmol/L)	0.6–2.8	0.05–0.5
Total Bilirubin (μmol/L)	0.1–3.7	0.06–0.12

Donkeys also have a different susceptibility to certain infectious agents and clinical manifestations when compared to horses. However, there are few scientific studies on pathogenesis, immune response and pathophysiology. Much of the available knowledge comes from the clinical experience of veterinarians, as well as knowledge, which is extrapolated from horses to donkeys [2].

Just like horses, donkeys and mules are also susceptible to infection by important virus such as equine infectious anemia (EIA), the eastern, western and venezuelan equine encephalomyelitis, Japanese encephalitis (JE), West Nile fever (WNF), equine viral arteritis (EVA), equine herpesvirus (EHV), equine influenza (EI), as well as the African horse sickness and rabies which are listed in the OIE Terrestrial Animal Health Code [13], and countries are required to report their occurrence accordingly. In addition to being relevant to the equine industry [14], some of them such as the Eastern, Western and Venezuelan equine encephalomyelitis, JE, WNF, rabies and Saint Louis encephalitis (SLE) [15,16] are of great importance for One Health. In addition, there are still unexplored viruses, such as the newly discovered *Nonprimate hepacivirus* (NPHV) [17], whose impact on equids is still poorly understood. Therefore, the main purpose of this article is to review the published literature and online resources (e.g., official donkey websites and non-governmental organizations (NGOs)) compiling and providing an update on the main viral diseases of equids on donkeys and mules.

2. Equine Infectious Anemia

Equine infectious anemia (EIA) is a worldwide disease caused by the *Equine infectious anemia virus* (EIAV), a *lentivirus* classified in the family *Retroviridae* that exclusively affects horses, mules, and donkeys [18]. The official EIA statistics on equids do not accurately report the prevalence of the disease in countries as they refer, and statistics are almost exclusively based on, laboratory tests carried out on the transit of animals and/or participation in events controlled by the Official Veterinary Services [19]. This most often involves animals of zootechnical interest value. It is believed that this estimate is even more compromised for mules and donkeys, which are typically animals of low value. Therefore, what is known about the occurrence of EIA in donkeys and mules comes from a limited number of studies that are summarized in Table 2.

Table 2. Studies on the occurrence of equine infectious anemia (EIA) in horses, donkeys, mules and zebras from 2009 to 2019.

Disease	Species	% Positivity	Location	References
EIA	Donkey Horse Mule Zebra	0% (0/1568) 0.2% (41/23971) 3.5% (27/767) 0% (0/3)	Italy	[20]
	Donkey Horse Mule	0.2% (1/662) 0% (0/289) 0% (0/51)	Ethiopia	[21]
	Donkey	3.27% (12/367)	Brazil	[22]
	Donkey Horse	8.28% (14/169) 3.17% (6/189)	Sudan	[23]

Blood from an infected horse, donkey and mule is the most important source of EIAV for transmission to susceptible animals. The most common natural transmission of EIAV is via blood-feeding insects of the family *Tabanidae* (e.g., horse flies and deer flies) [24]. Stable flies (*Stomoxys calrcitrans*) are also capable of transmitting the virus but they are less efficient than the tabanids. Because the virus does not replicate within the insect vector, insects only serve as mechanical vectors, transferring blood though their mouth parts [25]. The virus can be transmitted iatrogenically via contaminated needles or surgical instruments, as well as transfusion of contaminated blood products. The transmission of EIAV can also occur through contact of the infected blood with exposed wounds of a healthy animal [26]. In addition to these routes of transmission, some studies have shown the possibility of spreading EIAV by vertical transmission via the uterine/perinatal route (i.e., in utero, at parturition), or following the ingestion of infected colostrum, as well as via aerosols [24,27–29].

The EIA clinical course is well described in horses and ponies, with little information on donkeys and mules. It is known that the presentation of clinical signs can vary according to the virulence of the infecting strain, the amount of virus to which the animal was exposed and the carrier's immunological status [30,31]. In horses, three clinical phases of EIA have been described: acute, chronic and inapparent. The acute phase is characterized by viremia and thrombocytopenia, and occurs between 5 and 30 days after infection [18,32], they can also be accompanied by fever, lethargy, inappetence and in more severe cases may present petechiae and hemolytic anemia; however, the signs can also be mild or absent [33]. After the acute phase, the infected animal enters the chronic phase, characterized by recurrent cycles of viremia, weight loss, edema, anemia, thrombocytopenia and, less frequently, neurological clinical signs [18,33,34]. If the animal survives, the clinical signs of the chronic phase gradually decrease over a period of approximately one year [35,36]. After this period, about 90% of horses infected with EIAV evolve to the phase of inapparent carrier, where clinical signs are absent and viremia is undetectable in most cases [33,36].

Clinical, pathological, and laboratory findings of the mules show that they produce clinical signs similar to those seen in horses [35]. However, donkeys despite being susceptible to EIAV, have very low levels of viremia, which may explain the fact that they do not usually demonstrate clinical signs of the disease. Cook et al. [37] performed experimental infections of donkeys and horses with the pathogenic strain of EIAVpv strain (Pathogenic variant). All horses in the experiment showed clinical signs of EIA, while the donkeys remained asymptomatic during the 365 days of observation. Seroconversion of the animals occurred between 20- and 40-days post-infection for horses and donkeys. Interestingly, the viral load in donkeys was 100,000 times lower than in horses during the first three weeks of infection. In the case of infection with a highly virulent strain of EIAV (EIAV Wyoming), the difference in viral load was approximately 1000 times lower in donkeys than in horses, and seroconversion of the donkeys was detected two weeks after infection. In the same study, equine and donkey monocyte-derived macrophages (MDM) were equally susceptible to EIAV infection in vitro, suggesting that the clinical differences observed between these two species are not related to the permissiveness of the host cells.

A vaccine against EIA was developed in 1975 by the Harbin Veterinary Research Institute in China [38]. This vaccine was obtained through a three-step attenuation process. The first and second stages consisted of serial in vivo passage of the highly pathogenic strain $EIAV_{LN}$ (LN40) first in horses and then in donkeys, producing the strain $EIAV_{DV117}$ which was even more pathogenic. Thus, attenuation was obtained in the third stage, after 121 serial in vitro passages of the strain $EIAV_{DV117}$ in primary donkey macrophages ($EIAV_{DLV121}$). Vaccination of horses and donkeys with this cell culture adapted $EIAV_{DLV121}$ strain provided 85% and 100% protection, respectively, when challenged with the virulent virus [39].

Due to the cost and time spent isolating leukocytes, $EIAV_{DLV121}$ was cultured in FDD (fetal donkey dermal) cells and, after 13 passages, the new strain $EIAV_{FDDV13}$ showed the same protective efficiency as the previous strain ($EIAV_{DLV121}$) [39]. This viral attenuation process shows that the interaction of EIAV with horse and donkey cells is different, resulting in a distinct vaccinal protective efficacy between these species, which suggests that they may differ in susceptibility. After a previously vaccination with the attenuated strain ($EIAV_{DLV121}$), 25% of horses and only 5% of donkeys did not develop an effective immune response to EIAV when they were challenged with the pathogenic strain (LN40) [40]. To explain the greater susceptibility of horses even after vaccination, Yin et al., 2013 demonstrated that EIAV uses the cell translation machinery of the horses more efficiently than when compared to donkeys and this would increase viral proteins expression and consequently, increase the amount of virions circulating in the animal [41]. In addition, other studies have suggested that the RNA editing enzyme equine adenosine desaminase (eADAR1) upregulates EIAV in vitro by enhancement of viral protein expression associated with the LTR region and the RRE in the env region [42]. A genomic analysis revealed multiple mutations, specially A-to-G substitutions, in the genome of $EIAV_{DV117}$ strain, when compared to the parental strain $EIAV_{LN40}$. The mutations were found mostly at 5′TpA and 5′ApA dinucleotides, that are commonly targeted by ADAR1, therefore ADAR1 is most likely to be related in the EIAV adaptation from horses to donkeys [43].

Even though the induction of cellular and humoral immune responses has already been demonstrated in animals vaccinated with the Chinese vaccine [44], its effectiveness is not fully proven, especially when the animals are challenged with heterologous strains of EIAV. Therefore the efficiency of the vaccine would be compromised in other regions of the world, due to the genetic diversity of strains of EIAV [45,46]. An additional disadvantage of this vaccine is the impossibility of distinguishing between vaccinated and naturally infected animals, which can compromise control measures based on serological tests [24]. With the reported limitations on the vaccine developed in China, there is no vaccine or treatment in the world for EIA and, its control occurs through the identification, segregation, and/or euthanasia of EIAV-infected animals [44,47–49]. The official method for diagnosing EIA in several countries around the world is the agar gel immunodiffusion (AGID) technique, also known as the Coggins Test [50]. Some countries such as the USA, Brazil, and Italy use the enzyme-linked immunosorbent assay (ELISA) as a screening method due to its high sensitivity [24,51–53]. There is also a Western immunoblotting test, which can be used as a third confirmatory technique, but it is not commercially available and is used only in reference laboratories in the US and Italy [29].

One of the problems of using only serological tests for the diagnosis of EIA is that there is a "immunological window", that refers to the time after infection and before seroconversion. In horses and ponies, this period is usually less than 45 days, but periods of up to 157 days have already been reported for the AGID test [29]. Experimentally donkey seroconversion took twice as long as that for horses [37]. It is during the period of the "immunological window" that viral loads associated with blood can reach their highest levels, thus maximizing the risk of transmission [24]. There is also the hypothesis that EIAV can establish serologically silent infections similar to those seen with woodchuck hepatitis virus (WHV) and simian immunodeficiency virus (SIV) [54] that is, viruses and genetic material are identified, however by a mechanism not yet elucidated, viral replication does not stimulate detectable humoral immune response [55].

Scicluna et al. [53] observed an increase in the efficiency of the diagnosis of EIA when using three different methods in sequence, first the ELISA for screening, second the AGID to confirm the positive results in the ELISA, and third the Immunoblot (IB), to conclude on conflicting results between the two previous tests. It was observed that the frequency of samples with negative AGID and positive ELISA/immunoblot results was significantly higher in mules than in horses, which suggests a problem in the diagnosis of mules, which is important as this hybrid species has a role in the dissemination and maintenance of EIAV in the infected field herds.

Oliveira et al. [22] evaluated the seroprevalence of EIAV in a donkey population in Brazil, using the AGID, two ELISAs and the IB for discordant results. From the comparison between the tests, it was observed that the ELISA showed false-positive results, while the AGID was associated with false-negative results, which corroborates previous studies carried out in horses [29,53,56], emphasizing the need for a new diagnostic approach for EIAV research in equines, and particularly in donkeys, since they typically present an extremely weak reaction in AGID which is associated with false-negative results or results which are difficult to interpret, probably due to the delay in seroconverting or low levels of antibodies when compared to horses and ponies [37].

The other articles mentioned in Table 2 use only a single diagnostic test for EIA, the classic AGID [20,21,57] or ELISA [23]. Due to the limited techniques used in these surveys, the seropositivity found is likely to be overestimated in Sudan and underestimated in Italy, Bulgaria and Ethiopia. As reported above, although very specific, AGID is not very sensitive, increasing the number of false-negative results [53]. Therefore, the prevalence of studies that used only the AGID test is believed to be underestimated. The occurrence of EIA in Italy, Sudan, and Brazil was correlated with geographic location; regions less prone to the presence of natural fly vectors showed less seropositivity when compared to other regions within each country [20,22,23].

In circumstances where serology is not adequate, the use of the polymerase chain reaction (PCR) and even viral isolation are recommended, but these techniques have limitations. Viral isolation is extremely complex and rare, due to the difficulty, cost and time consuming for the cultivation of monocyte-derived macrophages, which prevents the routine use of this technique [24,58]. The molecular diagnosis by PCR is limited by the fact that the EIAV has a very complex global epidemiology, presenting a low level of identity between the known nucleotide sequences and their division into different clades [59], making it difficult to use universal PCR primers for its diagnosis. The effectiveness of molecular diagnosis in donkeys is still unknown, as there are no field isolates in this species that can serve as a basis for standardizing assays.

3. Equine Viral Arteritis

Equine viral arteritis (EVA) is a respiratory and reproductive disease that affects equides worldwide [60]. The etiological agent of EVA is the *Alphaarterivirus equid* (previously known as equine arteritis virus (EAV)), a single stranded, positive sense RNA virus that belongs to the family *Arteriviridae*, subfamily *Equarterivirinae* genus *Alphaartevirus* in the order *Nidovirales* [61].

Alphaarterivirus equid is transmitted via respiratory or venereal routes in horses, donkeys, and mules [11,62]. Horizontal transmission occurs via the respiratory tract after aerosolization of viral particles present in the respiratory secretions of horses with acute infection, facilitated by direct and/or close contact with susceptible animals [63,64]. Other body secretions and excretions from infected animals such as urine, aborted fetuses, and fetal membranes can also transmit the virus and play a significant role in precipitating new outbreaks of EVA [64,65]. Venereal transmission occurs exclusively via infected stallion semen in the acute or chronic phase [66].

Although *Alphaarterivirus equid* is present worldwide, there are few studies on the prevalence of this infection in donkeys and mules (Table 3). In horses, the prevalence of *Alphaarterivirus equid* varies between countries and among horses of different breeds and ages within the same country because of the management practices and genetics of host breeds that make stallions become long term *Alphaarterivirus equid* shedders [60,62]; these findings may also apply to donkeys. In the last 10 years, the presence of *Alphaarterivirus equid* in donkeys has been reported on the Asian (Eastern Anatolia, Turkey), European, and American continents (Central and South America). The highest reported prevalence was in Bulgaria (79.1%), and this study suggested that the high rate was due to a lack of control measures to prevent the disease [57]. In South Africa and Chile, *Alphaarterivirus equid* strains that infect donkeys have been reported [57]. Comparative genomic sequence and phylogenitc analyses of these strains with other equid *Alphaarterivirus equid* strains showed the existence of a new Alphaarterivirus genotypes in these regions [67,68]. In a study conducted by Bolfa et al. [69], only one donkey was positive for the presence of antibodies against *Alphaarterivirus equid*, and a possible explanation is that this animal could have been an asymptomatic carrier that could have been imported.

Table 3. Studies of the occurrence of *Alphaarterivirus equid*, including donkeys and/or mules, from 2009 to 2019.

Disease	Species	% Positivity	Location	References
EVA	Donkey	79.1% (152/192)	Bulgaria	[57]
	Donkey	20% (17/85)	Brazil	[70]
	Donkey	2 animals	Chile	[68]
	Donkey	2.5% (1/40)	West Indies	[69]
	Donkey	8.3% (19/227)	Turkey	[71]
	Horse	15% (29/193)		
	Donkey	3.46% (53/1532)	Turkey	[72]

There are a few studies which raise the hypothesis that donkeys are more resistant to the development of the disease caused by *Alphaarterivirus equid* than horses, presenting mostly inapparent infections or mild clinical signs [73,74]. In cases where there are evident clinical signs, donkeys usually present with fever, depression, ocular and nasal discharge [11]. Similar to equine stallions, where 10–70% become inapparent carriers and maintain viral spread through semen. All the evidence indicates that donkeys are mostly asymptomatic and also act as potential reservoirs of *Alphaarterivirus equid* and help to maintain and perpetuate virus in the equid population [11,62].

4. Equine Hepacivirus

Nonprimate Hepacivirus (NPHV), also known as Equine Hepacivirus, belongs to the family *Flaviviridae* and the species *Hepacivirus A*. Currently, the NPHV is described as the homologous virus phylogenetically closest to the *hepatitis C virus* (HCV) [75,76]. The NPHV was identified in horses in 2012, and they were considered the main natural host of the virus [17]. NPHV infection was identified in serum samples collected from 1974 to 2016 in donkeys from Germany, Spain, Bulgaria, Italy, France, and Mexico as well as in mules from Bulgaria, where the presence of antibodies was detected in 31.5% (278/882) of the tested animals and viral RNA in only 0.3% (3/882), who were also seropositive, suggesting a predominance of acute infection with rapid viral depletion [77]. This could explain why studies have failed to detect the virus in donkeys and mules in Brazil (n = 1 and 35) [78,79], China (n = 14) [80], United Kingdom (n = 16) [81] and Italy (n = 134) [82], since they used only PCR-based techniques to screen the virus and also a limited sample number.

A recently published work identified that a partial NPHV E2 nuccleotide sequence isolated from an infected donkey showed a 22% divergence between the virus sequenced from horses [83]. This indicates that the virus may acquire genetic variability when it crosses the species barrier. The accumulation of variations can produce antigenically distinct variants between horses and donkeys, thereby interfering with molecular diagnosis in donkeys.

NPHV is hepatotropic, and the infection is subclinical in horses. Although the viral load remained similar between horses and donkeys, a slight increase in liver enzymes was observed in horses, whereas in donkeys, enzyme concentrations remained within the reference range [76,77]. The absence of changes in donkeys may be a reflection of the time interval between the initial infection and the tests performed since over time, horses affected by the virus tend to overcome the infection and recover to normal functioning [76]. However, it could be speculated that donkeys are more resistant to NPHV and, therefore, clear the infection more quickly, as already observed in other diseases, such as EIA.

However, it is still not clear what the impact of this new virus is on the equine industry and, to date, NPHV poses no risk to public health [84]. Meanwhile, the use of equines as an alternative animal model to studies with HCV may help to elucidate the immunological mechanisms involved in viral depletion and possibly in the natural resistance of the asinine species, which may have considerable implications for the development of effective drugs and vaccines compared to other related viruses.

5. Equine Herpesvirus

Equine herpesviruses (EHVs) are pathogens that establish latent infections in their hosts and affect all members of the Equidae family worldwide. They are DNA enveloped viruses belonging to the family *Herpesviridae* [85]. There are currently nine identified equine herpesviruses (EHV-1 to EHV-9), six of which belong to the subfamily *Alphaherpesvirinae* and three to the subfamily *Gamaherpesvirinae*. There exist asinine herpesvirus (AHV) of both subfamilies (Table 4) [86,87].

EHV-1 and EHV-4 are clinically, economically, and epidemiologically the most relevant pathogens within this family, causing problems in the respiratory tract of horses worldwide [97]. Both have been associated with abortion and respiratory diseases in donkeys [98]. In addition, EHV-1 is also associated with neurological diseases in donkeys [11,99,100]. Although the equine herpesvirus myeloencephalopathy (EHM) caused by EHV-1 is uncommon, it has been described in donkeys and mules in outbreaks in Ethiopia between the years 2011 and 2013. EHM was fatal in donkeys, with deaths without apparent clinical signs [101]. These viruses are antigenically and genetically related, showing cross-reactions due to similarity between 55% and 96% in the amino acid sequence of surface glycoproteins [99]. EHV-1 and EHV-4 are distributed worldwide, in domestic equine species [102,103]. However, there are few studies on the presence of antibodies against herpesvirus in donkeys (Table 5).

Some of the studies listed in Table 5 describe a higher prevalence rate of equine herpesvirus in donkeys, suggesting that they may represent an important source of infection for other equines [21,101,104,110,111]. It was also suggested that older donkeys would have a higher infection incidence rate. Depending on the study, both sexes could be considered as correlated with the occurrence of the infection. The correlation between age and disease is easily explained, as older animals have a longer time of exposure to infection; therefore, they are more likely to be naturally challenged by EHVs and recover, increasing the prevalence of antibodies against the virus [101,104,105]. Regarding the correlation between sex and disease, there were differences in results among different studies. Bolfa et al. [69] suggested that males are more susceptible to infection by EHVs 1 and 4, while Lara et al. [70] and Negussie et al. [101] suggested a possible role of females in the "maintenance" of the disease in the herd, as they are more susceptible to EHV infection.

Table 4. A list of *Herpesviruses* that infect the family *Equidae*.

Virus	Synonym	Subfamily	Genus	Size of Genome (kpb)	Percent Identity	Natural Host	Disease
EHV-1	*Equine abortion virus*	α	*Varicellosvirus*	150	EHV-4 55 to 96%	*Equus caballus*	Respiratory, Abortion, Neurological
EHV-2	Old *equine eytomegalovirus*	γ	*Percavirus*	184	EHV-5 60%	*Equus caballus*	Rhinitis and Conjunctivitis
EHV-3	*Coital exanthema virus*	α	*Varicellosvirus*	151	EHV-1/EHV-4/EHV-8/EHV-9 63%	*Equus caballus*	Coital Exanthema
EHV-4	*Equine rhinopneumonitis virus*	α	*Varicellosvirus*	146	EHV-1 55 to 96%	*Equus caballus*	Respiratory
EHV-5	Old *equine cytomegalovirus*	γ	*Percavirus*	179	EHV-2 60%	*Equus caballus*	Equine Multinodular Pulmonary Fibrosis
EHV-6	*Asinine herpesvirus* 1	α	*Varicellosvirus*			*Equus asinus*	Coital Exanthema
EHV-7	*Asinine herpesvirus* 2	γ	*Rhadinovirus* *			*Equus asinus*	Undefined. Virus has been found in nasal secretion of donkeys and mules
EHV-8	*Asinine herpesvirus* 3	α	*Varicellosvirus*	149	EHV-1/EHV-9 89%/92%	*Equus asinus*	Rhinitis
EHV-9	*Gazelle herpesvirus*	α	*Varicellosvirus*	148	EHV-1 86 to 95%	*Equus grevyi*	Neurological
AHV-4	*Asinine herpesvirus* 4	γ	Unclassified			*Equus asinus*	Pneumonia
AHV-5	*Asinine herpesvirus* 5	γ	Unclassified			*Equus asinus*	Pneumonia, Equine Multinodular Pulmonary Fibrosis
AHV-6	*Asinine herpesvirus* 6	γ	Unclassified			*Equus asinus*	-
ZHV	*Zebra herpesvirus*	γ	Unclassified			*Equus zebra*	Pneumonia
WAHV	*Wildass herpesvirus*	γ	Unclassified			*Equus somalicus*	-

α: Alphaherpesvirinae; γ: Gammaherpesvirinae. * It may be a member of the genus but has not yet been approved as a species. Source [88–96].

Table 5. Studies on the occurrence of equine herpesvirus, including donkeys and/or mules, from 2009 to 2019.

Disease	Species	Agent and % Positivity	Location	References
EHV	Donkey	EHV-1/24.2% (31/128)	Turkey	[104]
	Horse	EHV-1/14.5% (42/290)		
	Mule	EHV-1/37.2% (32/86)		
	Donkey	EHV-1 and 4/0% (0/4)	Brazil	[105]
	Horse	EHV-1 and 4/18.7% (140/749)		
	Mule	EHV-1 and 4/6.8% (5/73)		
	Donkey	EHV-4/69.7% (134/192)	Bulgaria	[57]
	Donkey	EHV-1/20.2% (21/104) EHV-4/84.6% (88/104)	Ethiopia	[21]
	Horse	EHV-1/7% (7/100) EHV-4/91% (91/100)		
	Mule	EHV-1/0% (0/4) EHV-4 100% (4/4)		
	Donkey	EHV-1/33.3% (4/12) EHV-9/0% (0/12)	Tanzania and Namibia	[106]
	Somali wild ass	EHV-1/73.6% (14/19) EHV-9/5.2% (1/19)		
	Zebra	EHV-1/61.8% (55/89) EHV-9/29.2% (26/89)		
	Donkey	EHV-1/2 animals	EUA	[107]
	Horse	EHV-1/115 animals		
	Mule	EHV-1/2 animals		
	Donkey	EHV-1/51.85% (126/243) EHV-4/64.2% (156/243)	Turkey	[108]
	Horse	EHV-1/52.48% (222/423) EHV-4/83.69% (354/423)		
	Donkey	EHV-1 and 4/69.5% (57/82)	Sudan	[109]
	Horse	EHV-1 and 4/65.9% (83/126)		
	Donkey Mule	EHV-4/14.8% (16/108)	Iran	[110]
	Donkey	EHV-1/82 animals	Ethiopia	[101]
	Horse	EHV-1/6 animals		
	Mule	EHV-1/3 animals		
	Donkey	EHV-1 and 4/74.7% (201/269)	Ethiopia	[111]
	Horse	EHV-1 and 4/66.7% (64/96)		
	Mule	EHV-1 and 4/50% (6/12)		
	Donkey	EHV-1/10% (4/40) EHV-4/53% (21/40)	West Indies	[69]
	Horse	EHV-1/27.6% (39/140) EHV-4/90% (126/140)		
	Donkey	47% (40/85)	Brazil	[70]

In donkeys and mules, an infection with a gammaherpesviruses similar to *Asinine herpesvirus type 5* (AHV-5), isolated from a pharynx swab of a donkey with neurological disease in Ontario, Canada have been reported [112]. Beyond this, AHV type 4 and 5 are also considered potential causes of pulmonary fibrosis, which is a common disease in elderly donkeys [98]. AHV-5 has also been detected in donkeys with interstitial pneumonia, similar to AHV-4 [113]. *Equine herpesvirus type 7* (EHV-7), also known as *asinine herpesvirus type 2* (AHV-2), was isolated from the blood of a healthy donkey and nasal secretions of a mule after an outbreak of respiratory disease [114,115]. The first AHV-2 described in 1988 was isolated from a healthy donkey, but when inoculated into two newly weaned donkeys, it produced signs of acute rhinitis [115]. Another study showed that EHV-7 was recovered from nasal secretions in approximately 8% of healthy mules (n = 114) and donkeys (n = 13) [114]. *Asinine herpesvirus type* 3 (AHV-3) is a donkey virus that induces mild rhinitis in this species. Donkeys can also have lesions similar to the equine-coital rash, which can be caused by AHV-1 [116].

There are no specific therapeutic or biosafety procedures for AHVs. The measures that must be taken for donkeys infected with some herpesviruses must follow the protocols developed for the treatment of EHV in horses [98]. However, the unique physiology of donkeys must not be ignored. For example, respiratory infections can be aggravated because of hyperlipidemia, which is common in donkeys [11].

6. Flaviviral Encephalitis

The *West Nile virus* (WNV), *Saint Louis encephalitis virus* (SLEV) and *Japanese encephalitis virus* (JEV) are arboviruses belonging to the genus *Flavivirus* of the family *Flaviviridae* [117]. These viruses are transmitted by the bite of hematophagous mosquitoes and cause fatal neurological conditions in humans and horses [117]. In nature, the transmission cycle of the encephalitis, caused by those three viruses, is maintained primarily by biological vectors of genus *Culex*. Birds of the order Passeriformes may act as reservoirs and amplifiers of the WNV [118], just as water birds and pigs are potential reservoirs for JEV [119]. Humans and equines are considered accidental end-hosts and are not able to produce enough viremia to infect new vectors [117].

The neurologic clinical signs caused by these infections in horses are similar or identical to the signs induced by several other pathogens, including EHV-1 and *Rabies virus* [120]. Therefore, laboratory tests are crucial for an accurate diagnosis. Immunoenzymatic assays (i.e., ELISAs) are increasingly used because they are fast and inexpensive. However, most flaviviruses are antigenically related, which results in cross-reactivity between them, especially the SLEV, JEV, and WNV, which makes serological diagnosis more difficult [117,121]. It is often necessary to use additional confirmatory tests, especially in certain regions where there is co-circulation of those related viruses, with viral neutralization tests and plaque reduction tests (PRNT) being the most specific [119,122]. RT-PCR-based techniques can be used for direct detection of the virus and are highly specific. However, the transient nature of flavivirus viremia represents an important limitation for the use of this methodology. WNV infections, for example, have a single viremic phase of 4 to 6 days in horses. Therefore, due to the possibility of false-negative results by PCR, confirmation by serological tests is necessary, usually ELISAs for anti-viral IgM are used for this purpose [118,122].

Serological evidence of JEV infection has been reported in mules in the Himalayas and donkeys in Pakistan [123,124]. Data from the World Organization for Animal Health (OIE) show that JEV has already been identified in India, Japan, Korea, Bangladesh, Laos, Philippines, Timor-Leste, China, Papua New Guinea, and Australia [125]. However, there is no description of the affected species. JEV is a major public health problem in Asia as it is the leading cause of viral encephalitis in humans, with approximately 68,000 clinical cases per year [119]. Approximately 50% of these cases occur in China [126], where cases in equid are also frequent [125]. China is home to one of the largest populations of donkeys and mules in the world [5] but has shown a significant decline due to the deliberate culling of these species to obtain ejiao (Chinese gelatin made from donkey skin) raising the concern of experts [127]. The occurrence of infectious diseases can further aggravate this situation, as is the case of Japanese encephalitis, which has a high mortality rate, reaching up to 30–40% in more severe cases in horses, with impacts undescribed for other equids species [119].

The SLEV circulation seems to be restricted to the Americas, and serological studies in horses show its occurrence in Caribbean countries [128], Central America [129] and South America [121,130]. However, in donkeys and mules, the occurrence of the disease has only been described in Panama [129] and in two states in Brazil, Mato Grosso do Sul and Minas Gerais [131,132].

According to the OIE, the occurrence of WNV has already been reported in countries on all continents, except Antarctica, with emphasis on Israel, Hungary, Guatemala, Haiti, Canada, and the

USA, which has presented cases for several consecutive years [125]. WNV infection has been described in a donkey in the south of France with neurological signs, which showed a short period of remission, followed by severe liver failure [133]. The virus has already been identified in the central nervous system of symptomatic donkeys in Brazil, who had muscle tremors, dysphagia, anterior limb ataxia, lateral decubitus, and rowing movements in the first 24 h, followed by paralysis of the pelvic limbs, loss of sensation in the spine, and mandibular trismus [134]. In the USA, three mules with clinical signs and diagnosed with WNV survived, and none of them remained in decubitus in the course of the disease, while a donkey failed to recover [135]. Antibodies have also been found in clinically healthy animals in several countries (Table 6).

Table 6. *West Nile virus* (WNV) seroprevalence studies that include donkeys and/or mules.

Disease	Species	% Positivity	Location	References
WNV	Donkey Horse	20% (14/70) 0.8% (1/118)	Anatolian Province (Turkey)	[136]
	Donkey Horse	1.28% (3/234) 4.15% (27/650)	Turkey	[137]
	Equids [1]	28% (332/1189)	Tunisia	[138]
	Donkey Horse	8.6% (5/58) 11.5% (11/96)	Borno State (Nigeria)	[139]
	Donkey Horse	55.4% (249/449)	Punjab and Khyber Pakhtunkhwa Provinces (Pakistan)	[124]
	Donkey Horse	86.2% (25/29) 68.7% (44/64)	Northwest Senegal	[140]
	Donkey Horse	14.4% (33/222) 26.8% (19/71)	Northeast Algeria	[141]
	Donkey Horse	39.3% (50/127) 82.6% (380/460)	Palestine and Israel	[142]
	Donkey Horse	15% (6/40) 18.6% (26/140)	Leeward Islands (West Indies)	[69]
	Donkey Horse	12.7% (83/150) 20.7% (83/400)	Northern Egypt	[143]
	Donkey Horse/Pony	47.6% (10/21) 22.3% (69/309)	Guadaloupe Archipelago	[144]
	Donkey/Mule Horse	25% (4/16)	Chiapas and Puebla States (Mexico)	[145]
	Equids [1]	1.8% (4/217)	Mato Grosso State (Brazil)	[146]

[1] Authors did not specify the evaluated species.

Some authors suggest that donkeys and mules may be more resistant to WNV infection or may have milder clinical signs after exposure, even though severe signs may eventually occur [135,147]. Yildirim et al. [136] detected viral RNA in 28.5% (4/14) of seropositive donkeys, suggesting infection with no clinical signs. During an outbreak in Spain, WNV-positive donkeys and mules were detected in municipalities where clinical cases had not been reported, suggesting that there is greater geographical spread than previously thought and that these species can be used as sentinels for the WNV since this resistance to the virus supposedly allows the greater permanence of these animals in the herd [147].

Other studies have shown that the seroprevalence of WNV may be higher in donkeys than in horses (Table 6). Yildirim et al. [136] attributed this finding to the fact that the donkeys in the study were used mainly for the transport of water and therefore presented a greater risk of vector exposure. Davoust et al. [140] detected greater seroprevalence in donkeys than in horses, but with no statistically significant difference, while Bargoui et al. [138] did not observe a difference in the seroprevalence rate between both species, which was already expected since the animals were in similar environmental conditions and were used for the same purpose. On the other hand, some studies show a trend of greater seroprevalence in horses than in donkeys, but without statistical significance [124,143]. This is a complicated relationship to establish and a sampling discrepancy is common, and most studies do not fully explore the results obtained for donkeys and mules, while some do not differentiate between species in their data. It seems that both species have the same chance of being infected, but the rate of positivity can be influenced by factors such as geographical location, population size, and different uses of each animal species.

Neurological disorders are important health and economic threats to the equine industry. Publications from the last 20 years indicate that this morbidity represents the fifth largest cause of death among adult horses [148]. Nevertheless, little research has been carried out to evaluate the burden of neurological diseases in this species and even less in donkeys and mules, which are abundant in regions of the world that are conducive to the existence of infections due to high loads of arthropod vectors [116].

7. Equine Influenza

Influenza viruses makes up a large group of strains directly associated with severe respiratory infections in several species. *Influenza virus* is a member of the family *Orthomyxoviridae* and in the genus *Influenza* [149]. As the virus is airborne and is highly variable, it is constantly the target of concern related to the possible pandemics [150]. The *Influenza virus* has been classified into four types: A, B, C, and D, based on the matrix and nucleoprotein genes. Type A viruses infect animals and humans, whereas type B and C viruses infect only humans. Subtype D has been reported in pigs, cattle, sheep, and goats [151].

Equine influenza virus (EIV), classified as type A, is considered one of the most important viral respiratory pathogens, causing a disease with high morbidity. Equine influenza disease is mainly caused by the virus subtypes, H7N7, first recognized in 1956 [152] and H3N8, isolated for the first time in 1963 [153]. Since then, the increase in the transit of infected animals without clinical signs and which are subjected to inadequate quarantine procedures, has caused EIV to spread worldwide, except in a small number of island countries, including New Zealand and Iceland [154].

Once in the host, EIV has a tropism for the ciliated cells of the upper and lower respiratory tract, inducing necrosis of the epithelial cells, exudation of fluids rich in proteins, and agglomeration of the cilia [155]. The onset of clinical signs occurs between 5 and 14 days after infection. In horses, they are characterized by dry cough, fever, lethargy, anorexia, enlarged lymph nodes, tachycardia, hyperemia of the airways and conjunctival mucous membranes, serous nasal discharge, edema of the limbs, pain, muscle stiffness, and abortions [156]. Eventually, infected animals may develop more severe clinical signs, characterized by myocarditis and chronic obstructive pulmonary disease. Despite the high morbidity, the mortality rate is low and usually occurs in cases of pneumonia with sequelae [157].

Donkeys and mules, once affected, develop clinical manifestations similar to horses; however, in donkeys, the symptoms are more clinically severe with signs of typical broncho-interstitial pneumonia characterized by necrotizing bronchiolitis, hemorrhages, the presence of extremely swollen alveoli, and fibrinous exudation are commonly observed [158]. There is evidence from outbreaks in China between 1993 and 1994, that while mules and horses display mild clinical signs, donkeys infected under the same conditions of exposure to the virus were severely affected, and most of them died [149,159]. The greater susceptibility of donkeys to EIV infection is due to the greater propensity of these animals to develop bacterial bronchopneumonia [160]. Nevertheless, Rose et al. [161] attributed the greater severity of clinical signs seen in these animals, with the recurrent coinfection with pulmonary nematodes *Dictyocaulus arnfieldii*.

Although, approximately 50.5% of the world equine population corresponds to donkeys and mules [5], so far, infection by EIV in these animals is considered rare, even with the greater predisposition of donkeys to develop the severe form of the disease when infected. Epidemiological data regarding the presence of EIV in donkeys are limited to outbreaks of respiratory disease, caused by the H3N8 sub-lineage in Xinjuang in 2007, which affected approximately 13,600 animals [162], as well as in Shandong in 2017, limited to one property, and 300 seropositive animals were affected with a 25% mortality [158]. More recently, in 2020, in the city of Liaocheng, 120 unvaccinated donkeys showed an average seropositivity of 32.5% [163]. All of these outbreaks have been described in China's provinces, indicating that the EIV is a major threat to the country's large donkey farms.

Currently, vaccination is the most effective strategy, along with isolation measures, animal traffic control, and the adoption of basic biosafety measures to prevent EIV infections or limit their consequences [164,165]. However, as they are not mandatory, vaccination against EIV in herds, mainly of donkeys and mules, which are neglected species, is uncommon. However, we believe that the continuous circulation of EIV emphasizes the need for effective surveillance in herds of equines, which includes maintaining the updated vaccination schedule and applying diagnostic tests to detect subclinical cases before the animals move between different locations.

General information (taxonomy, natural host, transmission, diagnostic method and prevention) of the aforementioned viruses is summarized in Table 7.

8. Other Viral Diseases That Affect Donkeys

Currently, there are several diseases affecting members of the Equidae family worldwide considered of compulsory notification by the OIE; however only eleven are caused by viruses, and these include the equine infectious anemia, the equine encephalomyelitis caused by flaviviruses or alphaviruses, the equine influenza, infection with EHV-1 and *Alphaarterivirus equid*, as well as the African horse sickness and rabies [168]. Although several studies have demonstrated that donkeys and mules as well as horses are also affected by these diseases, almost all experimental studies, including epidemiological surveys, are carried out on horses, and there is little information about the pathogenesis of these diseases in donkeys and mules.

9. Rabies

Rabies is a zoonotic disease caused by *Rabies virus* (RABV), a highly neurotropic *Lyssavirus*, which belongs to the family *Rhabdoviridae*, order *Mononegavirales*, whose main transmission mechanism is through the bite of infected carnivorous animals (e.g., dogs, jackals, hyenas, and foxes) or blood-sucking bats [11].

Equidae are the animals most susceptible to RABV [169] and the first clinical signs usually appear between 2 and 9 weeks after infection and are initially characterized by fever, mild lameness, and cramps. However, with active replication of the virus, Equidae when infected can develop three distinct forms of rabies: the mute form characterized by muscle tremors, excessive salivation, ataxias, and depression [170], the paralytic form with ascending paralysis, loss of tone of the tail and anal sphincter [171] and the furious form, characterized by hydrophobia, photophobia, hypersensitivity to touch, difficulty in swallowing, aggressiveness, and self-mutilation [172]. In mules and donkeys, clinical signs such as abnormal vocalization, hyperesthesia, colic, self-bites, tendency to bite other animals, restlessness, and excessive salivation, are commonly observed, followed by a flaccid tail, phimosis, complete decubitus, and death [173].

Although some studies demonstrate that mainly working donkeys are an important source of transmission of RABV to humans, due to recurrent accidents involving deep wounds caused by bites which result in large scale loss of tissue, especially in children [174–176], the occurrence of rabies in donkeys is still limited to isolated cases in Canada, China and some countries in Africa and the Middle East [173,177–180].

Table 7. Summary of equine viral infectious diseases, listing the infectious agents, their taxonomic grouping, natural hosts, clinical characterization, transmission mechanisms, diagnosis methods and the precautions required to prevent their spread, as determined by World Organisation for Animal Health (OIE) terrestrial manual.

Virus	Synonym	Family	Genus	Genome		Natural Host	Disease	Transmission	Diagnostic Method	Prevention
				Type	Size (kbp)					
EIAV	*Equine infectious anemia virus*	*Retroviridae*	*Lentivirus*	ssRNA+	8.2	*Equus caballus Equus asinus*	Acute and Chronic Disease	Inoculation of contaminated blood by blood-feeding vectors and fomites	AGID, ELISA and IB *	Application of effective biosecurity measures and vector control
EAV	*Equine arteritis virus*	*Arteriviridae*	*Alphaartevirus*	ssRNA+	~12.7	*Equus caballus Equus asinus*	Respiratory and Reprodutive Disordrs	Respiratory or veneral routes	VI, RT-PCR, RT-qPCR, immunohistochemistry, or serological assays (ELISA, PRNT)	Application of effective biosecurity measures and vaccination
NPHV	*Nonprimate hepacivirus* or *Equine hepacivirus*	*Flaviviridae*	*Hepacivirus*	ssRNA+	9.5	*Equus caballus*	Subclinical Hepatitis	Blood/serum transfusion and possibly by vertical route	Unstandardized	Test of blood/sérum before transfusion
WNV	*West Nile virus*					Aquatic and passeriformes birds	Neurological Disease and Possibly Death			Vector control and vaccination (available for horses)
SLEV	*Saint Louis encephalitis virus*	*Flaviviridae*	*Flavivirus*	ssRNA+	~11.0	Birds	Neurological signs can be mild or severe. Death rarely ocorrs.	Bite of Culex spp.	PRNT, PCR and/or ELISA-IgM	Vector control. Vaccine is not available
JEV	*Japanese encephalitis virus*					Pigs and quatic birds	Neurological disease and possibly death			Vector control and vaccination (available for horses, swin and human)
IV	*Influenza virus*	*Orthomyxoviridae*	*Influenza*	ssRNA-	8 ssRNA-segments **	*Equus caballus Equus asinus*	Respiratory disese	Respiratory routes and fomites	Clinical diagnosis, VI, influenza A antigen detection, haemagglutinin inhibition, and qPCR	Application of effective biosecurity measures and vaccination

AGID: agar gel immunodiffusion; ELISA: enzyme-linked immunosorbent assay; IB: Western immunoblotting test; VI: virus isolation; PCR: polymerase chain reaction; RT-PCR: reverse transcription–PCR; RT-qPCR: real-time RT-PCR; qPCR: real-time quantitative PCR; PRNT: plaque reduction neutralization test. * Confirmatory technique, but it is not commercially available. ** EIV virion consists of eight single stranded negative sense RNA segments ranging from 866 to 2314 pb. Source: [24,62,117–119,155,166,167].

10. Alphaviral Encephalitis

The *Eastern equine encephalitis virus* (EEEV), *Western equine encephalitis virus* (WEEV), and *Venezuelan equine encephalitis virus* (VEEV) are enveloped, single-stranded, positive-sense RNA viruses that belong to the genus *Alphavirus* in the family *Togaviridae*, that cause neurological disease in equids (horses, mules, donkeys and zebras) and humans [181]. The EEEV and WEEV have been reported in several countries in the Americas, while VEE seems to be limited to South and Central Americas [182,183]. They are transmitted by mosquitoes, as biological vectors. Passerine birds (EEEV and WEEE) and rodents (VEEV) are the main reservoir hosts, with high viremia to infect vectors, while equids and humans are considered dead-end hosts, except for VEEV, where equids are efficient amplificators of the virus and can act as a source of infection for mosquito transmission [117,182].

These viruses affect the nervous system, so the clinical signs can be identical for EEV, VEEV and WEEV, with severe neurologic disease and frequently death [120]. The mortality rate of EEEV and VEEV infection can reach up to 90% in most of the equine cases, while WEEV is least lethal in horses, with a mortality rate of approximately 30% [184,185]. Donkeys and mules exhibit similar susceptibility and clinical signals to horses [116,186]. In 1943, cases of fatal VEEV had occurred among horses, asinines and mules in Trinidad. A suspension of donkey brain tissue was prepared and injected intra-cerebrally into guinea pigs, originating the virulent Trinidad Donkey strain, from which was developed the attenuated vaccinal TC-83 strain [117,186]. EEEV infection in donkeys and mules was reported in Brazil, with a fatal case in a donkey [187]. There has been a dramatic reduction in cases of WEEV during recent decades, with no outbreaks between equids reported since 1999 [183]. Even for EEEV and VEEV, reports on donkeys and mules are scarce. The neglect of these species can have negative impacts, especially regarding VEEV, since they also are important amplifiers of the virus. For this reason, donkeys and mules should be included in vaccination programs. This is more relevant in countries such as Brazil, which has one of the biggest populations of feral donkeys in the world [5].

11. African Horse Sickness

African horse sickness (AHS) is a disease caused by the *African horse sickness virus* (AHSV), an *Orbivirus* belonging to the *Reoviridae* family, non-contagious, transmitted by mosquitoes of the genus *Culicoides*. AHSV has a predilection for vascular endothelial cells, and equines when infected can develop up to four distinct forms of the disease: the cardiac form characterized mainly by edema, ocular and tongue hemorrhages, the pulmonary form where the infected animal has difficulty breathing, dyspnea, cough, and sweating, the mixed form that occurs in a combination of the cardiac and pulmonary forms and the form named sick fever characterized by moderate fever and some edema of the supraorbital fossae [188]. Although the risk of mortality in affected horses may reach 95% [189,190], experimental infections carried out on African donkeys have shown the presence of only minimal and mild histopathological lesions characterized by the accumulation of fluid in the peritoneal cavity, petechiae, and equimotic hemorrhages in the left hepatic ligament. Due to this, it is widely accepted that feral donkeys and mules are resistant and most of these animals, when affected, become asymptomatic carriers of the virus [191].

Currently, AHS is a disease restricted to sub-Saharan Africa and epidemiological data from Zimbabwe and Ethiopia, both African countries, have a prevalence ranging from 59.3% to 75% in donkeys as well as 55.5% in mules [192,193]. Although typically limited to North Africa and Middle Eastern countries, epizootic events affected Spain in 1987, after the importation of several infected zebras with no apparent clinical signs. However, in 1988, the disease reemerged in the country and killed about 13,000 horses. Studies have indicated that viral reemergence in Spain occurred due to the clinical resistance of mules and donkeys with no apparent clinical signs and inaccurate detection which facilitated disease spread in the region [194]. Recently, a new outbreak of AHS was reported in different provinces of Thailand, infecting about 422 animals and leading to 386 deaths. Although AHS is not a transmissible disease, possibly the number of infected animals is still increasing, due to the large population of insect vectors in that region [195], representing a significant threat to other

Asian countries, including China, which has the largest equids population in the world [196] and about 6.8 million animals without a history of AHS vaccination [197].

Although the presence of infections caused by RABV and AHSV are compulsory notifications, the low economic importance of donkeys and mules, when compared to horses, make studies on the pathogenesis of diseases that affect the Equidae family, increasingly scarce for these species and, consequently, the world population of 50.4 million donkeys and 8.5 million mules poses a significant risk, acting as a reservoir for the maintenance and transmission of important pathogenic viruses. Considering their unique physiology and their natural resistance to the development of clinical signs when affected by some diseases, it would be important for donkeys and mules to have the same focus of interest when it comes to infection since they are abundant in number and have a close coexistence with horses in various regions of the world.

12. Conclusions

Most of the viral diseases affecting the family *Equidae* are considered compulsory to be reported to the OIE [168]. In the absence of an effective vaccine, the control of an infectious disease is dependent on breaking the transmission cycle through the use of highly sensitive and specific diagnostic methods capable of detecting positive animals, followed by segregation of them from the rest of the herd, and reporting to state and/or government authorities.

Donkeys and mules are susceptible to infection by viruses such as EIAV, *Alphaarterivirus equid*, NPHV, EHVs, WNV, SLEV, JEV, EIV, RABV, EEEV, WEEV, VEEV, and AHS. However, in many cases, when infected, they show resistance/low susceptibility to the development of clinical signs. In this situation, infected animals are hardly identified and remain for long periods as possible sources of infection for horses and other species, including humans. In addition, with regard to the natural "resistance" of donkeys and mules linked to their growing devaluation, there is a significant deficit of studies related to the epidemiology and pathogenesis of viral infectious diseases in these animals. As a result, the control of infectious diseases in donkeys and mules is most often compromised. Furthermore, it would be important to create robust funding for studies of infectious diseases in donkeys and mules with the development of diagnostic tests specific to them, when necessary, and that the OIE and regional agencies in countries where donkeys and mules are abundant adopt preventive measures against the viral spread in these animals, with compulsory disease notification in these populations.

Author Contributions: Writing—original draft preparation, R.J.F.C., B.L.B., C.F.R.; writing—review and editing, R.J.F.C., B.L.B., C.F.R., J.K.P.d.R., S.M.S., U.B.R.B.; visualization, R.J.F.C., B.L.B., C.F.R., J.K.P.d.R., S.M.S., U.B.R.B.; supervision, J.K.P.d.R.; project administration, J.K.P.d.R. All authors have read and agreed to the published version of the manuscript.

Funding: This work received no external funding.

Acknowledgments: This work was supported by Conselho Nacional de Desenvolvimento Científco e Tecnológico (CNPq) and Coordenação de Aperfeiçoamento de Pessoal de Nível Superior—Brazil (CAPES). JKPR is CNPq fellowship recipient.

Conflicts of Interest: The authors declare no conflict of interest.

References

1. Jónsson, H.; Schubert, M.; Seguin-Orlando, A.; Ginolhac, A.; Petersen, L.; Fumagalli, M.; Albrechtsen, A.; Petersen, B.; Korneliussen, T.S.; Vilstrup, J.T.; et al. Speciation with gene flow in equids despite extensive chromosomal plasticity. *Proc. Natl. Acad. Sci. USA* **2014**, *111*, 18655–18660. [CrossRef] [PubMed]
2. Burden, F.; Thiemann, A. Donkeys are different. *J. Equine Vet. Sci.* **2015**, *35*, 376–382. [CrossRef]
3. Matthews, N.; van Loon, J.P.A.M. Anaesthesia and analgesia of the donkey and the mule. *Equine Vet. Educ.* **2013**, *25*, 47–51. [CrossRef]
4. Matthews, N.S.; Taylor, T.S.; Hartsfield, S.M. Anaesthesia of donkeys and mules. *Equine Vet. Educ.* **2010**, *15*, 102–107. [CrossRef]
5. FAO. FAOSTAT. Available online: http://www.fao.org/faostat/en/#data/QA (accessed on 11 May 2020).

6. Toribio, R.E. Dear Donkey and Mule: You Deserve More Appreciation and Better Medicine. *Vet. Clin. N. Am. Equine Pract.* **2019**, *35*, 13–14. [CrossRef]
7. Camillo, F.; Rota, A.; Biagini, L.; Tesi, M.; Fanelli, D.; Panzani, D. The Current Situation and Trend of Donkey Industry in Europe. *J. Equine Vet. Sci.* **2018**, *65*, 44–49. [CrossRef]
8. Orlando, L.; Ginolhac, A.; Zhang, G.; Froese, D.; Albrechtsen, A.; Stiller, M.; Schubert, M.; Cappellini, E.; Petersen, B.; Moltke, I.; et al. Recalibrating equus evolution using the genome sequence of an early Middle Pleistocene horse. *Nature* **2013**, *499*, 74–78. [CrossRef]
9. Alberghina, D.; Fazio, F.; Arfuso, F.; Scianò, S.; Zumbo, A.; Piccione, G. Reference intervals of serum protein concentrations from clinically healthy female Ragusana Donkeys (*Equus asinus*) determined by cellulose acetate electrophoresis. *J. Equine Vet. Sci.* **2013**, *33*, 433–436. [CrossRef]
10. Mendoza, F.J.; Perez-Ecija, R.A.; Toribio, R.E.; Estepa, J.C. Thyroid hormone concentrations differ between donkeys and horses. *Equine Vet. J.* **2013**, *45*, 214–218. [CrossRef]
11. Evans, L.; Crane, M. (Eds.) *The Clinical Companion of the Donkey*, 1st ed.; The Donkey Sanctuary: Devon, UK, 2018.
12. Dunkel, B. Disorders of the Hematopoietic System. In *Equine Internal Medicine*; Reed, S.M., Bayly, W.M., Sellon, D.C., Eds.; Elsevier: St. Louis, MO, USA, 2018; pp. 991–1028.
13. OIE. World Organisation for Animal Health. Terrestrial Animal Health Code. Available online: https://www.oie.int/index.php?id=169&L=0&htmfile=chapitre_diagnostic_tests.htm (accessed on 8 November 2020).
14. Timoney, P.J. Factors influencing the international spread of equine diseases. *Vet. Clin. N. Am. Equine Pract.* **2000**, *16*, 537–551. [CrossRef]
15. Arbovirus Catalog—CDC Division of Vector-Borne Diseases (DVBD). Available online: https://wwwn.cdc.gov/arbocat/ (accessed on 8 November 2020).
16. Center for Disease Control and Prevention Rabies (CDC). National Center for Emerging and Zoonotic Infectious Diseases (NCEZID), Division of High-Consequence Pathogens and Pathology. 2019. Available online: https://www.cdc.gov/rabies/ (accessed on 8 November 2020).
17. Burbelo, P.D.; Dubovi, E.J.; Simmonds, P.; Medina, J.L.; Henriquez, J.A.; Mishra, N.; Wagner, J.; Tokarz, R.; Cullen, J.M.; Iadarola, M.J.; et al. Serology-Enabled Discovery of Genetically Diverse Hepaciviruses in a New Host. *J. Virol.* **2012**, *86*, 6171–6178. [CrossRef] [PubMed]
18. Issel, C.J.; Coggins, L. Equine infectious anemia: Current knowledge. *J. Am. Vet. Med. Assoc.* **1979**, *174*, 727–733. [PubMed]
19. Almeida, V.M.A.; Gonçalves, V.S.P.; Martins, M.F.; Haddad, J.P.A.; Dias, R.A.; Leite, R.C.; Reis, J.K.P. Anemia infecciosa eqüina: Prevalência em eqüídeos de serviço em Minas Gerais. *Arq. Bras. Med. Vet. Zootec.* **2006**, *58*, 141–148. [CrossRef]
20. Maresca, C.; Scoccia, E.; Faccenda, L.; Zema, J.; Costarelli, S. Equine Infectious Anemia: Active Surveillance in Central Italy 2007–2009. *J. Equine Vet. Sci.* **2012**, *32*, 596–598. [CrossRef]
21. Getachew, M.; Alemayehu, F.; Chala, C.; Amare, B.; Kassa, D.; Burden, F.; Wernery, R.; Wernery, U. A cross-sectional sero-survey of some infectious diseases of working equids in Central Ethiopia. *J. Vet. Med. Anim. Health* **2014**, *6*, 231–238. [CrossRef]
22. Oliveira, F.G.; Cook, R.F.; Naves, J.H.F.; Oliveira, C.H.S.; Diniz, R.S.; Freitas, F.J.C.; Lima, J.M.; Sakamoto, S.M.; Leite, R.C.; Issel, C.J.; et al. Equine infectious anemia prevalence in feral donkeys from Northeast Brazil. *Prev. Vet. Med.* **2017**, *140*, 30–37. [CrossRef]
23. Wegdan, H.A.; Sahar, M.E.; Ballal, A.; Intisar, K.S.; Shaza, M.M.; Algezoli, O.A.; Ihsan, H.A.; Baraa, A.M.; Taha, K.M.; Nada, E.M.; et al. Sero Prevalence of Equine Infectious Anemia (EIA) Virus in Selected Regions in Sudan. *Microbiol. Res. J. Int.* **2017**, 1–6. [CrossRef]
24. Cook, R.F.; Leroux, C.; Issel, C.J. Equine infectious anemia and equine infectious anemia virus in 2013: A review. *Vet. Microbiol.* **2013**, *167*, 181–204. [CrossRef]
25. Williams, D.L.; Issel, C.J.; Steelman, C.D.; Adams, W.V., Jr.; Benton, C.V. Studies with equine infectious anemia virus: Transmission attempts by mosquitoes and survival of virus on vector mouthparts and hypodermic needles, and in mosquito tissue culture. *Am. J. Vet. Res.* **1981**, *42*, 1469–1473.
26. Borges, A.M.C.M.; Silva, L.G.; Nogueira, M.F.; Oliveira, A.C.S.; Segri, N.J.; Ferreira, F.; Witter, R.; Aguiar, D.M. Prevalence and risk factors for Equine Infectious Anemia in Poconé municipality, northern Brazilian Pantanal. *Res. Vet. Sci.* **2013**, *95*, 76–81. [CrossRef]

27. Bolfa, P.; Nolf, M.; Cadoré, J.L.; Catoi, C.; Archer, F.; Dolmazon, C.; Mornex, J.F.; Leroux, C. Interstitial lung disease associated with Equine Infectious Anemia Virus infection in horses. *Vet. Res.* **2013**, *44*. [CrossRef] [PubMed]
28. Bueno, B.L.; Câmara, R.J.F.; Moreira, M.V.L.; Galinari, G.C.F.; Souto, F.M.; Victor, R.M.; Bicalho, J.M.; Ecco, R.; dos Reis, J.K.P. Molecular detection, histopathological analysis, and immunohistochemical characterization of equine infectious anemia virus in naturally infected equids. *Arch. Virol.* **2020**. [CrossRef] [PubMed]
29. Cullinane, A.; Quinlivan, M.; Nelly, M.; Patterson, H.; Kenna, R.; Garvey, M.; Glldea, S.; Lyons, P.; Flynn, M.; Galvin, P.; et al. Diagnosis of equine infectious anaemia during the 2006 outbreak in Ireland. *Vet. Rec.* **2007**, *161*, 647–652. [CrossRef] [PubMed]
30. Cook, R.F.; Cook, S.J.; Berger, S.L.; Leroux, C.; Ghabrial, N.N.; Gantz, M.; Bolin, P.S.; Mousel, M.R.; Montelaro, R.C.; Issel, C.J. Enhancement of equine infectious anemia virus virulence by identification and removal of suboptimal nucleotides. *Virology* **2003**, *313*, 588–603. [CrossRef]
31. Russell, K.E.; Walker, K.M.; Miller, R.T.; Sellon, D.C. Hyperglobulinemia and lymphocyte subset changes in naturally infected, inapparent carriers of equine infectious anaemia virus. *Am. J. Vet. Res.* **1998**, *59*, 1009–1015.
32. Sellon, D.C. Equine Infectious Anemia. *Vet. Clin. N. Am. Equine Pract.* **1993**, *9*, 321–336. [CrossRef]
33. Issel, C.J.; Cook, R.F.; Mealey, R.H.; Horohov, D.W. Equine infectious anemia in 2014: Live with it or eradicate it? *Vet. Clin. N. Am. Equine Pract.* **2014**, *30*, 561–577. [CrossRef]
34. Leroux, C.; Cadoré, J.-L.; Montelaro, R.C. Equine Infectious Anemia Virus (EIAV): What has HIV?s country cousin got to tell us? *Vet. Res.* **2004**, *35*, 485–512. [CrossRef]
35. Spyrou, V.; Papanastassopoulou, M.; Psychas, V.; Billinis, C.; Koumbati, M.; Vlemmas, J.; Koptopoulos, G. Equine infectious anemia in mules: Virus isolation and pathogenicity studies. *Vet. Microbiol.* **2003**, *95*, 49–59. [CrossRef]
36. Dong, J.B.; Zhu, W.; Cook, F.R.; Goto, Y.; Horii, Y.; Haga, T. Development of a nested PCR assay to detect equine infectious anemia proviral DNA from peripheral blood of naturally infected horses. *Arch. Virol.* **2012**, *157*, 2105–2111. [CrossRef]
37. Cook, S.J.; Cook, R.F.; Montelaro, R.C.; Issel, C.J. Differential responses of *Equus caballus* and *Equus asinus* to infection with two pathogenic strains of equine infectious anemia virus. *Vet. Microbiol.* **2001**, *79*, 93–109. [CrossRef]
38. Hvri-Harbin Veterinary Research Institute, CAAS. Available online: http://www.hvri.ac.cn/en/aboutus/athvri/index.htm (accessed on 12 May 2020).
39. Shen, R.X.; Xu, Z.D.; HE YS, Z.S. Development of a EIAV donkey leucocyte attenuated vaccine. *China Agric. Sci.* **1979**, *4*, 1–15.
40. Wang, X.F.; Lin, Y.Z.; Li, Q.; Liu, Q.; Zhao, W.W.; Du, C.; Chen, J.; Wang, X.; Zhou, J.H. Genetic Evolution during the development of an attenuated EIAV vaccine. *Retrovirology* **2016**, *13*, 9. [CrossRef] [PubMed]
41. Yin, X.; Lin, Y.; Cai, W.; Wei, P.; Wang, X. Comprehensive analysis of the overall codon usage patterns in equine infectious anemia virus. *Virol. J.* **2013**, *10*, 356. [CrossRef] [PubMed]
42. Tang, Y.D.; Na, L.; Fu, L.H.; Yang, F.; Zhu, C.H.; Tang, L.; Li, Q.; Wang, J.Y.; Li, Z.; Wang, X.F.; et al. Double-stranded RNA-specific adenosine deaminase 1 (ADAR1) promotes EIAV replication and infectivity. *Virology* **2015**, *476*, 364–371. [CrossRef] [PubMed]
43. Tang, Y.D.; Zhang, X.; Na, L.; Wang, X.F.; Fu, L.H.; Zhu, C.H.; Wang, X.; Zhou, J.H. Double-stranded-RNA-specific adenosine deaminase 1 (ADAR1) is proposed to contribute to the adaptation of equine infectious anemia virus from horses to donkeys. *Arch. Virol.* **2016**, *161*, 2667–2672. [CrossRef]
44. Wang, H.N.; Rao, D.; Fu, X.Q.; Hu, M.M.; Dong, J.G. Equine infectious anemia virus in China. *Oncotarget* **2018**, *9*, 1356–1364. [CrossRef]
45. Meng, Q.; Lin, Y.; Ma, J.; Ma, Y.; Zhao, L.; Li, S.; Liang, H.; Zhou, J.; Shen, R.; Zhang, X.; et al. A pilot study on an attenuated Chinese EIAV vaccine inducing broadly neutralizing antibodies. *Arch. Virol.* **2011**, *156*, 1455–1462. [CrossRef]
46. Lin, Y.Z.; Shen, R.X.; Zhu, Z.Y.; Deng, X.L.; Cao, X.Z.; Wang, X.F.; Ma, J.; Jiang, C.G.; Zhao, L.P.; Lv, X.L.; et al. An attenuated EIAV vaccine strain induces significantly different immune responses from its pathogenic parental strain although with similar in vivo replication pattern. *Antivir. Res.* **2011**, *92*, 292–304. [CrossRef]
47. Italy Ordinanza Ministeriale 14 November 2006: Disposizioni Urgenti in Materia di Sorveglianza Dell'anemia Infettiva Degli Equidi. Available online: https://www.gazzettaufficiale.it/atto/serie_generale/

caricaDettaglioAtto/originario?atto.dataPubblicazioneGazzetta=2006-12-07&atto.codiceRedazionale=06A11016&elenco30giorni=false (accessed on 8 November 2020).
48. Brazil Portaria no. 378, de 17 de dezembro de 2014. In *Diário Oficial da União*; Federal Government of Brazil: Brasilia, Brazil, 2004.
49. USDA. Equine Infectious Anemia: Uniform Methods and Rules. Available online: https://aws.state.ak.us/OnlinePublicNotices/Notices/Attachment.aspx?id=106877 (accessed on 8 November 2020).
50. WHO. Equine Infectious Anemia. In *Manual of Diagnostic Tests and Vaccines for Terrestrial Animals*; Office International des Epizooties: Paris, France, 2019.
51. *Brasil Normas para Credenciamento e Monitoramento de Laboratórios de Diagnóstico de Anemia Infecciosa Equina*; Diário Oficial da República Federativa do Brasil; Secretaria de Defesa: Brasilia, Brazil, 2014; p. 136.
52. Issel, C.J.; Cook, R.F. A review of techniques for the serologic diagnosis of equine infectious anemia. *J. Vet. Diagn. Investig.* **1993**, *5*, 137–141. [CrossRef]
53. Scicluna, M.T.; Issel, C.J.; Cook, F.R.; Manna, G.; Cersini, A.; Rosone, F.; Frontoso, R.; Caprioli, A.; Antonetti, V.; Autorino, G.L. Is a diagnostic system based exclusively on agar gel immunodiffusion adequate for controlling the spread of equine infectious anaemia? *Vet. Microbiol.* **2013**, *165*, 123–134. [CrossRef] [PubMed]
54. Ricotti, S.; Garcia, M.I.; Veaute, C.; Bailat, A.; Lucca, E.; Cook, R.F.; Cook, S.J.; Soutullo, A. Serologically silent, occult equine infectious anemia virus (EIAV) infections in horses. *Vet. Microbiol.* **2016**, *187*, 41–49. [CrossRef] [PubMed]
55. Cook, R.F.; Barrandeguy, M.; Lee, P.-Y.A.; Tsai, C.-F.; Shen, Y.-H.; Tsai, Y.-L.; Chang, H.-F.G.; Wang, H.-T.T.; Balasuriya, U.B.R. Rapid detection of equine infectious anaemia virus nucleic acid by insulated isothermal RT-PCR assay to aid diagnosis under field conditions. *Equine Vet. J.* **2018**, *51*, 489–494. [CrossRef] [PubMed]
56. Issel, C.J.; Scicluna, M.T.; Cook, S.J.; Cook, R.F.; Caprioli, A.; Ricci, I.; Rosone, F.; Craigo, J.K.; Montelaro, R.C.; Autorino, G.L. Challenges and proposed solutions for more accurate serological diagnosis of equine infectious anaemia. *Vet. Rec.* **2013**, *172*, 210. [CrossRef]
57. Chenchev, I.; Rusenova, N.; Sandev, N. Seroepidemiological studies of donkeys' blood for detection of some virus infections on ungulates. *Trakia J. Sci.* **2011**, *9*, 82–86.
58. OIE. World Organisation for Animal Health. Equine Viral Arteritis (Infection with Equine Arteritis Virus). In *Manual of Diagnostic Tests and Vaccines for Terrestrial Animals*; OIE: Paris, France, 2018; pp. 1334–1349.
59. Cook, S.J.; Li, G.; Zheng, Y.; Willand, Z.A.; Issel, C.J.; Cook, R.F. Molecular Characterization of the Major Open Reading Frames (ORFs) and Enhancer Elements from Four Geographically Distinct North American Equine Infectious Anemia Virus (EIAV) Isolates. *J. Equine Vet. Sci.* **2020**, *85*, 102852. [CrossRef]
60. Balasuriya, U.B.R.; Go, Y.Y.; MacLachlan, N.J. Equine arteritis virus. *Vet. Microbiol.* **2013**, *167*, 93–122. [CrossRef]
61. ICTV. ICTV Taxonomy History: Alphaarterivirus Equid. Available online: https://talk.ictvonline.org/taxonomy/p/taxonomy-history?taxnode_id=201901827 (accessed on 8 September 2020).
62. Balasuriya, U.B.R.; Carossino, M.; Timoney, P.J. Equine viral arteritis: A respiratory and reproductive disease of significant economic importance to the equine industry. *Equine Vet. Educ.* **2018**, *30*, 497–512. [CrossRef]
63. Balasuriya, U.B.R.; Snijder, E.J.; Van Dinten, L.C.; Heidner, H.W.; Wilson, W.D.; Hedges, J.F.; Hullinger, P.J.; MacLachlan, N.J. Equine arteritis virus derived from an infectious cDNA clone is attenuated and genetically stable in infected stallions. *Virology* **1999**, *260*, 201–208. [CrossRef]
64. McCollum, W.H.; Prickett, M.E.; Bryans, J.T. Temporal distribution of equine arteritis virus in respiratory mucosa, tissues and body fluids of horses infected by inhalation. *Res. Vet. Sci.* **1971**, *12*, 459–464. [CrossRef]
65. Guthrie, A.J.; Howell, P.G.; Hedges, J.F.; Bosman, A.-M.; Balasuriya, U.B.R.; McCollum, W.H.; Timoney, P.J.; MacLachlan, N.J. Lateral transmission of equine arteritis virus among Lipizzaner stallions in South Africa. *Equine Vet. J.* **2010**, *35*, 596–600. [CrossRef] [PubMed]
66. Timoney, P.J.; McCollum, W.H.; Murphy, T.W.; Roberts, A.W.; Willard, J.G.; Carswell, G.D. The carrier state in equine arteritis virus infection in the stallion with specific emphasis on the venereal mode of virus transmission. *J. Reprod. Fertil. Suppl.* **1987**, *35*, 95–102. [PubMed]
67. Stadejek, T.; Mittelholzer, C.; Oleksiewicz, M.B.; Paweska, J.; Belák, S. Highly diverse type of equine arteritis virus (EAV) from the semen of a South African donkey: Short communication. *Acta Vet. Hung.* **2006**, *54*, 263–270. [CrossRef] [PubMed]

68. Rivas, J.; Neira, V.; Mena, J.; Brito, B.; Garcia, A.; Gutierrez, C.; Sandoval, D.; Ortega, R. Identification of a divergent genotype of equine arteritis virus from South American donkeys. *Transbound. Emerg. Dis.* **2017**, *64*, 1655–1660. [CrossRef] [PubMed]
69. Bolfa, P.; Jeon, I.; Loftis, A.; Leslie, T.; Marchi, S.; Sithole, F.; Beck, C.; Lecollinet, S.; Zientara, S.; Hans, A.; et al. Detection of West Nile Virus and other common equine viruses in three locations from the Leeward Islands, West Indies. *Acta Trop.* **2017**, *174*, 24–28. [CrossRef] [PubMed]
70. De Souza Hunold Lara, M.D.C.C.; Villalobos, E.M.C.; Cunha, E.M.S.; De Oliveira, J.V.; de Castro Nassar, A.F.; Silva, L.M.P.; Okuda, L.; Nogueira, A.H.C.; Cunha, M.S.; Marques, E.C.; et al. Occurrence of viral diseases in donkeys (*Equus asinus*) in São Paulo State, Brazil. *Braz. J. Vet. Res. Anim. Sci.* **2017**, *54*, 154. [CrossRef]
71. Gür, S.; Gürçay, M.; İrehan, B.; Turan, T. Serological investigation of equine viral arteritis infection in donkeys and horses in the Eastern Anatolia Region. *Harran Üniver. Vet. Fakültesi Derg.* **2018**, *7*, 186–191. [CrossRef]
72. Gür, S.; İrehan, B.; Gürçay, M.; Turan, T. Serological investigation of equine viral arteritis in donkeys in eastern and south-eastern anatolia regions of Turkey. *Acta Vet. Brno* **2019**, *88*, 385–391. [CrossRef]
73. McCollum, W.H.; Timoney, P.J.; Tengelsen, L.A. Clinical, virological and serological responses of donkeys to intranasal inoculation with the KY-84 strain of equine arteritis virus. *J. Comp. Pathol.* **1995**, *112*, 207–211. [CrossRef]
74. Paweska, J.T.; Binns, M.M.; Woods, P.S.A.; Chirnside, E.D. A survey for antibodies to equine arteritis virus in donkeys, mules and zebra using virus neutralisation (VN) and enzyme linked immunosorbent assay (ELISA). *Equine Vet. J.* **1997**, *29*, 40–43. [CrossRef]
75. ICTV. Genus: Hepacivirus-Flaviviridae-Positive-Sense RNA Viruses—International Committee on Taxonomy of Viruses (ICTV). Available online: https://talk.ictvonline.org/ictv-reports/ictv_online_report/positive-sense-rna-viruses/w/flaviviridae/362/genus-hepacivirus (accessed on 5 June 2020).
76. Pfaender, S.; Cavalleri, J.M.V.; Walter, S.; Doerrbecker, J.; Campana, B.; Brown, R.J.P.; Burbelo, P.D.; Postel, A.; Hahn, K.; Anggakusuma; et al. Clinical course of infection and viral tissue tropism of Hepatitis C Virus-like nonprimate hepaciviruses in horses. *Hepatology* **2015**, *61*, 447–459. [CrossRef] [PubMed]
77. Walter, S.; Rasche, A.; Moreira-Soto, A.; Pfaender, S.; Bletsa, M.; Corman, V.M.; Aguilar-Setien, A.; García-Lacy, F.; Hans, A.; Todt, D.; et al. Differential Infection Patterns and Recent Evolutionary Origins of Equine Hepaciviruses in Donkeys. *J. Virol.* **2017**, *91*, e01711–e01716. [CrossRef] [PubMed]
78. Figueiredo, A.S.; Lampe, E.; Espírito-Santo, M.P.; do Mello, F.C.A.; Almeida, F.Q.; Lemos, E.R.S.; Godoi, T.L.O.S.; Dimache, L.A.G.; Santos, D.R.L.; Villar, L.M. Identification of two phylogenetic lineages of equine hepacivirus and high prevalence in Brazil. *Vet. J.* **2015**, *206*, 414–416. [CrossRef] [PubMed]
79. Ribeiro, D.B.; Silva, J.S.; Moraes, L.A.; Ribeiro, A.S.S.; Assunção, P.W.L. Hepacivirus infection in domestic horses, Brazil, 2011–2013. *Emerg. Infect. Dis.* **2014**, *20*, 2180–2182. [CrossRef]
80. Lu, G.; Sun, L.; Xu, T.; He, D.; Wang, Z.; Ou, S.; Jia, K.; Yuan, L.; Li, S. First description of hepacivirus and pegivirus infection in domestic Horses in China: A study in guangdong province, heilongjiang province and Hong Kong district. *PLoS ONE* **2016**, *11*, e155662. [CrossRef] [PubMed]
81. Lyons, S.; Kapoor, A.; Sharp, C.; Schneider, B.S.; Wolfe, N.D.; Culshaw, G.; Corcoran, B.; McGorum, B.C.; Simmonds, P. Nonprimate hepaciviruses in domestic horses, United Kingdom. *Emerg. Infect. Dis.* **2012**, *18*, 1976–1982. [CrossRef]
82. Elia, G.; Lanave, G.; Lorusso, E.; Parisi, A.; Cavaliere, N.; Patruno, G.; Terregino, C.; Decaro, N.; Martella, V.; Buonavoglia, C. Identification and genetic characterization of equine hepaciviruses in Italy. *Vet. Microbiol.* **2017**, *207*, 239–247. [CrossRef]
83. Schlottau, K.; Fereidouni, S.; Beer, M.; Hoffmann, B. Molecular identification and characterization of nonprimate hepaciviruses in equines. *Arch. Virol.* **2019**, *164*, 391–400. [CrossRef]
84. Pfaender, S.; Walter, S.; Todt, D.; Behrendt, P.; Doerrbecker, J.; Wölk, B.; Engelmann, M.; Gravemann, U.; Seltsam, A.; Steinmann, J.; et al. Assessment of cross-species transmission of hepatitis C virus-related non-primate hepacivirus in a population of humans at high risk of exposure. *J. Gen. Virol.* **2015**, *96*, 2636–2642. [CrossRef]
85. ICTV. Herpesviridae—dsDNA Viruses—dsDNA Viruses. Available online: https://talk.ictvonline.org/ictv-reports/ictv_9th_report/dsdna-viruses-2011/w/dsdna_viruses/91/herpesviridae (accessed on 11 May 2020).
86. Allen, G.P.; Kydd, J.H.; Slater, J.D.; Smith, K.C. *Equid Herpesvirus-1 (EHV-1) and -4 (EHV-4) Infections*, 2nd ed.; Oxford Press: Cape Town, South Africa, 2004.

87. Slater, J. Equine herpesviruses. In *Equine Infectious Diseases*; WB Saunders: Philadelphia, PA, USA, 2014; pp. 151–168.
88. Fortier, G.; van Erck, E.; Pronost, S.; Lekeux, P.; Thiry, E. Equine gammaherpesviruses: Pathogenesis, epidemiology and diagnosis. *Vet. J.* **2010**, *186*, 148–156. [CrossRef]
89. Fukushi, H.; Yamaguchi, T.; Yamada, S. Complete Genome Sequence of Equine Herpesvirus Type 9. *J. Virol.* **2012**, *86*, 13822. [CrossRef]
90. Garvey, M.; Suárez, N.M.; Kerr, K.; Hector, R.; Moloney-Quinn, L.; Arkins, S.; Davison, A.J.; Cullinane, A. Equid herpesvirus 8: Complete genome sequence and association with abortion in mares. *PLoS ONE* **2018**, *13*, e192301. [CrossRef] [PubMed]
91. Izume, S.; Kirisawa, R.; Ohya, K.; Ohnuma, A.; Kimura, T.; Omatsu, T.; Katayama, Y.; Mizutani, T.; Fukushi, H. The full genome sequences of 8 equine herpesvirus type 4 isolates from horses in Japan. *J. Vet. Med. Sci.* **2017**, *79*, 206–212. [CrossRef] [PubMed]
92. King, A.M.; Lefkowitz, E.; Adams, M.J.; Carstens, E.B. (Eds.) Family Herpesviridae. In *Virus Taxonomy: Ninth Report of the International Committee on Taxonomy of Viruses*; Elsevier: Amsterdam, The Netherlands, 2011; pp. 111–122.
93. Sijmons, S.; Vissani, A.; Tordoya, M.S.; Muylkens, B.; Thiry, E.; Maes, P.; Matthijnssens, J.; Barrandeguy, M.; Van Ransta, M. Complete genome sequence of equid herpesvirus 3. *Genome Announc.* **2014**, *2*. [CrossRef] [PubMed]
94. Wilkie, G.S.; Kerr, K.; Stewart, J.P.; Studdert, M.J.; Davison, A.J. Genome sequences of equid herpesviruses 2 and 5. *Genome Announc.* **2016**, *3*. [CrossRef] [PubMed]
95. Back, H.; Kendall, A.; Grandón, R.; Ullman, K.; Treiberg-Berndtsson, L.; Ståhl, K.; Pringle, J. Equine multinodular pulmonary fibrosis in association with asinine herpesvirus type 5 and equine herpesvirus type 5: A case report. *Acta Vet. Scand.* **2012**, *54*, 57. [CrossRef] [PubMed]
96. Scheurer, L.; Bachofen, C.; Herteman, N.; Hilbe, M.; Wolfer, N.; Schoster, A. A case series highlighting the role of different gamma-herpesviruses in Equine Multinodular Pulmonary Fibrosis. *Zur. Open Repos. Arch.* **2020**, 245–256. [CrossRef]
97. Patel, J.R.; Heldens, J. Equine herpesviruses 1 (EHV-1) and 4 (EHV-4)—Epidemiology, disease and immunoprophylaxis: A brief review. *Vet. J.* **2005**, *170*, 14–23. [CrossRef]
98. Mendoza, F.J.; Toribio, R.E.; Perez-Ecija, A. Donkey Internal Medicine—Part II: Cardiovascular, Respiratory, Neurologic, Urinary, Ophthalmic, Dermatology, and Musculoskeletal Disorders. *J. Equine Vet. Sci.* **2018**, *65*, 86–97. [CrossRef]
99. Kapoor, S.; Sharma, H.; Singh, M.; Ranjan, K.; Patel, S.V.; Kumar, P.; Kumari, A.; Khirbat, R. Equine Herpesviruses: A Brief Review. *Adv. Anim. Vet. Sci.* **2014**, *2*, 46–54. [CrossRef]
100. Thiemann, A.K. Respiratory disease in the donkey. *Equine Vet. Educ.* **2012**, *24*, 469–478. [CrossRef]
101. Negussie, H.; Gizaw, D.; Tessema, T.S.; Nauwynck, H.J. Equine Herpesvirus-1 Myeloencephalopathy, an Emerging Threat of Working Equids in Ethiopia. *Transbound Emerg. Dis.* **2017**, *64*, 389–397. [CrossRef] [PubMed]
102. Mori, E.; Maria Do Carmo, C.S.H.L.; Cunha, E.M.S.; Villalobos, E.M.C.; Mori, C.M.C.; Soares, R.M.; Brandão, P.E.; Fernandes, W.R.; Richtzenhain, L.J. Molecular characterization of Brazilian equid herpesvirus type 1 strains based on neuropathogenicity markers. *Braz. J. Microbiol.* **2015**, *46*, 565–570. [CrossRef] [PubMed]
103. Pusterla, N.; Hussey, G.S. Equine herpesvirus 1 myeloencephalopathy. *Vet. Clin. N. Am. Equine Pract.* **2014**, *30*, 489–506. [CrossRef]
104. Ataseven, V.S.; Dağalp, S.B.; Güzel, M.; Başaran, Z.; Tan, M.T.; Geraghty, B. Prevalence of equine herpesvirus-1 and equine herpesvirus-4 infections in equidae species in Turkey as determined by ELISA and multiplex nested PCR. *Res. Vet. Sci.* **2009**, *86*, 339–344. [CrossRef]
105. De Souza Hunold Lara, M.d.C.C.; Torelli, C.S.; Cunha, E.M.S.; Villalobos, E.M.C.; Cunha, M.S.; Bello, A.C.P.P.; Cunha, A.P.; Reis, J.K.P.; Leite, R.C.; Mori, E. Serological survey of equine herpesvirus infection in Minas Gerais state. *Braz. J. Vet. Res. Anim. Sci.* **2010**, *47*, 352–356. [CrossRef]
106. Abdelgawad, A.; Hermes, R.; Damiani, A.; Lamglait, B.; Czirják, G.Á.; East, M.; Aschenborn, O.; Wenker, C.; Kasem, S.; Osterrieder, N.; et al. Comprehensive Serology Based on a Peptide ELISA to Assess the Prevalence of Closely Related Equine Herpesviruses in Zoo and Wild Animals. *PLoS ONE* **2015**, *10*, e138370. [CrossRef]

107. Pusterla, N.; Mapes, S.; Akana, N.; Barnett, C.; Mackenzie, C.; Gaughan, E.; Craig, B.; Chappell, D.; Vaala, W. Prevalence factors associated with equine herpesvirus type 1 infection in equids with upper respiratory tract infection and/or acute onset of neurological signs from 2008 to 2014. *Vet. Rec.* **2016**, *178*, 70. [CrossRef]
108. Yildirim, Y.; Yilmaz, V.; Kirmizigul, A.H. Equine herpes virus type 1 (EHV-1) and 4 (EHV-4) infections in horses and donkeys in northeastern Turkey. *Iran. J. Vet. Res.* **2015**, *16*, 341–344. [CrossRef]
109. Wegdan, H.; Intisar, K.; Shaza, M.; Algezoli, O.; Ballal, A.; Ihsan, H.; Sahar, M.; Baraa, A.; Manal, H.; Muna, E.; et al. Serological Detection of Equine Herpes Virus (EHV) Type 1 and 4 in Sudan. *Br. Microbiol. Res. J.* **2016**, *14*, 1–6. [CrossRef]
110. Ali, S.; Mohammadi, G. Detection of equine herpesvirus-1 and equine herpesvirus-4 in mules and donkeys by real time PCR. *Vet. Pract.* **2016**, *17*, 160–163.
111. Mekonnen, A.; Eshetu, A.; Gizaw, D. Equine herpesvirus 1 and/or 4 in working equids: Seroprevalence and risk factors in North Shewa Zone, Ethiopia. *Ethiop. Vet. J.* **2017**, *21*, 28. [CrossRef]
112. Vengust, M.; Wen, X.; Bienzle, D. Herpesvirus-associated neurological disease in a donkey. *J. Vet. Diagn. Investig.* **2008**, *20*, 820–823. [CrossRef] [PubMed]
113. Kleiboeker, S.B.; Schommer, S.K.; Johnson, P.J.; Ehlers, B.; Turnquist, S.E.; Boucher, M.; Kreeger, J.M. Association of two newly recognized herpesviruses with interstitial pneumonia in donkeys (*Equus asinus*). *J. Vet. Diagn. Investig.* **2002**, *14*, 273–280. [CrossRef] [PubMed]
114. Bell, S.A.; Pusterla, N.; Balasuriya, U.B.R.; Mapes, S.M.; Nyberg, N.L.; MacLachlan, N.J. Isolation of a gammaherpesvirus similar to asinine herpesvirus-2 (AHV-2) from a mule and a survey of mules and donkeys for AHV-2 infection by real-time PCR. *Vet. Microbiol.* **2008**, *130*, 176–183. [CrossRef] [PubMed]
115. Browning, G.F.; Ficorilli, N.; Studdert, M.J. Asinine herpesvirus genomes: Comparison with those of the equine herpesviruses. *Arch. Virol.* **1988**, *101*, 183–190. [CrossRef]
116. Barrandeguy, M.E.; Carossino, M. Infectious Diseases in Donkeys and Mules: An Overview and Update. *J. Equine Vet. Sci.* **2018**, *65*, 98–105. [CrossRef]
117. Go, Y.Y.; Balasuriya, U.B.R.; Lee, C. Zoonotic encephalitides caused by arboviruses: Transmission and epidemiology of alphaviruses and flaviviruses. *Clin. Exp. Vaccine Res.* **2014**, *3*, 58–77. [CrossRef]
118. Angenvoort, J.; Brault, A.C.; Bowen, R.A.; Groschup, M.H. West Nile viral infection of equids. *Vet. Microbiol.* **2013**, *167*, 168–180. [CrossRef]
119. World Organization For Animal Health (OIE). Japanese Encephalitis. In *OIE Technical Disease Cards*; World Organization for Animal Health: Paris, France, 2019.
120. Aleman, M.; Nout-Lomas, Y.S.; Reed, S.M. Disorders of the Neurologic System. In *Equine Internal Medicine*; Reed, S.M., Bayly, W.M., Sellon, D.C., Eds.; Elsevier: St. Louis, MO, USA, 2018; pp. 580–708. ISBN 9780323443296.
121. Tauro, L.; Marino, B.; Diaz, L.A.; Lucca, E.; Gallozo, D.; Spinsanti, L.; Contigiani, M. Serological detection of St. Louis encephalitis virus and West Nile virus in equines from Santa Fe, Argentina. *Mem. Inst. Oswaldo Cruz* **2012**, *107*, 553–556. [CrossRef]
122. Barba, M.; Fairbanks, E.L.; Daly, J.M. Equine viral encephalitis: Prevalence, impact, and management strategies. *Vet. Med. Res. Rep.* **2019**, *10*, 99–110. [CrossRef] [PubMed]
123. Chakravarti, S.K.; Sardana, D.N.; Mukherjee, K.K.; Mitra, A.C.; Chakrabarty, M.S. Serological evidence of infection with Japanese encephalitis virus in mules of eastern Himalayan region. *Indian J. Med. Res.* **1981**, *73*, 4–7. [PubMed]
124. Zohaib, A.; Saqib, M.; Beck, C.; Hussain, M.H.; Lowenski, S.; Lecollinet, S.; Sial, A.; Asi, M.N.; Mansoor, M.K.; Saqalein, M.; et al. High prevalence of West Nile virus in equines from the two provinces of Pakistan. *Epidemiol. Infect.* **2014**, *143*, 1931–1935. [CrossRef]
125. OIE. OIE World Animal Health Information System. Available online: https://www.oie.int/wahis_2/public/wahid.php/Diseaseinformation/reportarchive (accessed on 10 May 2020).
126. Campbell, G.L.; Hills, S.L.; Fischer, M.; Jacobson, J.A.; Hoke, C.H.; Hombach, J.M.; Marfin, A.A.; Solomon, T.; Tsai, T.F.; Tsu, V.D.; et al. Estimated global incidence of Japanese encephalitis: A systematic review. *Bull. World Health Organ.* **2011**, *89*, 766–774. [CrossRef] [PubMed]
127. Zhu, M.; Weng, Q. China: Will the donkey become the next pangolin? *Equine Vet. J.* **2018**, *50*, 276. [CrossRef] [PubMed]

128. Thompson, N.N.; Auguste, A.J.; Coombs, D.; Blitvich, B.J.; Carrington, C.V.F.; Rosa, A.P.T.; Wang, E.; Chadee, D.D.; Drebot, M.A.; Tesh, R.B.; et al. Serological evidence of Flaviviruses and Alphaviruses in Livestock and Wildlife in Trinidad. *Vector-Borne Zoonotic Dis.* **2012**, *12*, 969–978. [CrossRef]
129. Carrera, J.-P.; Bagamian, K.H.; Rosa, A.P.T.; Wang, E.; Beltran, D.; Gundaker, N.D.; Armien, B.; Arroyo, G.; Sosa, N.; Pascale, J.M.; et al. Human and equine infection with alphaviruses and flaviviruses in Panamá during 2010: A cross-Sectional study of household contacts during an encephalitis outbreak. *Am. J. Trop. Med. Hyg.* **2018**, *98*, 1798–1804. [CrossRef]
130. Burgueño, A.; Spinsanti, L.; Díaz, L.A.; Rivarola, M.E.; Arbiza, J.; Contigiani, M.; Delfraro, A. Seroprevalence of St. Louis encephalitis virus and West Nile virus (Flavivirus, Flaviviridae) in horses, Uruguay. *BioMed. Res. Int.* **2013**, *2013*, 582957. [CrossRef]
131. Costa, G.B.; Marinho, P.E.S.; Vilela, A.P.P.; Saraiva-Silva, A.T.; Crispim, A.P.C.; Borges, I.A.; Dutra, A.G.S.; Lobato, Z.I.P.; Dos Reis, J.K.P.; De Oliveira, D.B.; et al. Silent circulation of the saint louis encephalitis virus among humans and equids, Southeast Brazil. *Viruses* **2019**, *11*, 1029. [CrossRef]
132. Pauvolid-Corrêa, A.; Campos, Z.; Juliano, R.; Velez, J.; Nogueira, R.M.R.; Komar, N. Serological Evidence of Widespread Circulation of West Nile Virus and Other Flaviviruses in Equines of the Pantanal, Brazil. *PLoS Negl. Trop. Dis.* **2014**, *8*, e2706. [CrossRef]
133. Murgue, B.; Murri, S.; Zientara, S.; Durand, B.; Durand, J.; Zeller, H. West Nile Outbreak in Horses in Southern France, 2000: The Return after 35 Years. *Emerg. Infect. Dis.* **2001**, *7*, 692–696. [CrossRef] [PubMed]
134. Silva, A.S.G.; Matos, A.C.D.; da Cunha, M.A.C.R.; Rehfeld, I.S.; Galinari, G.C.F.; Marcelino, S.A.C.; Saraiva, L.H.G.; da Martins, N.R.S.; de Maranhão, R.P.A.; Lobato, Z.I.P.; et al. West Nile virus associated with equid encephalitis in Brazil, 2018. *Transbound. Emerg. Dis.* **2019**, *66*, 445–453. [CrossRef] [PubMed]
135. Salazar, P.; Traub-Dargatz, J.L.; Morley, P.S.; Wilmot, D.D.; Steffen, D.J.; Cunningham, W.E.; Salman, M.D. Outcome of equids with clinical signs of West Nile virus infection and factors associated with death. *J. Am. Vet. Med. Assoc.* **2004**, *225*, 267–274. [CrossRef] [PubMed]
136. Yildirim, Y.; Yilmaz, V.; Yazici, K.; Özic, C.; Ozkul, A. Molecular and serological investigation of West Nile virus (WNV) infection in donkeys, horses and native geese in Turkey. *Rev. Med. Vet.* **2018**, *169*, 87–92.
137. Mehmet, K.; Sibel, G.; Orhan, Y.; Nuri, M.; Sibel, Y.; Sibel, H.; Oya, B.; Metin, G. Serological investigation of West Nile Virus infection in domestic horses and donkeys in Turkey. *Pak. Vet. J.* **2017**, *37*, 51–54.
138. Bargaoui, R.; Lecollinet, S.; Lancelot, R. Mapping the Serological Prevalence Rate of West Nile fever in Equids, Tunisia. *Transbound. Emerg. Dis.* **2013**, *62*, 55–66. [CrossRef] [PubMed]
139. Baba, S.S.; NNnadi, O.D.; Hamman, K.D.; Saidu, A.; El-Yuguda, A.; Oderinde, B.S. Preliminary Study on the Prevalence of West Nile Virus Antibody among Horses, Donkeys and Camels in Borno State, Nigeria. *J. Appl. Virol.* **2014**, *3*, 39–45. [CrossRef]
140. Davoust, B.; Maquart, M.; Roqueplo, C.; Gravier, P.; Sambou, M.; Mediannikov, O.; Leparc-Goffart, I. Serological Survey of West Nile Virus in Domestic Animals from Northwest Senegal. *Vector-Borne Zoonotic Dis.* **2016**, *16*, 359–361. [CrossRef]
141. Lafri, I.; Prat, C.M.; Bitam, I.; Gravier, P.; Besbaci, M.; Zeroual, F.; Ben-Mahdi, M.H.; Davoust, B.; Leparc-Goffart, I. Seroprevalence of West Nile virus antibodies in equids in the North-East of Algeria and detection of virus circulation in 2014. *Comp. Immunol. Microbiol. Infect. Dis.* **2017**, *50*, 8–12. [CrossRef]
142. Azmi, K.; Tirosh-Levy, S.; Manasrah, M.; Mizrahi, R.; Nasereddin, A.; Al-Jawabreh, A.; Ereqat, S.; Abdeen, Z.; Lustig, Y.; Gelman, B.; et al. West Nile Virus: Seroprevalence in Animals in Palestine and Israel. *Vector-Borne Zoonotic Dis.* **2017**, *17*, 558–566. [CrossRef]
143. Selim, A.; Radwan, A.; Arnaout, F.; Khater, H. The Recent Update of the Situation of West Nile Fever among Equids in Egypt after Three Decades of Missing Information. *Pak. Vet. J.* **2020**, *8318*. [CrossRef]
144. Pradel, J.; Monfray, K.C.; Molia, S.; Vachiéry, N.; Rousteau, A.; Imbert, D.; Martinez, D.; Sabatier, P.; Lefrançois, T. Risk factors for West Nile virus seropositivity of equids in Guadeloupe. *Prev. Vet. Med.* **2009**, *92*, 71–78. [CrossRef] [PubMed]
145. Alonso-Padilla, J.; Loza-Rubio, E.; Escribano-Romero, E.; Córdoba, L.; Cuevas, S.; Mejía, F.; Calderón, R.; Milián, F.; da Rosa Travassos, A.; Weaver, S.C.; et al. The continuous spread of West Nile virus (WNV): Seroprevalence in asymptomatic horses. *Epidemiol. Infect.* **2009**, *137*, 1163–1168. [CrossRef] [PubMed]
146. Ometto, T.; Durigon, E.L.; de Araujo, J.; Aprelon, R.; Aguiar, D.M.; Cavalcante, G.T.; Melo, R.M.; Levi, J.E.; de Azevedo Júnior, S.M.; Petry, M.V.; et al. West nile virus surveillance, Brazil, 2008–2010. *Trans. R. Soc. Trop. Med. Hyg.* **2013**, *107*, 723–730. [CrossRef] [PubMed]

147. García-Bocanegra, I.; Arenas-Montes, A.; Jaén-Téllez, J.A.; Napp, S.; Fernández-Morente, M.; Arenas, A. Use of sentinel serosurveillance of mules and donkeys in the monitoring of West Nile virus infection. *Vet. J.* **2012**, *194*, 262–264. [CrossRef]
148. Lecollinet, S.; Pronost, S.; Coulpier, M.; Beck, C.; Gonzalez, G.; Leblond, A.; Tritz, P. Viral equine encephalitis, a growing threat to the horse population in Europe? *Viruses* **2020**, *12*, 23. [CrossRef]
149. Timoney, P.J. Equine influenza. *Comp. Immunol. Microbiol. Infect. Dis.* **1996**. [CrossRef]
150. Tellier, R. Review of aerosol transmission of influenza A virus. *Emerg. Infect. Dis.* **2006**, *12*, 1657–1662. [CrossRef]
151. Ferguson, L.; Olivier, A.K.; Genova, S.; Epperson, W.B.; Smith, D.R.; Schneider, L.; Barton, K.; McCuan, K.; Webby, R.J.; Wan, X.-F. Pathogenesis of Influenza D Virus in Cattle. *J. Virol.* **2016**, *90*, 5636–5642. [CrossRef]
152. Webster, R.G.; Bean, W.J.; Gorman, O.T.; Chambers, T.M.; Kawaoka, Y. Evolution and ecology of influenza a viruses. *Microbiol. Rev.* **1992**, *56*, 153–174. [CrossRef]
153. Oxburgh, L.; Berg, M.; Klingeborn, B.; Emmoth, E.; Linné, T. Evolution of H3N8 equine influenza virus from 1963 to 1991. *Virus Res.* **1994**, *34*, 153–165. [CrossRef]
154. Cullinane, A.; Newton, J.R. Equine influenza—A global perspective. *Vet. Microbiol.* **2013**, *167*, 205–214. [CrossRef] [PubMed]
155. Singh, R.K.; Dhama, K.; Karthik, K.; Khandia, R.; Munjal, A.; Khurana, S.K.; Chakraborty, S.; Malik, Y.S.; Virmani, N.; Singh, R.; et al. A comprehensive review on equine influenza virus: Etiology, epidemiology, pathobiology, advances in developing diagnostics, vaccines, and control strategies. *Front. Microbiol.* **2018**, *9*, 1941. [CrossRef] [PubMed]
156. Van Maanen, C.; Cullinane, A. Equine influenza virus infections: An update. *Vet. Q.* **2002**, *24*, 79–94. [CrossRef] [PubMed]
157. Landolt, G.A.; Olsen, C.W. Up to new tricks—A review of cross-species transmission of influenza A viruses. *Anim. Health Res. Rev.* **2007**, *8*. [CrossRef] [PubMed]
158. Yang, H.; Xiao, Y.; Meng, F.; Sun, F.; Chen, M.; Cheng, Z.; Chen, Y.; Liu, S.; Chen, H. Emergence of H3N8 equine influenza virus in donkeys in China in 2017. *Vet. Microbiol.* **2018**, *214*, 1–6. [CrossRef] [PubMed]
159. Shortridge, K.E. Influenza, Hong Kong, and the China Traveler. *J. Travel Med.* **1995**, *2*, 257–259. [CrossRef]
160. Thiemann, A.K.; Bell, N.J. The Peculiarities of Donkey Respiratory Disease. *Equine Respir. Dis.* **2001**. [CrossRef]
161. Rose, M.A.; Round, M.C.; Beveridge, W.I. Influenza in horses and donkeys in Britain, 1969. *Vet. Rec.* **1970**. [CrossRef]
162. Qi, T.; Guo, W.; Huang, W.; Dai, L.; Zhao, L.; Li, H.; Li, X.; Zhang, X.; Wang, Y.; Yan, Y.; et al. Isolation and genetic characterization of H3N8 equine influenza virus from donkeys in China. *Vet. Microbiol.* **2010**, *144*, 455–460. [CrossRef]
163. Yu, Y.; Sun, X.; Xia, N.; Zhang, J.; Liu, W. Detection of the epidemic of the H3N8 subtype of the equine influenza virus in large-scale donkey farms. *Int. J. Vet. Sci. Med.* **2020**, *8*, 26–30. [CrossRef]
164. Pica, N.; Palese, P. Toward a Universal Influenza Virus Vaccine: Prospects and Challenges. *Annu. Rev. Med.* **2013**, *64*, 189–202. [CrossRef] [PubMed]
165. Wong, S.S.; Webby, R.J. Traditional and new influenza vaccines. *Clin. Microbiol. Rev.* **2013**, *26*, 476–492. [CrossRef] [PubMed]
166. Reed, S.M.; Bayly, W.M.; Sellon, D.C. (Eds.) *Equine Internal Medicine*, 4th ed.; Elsevier: St. Louis, MO, USA, 2018.
167. King, A.M.; Lefkowitz, E.; Adams, M.J.; Carstens, E.B. (Eds.) *Virus Taxonomy: Ninth Report of the International Committee on Taxonomy of Viruses*, 1st ed.; Elsevier: Amsterdam, The Netherlands, 2011.
168. OIE. OIE-Listed Diseases 2020: OIE—World Organisation for Animal Health. Available online: https://www.oie.int/animal-health-in-the-world/oie-listed-diseases-2020/ (accessed on 29 September 2020).
169. Johnson, N.; Phillpotts, R.; Fooks, A.R. Airborne transmission of lyssaviruses. *J. Med. Microbiol.* **2006**, *55*, 785–790. [CrossRef]
170. Green, S.L. Rabies. *Vet. Clin. N. Am. Equine Pract.* **1997**, *13*, 1–11. [CrossRef]
171. Hemachudha, T.; Tirawatnpong, S.; Phanthumchinda, K. Seizures as the initial manifestation of ocular paralytic rabies. *J. Neurol. Neurosurg. Psychiatry* **1989**, *52*, 808–810. [CrossRef]
172. Hudson, L.C.; Weinstock, D.; Jordan, T.; Bold-Fletcher, N.O. Clinical presentation of experimentally induced rabies in horses. *J. Vet. Med. Ser. B* **1996**, *43*, 277–285. [CrossRef]

173. Numan, M.; Qureshi, Z.A.; Shauket, M.; Hashmi, H.A.; Iqbal, M.; Gill, Z.J.; Habib, M.; Siddique, M. Rabies out-break in mules at Mansehra, Pakistan. *Res. Vet. Sci.* **2011**, *90*, 160–162. [CrossRef]
174. Droussi, H.; El Amrani, D.; Dlimi, M.; Elatiqi, K.; Boukind, S.; Exadaktylos, A.K.; Weiner, S.; Ettalbi, S. Domestic donkey (*Equus africanus* asinus) bites: An unusual aetiology of severe scalp injuries in Morocco. *Afr. J. Emerg. Med.* **2014**, *4*, e5–e8. [CrossRef]
175. Feng, Y.; Shi, Y.; Yu, M.; Xu, W.; Gong, W.; Tu, Z.; Ding, L.; He, B.; Guo, H.; Tu, C. Livestock rabies outbreaks in Shanxi province, China. *Arch. Virol.* **2016**, *161*, 2851–2854. [CrossRef]
176. Sheth, S.S.; Gharpure, V.; Nair, K.; Nadkarni, U.B.; Deshmukh, C.T.; Desai, A.P.; Jain, M.K.; Shah, M.D. Fatal rabies encephalomyelitis despite chick embryo vaccine prophylaxis. *Indian Pediatr.* **1992**, *29*, 766–769. [PubMed]
177. Badiali, L.; Ferris, D.H. A preliminary report on rabies in suspected equine encephalomyelitis cases in the United Arab Republic. *Bull. World Health Organ.* **1966**, *34*, 797–798. [PubMed]
178. Fekadu, M. Rabies in Ethiopia. *Am. J. Epidemiol.* **1982**, *115*, 266–273. [CrossRef] [PubMed]
179. Kariuki, D.P.; Ngulo, W.K. Epidemiology of Animal Rabies in Kenya (1900–1983). *Rabies Trop.* **1985**, 451–464. [CrossRef]
180. Tabel, H.; Corner, A.H.; Webster, W.A.; Casey, C.A. History and epizootiology of rabies in Canada. *Canad. Vet. J.* **1974**, *15*, 271–281.
181. Stromberg, Z.R.; Fischer, W.; Bradfute, S.B.; Kubicek-Sutherland, J.Z.; Hraber, P. Vaccine advances against venezuelan, eastern, and western equine encephalitis viruses. *Vaccines* **2020**, *8*, 273. [CrossRef]
182. Spickler, A.R. Eastern, Western and Venezuelan Equine Encephalomyelitis. Available online: http://www.cfsph.iastate.edu/DiseaseInfo/factsheets.php (accessed on 7 September 2020).
183. World Organization For Animal Health (OIE). Equine Encephalomyelitis. In *Terrestrial Manual 2019*; World Organization for Animal Health: Paris, France, 2019.
184. Franklin, R.P.; Kinde, H.; Jay Laura, D.; Kramer, M.T.; Green, E.G.N.; Chiles, R.E.; Ostlund, E.; Husted, S.; Smith, J.; Parker, M.D. Eastern equine encephalomyelitis virus infection in a horse from California. *Emerg. Infect. Dis.* **2002**, *8*, 283–288. [CrossRef]
185. The Center for Food Security & Public Health. Equine Encephalitis. Available online: http://www.cfsph.iastate.edu/FastFacts/pdfs/easter_wester_venezuelan_equine_encephalomyelitis_F.pdf (accessed on 8 September 2020).
186. Randall, R.; Mills, J.W. Fatal Encephalitis in Man Due to the Venezuelan Virus of Equine Encephalomyelitis in Trinidad. Available online: https://www.cabdirect.org/cabdirect/abstract/19441000327 (accessed on 8 September 2020).
187. Silva, M.L.C.R.; Galiza, G.J.N.; Dantas, A.F.M.; Oliveira, R.N.; Iamamoto, K.; Achkar, S.M.; Riet-Correa, F. Outbreaks of eastern equine encephalitis in northeastern Brazil. *J. Vet. Diagn. Investig.* **2011**, *23*, 570–575. [CrossRef]
188. Thompson, G.M.; Jess, S.; Murchie, A.K. A review of african horse sickness and its implications for Ireland. *Ir. Vet. J.* **2012**, *65*, 1–8. [CrossRef]
189. Meiswinkel, R.; Paweska, J.T. Evidence for a new field Culicoides vector of African horse sickness in South Africa. *Prev. Vet. Med.* **2003**, *60*, 243–253. [CrossRef]
190. Zientara, S.; Weyer, C.T.; Lecollinet, S. African horse sickness. *OIE Rev. Sci. Tech.* **2015**. [CrossRef]
191. Mellor, P.S.; Baylis, M.; Hamblin, C. *African Horse Sickness*, 1st ed.; Mertens, P.P., Ed.; Springer: New York, NY, USA, 1998; Volume 53, ISBN 9788578110796.
192. Gordon, S.J.G.; Bolwell, C.; Rogers, C.W.; Musuka, G.; Kelly, P.; Guthrie, A.; Mellor, P.S.; Hamblin, C. The sero-prevalence and sero-incidence of African horse sickness and equine encephalosis in selected horse and donkey populations in Zimbabwe. *Onderstepoort J. Vet. Res.* **2017**, *84*, 1–5. [CrossRef] [PubMed]
193. Teshome, M.; Addis, M.; Temesgen, W. Seroprevalence and risk factors of African horse sickness in mules and donkeys in selected sites of West Amhara Region, Ethiopia. *African J. Microbiol. Res.* **2012**, *6*. [CrossRef]
194. Rodriguez, M.; Hooghuis, H.; Castaño, M. African horse sickness in Spain. *Vet. Microbiol.* **1992**. [CrossRef]
195. Lu, G.; Pan, J.; Ou, J.; Shao, R.; Hu, X.; Wang, C.; Li, S. African horse sickness: Its emergence in Thailand and potential threat to other Asian countries. *Transbound. Emerg. Dis.* **2020**. [CrossRef] [PubMed]
196. Gilbert, M.; Nicolas, G.; Cinardi, G.; Van Boeckel, T.P.; Vanwambeke, S.O.; Wint, G.R.W.; Robinson, T.P. Global distribution data for cattle, buffaloes, horses, sheep, goats, pigs, chickens and ducks in 2010. *Sci. Data* **2018**, *5*, 180227. [CrossRef]

197. National Bureau of Statistics of China. Latest Release. Available online: http://www.stats.gov.cn/tjsj/zxfb/index_1.html (accessed on 7 September 2020).

Publisher's Note: MDPI stays neutral with regard to jurisdictional claims in published maps and institutional affiliations.

animals

MDPI

Article

Donkey Epididymal Transport for Semen Cooling and Freezing

Yamilka Lago-Alvarez [1,†], Giorgia Podico [1,†], Lorenzo G. Segabinazzi [1], Lais L. Cunha [1], Leonardo Barbosa [1], Carolyn E. Arnold [2], Fabio S. Lima [1], Luise T. King [3], Amy K. McLean [4] and Igor F. Canisso [1,*]

1 Department of Veterinary Clinical Medicine, College of Veterinary Medicine, University of Illinois, Urbana, IL 61801, USA; yla3@cornell.edu (Y.L.-A.); gpodico@illinois.edu (G.P.); lgseg@hotmail.com (L.G.S.); laisl@illinois.edu (L.L.C.); lbarbosa@illinois.edu (L.B.); falima@ucdavis.edu (F.S.L.)

2 Department of Large Animal Clinical Sciences, College of Veterinary Medicine and Biomedical Sciences, Texas A&M University, College Station, TX 77840, USA; cearnold@tamu.edu

3 Department of Veterinary Clinical Medicine, University of Arizona, Oro Valley, AZ 85704, USA; tluiseking@arizona.edu

4 Department of Animal Sciences, University of California, Davis, CA 95161, USA; acmclean@ucdavis.edu

* Correspondence: canisso@illinois.edu

† These authors contributed equally.

Received: 31 October 2020; Accepted: 17 November 2020; Published: 25 November 2020

Simple Summary: In the event of death, euthanasia, or forceful castration for medical reasons, epididymal semen harvesting represents the last opportunity to preserve the genetic material of valuable sires. However, this technique has yet to be tested in donkeys. Three experiments were carried out to assess epididymal semen cooling and freezing in donkeys. In experiment 1, semen cooling and freezing were conducted immediately after castration, and in experiments 2 and 3, epididymides were shipped overnight, and then epididymal semen cooled and frozen. Results showed that cooling of epididymal semen up to 24 h after harvesting did not affect motility parameters or plasma membrane integrity. Collectively, the post-thaw results revealed low motility parameters across groups; At the same time, the plasma membrane integrity did not reflect this trend, and the values remained high, suggesting that there was a lack of epididymal sperm activation after freezing. In summary, freshly harvested and cooled-shipped epididymal donkey semen had satisfactory semen parameters. New studies need to address donkey epididymal semen fertility in mares and jennies.

Abstract: The objectives of this study were to assess the cooling and freezing of donkey epididymal semen harvested immediately after castration (Experiment 1, $n = 4$) or after the shipment (24 or 48 h) of epididymides attached to testicles (Experiment 2, $n = 14$) or dissected apart (Experiment 3, $n = 36$). In each experiment, semen was frozen immediately (Non-Centrif) in an egg yolk-based semen extender (EY) or after processing through cushion-centrifugation (Centrif) while extended in a skim milk-based extender (SC). In all three experiments, cooled, pre-freeze, and post-thaw epididymal semen was assessed for total motility (TM), progressive motility (PM), plasma membrane integrity (PMI), and high mitochondrial membrane potential (HMMP). Data were analyzed with R using mixed models and Tukey's test as posthoc. Results showed that the cooling of epididymal semen up to 24 h after harvesting did not affect motility parameters or plasma membrane integrity; furthermore, in Experiment 3, the post-thaw evaluation of both Centrif and Non-Centrif achieved similar TM and PM. Collectively, the post-thaw results revealed low motility parameters across groups; while, the PMI and HMMP did not reflect this trend, and the values remained high, suggesting that there was a lack of epididymal sperm activation with either centrifugation or extenders. In summary, freshly harvested and cooled-shipped and cooled semen had satisfactory semen parameters. Future studies need to address donkey epididymal semen fertility in mares and jennies.

Keywords: epididymis; cooled-shipped epididymides; castration; testis; semen cryopreservation

1. Introduction

In the American continent, donkey-jacks (jacks, Equus asinus) are primarily used for breeding mares (Equus caballus) to produce mules (Equus mulus) [1,2]. Top mule producers are increasingly prized animals in the donkey and mule show industry in the United States and Brazil [1,3]. Most recently, there has been a massive increase in the number of breeding programs to produce donkey-hide gelatin (ejiao, colla cori asini), a collagen product extracted from the skin, which is the central pillar of traditional Chinese medicine [2]. The increased value of show mule sires and interest for donkeys to produce ejiao has generated an exponential interest in donkey semen freezing [2].

While much progress has been made in recent years to cryopreserve ejaculated jack semen [2,4,5], harvesting epididymal semen has yet to be investigated in this species. In the event of death, euthanasia, or forceful castration for medical reasons (e.g., trauma to the scrotum, life-threatening conditions, and testicular neoplasia), epididymal semen harvesting represents the last opportunity to preserve the genetic material of valuable sires [6]. Since most jacks are located far away from laboratories capable of harvesting and freezing epididymal semen, cooling and shipping the epididymides alone or attached to testicles could prove advantageous. In horses, epididymal shipment for semen freezing has been successfully carried out by different groups [7–10]; however, to our knowledge, this approach has not been investigated in donkeys.

Furthermore, the preservation of epididymal spermatozoa could be a valuable tool to preserve the genetics of endangered domestic donkey breeds (e.g., Amiata, Catalan, Andalusian, Baudet Du Poitou, and Martina Franca) and endangered wild donkey species, such as the African wild ass (Equus africanus), Somali wild ass (Equus africanus somaliensis), and Asiatic wild ass (Equus hemionus) [11,12]. To prevent the imminent extinction of these animals, decisive conservation measures are needed to preserve the genetic material of endangered breeds and species [12]. Historically, routine semen collections of captive equids in zoological collections have been performed via electroejaculation under chemical restraint [12]. However, when ejaculated sperm are not available due to death or euthanasia for medical reasons, epididymal semen harvesting could be used as an alternative method. Despite having valuable species in the collection, zoos across the world are often not equipped or do not have the expertise to process epididymal semen for cryopreservation. Therefore, zoos could benefit from shipping epididymides to referral centers acquainted with freezing epididymal semen.

Numerous harvesting methods of epididymal semen have been evaluated and optimized in the domesticated species [8,13–16]. In horses, the methods available for epididymal semen harvesting include direct aspiration, slicing, and flotation, or retrograde flushing of the cauda epididymis; [8,17]. Although there is no consensus regarding the most suitable method, the latter appears to be the most widely used [10]. Furthermore, if the retrograde flushing is performed with a freezing extender, the semen can be directly frozen and requires no further processing; however, if the retrograde flushing is performed with a cooling extender, the harvested semen needs to be centrifuged, the supernatant discarded, and the pellet resuspended in a semen freezing extender. While some authors advocate for the direct method [8], others advocate for the other approach, considering centrifugation a necessary step to activate epididymal sperm motility [18,19]. The only study comparing the two approaches failed to find any differences in post-thaw semen parameters for stallion epididymal semen processed via either approach [6]. The use of centrifugation is controversial in harvested epididymal semen; since epididymal sperm usually yields high sperm concentration, there is no need to concentrate sperm and remove seminal plasma [6]. In the stallion, semen centrifugation can lead to a sperm loss from 10 to 45% [20,21]. Therefore, the objectives of the present study were to compare the cooling and freezing ability of donkey epididymal semen obtained either from retrograde flushing and direct freezing or retrograde flushing and then cooling and freezing after cushion centrifugation. In addition, this study aimed to assess donkey epididymal semen cooling and freezing harvested immediately after castration and on cooled shipped epididymides attached to testicles or dissected apart during transport. We hypothesized that the semen quality parameters of donkey epididymal sperm result in equivalent post-cooling and post-thaw semen parameters irrespective of the method used for sperm harvesting, processing, and storage time.

2. Materials and Methods

Three experiments were conducted in the present study, from October 2019 to February 2020. The donkey epididymal semen was harvested immediately after castration (Experiment 1) or after the epididymides were shipped (Experiments 2 and 3). In Experiment 2, epididymides attached to the testicles were shipped in eight passive semen cooling containers (Equitainer II; Hamilton Research, Inc., Ipswich, MA, USA). In Experiment 3, epididymides dissected away from the testes were shipped in twelve passive cooling containers (Botuflex, Botupharma USA, Phoenix, AZ, USA). In Experiments 2 and 3, subsets of the passive cooling containers were randomly chosen and processed after arrival (cooled-shipped 24 h, $n = 8$ and $n = 20$ pairs for Experiments 2 and 3, respectively) and the next day (cooled-shipped 48 h, $n = 6$ and $n = 16$ pairs for Experiments 2 and 3, respectively).

2.1. *Experiment 1: On-Site Epididymal Semen Cooling and Freezing*

2.1.1. Animals and Castration

Four privately-owned jacks (2 American Mammoth and 2 Miniature, ranging from 2 to 10 years old, bodyweights in the range of 95–350 kg) were enrolled in this study. The jacks belonged to four different clients and were presented for routine castration at the University of Illinois Veterinary Teaching Hospital, Urbana, IL (40.1206°N, 88.2073°W). The owners reported that their diets consisted of mixed alfalfa–grass hay and no grain supplementation. This experiment was carried out to determine whether the techniques used to process horse epididymal semen could be effectively applied to donkeys. For each jack, a full physical examination including palpation of the scrotal content was performed prior to castration to exclude gross scrotal abnormalities (e.g., hernias, no adhesions, or lacerations and confirmation of two descending testicles).

Standard intravenous general anesthesia was induced with triple-drip, which includes xylazine, midazolam, and ketamine [22]. Once under general anesthesia, routinely closed castration was performed through a scrotal approach using Reimer's emasculators as previously described [22]. In addition, for each testicle, two mosquito hemostats were placed at the vas deferens distal to the emasculator to minimize the loss of semen. The cord was transected immediately distal to the emasculator and proximal to the mosquito hemostats, removing the testicle and a portion of the spermatic cord. Client consent was obtained from all four owners.

2.1.2. Epididymal Semen Harvesting

The epididymides were rinsed with room temperature sterile Lactated Ringer's Solution and wiped with cotton gauze to remove any contamination from blood or debris. The epididymides were dissected from the testes, as previously described [6]. Briefly, the vas deferens and tail of the epididymides were dissected free of blood vessels and connective tissue working distally from the vas deferens toward the body of the epididymis. Then, the tail of the epididymis was transected with a straight-bladed Mayo scissor. Each epididymal tail was weighed on a scale before further processing. An intravenous catheter (16 Ga × 13 cm, MilaCath®) was inserted into the proximal end of the ductus deferens and slowly flushed with 5–10 mL of a cooling extender based on sodium caseinate cyclodextrin–cholesterol loaded (SC; BotuSemen Gold, Botupharma USA, Phoenix, AZ, USA) or egg yolk-based semen freezing extender (EY; Botucrio, Botupharma USA) as previously described [6]. One epididymis of each pair was submitted to retrograde flushing with EY, while the contralateral epididymis was flushed with SC. Once the tail of the epididymis was fully distended, it was cut at its most distal portion with a straight-bladed Mayo scissor, and the fluid containing the semen allowed to outflow into a 50 mL conical tube (Thermo Scientific™, Nunc™, Rochester, NY, USA) [6].

2.1.3. Epididymal Semen Processing

The semen recovered from the epididymides flushed with the EY extender was further extended at 100 million sperm/mL, and a portion was frozen immediately without centrifugation (Non-Centrif-0 h) (as described in Section 2.1.4) or packed into an 18 oz disposable plastic bag (Whirl-Pak®, Nasco Inc., Fort Atkinson, WI, USA) and maintained in a passive cooling semen container (BotuFlex, Botupharma USA) for 24 h before being frozen in a similar manner (Non-Centrif-24 h). The contralateral epididymides flushed with SC extender were extended at 100 million sperm/mL and stored in a passive cooling container (BotuFlex) for 24 h (Centrif-24 h), or cushion-centrifuged and then re-extended with the same extender and then stored in the same container for 24 h (C-Centrif-24 h) or resuspended in the EY extender (Centrif-0 h) and frozen. Cushion centrifugation was performed as previously described [6]. Briefly, the extended semen was loaded in 50 mL conical tubes (Thermo Scientific™, Nunc™) with the addition of 1 mL cushion fluid (Red-Cushion, Botupharma USA) placed at the bottom of the tube with a blunted spinal needle (18 Ga × 13.5 cm). Centrifugation was performed at 1000× *g* × 20 min at room temperature.

Following centrifugation, the supernatant and cushion solution was discarded. The concentration of the remaining pellet was assessed as described below, and the semen cushion-centrifuged was either re-extended in the same SC extender at 100 million sperm/mL or in the EY extender at 100 million sperm/mL. The semen extended in the EY extender was frozen immediately. After 24 h of cooled storage, semen was cushion-centrifuged, resuspended in the EY extender, and then frozen. Sperm motility parameters, plasma membrane integrity (PMI), and high-mitochondrial membrane potential (HMMP) were assessed immediately before (pre-cooling) and after 24 h (post-cooling) of cooled-storage in a passive cooling semen container, and before and after cushion centrifugation, as described in Sections 2.4 and 2.5.

2.1.4. Epididymal Semen Freezing

Once semen was finally extended with the EY extender at 100 million sperm/mL, it was manually loaded in 0.5 mL straws and sealed with a porTable Straw sealer (UltraSeal21TM, Minitube of America, Vernon, WI, USA). Once the straws were sealed, they were placed in a cold room at 5 °C for 20 min. The isothermal box (Lifoam™, Fishers, IN, USA) of 42 L capacity was filled with a depth of 6 cm of liquid nitrogen. The straws were placed horizontally on a rack at 3 cm above the liquid nitrogen for 20 min in the airtight isothermal box. Subsequently, the straws were immersed in liquid nitrogen [8]. Then, the straws were loaded into canes and transferred to liquid nitrogen tanks until further analyses. The thawing of semen was performed by placing one straw at the time in a water bath at 38 °C for 60 s. Then, the samples were assessed for either treatment to determine the sperm motility parameters, PMI, and HMMP, as described in Sections 2.4 and 2.5.

2.2. Experiment 2. Cooled-Transported Testicles and Epididymides for Semen Cooling and Freezing

2.2.1. Animals and Castration

Fourteen feral small standard jacks (ranging from 4 to 13 years old, bodyweights in the range of 150–200 kg) maintained in corrals at the Bureau of Land Management facility in Florence, AZ (33.0315° N, 111.3873° W) were enrolled in this experiment. The agency gave us permission to use the samples harvested from these animals. This experiment was conducted to determine whether harvesting the testicles and epididymides under field conditions and shipping them to a specialized laboratory could be an alternative approach to cryopreserve epididymal donkey semen. The donkeys were housed in small groups and fed alfalfa and bermudagrass hay. All jacks were gathered at least six months from the states of Arizona, Nevada, and California. Physical examination was performed, which included the confirmation of two descending testicles with no adhesions, abrasions, or lacerations, and only jacks with unremarkable physical examinations were enrolled in the study.

All jacks were anesthetized with the intravenous administration of detomidine, butorphanol, and ketamine [22]. Once under general anesthesia, open castration was performed using Henderson's instrument, as previously described [22]. For each testicle, ligations were placed at the vas deferens immediately after the cord transection. The testes attached to the epididymis and spermatic cord were maintained in an isothermal box for about one hour until processing for shipping.

2.2.2. Epididymal Processing and Shipping

The epididymides attached to the testicles were rinsed and packed in a disposable, plastic Whirl-Pak® bag containing fresh skim-milk based extender (25 mL, Botusemen, Botupharma USA, Phoenix, AZ, USA), and shipped overnight in a passive cooling device (Equitainer II; Hamilton Research, Inc.) to the University of Illinois Veterinary Teaching Hospital, Urbana, IL, USA for semen cooling and freezing. The Equitainer cans were deep-frozen at −20 °C for at least 24 h. Upon arrival, each Equitainer contained one to two pairs of testicles with the epididymides attached, individually packed in Whirl-Pak® without an isothermolizer cup.

2.2.3. Epidydimal Semen Processing, Cooling, and Freezing

The samples ($n = 14$ pairs) were randomly divided and either processed after arrival (Cooled-shipped 24 h, $n = 8$) or the next day (Cooled-shipped 48 h, $n = 6$). Once each passive cooling device was opened, the extender's temperature where the sample was submerged was assessed. The epididymides were dissected from the testes as previously described [6]. Briefly, the epididymides were rinsed with room temperature sterile Lactated Ringer's Solution and wiped with cotton gauze to remove any contamination from blood or debris. Each epididymal tail was weighed before processing. Thereafter, retrograde epididymal flushing was performed as aforementioned in Experiment 1—Section 2.1.2. The lumen of the ductus deferens was cannulated with a 16 Ga × 13 cm (MilaCath®) and flushed with 5–10 mL of the SC or EY extenders. Further semen processing and freezing were performed as described in Experiment 1—Sections 2.1.3 and 2.1.4. The assessment of sperm motility parameters, PMI and HMMP, was performed for all samples before (pre-cooling) and after cooling (post-cooling) and pre-freezing and post-thaw, as described in Sections 2.4 and 2.5.

2.3. *Experiment 3: Cooled-Transported Epididymides for Semen Cooling and Freezing*

2.3.1. Animals and Castration

Thirty-six feral small standard jacks (ranging from 2 to 17 years old, 100–300 kg in body weight) were housed in a 172-acre ranch in San Angelo, TX, USA (31.4638°N, 100.4370°W) at the Peaceful Valley Donkey Rescue. This experiment was performed to determine the feasibility of dissecting the epididymides away from the testicles and shipping overnight in a passive cooling container to a referral laboratory equipped to perform semen cryopreservation and to investigate the epididymal semen harvesting and processing techniques with a larger number of animals. The jacks were maintained in a herd of approximately 1000 donkeys, and their diet consisted of free access to coastal bermudagrass, Sudan grass hay, and the grain supplementation of ADM® Sweet UniqueTM 14% (ADM Animal Nutrition™, Quincy, IL, USA). The rescue facility granted us full access to the samples and records to conduct the present study. All jacks were gathered from Goldstone, Fort Irwin, and Butte Valley in California. Only jacks with unremarkable physical examinations were enrolled in this study.

Each donkey was placed under general anesthesia by administering intravenous xylazine and ketamine [22]. Prior to the aseptic preparation of the surgical site, both testicles were evaluated for gross abnormalities. Following anesthetic induction and aseptic preparation of the scrotum, each donkey was castrated with an open technique using a Serra and/or Reimer emasculators as previously described [22]. For each testicle, ligation of the vas deferens was performed while maintaining the longest portion to prevent loss of semen from the epididymis and deferent duct before the spermatic cord was transected. The samples were maintained in an isothermal box until processing within one hour of surgical removal.

2.3.2. Epididymal Processing and Shipping

The epididymides were dissected away from the testes and packed in a disposable, plastic Whirl-Pak® bag containing fresh skim-milk based extender (25 mL, Botusemen), and cooled-shipped in a passive cooling device overnight (Botuflex). The Botuflex consists of two ice packs placed on each side of the device. Ice packs were deep-frozen at −20 °C for 24 h. Three pairs of epididymides individually packed in disposable plastic bags (Whirl-Pak®) were placed in the well of the container.

2.3.3. Epidydimal Semen Processing, Cooling, and Freezing

Samples were randomly processed after arrival (cooled-shipped 24 h $n = 20$ pairs) or in the next day (Cooled-shipped 48 h $n = 16$ pairs). The epididymides were rinsed with Lactated Ringer's Solution at room temperature and wiped with cotton gauze to remove any contamination from blood or debris. Each epididymal tail was weighed before processing. Thereafter, retrograde epididymal flushing was performed as aforementioned in Experiment 1—Section 2.1.2. The lumen of the ductus deferens was cannulated with a 16 Ga × 13 cm (MilaCath®) and flushed with 5–10 mL of the SC extender or EY extenders. Further semen processing and freezing were performed as described in Experiment 1—Sections 2.1.3 and 2.1.4. The assessment of sperm motility parameters, PMI and HMMP was performed for all samples before and after cooling and freezing as described in Sections 2.4 and 2.5.

2.4. Assessment of Sperm Concentration and Motility

Across the three experiments, sperm concentration was determined using an automated cell counter (Nucleocounter® SP-100™, Chemometec, Denmark) following the manufacturer's instructions. Briefly, 50 µL of semen was diluted in 5 mL of lysis buffer (Reagent S100, Chemometec, Denmark) and loaded into the cassettes before the assessment.

Throughout the three experiments, the assessment of the sperm motility parameters was performed using computer-assisted sperm analysis (CASA) with default settings recommended by the manufacturer (Spermvision, Minitube of America, Verona, WI, USA) for equine sperm. The preset values for the CASA were: static cell area 14–100 µm^2, straightness threshold for progressive motility 90%, average path velocity threshold for static cell <9.5 µm/s, light-emitting diode illumination intensity 180–255. Each sample was incubated for 10 min at 38 °C before each evaluation. A small aliquot (10 µL) of extended semen was placed on a pre-heated slide with a coverslip for the assessments. Motility parameters assessed included the total percent of sperm motility (TM), progressive sperm motility (PM), curvilinear velocity (VCL, µm/s), average path velocity (VAP, µm/s), and straight-line velocity (VSL, µm/s).

2.5. Flow Cytometry Analysis

The evaluation of PMI and HMMP was conducted using a spectral flow cytometer, as previously described [23]. Briefly, the staining solution of Zombie Green dye (#423112 Biolegend, San Diego, CA, USA) was freshly prepared with 100 µL of DMSO added to each vial of dye; similarly, MitoTracker Deep Red FM (M22426, Molecular Probes, Eugene, OR, USA) stock solution was prepared by adding DMSO to create a 10 µM solution. The stock solution was aliquoted and frozen at −20 °C until it was used.

One milliliter containing 50 million sperm/mL was centrifuged (600× g × 10 min) and then resuspended in PBS to a concentration of 3–5 million sperm/mL. Subsequently, a 100 µL aliquot of this solution was stained with both dyes (1 µL of Zombie Green and 1 µL MitoTracker Deep Red). After mixing, the sample was incubated for 30 min at room temperature in the dark. The incubation was followed by centrifugation (400× g × 5 min). The supernatant was discarded, and each pellet was fixed with 500 µL of 2% buffered formalin and stored in the dark until flow cytometric evaluation. Before the flow cytometric analysis, samples were washed with 1 mL of PBS, centrifuged at 400× g × 5 min, and resuspended in PBS (250 μL). The analyses of the stained samples were conducted using a full-spectrum detector based (filter-less) Cytek Aurora Flow Cytometer (Cytek Biosciences Inc., Fremont, CA, USA). The analysis was concluded when at least 10,000 fluorescent gated events were recorded. Zombie Green was excited and detected with a 488 nm fluorescence detector, whereas MitoTracker Deep Red was excited with a 644/665 nm detector. Unstained and single-stained controls were used to unmix the signals. As previously described [23], four subpopulations of sperm were identified. The populations of sperm with intact (low Zombie Green signal) or damaged (high Zombie Green signal) plasma membrane were subdivided into low or high mitochondrial membrane potential based on the intensity of the signal given by Mitotracker Deep Red staining. Debris was manually excluded based on the minimal emitted fluorescence. Data from the flow cytometer were exported and analyzed with FlowJo (FlowJo v. 10 Software, Ashland, OR, USA). The PMI and HMMP potential were accounted for comparisons across groups.

2.6. Statistical Analysis

Assumptions of linearity, homogeneity of variance, and normality of the residuals were tested with plots, Levene's, and Shapiro–Wilk test. Data analyses were carried out with RStudio v 3.2.3 (RStudio Team, Boston, MA). Data were analyzed by mixed models with cooled-storage, shipment duration, semen freezing, and the type of extender considered as fixed effects and the individual donkey as a random effect. Tukey's test was used for post-hoc comparisons. Statistical significance was set at $p < 0.05$. A statistically significant tendency was determined with $0.05 < p < 0.1$. All data are presented as mean ± SEM.

3. Results

3.1. Experiment 1. On-Site Epididymal Semen Cooling and Freezing

The retrograde flushing of the epididymal tails was successfully performed on all four pairs. None of the epididymides had appreciable gross lesions. The total number of sperm recovered from each pair was 9.6 ± 2.2 billion (range 5.1–18.1 billion) (Supplementary Materials Table S1). The average weight of each epididymis was 16.7 ± 2.5 g (range 10.1–25.1 g) (Supplementary Materials Table S1).

3.1.1. Cooling

There were no differences in any of the parameters (TM, PM, VAP, VCL, VSL, PMI, and HMMP) measured at any timepoints during cooling or between groups (Table 1) ($p > 0.05$). The progressive motility of the donkey semen harvested in the EY extender immediately after castration tended ($p = 0.08$) to yield greater values compared to the semen recovered in the SC extender (Table 1).

3.1.2. Freezing–Thawing

There was a reduction in post-thaw TM and PM in comparison with pre-freeze values for all groups ($p < 0.05$) (Table 2). Groups Non-Centrif-24 h and Centrif-0 h had a reduction in post-thaw PMI in comparison to the pre-freeze values ($p < 0.05$) (Table 2). Post-thaw HMMP was reduced in comparison to the pre-freeze values for the Non-Centrif-0 h group ($p < 0.05$) but not different for the remaining groups ($p > 0.05$) (Table 2). There were no differences between the groups for the post-thaw results of TM, PM, PMI, and HMMP ($p > 0.05$) (Table 2). Post-thaw, there was a reduction in VAP, VCL, and VSL in comparison to the pre-freeze values for the groups Non-Centrif-0 h and Centrif-0 h ($p < 0.05$) (Table 2). Pre-freeze values of VCL, VSL, and VAP were greater at 0 h than the ones at 24 h groups (Non-Centrif-0 h vs. Non-Centrif-24 h and Centrif-0 h and Centrif-24 h) ($p < 0.05$) (Table 2).

Table 1. Donkey epididymal semen parameters were assessed immediately after semen harvesting (pre-cooling) or after cooling for 24 h (post-cooling). Harvesting was performed within 1 h post-castration (n = 4 pairs). Each epididymal pair was submitted to retrograde flushing with a freezing egg yolk-based extender (EY) or a cooling extender based on sodium caseinate cholesterol-loaded cyclodextrin (SC). Thereafter, the semen extended in EY was cooled for 24 h, whereas the semen extended in SC was cooled for 24 h, or cushion-centrifuged, re-extended with SC, and then cooled for 24 h (C-SC). Data are expressed as mean ± SEM.

	EY		SC		C-SC	
	Pre-Cooling	**Post-Cooling**	**Pre-Cooling**	**Post-Cooling**	**Pre-Cooling**	**Post-Cooling**
TM	88.0 ± 0.6	71.5 ± 4.4	47.0 ± 15.6	51.6 ± 8.6	57.2 ± 13.3	54.4 ± 8.8
PM	81.8 ± 0.9	63.8 ± 4.1	37.4 ± 14.7	39.3 ± 8.5	44.7 ± 13.6	40.1 ± 10.4
VSL	66.8 ± 3.3	55.1 ± 7.1	55.3 ± 8.6	43.6 ± 2.4	52.7 ± 6.0	38.8 ± 1.4
VCL	153.5 ± 4.8	123.5 ± 9.3	135.6 ± 13.4	118.5 ± 11.8	131.9 ± 11.8	112.5 ± 7.2
VAP	82.0 ± 2.7	65.2 ± 6.9	67.9 ± 7.9	56.4 ± 3.0	65.5 ± 6.6	50.8 ± 0.9
PMI	91.1 ± 0.8	84.0 ± 4.5	86.9 ± 3.6	80.2 ± 8.7	86.3 ± 0.7	86.2 ± 0.7
HMMP	94.0 ± 1.8	93.5 ± 1.9	89.0 ± 2.0	93.7 ± 1.4	91.5 ± 1.5	91.9 ± 0.9

TM, total motility (%); **PM**, progressive motility (%); **VAP**, average path velocity (μm/s); **VCL**, curvilinear velocity (μm/s); **VSL**, straight-line velocity (μm/s); **PMI**, plasma membrane integrity (%); **HMMP**, sperm with intact plasma membrane and high mitochondrial membrane potential (%).

Table 2. Pre-freeze and post-thaw parameters of the epididymal semen obtained from freshly castrated donkeys (n = 4 pairs). Each epididymal pair was submitted to retrograde flushing with a semen freezing egg yolk-based extender (EY), or a cooling extender based on sodium caseinate cholesterol-loaded cyclodextrin (SC). Thereafter, the semen extended in EY was frozen immediately after recovery (**Non-Centrif-0 h**) or cooled for 24 h and then frozen (**Non-Centrif-24 h**). The semen extended in **SC** was cushion-centrifuged, resuspended in EY and then frozen (**Centrif-0 h**); or cooled for 24 h before being frozen (**Centrif-24 h**); or cushion-centrifuged, re-extended with SC, and then cooled for 24 h, re-centrifuged and extended **in EY** and then frozen (**C-Centrif-24 h**). Samples were assessed **pre-freezing** and **post-freezing**. Data are expressed as mean ± SEM.

		Non-Centrif		Centrif		C-Centrif
		0 h	**24 h**	**0 h**	**24 h**	**24 h**
Pre-Freezing	**TM**	88.0 ± 0.6 ^{A}	71.5 ± 4.4 ^{A}	86.4 ± 4.4 ^{Aa}	72.7 ± 4.6 ^{Ab}	75.3 ± 3.7 ^{A}
	PM	81.8 ± 0.9 ^{A}	63.8 ± 4.1 ^{A}	80.2 ± 7.4 ^{Aa}	64.0 ± 5.2 ^{Ab}	64.4 ± 6.8 ^{A}
	VSL	66.8 ± 3.3 ^{Aa}	55.1 ± 7.1	67.6 ± 10.1 ^{aA}	44.8 ± 0.8 ^{b}	46.6 ± 2.2 ^{b}
	VCL	153.5 ± 4.8 ^{aA}	123.5 ± 9.3	146.8 ± 16.1 ^{aA}	117.1 ± 1.2 ^{b}	121.8 ± 5.8 ^{b}
	VAP	82.0 ± 2.7 ^{aA}	65.2 ± 6.9	82.2 ± 8.8 ^{aA}	57.6 ± 1.0 ^{b}	58.6 ± 3.3 ^{b}
	PMI	91.1 ± 0.8	84.0 ± 4.5 ^{A}	83.9 ± 6.0 ^{A}	88.2 ± 0.9	86.6 ± 0.9
	HMMP	93.8 ± 1.8	93.5 ± 1.9	94.5 ± 0.8 ^{A}	93.2 ± 3.8	92.9 ± 3.9
Post-Freezing	**TM**	35.2 ± 1.7 ^{B}	31.9 ± 7.7 ^{B}	22.3 ± 5.3 ^{B}	39.8 ± 8.2 ^{B}	36.2 ± 2.2 ^{B}
	PM	26.9 ± 1.3 ^{B}	24.6 ± 6.8 ^{B}	15.4 ± 3.4 ^{B}	31.3 ± 5.9 ^{B}	27.0 ± 1.8 ^{B}
	VSL	40.2 ± 7.3 ^{B}	39.1 ± 0.4	40.3 ± 0.2 ^{B}	36.9 ± 2.5	37.1 ± 0.7
	VCL	111.1 ± 16.6 ^{B}	99.8 ± 4.9	118.4 ± 1.2 ^{B}	93.82 ± 2.2	97.3 ± 9.1
	VAP	52.0 ± 7.0 ^{aB}	48.0 ± 0.6	52.5 ± 4.0 ^{aB}	45.2 ± 2.0 ^{b}	45.3 ± 1.3 ^{b}
	PMI	70.0 ± 8.6	56.7 ± 10.6 ^{B}	47.2 ± 7.3 ^{B}	68.8 ± 2.2	67.9 ± 3.2
	HMMP	90.3 ± 0.9	91.0 ± 1.9	78.9 ± 3.7 ^{B}	89.5 ± 4.2	87.6 ± 1.9

TM, total motility (%); **PM**, progressive motility (%); **VAP**, average path velocity (μm/s); **VCL**, curvilinear velocity (μm/s); **VSL**, straight-line velocity (μm/s); **PMI**, plasma membrane integrity (%); **HMMP**, sperm with intact plasma membrane and high mitochondrial membrane potential (%). Different superscripts denote differences within (^{AB}) or between (^{ab}) columns. ($p < 0.05$).

3.2. Experiment 2. Cooled-Transported Testicles and Epididymides for Semen Cooling and Freezing

The total number of sperm recovered was 6.4 ± 0.7 billion (range 0.4–16.2 billion) (Supplementary Materials Table S2). One donkey was excluded from the study due to insufficient sperm counts to assess all the different endpoints. The average weight for each epididymis was 25.1 ± 2.0 g (range 10.8–56.7 g) (Supplementary Materials Table S2). None of the epididymides had gross morphological abnormalities. The average temperature of the samples upon arrival was 7.6 ± 0.7 °C (range 5–9.7 °C, cooled-shipped 24 h), and 15.5 ± 0.4 °C (range 14–17 °C, cooled-shipped 48 h).

3.2.1. Cooling

There were no differences across groups for any of the endpoints (TM, PM, PMI, HMMP, VCL, VSL, and VAP) assessed pre- and post-cooling or between samples allocated to the cooled-shipped 24 h and the cooled-shipped 48 h ($p > 0.05$) (Table 3).

Table 3. Sperm velocity and viability parameters of donkey epididymal semen cooling harvested from cooled shipped epididymides. The scrotal content was shipped in a passive cooling semen container. The epididymides were processed upon arrival ($n = 8$ pairs, **cooled-shipped 24 h**) or the following day ($n = 6$ pairs, **cooled-shipped 48 h**). All epididymides were kept in the passive cooling device until processing. For each of the time points, one epididymis of each pair was submitted to retrograde flushing with a freezing egg yolk-based extender (**EY**), while the contralateral epididymis was flushed with a cooling extender based on sodium caseinate cholesterol-loaded cyclodextrin (**SC**). Thereafter, the semen extended in EY was cooled for 24 h; whereas the semen extended in **SC** was cooled for 24 h, or cushion-centrifuged, re-extended with SC, and then cooled for 24 h (**C-SC**). Semen was assessed immediately after harvesting (**pre-cooling**) and after cooling for 24 h (**post-cooling**). Data are expressed as the mean ± SEM.

		Cooled-Shipped 24 h			**Cooled-Shipped 48 h**		
		EY	SC	C-SC	EY	SC	C-SC
Pre-Cooling	**TM**	67.7 ± 7.5	55.6 ± 7.6	65.6 ± 7.4	67.6 ± 5.7	58.3 ± 12.7	73.3 ± 7.1
	PM	58.0 ± 8.2	44.2 ± 7.0	59.5 ± 7.9	58.8 ± 6.6	52.1 ± 11.9	66.0 ± 6.7
	VSL	58.8 ± 4.3	49.1 ± 3.9	66.9 ± 3.8	55.0 ± 1.4	59.7 ± 5.0	68.0 ± 4.0
	VCL	134.5 ± 8.8	121.0 ± 4.4	149.2 ± 5.0	143.2 ± 3.2	146.5 ± 7.2	148.8 ± 2.6
	VAP	71.7 ± 4.6	62.0 ± 3.3	80.7 ± 4.0	64.7 ± 5.5	69.2 ± 3.2	68.8 ± 2.8
	PMI	78.5 ± 8.9	87.0 ± 2.2	83.8 ± 2.4	82.1 ± 10.0	85.7 ± 2.7	86.8 ± 1.1
	HMMP	94.2 ± 1.6	88.9 ± 2.4	91.2 ± 1.5	75.2 ± 18.8	89.0 ± 3.4	88.2 ± 2.5
Post-Cooling	**TM**	64.0 ± 6.2	40.0 ± 8.2	44.0 ± 7.6	71.3 ± 1.3	60.4 ± 9.7	60.8 ± 8.6
	PM	55.8 ± 8.3	33.8 ± 7.7	37.1 ± 7.6	64.1 ± 1.5	53.8 ± 9.2	56.1 ± 7.2
	VSL	50.8 ± 5.3	56.6 ± 3.3	57.4 ± 3.3	55.1 ± 7.7	61.9 ± 2.2	61.9 ± 2.4
	VCL	138.1 ± 13.5	136.7 ± 4.0	135 ± 1.9	133.4 ± 19.0	145.1 ± 3.1	139.1 ± 5.0
	VAP	70.6 ± 2.1	75.4 ± 4.8	83.5 ± 4.1	71.4 ± 7.6	76.0 ± 1.4	73.1 ± 2.3
	PMI	84.8 ± 6.1	82.9 ± 4.1	78.6 ± 4.1	90.2 ± 1.3	83.4 ± 2.5	75.8 ± 3.4
	HMMP	88.6 ± 2.9	86.9 ± 2.1	78.9 ± 7.6	91.5 ± 4.7	88.3 ± 1.7	84.3 ± 4.2

TM, total motility (%); **PM**, progressive motility (%); **VAP**, average path velocity (μm/s); **VCL**, curvilinear velocity (μm/s); **VSL**, straight-line velocity (μm/s); **PMI**, plasma membrane integrity (%); **HMMP**, sperm with intact plasma membrane and high mitochondrial membrane potential (%).

3.2.2. Freezing–Thawing

There were no differences between the cooled-shipped 24 h vs. the cooled-shipped 48 h ($p > 0.05$) (Table 4). Post-thaw, there was an overall reduction in TM and PM across the groups in comparison to the pre-freeze values (Table 4) ($p < 0.05$). However, there were no differences between groups for pre-freeze and post-thaw TM and PM ($p > 0.05$). Pre- and post-freezing VCL and VSL did not differ across groups ($p > 0.05$) (Table 4). Post-freezing VAP decreased in comparison to pre-freeze for the Non-Centrif-24 h group ($p < 0.05$), but it was not different for other groups ($p > 0.05$) (Table 4). There was a reduction of HMMP in post-thaw samples in comparison to pre-freeze values for group Centrif-24 h ($p < 0.05$), but it was not different between other groups ($p > 0.05$) (Table 4).

Table 4. Pre- and post-freezing parameters of the epididymal semen obtained from donkey cooled-shipped epididymides attached to the testes. All epididymides were kept in the passive cooling device until processing. The epididymides were processed upon arrival ($n = 8$ pairs) (**cooled-shipped 24 h**) or the following day ($n = 6$ pairs) (**cooled-shipped 48 h**). One epididymis of each pair was submitted to retrograde flushing with a semen freezing egg yolk-based extender (EY), while the contralateral was flushed with a cooling extender based on sodium caseinate cholesterol-loaded cyclodextrin (SC). Thereafter, the semen extended in EY was frozen immediately after recovery (**Non-Centrif-0 h**) or cooled for 24 h and then frozen (**Non-Centrif-24 h**). The semen extended in SC was cushion-centrifuged, resuspended in EY and then frozen (**Centrif-0 h**); or cooled for 24 h before being frozen (**Centrif-24 h**); or cushion-centrifuged, re-extended with SC, and then cooled for 24 h, re-centrifuged and extended in EY and then frozen (**C-Centrif-24 h**). Data are expressed as mean ± SEM.

		Cooled-Shipped 24 h					Cooled-Shipped 48 h				
		Non-Centrif		Centrif		C-Centrif	Non-Centrif		Centrif		C-Centrif
		0 h	24 h	0 h	24 h	24 h	0 h	24 h	0 h	24 h	24 h
Pre-Freezing	**TM**	67.7 ± 7.5 ^{A}	64.0 ± 6.2 ^{A}	74.6 ± 5.3 ^{A}	68.3 ± 6.4 ^{A}	59.5 ± 6.8 ^{A}	67.6 ± 5.7 ^{A}	71.3 ± 1.3 ^{A}	72.7 ± 8.8 ^{A}	64.3 ± 7.6 ^{A}	64.4 ± 9.0 ^{A}
	PM	58.0 ± 8.2 ^{A}	55.8 ± 8.4 ^{A}	68.6 ± 5.5 ^{A}	61.2 ± 7.9 ^{A}	50.0 ± 8.9 ^{A}	58.8 ± 6.6	64.1 ± 1.5	67.0 ± 8.8	57.8 ± 7.8	57.9 ± 8.8
	VSL	58.8 ± 4.3	50.9 ± 5.3	60.0 ± 2.8	54.5 ± 5.7	54.5 ± 4.5	55.0 ± 1.4	55.1 ± 7.7	56.6 ± 2.8	53.2 ± 5.8	53.7 ± 5.3
	VCL	134.5 ± 8.8	138.1±13.5	134.1 ± 6.9	131.2 ± 6.6	123.8 ± 5.9	143.2 ± 3.2	133.4 ± 19.1	132.6 ± 4.7	126.6 ± 7.3	121.7 ± 9.3
	VAP	71.7 ± 4.6	64.6 ± 5.5	71.6 ± 2.9	67.9 ± 5.7	65.8 ± 4.4	70.6 ± 2.1	71.4 ± 7.6	70.6 ± 3.8	64.9 ± 6.3	64.5 ± 6.1
	PMI	78.5 ± 8.9	84.8 ± 6.1	87.6 ± 3.8	82.6 ± 7.9	85.0 ± 1.7	82.1 ± 10.0	90.2 ± 1.3	90.0 ± 4.2	85.8 ± 1.1	74.3 ± 6.6
	HMMP	94.2 ± 1.6	88.6 ± 2.9	87.8 ± 3.9	72.5 ± 8.2	86.5 ± 2.7	75.2 ± 18.8	91.5 ± 4.7	91.8 ± 1.8	88.2 ± 4.6	82.7 ± 5.2
Post-Freezing	**TM**	22.8 ± 4.1 ^{B}	19.6 ± 4.9 ^{B}	30.8 ± 4.3 ^{B}	22.1 ± 2.9 ^{B}	25.2 ± 4.0 ^{B}	20.4 ± 3.0 ^{B}	25.1 ± 4.2 ^{B}	25.6 ± 3.4 ^{B}	26.0 ± 0.1 ^{B}	20.2 ± 3.5 ^{B}
	PM	15.2 ± 4.1 ^{B}	14.5 ± 4.7 ^{B}	21.1 ± 3.5 ^{B}	12.6 ± 2.3 ^{B}	13.9 ± 3.0 ^{B}	11.5 ± 3.1 ^{B}	15.1 ± 3.4 ^{B}	15.4 ± 3.4 ^{B}	13.4 ± 1.1 ^{B}	10.5 ± 2.3 ^{B}
	VSL	40.6 ± 2.8	42.5 ± 4.2	46.1 ± 1.5	36.1 ± 3.1	40.3± 2.6	45.7 ± 1.4	41.0 ± 4.1	40.6 ± 2.4	38.1 ± 0.3	38.9 ± 1.5
	VCL	100.8 ± 6.5	100.6 ± 12.8	118.0 ± 3.2	99.7 ± 4.8	107.9 ± 3.2	127.4 ± 3.1	116.6 ± 6.5	112.5 ± 5.3	103.5 ± 1.4	109.5 ± 3.5
	VAP	50.2 ± 2.8	51.3 ± 5.0	56.3 ± 1.3	45.7 ± 3.3	51.1 ± 2.5	59.2 ± 1.5	51.8 ± 4.4	50.7 ± 3.2	46.6 ± 0.7	49.4 ± 2.1
	PMI	46.5 ± 5.4	41.0 ± 9.4	63.6 ± 4.3	59.2 ± 3.3	57.7 ± 3.2	51.7 ± 4.3	46.6 ± 9.7	46.3 ± 5.4	60.0 ± 4.7	58.7 ± 7.4
	HMMP	87.5 ± 3.2	87.3 ± 3.0	91.5 ± 1.6	94.0 ± 1.1	91.7 ± 1.7	86.6 ± 4.6	90.3 ± 2.5	90.0 ± 5.0	91.6 ± 6.5	72.5 ± 21.5

TM, total motility (%); **PM**, progressive motility (%); **VAP**, average path velocity (μm/s); **VCL**, curvilinear velocity (μm/s); **VSL**, straight-line velocity (μm/s); **PMI**, plasma membrane integrity (%); **HMMP**, sperm with intact plasma membrane and high mitochondrial membrane potential (%). Different superscripts denote differences within (^{AB}) columns. ($p < 0.05$).

3.3. Experiment 3. Cooled-Transported Epididymides Tails for Semen Cooling and Freezing

The retrograde flushing of the epididymis tail was successfully performed on all thirty-six pairs; total sperm count average was 8.4 ± 0.3 billion (range 0.4–19 billion) from each epididymis. The average weight of each epididymis was 28.5 ± 0.6 g (range 20.8–45 g) (Supplementary Materials Table S3). The average temperature of the samples upon arrival was 9.4 ± 0.5 °C (range 7–13 °C, cooled-shipped 24 h), and 17.2 ± 0.5 °C (range 14–20 °C, cooled-shipped 48 h).

3.3.1. Cooling

Semen harvested in the EY extender had greater TM and PM than SC and C-SC for both cooled-shipped 24 h and cooled-shipped 48 h ($p < 0.05$) (Table 5). It is noteworthy that PM in the EY group increased after 24 h of cooling in a passive-cooling device for both cooled-shipped 24 h and cooled-shipped 48 h ($p < 0.05$) (Table 5).

Table 5. Motility and viability parameters of the donkey epididymal semen harvested from cooled-shipped epididymides dissected away from the testes. The epididymides were processed upon arrival (n = 20 pairs, **cooled-shipped 24 h**) or the following day post-arrival (n = 16 pairs, **cooled-shipped 48 h**). All epididymides were kept in the passive cooling device until processing. For each of the time points, each epididymal pair was submitted to retrograde flushing with a freezing egg yolk-based extender (**EY**), or a cooling extender based on sodium caseinate cholesterol-loaded cyclodextrin (**SC**). Thereafter, the semen extended in EY was cooled for 24 h; whereas the semen extended in SC was cooled for 24 h, or cushion-centrifuged, re-extended with SC, and then cooled for 24 h (**C-SC**). Semen was assessed immediately after harvesting (**pre-cooling**) and after cooling for 24 h (**post-cooling**). Data are expressed as the mean ± SEM.

		Cooled-Shipped 24 h			Cooled-Shipped 48 h		
		EY	SC	C-SC	EY	SC	C-SC
Pre-Cooling	**TM**	43.0 ± 3.9 ^{a}	22.3 ± 4.5 ^{b}	25.9 ± 5.2 ^{b}	43.3 ± 4.7 ^{a}	21.7 ± 3.1 ^{b}	18.7 ± 2.4 ^{b}
	PM	31.7 ± 3.6 ^{aA}	15.3 ± 4.2 ^{b}	16.6 ± 5.0 ^{b}	32.8 ± 4.7 ^{aA}	12.9 ± 2.3 ^{b}	10.2 ± 1.8 ^{b}
	VSL	44.0 ± 1.9	35.0 ± 4.0	38.1 ± 3.1	45.0 ± 2.1	38.3 ± 2.7	45.7 ± 3.6
	VCL	107.5 ± 3.8 ^{a}	79.3 ± 9.1 ^{b}	98.4 ± 5.5	114.1 ± 5.1	97.8 ± 6.2	105.0 ± 6.8
	VAP	53.6 ± 2.1	42.6 ± 4.6	46.8 ± 3.4	56.8 ± 2.7	48.6 ± 3.1	54.9 ± 4.0
	PMI	76.0 ± 4.2	72.5 ± 4.0 ^{a}	74.0 ± 2.5	57.5 ± 5.0 ^{A}	54.9 ± 6.2 ^{b}	44.8 ± 4.5
	HMMP	80.3 ± 5.4	82.9 ± 4.7	77.3 ± 6.8	84.3 ± 4.5	83.6 ± 3.6	83.0 ± 3.9
Post-Cooling	**TM**	56.8 ± 5.9	21.9 ± 3.8	25.6 ± 4.5	53.6 ± 5.2	24.3 ± 2.8	23.0 ± 2.9
	PM	48.1 ± 5.7 ^{B}	13.8 ± 3.1	15.1 ± 3.7	44.1 ± 4.9 ^{B}	15.4 ± 2.3	11.7 ± 2.0
	VSL	44.9 ± 2.3	36.2 ± 2.3	35.8 ± 2.2	42.6 ± 1.6	36.1 ± 2.7	43.5 ± 3.8
	VCL	111.0 ± 5.0	83.4 ± 6.6	94.2 ± 4.0	110.2 ± 5.1	87.9 ± 6.1	100.8 ± 3.0
	VAP	56.9 ± 2.8	44.5 ± 2.7	44.8 ± 2.6	54.5 ± 2.4	45.6 ± 3.1	52.8 ± 3.5
	PMI	78.9 ± 4.2	65.8 ± 6.1	72.0 ± 5.8	77.6 ± 2.9 ^{B}	61.9 ± 4.3	62.5 ± 2.4
	HMMP	88.9 ± 2.5	75.8 ± 6.7	83.8 ± 5.2	87.6 ± 2.1	81.8 ± 2.6	78.2 ± 5.4

TM, total motility (%); **PM**, progressive motility (%); **VAP**, average path velocity (μm/s); **VCL**, curvilinear velocity (μm/s); **VSL**, straight-line velocity (μm/s); **PMI**, plasma membrane integrity (%); **HMMP**, sperm with intact plasma membrane and high mitochondrial membrane potential (%). Different superscripts denote differences within (^{AB}) or between (^{ab}) columns. ($p < 0.05$).

There were no differences in PMI for any of the groups for cooled-shipped 24 h ($p > 0.05$) (Table 5). In the cooled-shipped 48 h, PMI increased after cooling in comparison to pre-cooling for EY ($p < 0.05$). In addition, after centrifugation (SC), the PMI was lower in cooled-shipped 48 h than in the cooled-shipped 24 h ($p = 0.0006$) (Table 5). There were no differences between groups for VAP and VSL across timepoints and processing ($p > 0.05$), whereas the VCL was higher in the EY semen than in the SC semen cooled-shipped for 24 h ($p < 0.05$) (Table 5).

3.3.2. Freezing–Thawing

There was a reduction in post-thaw TM and PM across groups and cooled-shipped 24 h and cooled-shipped 48 h ($p < 0.05$) (Table 6). In cooled-shipped 24 h, the pre-freeze values of PM in Non-Centrif-24 h were greater than the one in Non-Centrif-0 h ($p < 0.05$) (Table 6). The pre-freeze values of PM in Non-Centrif-24 h was greater than those in Centrif-24 h in cooled-shipped 24 h ($p < 0.05$) (Table 6).

Table 6. Pre- and post-freezing parameters of the epididymal semen obtained from donkey cooled-shipped epididymides (**cooled-shipped 24 h**) dissected from the testes. All epididymides were kept in the passive cooling device until processing. The epididymides were processed upon arrival (n = 20 pairs, **cooled-shipped 24 h**) or the following day post-arrival (n = 16 pairs, **cooled-shipped 48 h**). Each epididymal pair was submitted to retrograde flushing with a semen freezing egg yolk-based extender (EY), or a cooling extender based on sodium caseinate cholesterol-loaded cyclodextrin (SC). Thereafter, the semen extended in EY was frozen immediately after recovery (**Non-Centrif-0 h**) or cooled for 24 h and then frozen (**Non-Centrif-24 h**). The semen extended in SC was cushion-centrifuged, resuspended in EY and then frozen (**Centrif-0 h**); or cooled for 24 h before being frozen (**Centrif-24 h**); or cushion-centrifuged, re-extended with SC, and then cooled for 24 h, re-centrifuged and extended in EY and then frozen (**C-Centrif-24 h**). Data are expressed as the mean ± SEM.

		Cooled-Shipped 24 h					Cooled-Shipped 48 h				
		Non-Centrif		Centrif		C-Centrif	Non-Centrif		Centrif		C-Centrif
		0 h	24 h	0 h	24 h	24 h	0 h	24 h	0 h	24 h	24 h
Pre-Freezing	**TM**	43.0 ± 3.9 ^{A}	56.8 ± 5.9 ^{A}	43.9 ± 4.1 ^{A}	43.9 ± 4.2 ^{A}	50.8 ± 4.8 ^{A}	43.3 ± 4.7 ^{A}	53.6 ± 5.2 ^{Aa}	49.0 ± 4.1 ^{A}	36.0 ± 4.8 ^{Ab}	39.7 ± 3.9 ^{A}
	PM	31.7 ± 3.6 ^{aA}	48.0 ± 5.7 ^{Ab}	33.2 ± 4.3 ^{A}	30.7 ± 4.0 ^{Aa}	37.2 ± 4.7 ^{A}	32.8 ± 4.7 ^{A}	44.1 ± 4.9 ^{Aa}	39.6 ± 4.3 ^{A}	27.1 ± 4.1 ^{Ab}	27.0 ± 3.8 ^{A}
	VSL	44.0 ± 1.9	44.9 ± 2.3 ^{A}	44.8 ± 2.5 ^{Aa}	37.0 ± 1.5 ^{Ab}	40.3 ± 1.9	45.2 ± 2.1 ^{A}	42.6 ± 1.6 ^{A}	47.5 ± 2.1 ^{A}	35.8 ± 2.3	36.3 ± 1.8
	VCL	107.5 ± 3.8 ^{A}	111.0 ± 5.0 ^{A}	108.8 ± 4.7 ^{Aa}	101.6 ± 3.7 ^{Ab}	105.7 ± 3.3	114.1 ± 5.1	110.2 ± 5.1 ^{A}	114.7 ± 4.5 ^{a}	93.0 ± 7.1 ^{Bb}	95.2 ± 4.0 ^{Bb}
	VAP	53.6 ± 2.1	56.9 ± 2.7 ^{A}	54.6 ± 2.7 ^{A}	48.3 ± 2.0 ^{B}	51.2 ± 2.2	56.8 ± 2.7 ^{A}	54.5 ± 2.4 ^{A}	58.9 ± 2.5 ^{Aa}	46.4 ± 2.9 ^{b}	46.7 ± 2.1 ^{b}
	PMI	76.0 ± 4.2	76.0 ± 4.2 ^{A}	76.8 ± 2.4 ^{A}	68.7 ± 7.0 ^{A}	72.8 ± 9.3	57.5 ± 5.0 ^{a}	77.6 ± 2.9 ^{b}	61.9 ± 3.8 ^{A}	69.1 ± 4.6 ^{A}	73.6 ± 2.7 ^{A}
	HMMP	80.3 ± 5.4 ^{a}	80.3 ± 5.4 ^{a}	52.8 ± 10.1 ^{b}	76.4 ± 7.5 ^{a}	91.0 ± 2.0 ^{a}	84.3 ± 4.5	87.6 ± 2.1	76.2 ± 5.7	83.0 ± 5.4	90.3 ± 1.6
Post-Freezing	**TM**	17.6 ± 1.3 ^{B}	22.5 ± 2.3 ^{B}	20.4 ± 2.0 ^{B}	15.7 ± 2.1 ^{B}	20.3 ± 2.0 ^{B}	13.3 ± 1.2 ^{B}	19.8 ± 2.1 ^{B}	20.0 ± 2.2 ^{B}	14.2 ± 1.6 ^{B}	17.5 ± 1.7 ^{B}
	PM	10.6 ± 1.3 ^{B}	15.1 ± 2.2 ^{B}	11.0 ± 1.5 ^{B}	8.0 ± 1.8 ^{B}	10.5 ± 1.6 ^{B}	6.9 ± 1.0 ^{B}	12.9 ± 2.1 ^{B}	11.9 ± 1.5 ^{B}	6.7 ± 1.2 ^{B}	8.5 ± 1.1 ^{B}
	VSL	36.3 ± 1.7 ^{B}	33.6 ± 1.7 ^{B}	33.9 ± 1.6 ^{B}	28.6 ± 2.0 ^{B}	34.0 ± 1.3	33.9 ± 1.4 ^{B}	34.0 ± 1.3 ^{B}	34.7 ± 0.9 ^{B}	30.6 ± 1.5	32.4 ± 0.9
	VCL	88.9 ± 3.7	84.4 ± 4.6 ^{B}	87.7 ± 4.4 ^{B}	72.0 ± 7.2 ^{B}	92.9 ± 2.1	94.5 ± 3.7	81.7 ± 4.6 ^{B}	95.4 ± 3.1	80.4 ± 4.6	90.9 ± 3.7
	VAP	44.1 ± 1.9 ^{B}	41.1 ± 1.8 ^{B}	41.4 ± 1.8 ^{B}	35.5 ± 2.7 ^{B}	42.2 ± 1.3	42.4 ± 1.5 ^{B}	40.7 ± 1.6 ^{B}	44.0 ± 1.2 ^{B}	38.8 ± 1.5	42.5 ± 1.2
	PMI	44.8 ± 3.1 ^{B}	41.0 ± 2.1 ^{B}	53.6 ± 3.1 ^{B}	41.4 ± 4.8 ^{B}	53.3 ± 2.9	44.7 ± 2.7	49.3 ± 3.8	50.3 ± 2.3 ^{B}	47.6 ± 3.7 ^{B}	53.4 ± 2.5 ^{B}
	HMMP	83.0 ± 2.0	84.4 ± 2.9	86.4 ± 1.9	79.1 ± 4.8	83.7 ± 4.7	82.6 ± 2.7	90.7 ± 1.5	86.8 ± 2.2	80.3 ± 6.2	87.0 ± 2.2

TM, total motility (%); **PM**, progressive motility (%); **VAP**, average path velocity (µm/s); **VCL**, curvilinear velocity (µm/s); **VSL**, straight-line velocity (µm/s); **PMI**, plasma membrane integrity (%); **HMMP**, sperm with intact plasma membrane and high mitochondrial membrane potential (%). Different superscripts denote differences within (^{AB}) or between (^{ab}) columns. ($p < 0.05$).

Post-thaw velocity parameters (VSL, VAP, VCL) were reduced in cooled-shipped 24 h across groups in comparison to the pre-freeze values ($p < 0.05$), except for group C-Centrif-24 h which maintained its pre-freeze values ($p > 0.05$) (Table 6). Pre-freezing values of VSL and VCL were greater in Centrif-0 h than in Centrif-24 h ($p < 0.05$). Post-thaw PMI decreased in comparison to pre-freeze values for all groups ($p < 0.05$), except for semen in C-Centrif-24 h group, which maintained its pre-freezing values ($p > 0.05$) (Table 6). The population of sperm HMMP did not change from pre-freeze to post-thaw ($p > 0.05$) (Table 6). Pre-freeze HMMP was higher in Non-Centrif-0 h than semen in Centrif-0 h ($p < 0.05$) (Table 6). Moreover, cooled-shipped 24 h pre-freezing values of semen extended in Centrif-0 h group had less sperm with HMMP than any of the other groups at 24 h ($p < 0.05$) (Table 6).

Cooled-shipped 48 h pre-freeze values of TM and PM were greater in Non-Centrif-24 h than in Centrif-24 h ($p < 0.05$). Post-thaw VSL and VAP decreased in comparison with pre-freeze values in groups Non-Centrif-0 h, Centrif-0 h, and Non-Centrif-24 h. Moreover, post-thaw VCL decreased in group Non-Centrif-24 h in comparison with pre-freeze values ($p < 0.05$) (Table 6). Pre-freeze values of VCL and VAP were greater in the Centrif-0 h group than groups Centrif-24 h and C-Centrif-24 h ($p < 0.05$). The values of PMI pre-freeze of semen extended in Non-Centrif-24 h was greater than in Non-Centrif-0 h ($p < 0.05$) (Table 6). Post-thaw HMMP did not change after freezing with any extender or methods or processing ($p > 0.05$) (Table 6).

4. Discussion

The present study was designed to assess donkey epididymal semen cooling and freezing, from freshly harvested and cooled-shipped specimens. The first experiment was conducted to determine whether freshly castrated donkeys could have epididymal semen harvested, cooled, and then frozen. The second experiment was conducted to determine the feasibility of collecting, packaging, and shipping testes attached to the epididymis and spermatic cord to a referral laboratory for epididymal semen harvesting, cooling, and freezing. This design was implemented to simulate a scenario when a layperson or an inexperienced practitioner was unable to effectively dissect the epididymides from the testes before shipping to a referral laboratory. Additionally, a subset of the shipment was processed after delivery (cooled-shipped 24 h), and another subset was processed the next day (cooled-shipped 48 h). The delay in processing the samples was created to mimic common problems encountered in clinical practice. It has been estimated that about 5% of the semen shipments in the United States may be delayed for various reasons. Furthermore, animals located in rural and remote indigenous locations cannot have their testes and/or epididymis timely shipped to a referral laboratory for further processing. Experiment three was conducted with a larger sample size to determine the feasibility of dissecting the epididymal tail, packaging, shipping, and further processing for cooling and freezing. In addition, shipping the epididymides dissected from the testes minimizes the potential for blood contamination and debris, decreases the amount of space occupied in a passive cooling container and shipping charges. In this experiment, a similar approach for delayed processing (cooled-shipped 24 h vs. 48 h) was performed.

In all three experiments, the concept of cooling donkey semen after harvesting was assessed to simulate cooled-shipped semen for potential artificial insemination with cooled semen or as a means to pre-process semen before freezing. Cooling epididymal semen after harvesting could be a useful strategy if a popular jack needs to be castrated or euthanized in the middle of the breeding season when there are mares or jennies lined up to be inseminated. Breeding mares or jennies with cooled semen could circumvent the problems with frozen semen (e.g., frequent follicular checks, low fertility, and excessive post-breeding endometrial inflammation) [24]. Interestingly, the cooling extender used in the present study contains cholesterol-loaded cyclodextrin (CLC). Although this compound had not been tested yet to cool donkey epididymal semen, satisfactory fertility rates have been obtained in mares inseminated with donkey ejaculated semen extended in the CLC SC extender (Papa FO, personal communication). In addition, CLC has been reported to improve the sperm viability of post-thawed donkey semen with no detrimental effect on fertility rates in mares [25]. The addition of CLC improved semen quality and fertility of cooled stallion semen [25,26]. The CLC has also been reported in one study to decrease cooling-induced damages and to improve the semen freezing ability of stallion's epididymal sperm [27].

Cooling epididymal semen before freezing was assessed as a strategy herein; the authors were interested in answering whether donkey semen could be cooled overnight before being frozen the next day. Such a strategy could prove to be useful in clinical practice. For instance, if a donkey needs to have epididymal semen harvested and processed outside regular working hours, the findings of the present study suggest that semen can be harvested, cooled, and frozen the next day. Some mature jacks have a large amount of semen, which requires a long processing time. The jacks used in the present study were in general, of the small frame (except two among the young animals used in Experiment 1); however, some yielded close to 40 billion sperm. Larger breeds such as Mammoth, Poitou, and Catalan can easily surpass that amount [2]. While the fertility of this type of semen has not been assessed, the present study provided an introductory concept for its use and future studies.

Across experiments, epididymal semen cooling displayed satisfactory parameters, and semen flushed with EY tended to yield superior results than semen flushed with SC. The use of a freezing extender for cooling semen may come as a surprise to investigators as it has been proposed that stallion and donkey semen are sensitive to glycerol and high concentrations of this cryoprotectant affect the fertility of semen extended in milk-based or milk-based protein extenders [28]. Vidament and collaborators proposed that donkey semen extended with a milk-based extender containing glycerol used to breed mares is less sensitive to glycerol-toxicity than the same semen used to breed jennies, as the final concentration affecting the fertility of donkey semen in mares was ~5% and approximately 2% in jennies [28–30]. The EY extender used in the present study contained 1% of glycerol and 4% of formamide, approximately half of the toxic glycerol concentration proposed by Vidament et al. [28]. A report comparing ejaculated donkey semen extended in EY, SC and milk-based extender cooled for 24 h demonstrated that the EY extender resulted in superior sperm motility parameters [31].

A study demonstrated that epididymal semen has superior cooling ability than ejaculated stallion semen; this does not seem to be the case in donkeys [32]. While the present study did not compare donkey ejaculated sperm with donkey epididymal semen, the overall results of the semen parameters appear inferior to that expected with cooled ejaculated donkey semen [2]. However, it remains to be determined if cooled donkey epididymal semen extended in the EY used in the present study can result in satisfactory fertility when used to inseminate mares and jennies. It is also unknown if the same formula of the EY extender used herein deprived of glycerol and formamide would result in superior results than those obtained herein and by previous authors. Finally, donkey semen extended in EY-based extender seems to behave similarly to ruminant semen regarding glycerol exposure during cooling.

While the present design was not set to compare semen in the three different experiments, a subjective comparison across experiments seems to trend towards a reduction in TM and PM between the freshly harvested epididymal semen and the cooled-shipped semen. In stallions, it has been shown that epididymal semen can be stored at 5 °C for up to 96 h before being processed [19]. However, the apparent differences could have also been due to the different sample sizes, husbandry practices, levels of stress, weather, and other potential confounding factors not identified. The shipment of the entire scrotal content (Experiment 2) rather than the dissected epididymides (Experiment 3) seemed to yield superior sperm parameters; however, the comparison was not included in the statistical analyses. Several variables can be accounted for the different results. The temperature for both passive cooling containers was equivalent both on the day of arrival and 24 h later. The donkeys included in Experiment 2 appeared to be more mature than the donkeys used in Experiment 3 and were more likely to be kept in a more stable environment. While a cross-over design involving both passive cooling containers were not tested, the authors believe that the apparent different results cannot be accounted for by the type of container. It is also unclear whether this difference is associated with sample preparation. Upon arrival, both types of containers had a similar internal temperature; due to each container's different features, it is reasonable to suggest that the cooling curve might have been different. In Experiment 2, the testes were placed in direct contact with the iced can, without the use of an isothermolizer cup. It is reasonable to assume that the presence of the testicles inside the shipper could have mitigated the reduction temperature, creating a less steep cooling curve. Further investigations are needed to evaluate the effect of different containers and cooling curves on the quality of the semen harvested from the epididymis.

The results of the present study with the donkey epididymal semen parameters were lower than those reported in the literature with raw or frozen-thawed donkey ejaculated semen [2]. Interestingly, there was an overall high percentage of PMI with HMMP. Across mammalian species, sperm stored at the tail of the epididymis is immotile due to plasma membrane-bound proteins that prevent premature motility and loss of viability [33,34]. In stallions, the EY extender used in the present study was shown to activate the motility of epididymal semen [8]; thus, the protocol of directly harvesting epididymal semen with this extender became popular in equine clinical practice and has been used in numerous investigations [6,32]. Semen centrifugation has been used as a means to potentially activate epididymal sperm motility in mice by removing plasma membrane-bound proteins [35]. In addition, centrifugation is a mandatory step for the cryopreservation of ejaculated semen of a jack or stallion to improve sperm interaction with cryoprotectants by removing the seminal plasma [2,9]. However, sperm harvested from the epididymis does not have seminal plasma; therefore, the need for centrifugation is questionable, as a very high sperm concentration is often yielded. Some authors do not recommend centrifugation before freezing epididymal semen in stallions [6,18]. One study found no differences in the post-thaw parameters for semen flushed and frozen with the EY extender used herein or semen flushed with a milk-based extender, cushion-centrifuged and then extended and frozen in the same EY extender [6]. Our results in the present experiment appear to concur with the latter study; in fact, it appears to suggest that centrifugation may not be beneficial for donkey epididymal semen cooling or freezing. However, since we did not centrifuge semen flushed with the EY extender, it is possible that some of the differences observed herein could have been simply the difference between extenders rather than centrifugation. In addition, since semen freezing extenders are more expensive than cooling extenders, the use of EY could be cost-prohibitive in practice; thus, this approach was not tested. In addition, the sperm loss associated with traditional centrifugation (600 g × 10 min, ~40%) or cushioned-centrifugation (1000 g × 20 min, ~10%) may be discouraging in using either approach as the last chance to cryopreserve semen from a valuable stallion [21].

5. Conclusions

In conclusion, freshly harvested, cooled-shipped, and cooled semen had satisfactory semen parameters. The post-thaw results revealed low motility parameters across any type of extender or methods of processing; unexpectedly, the HMMP and PMI did not reflect this trend, and the values remained high, suggesting that there was a lack of epididymal sperm activation with either centrifugation or extenders. While fertility was not tested in the present study, our in vitro results with cooled epididymal semen are encouraging and suggest that donkey epididymal semen may have satisfactory fertility. New studies need to address the fertility of donkey epididymal semen in mares and jennies.

Supplementary Materials: The following are available online at http://www.mdpi.com/2076-2615/10/12/2209/s1, Table S1: Donkey ages, tail of epididymis weight, and total sperm recovery harvested freshly after castration; Table S2: Donkey age, tail of epididymis weight, total sperm recovery harvested and temperature of the container at the opening. Scrotal content was shipped in passive cooling semen containers. Donkeys 1–8 were processed after arrival (Cooled-shipped 24 h), whereas donkeys 9–14 were processed 24 h later (Cooled-shipped 48 h); Table S3: Donkey age, tail of epididymis weight, total sperm recovery harvested, and temperature of the container at opening. Epididymides were shipped in passive cooling semen containers. Donkeys 1–20 were processed after arrival (Cooled-shipped 24 h), whereas donkeys 21–36 were processed 24 h later (Cooled-shipped 48 h).

Author Contributions: Conceptualization, I.F.C.; methodology, I.F.C., Y.L.-A., G.P., L.G.S., L.L.C., L.B.; formal analysis, G.P.; resources, C.E.A., L.T.K., A.K.M.; data curation, G.P., Y.L.A.; writing—original draft preparation, G.P., Y.L.A., L.G.S.; writing—review and editing, I.F.C.; supervision, I.F.C., F.S.L.; project administration, I.F.C., F.S.L.; funding acquisition, I.F.C. All authors have read and agreed to the published version of the manuscript.

Funding: This study was funded by the Department of Veterinary Clinical Medicine, University of Illinois.

Acknowledgments: The authors are grateful to the clients, B.L.M., and Peaceful Valley, for allowing us to their animals. Botupharma USA is thanked for donating the extenders and passive cooling semen containers used in this study.

Conflicts of Interest: The authors declare no conflict of interest.

References

1. Canisso, I.F.; Mcdonnell, S.M. Donkey breeding behavior with an emphasis on the Pêga breed. *Vet. Care Donkeys Int. Vet. Inf. Serv.* **2010**, *20*, 0310.
2. Canisso, I.F.; Panzani, D.; Miró, J.; Ellerbrock, R.E. Key Aspects of Donkey and Mule Reproduction. *Vet. Clin. N. Am. Equine Pract.* **2019**, *35*, 607–642. [CrossRef] [PubMed]
3. Canisso, I.F.; Davies Morel, M.C.G.; McDonnell, S. Strategies for the management of donkey jacks in intensive breeding systems. *Equine Vet. Educ.* **2009**, *21*, 652–659. [CrossRef]
4. Canisso, I.F.; Carvalho, G.R.; Morel, M.D.; Ker, P.G.; Rodrigues, A.L.; Silva, E.C.; Coutinho da Silva, M.A. Seminal parameters and field fertility of cryopreserved donkey jack semen after insemination of horse mares. *Equine Vet. J.* **2011**, *43*, 179–183. [CrossRef]
5. Rota, A.; Panzani, D.; Sabatini, C.; Camillo, F. Donkey jack (*Equus asinus*) semen cryopreservation: Studies of seminal parameters, post-breeding inflammatory response, and fertility in donkey jennies. *Theriogenology* **2012**, *78*, 1846–1854. [CrossRef]
6. Ellerbrock, R.E.; Prell, M.J.; Stewart, J.L.; Bojko, M.S.; Lima, F.S.; Canisso, I.F. Comparison of Centrifugation and Noncentrifugation Methods to Cryopreserve Stallion Epididymal Semen. *J. Equine Vet. Sci.* **2017**, *50*, 27–32. [CrossRef]
7. Braun, J.; Sakai, M.; Hochi, S.; Oguri, N. Preservation of ejaculated and epididymal stallion spermatozoa. *Theriogenology* **1994**, *41*, 809–818. [CrossRef]
8. Papa, F.O.; Melo, C.M.; Fioratti, E.G.; Dell'Aqua, J.A.; Zahn, F.S.; Alvarenga, M.A. Freezing of stallion epididymal sperm. *Anim. Reprod. Sci.* **2008**, *107*, 293–301. [CrossRef]
9. Amann, R.P.; Pickett, B.W. Principles of cryopreservation and a review of cryopreservation of stallion spermatozoa. *J. Equine Vet. Sci.* **1987**, *7*, 145–173. [CrossRef]
10. Bruemmer, J.E. Collection and Freezing of Epididymal Stallion Sperm. *Vet. Clin. N. Am. Equine Pract.* **2006**, *22*, 677–682. [CrossRef]
11. Camillo, F.; Rota, A.; Biagini, L.; Tesi, M.; Fanelli, D.; Panzani, D. The Current Situation and Trend of Donkey Industry in Europe. *J. Equine Vet. Sci.* **2018**, *65*, 44–49. [CrossRef]
12. Adams, G.P.; Ratto, M.H.; Collins, C.W.; Bergfelt, D.R. Artificial insemination in South American camelids and wild equids. *Theriogenology* **2009**, *71*, 166–175. [CrossRef] [PubMed]
13. Beek, J.; Maes, D.; Nauwynck, H.; Piepers, S.; van Soom, A. A critical assessment of the effect of serine protease inhibitors on porcine fertilization and quality parameters of porcine spermatozoa in vitro. *Reprod. Biol.* **2015**, *15*, 9–19. [CrossRef] [PubMed]
14. Herold, F.C.; Aurich, J.E.; Gerber, D. Epididymal sperm from the African buffalo (*Syncerus caffer*) can be frozen successfully with AndroMed® and with Triladyl™ but the addition of bovine seminal plasma is detrimental. *Theriogenology* **2004**, *61*, 715–724. [CrossRef]
15. Mamani-Mango, G.; Gonzales, M.M.; Hidalgo, M.R.; Mallma, J.M.; Béjar, J.R.; Palma, V.R.; Mellisho Salas, E. Effect of Extender and Freezing Rate on Quality Parameters and in Vitro Fertilization Capacity of Alpaca Spermatozoa Recovered from *Cauda Epididymis*. *Biopreserv. Biobank* **2019**, *17*, 39–45. [CrossRef]

16. Vilela, C.G.; Marquez, J.M.; Graham, J.K.; Barfield, J.P. Cryopreservation of bison epididymal sperm: A strategy for improving post-thaw quality when collecting sperm in field conditions. *Theriogenology* **2017**, *89*, 155–161. [CrossRef]
17. Miró, J.; Morató, R.; Vilagran, I.; Taberner, E.; Bonet, S.; Yeste, M. Preservation of Epididymal Stallion Sperm in Liquid and Frozen States: Effects of Seminal Plasma on Sperm Function and Fertility. *J. Equine Vet. Sci.* **2020**, *88*, 102940. [CrossRef]
18. Stawicki, R.J. How to Process Stallion Epididymal Sperm in Your Practice. *Am. Assoc. Equine Pract.* **2015**, *61*, 485–489.
19. Stawicki, R.J.; McDonnell, S.M.; Giguère, S.; Turner, R.M. Pregnancy outcomes using stallion epididymal sperm stored at 5 °C for 24 or 48 h before harvest. *Theriogenology* **2016**, *85*, 698–702. [CrossRef]
20. Waite, J.A.; Love, C.C.; Brinsko, S.P.; Teague, S.R.; Salazar, J.L.; Mancill, S.S.; Varner, D.D. Factors impacting equine sperm recovery rate and quality following cushioned centrifugation. *Theriogenology* **2008**, *70*, 704–714. [CrossRef]
21. Bliss, S.B.; Voge, J.L.; Hayden, S.S.; Teague, S.R.; Brinsko, S.P.; Love, C.C.; Varner, D.D. The impact of cushioned centrifugation protocols on semen quality of stallions. *Theriogenology* **2012**, *77*, 1232–1239. [CrossRef] [PubMed]
22. Auer, J.; Stick, J. *Equine Surgery*; Elsevier: Amsterdam, The Netherlands, 2012. [CrossRef]
23. Peña, F.J.; Ball, B.A.; Squires, E.L. A New Method for Evaluating Stallion Sperm Viability and Mitochondrial Membrane Potential in Fixed Semen Samples. *Cytom. Part B Clin. Cytom.* **2018**, *94*, 302–311. [CrossRef] [PubMed]
24. Oliveira, R.R.; Rates, D.M.; Pugliesi, G.; Ker, P.G.; Arruda, R.P.; Moraes, E.A.; Carvalho, G.R. Use of cholesterol-loaded cyclodextrin in donkey semen cryopreservation improves sperm viability but results in low fertility in mares. *Reprod. Domest. Anim.* **2014**, *49*, 845–850. [CrossRef] [PubMed]
25. Novello, G.; Podico, G.; Segabinazzi, L.G.T.M.; Lima, F.S.; Canisso, I.F. Stallion Semen Cooling Using Native Phosphocaseinate-based Extender and Sodium Caseinate Cholesterol-loaded Cyclodextrin-based Extender. *J. Equine Vet. Sci.* **2020**, *92*. [CrossRef] [PubMed]
26. Hartwig, F.P.; Lisboa, F.P.; Hartwig, F.P.; Monteiro, G.A.; Maziero, R.R.D.; Freitas-Dell'Aqua, C.P.; Alvarenga, M.A.; Papa, F.O.; Dell'Aqua, J.A., Jr. Use of cholesterol-loaded cyclodextrin: An alternative for bad cooler stallions. *Theriogenology* **2014**, *81*, 340–346. [CrossRef] [PubMed]
27. Pamornsakda, T.; Pojprasath, T.; Suwimonteerabutr, J.; Tharasanit, T. Effects of cholesterol-loaded cyclodextrins on the quality of frozen-thawed equine epididymal sperm. *Cryobiology* **2011**, *63*, 90–95. [CrossRef] [PubMed]
28. Vidament, M.; Vincent, P.; Yvon, J.M.; Bruneau, B.; Martin, F.X. Glycerol in semen extender is a limiting factor in the fertility in asinine and equine species. *Anim. Reprod. Sci.* **2005**, *89*, 302–305. [PubMed]
29. Vidament, M.; Vincent, P.; Martin, F.X.; Magistrini, M.; Blesbois, E. Differences in ability of jennies and mares to conceive with cooled and frozen semen containing glycerol or not. *Anim. Reprod. Sci.* **2009**, *112*, 22–35. [CrossRef]
30. Saragusty, J.; Lemma, A.; Hildebrandt, T.B.; Göritz, F. Follicular size predicts success in artificial insemination with frozen-Thawed sperm in donkeys. *PLoS ONE* **2017**, *12*, e0175637. [CrossRef]
31. Rodrigues, L.T.; Canuto, L.E.F.; Oliveira, S.N.; Andrade, L.R.P.; Alvarenga, M.A.; Dell'Aqua, J.A.; Canisso, I.F.; Papa, F.O. Skim Milk, Casein Based Extender or Botucrio in Cooled Semen Asinus (*Equus asinus*). *J. Equine Vet. Sci.* **2018**, *66*, 84. [CrossRef]
32. Monteiro, G.A.; Guasti, P.N.; Hartwig, F.P.; Dellaqua, J.A.; Alvarenga, M.A.; Papa, F.O. Cooling of ejaculated and epididymal stallion sperm. *Arq. Bras. Med. Vet. Zootec.* **2013**, *65*, 681–686. [CrossRef]
33. Colenbrander, B.; Brouwers, J.F.H.M.; Neild, D.M.; Stout, T.A.E.; da Silva, P.; Gadella, B.M. Capacitation dependent lipid rearrangements in the plasma membrane of equine sperm. *Theriogenology* **2002**, *58*, 341–345. [CrossRef]

34. Combes, G.; Varner, D.; Schroeder, F.; Burghardt, R.; Blanchard, T. Effect of cholesterol on the motility e plasma membrane integrity of frozen equine spermatozoa after thawing. *J. Reprod. Fertil. Suppl.* **2000**, *56*, 127–132.
35. Carneiro, J.A.M.; Monteiro, G.A.; Martin, I.; Maziero, R.R.D.; Sancler-Silva, Y.F.R.; de Freitas-Dell'Aqua, C.P.; Guasti, P.N.; Bandeira, R.; Hartwig, F.P.; Ignácio, F.S.; et al. Heterologous Oviductal Cells Binding Capacity of Cryopreserved Equine Ejaculated and Epididymal Spermatozoa. *J. Equine Vet. Sci.* **2017**, *59*, 40–48. [CrossRef]

animals

MDPI

Article

Comparison of the Surface Thermal Patterns of Horses and Donkeys in Infrared Thermography Images

Małgorzata Domino [1,*], Michał Romaszewski [2], Tomasz Jasiński [1] and Małgorzata Maśko [3]

1 Veterinary Research Centre and Center for Biomedical Research, Department of Large Animal Diseases and Clinic, Institute of Veterinary Medicine, Warsaw University of Life Sciences (WULS–SGGW), 02-787 Warsaw, Poland; tomasz_jasinski@sggw.edu.pl

2 Institute of Theoretical and Applied Informatics, Polish Academy of Sciences, 44-100 Gliwice, Poland; mromaszewski@iitis.pl

3 Department of Animal Breeding, Institute of Animal Science, Warsaw University of Life Sciences (WULS–SGGW), 02-787 Warsaw, Poland; malgorzata_masko@sggw.edu.pl

* Correspondence: malgorzata_domino@sggw.edu.pl; Tel.: +48-512-388-517

Received: 14 October 2020; Accepted: 18 November 2020; Published: 24 November 2020

check for updates

Simple Summary: In this study, the thermal patterns of horses and donkeys in infrared thermography (IRT) images are analyzed and compared. Thermal patterns are defined as statistically significant differences between groups of regions of interest (ROIs) corresponding to underlying large muscles. The dataset used in the experiments consists of images of healthy and rested animals: sixteen horses and eighteen donkeys. Thermal patterns between species are compared, and the results are discussed along with special cases of animals identified as outliers. The results support the thesis about the similarities in the thermal patterns of horses and donkeys.

Abstract: Infrared thermography (IRT) is a valuable diagnostic tool in equine veterinary medicine; however, little is known about its application to donkeys. This study aims to find patterns in thermal images of donkeys and horses and determine if these patterns share similarities. The study is carried out on 18 donkeys and 16 horses. All equids undergo thermal imaging with an infrared camera and measurement of the skin thickness and hair coat length. On the class maps of each thermal image, fifteen regions of interest (ROIs) are annotated and then combined into 10 groups of ROIs (GORs). The existence of statistically significant differences between surface temperatures in GORs is tested both "globally" for all animals of a given species and "locally" for each animal. Two special cases of animals that differed from the rest are also discussed. The results indicate that the majority of thermal patterns are similar for both species; however, average surface temperatures in horses (22.72 ± 2.46 °C) are higher than in donkeys (18.88 ± 2.30 °C). This could be related to differences in the skin thickness and hair coat. The patterns of both species are associated with GORs, rather than with an individual ROI, and there is a higher uniformity in the donkeys' patterns.

Keywords: infrared thermography; equids; thermal patterns; surface temperature; skin thickness; hair coat

1. Introduction

Infrared thermography (IRT) is a non-invasive imaging technique that allows for the detection of radiant energy emitted by any object with a temperature above absolute zero. The radiated power detected by the thermal camera in the infrared spectrum is proportional to the fourth power of the object's absolute temperature, and it is used to calculate the temperature of the target, e.g., the surface of the animal's body. Infrared radiation is often presented as a thermogram, which is an image where the color gradient corresponds to the distribution of surface temperatures [1]. Furthermore, the relationship of temperature gradients may create specific thermal patterns, which may be used, e.g., for assessing the influence of load on saddle fit in horses [2] or the horses' response to the training [3].

IRT has been used as a diagnostic tool in equine veterinary medicine since the mid-1960s, particularly in the field of orthopedics, in the management of lameness [4–7]. The surface temperature changes, reflecting heat emitted from overloaded or injured tissue, are considered a valuable indicator for identifying areas of inflammation and blood flow alterations [8,9]. This allows for the detection of temperature changes before they can be detected by palpation [10,11] and before the onset of other clinical signs of injury [11,12]. IRT also enables the identification of continuing subclinical changes and allows the verification of the complete clinical healing and, hence, if the horse may return to exercise after required rest [13]. In recent studies, IRT was also applied to interpret changes in the surface temperatures of the thoracic region in the case of back pain diagnosis of equine athletes [14,15], as well as the results of the impact of a load on a saddle [2] or incorrect saddle fit [16]. Moreover, the usefulness of equine IRT in the assessment of transient stress response during training [17,18] and competitive sport [19,20] has been demonstrated. Equine IRT seems to be highly related to thermoregulation, the increase in blood flow due to exercise [18], and the blood concentration of metabolic biochemical measurements [21,22]. During physical exercise, metabolic heat production increases as exercise intensity increases [23], and only a quarter of the energy used by a muscle is converted to mechanical energy. The remaining three quarters are dissipated as heat [24]. Therefore, the radiant energy emitted from the horse's skin surface may be found as a product of basic metabolic processes, exercise, and pathological conditions. However, it should be kept in mind that the temperature measured from the body surface is related not only to the above internal conditions, but also to the thermal properties of the skin and hair coat and the thermal gradient between the skin surface and the environment [25,26].

It is easy to see that IRT is widespread in the equestrian industry as a valuable tool to monitor the underlying circulation, tissue metabolism, and local blood flow in response to different physiological, pathological, or environmental conditions. However, little or no attention has been paid to the application of IRT in donkeys. The only work the authors are aware of is the study of the effects of season and age on the daily rhythmicity of rectal temperature and body surface temperature during the cold-dry and hot-dry seasons in a tropical savannah [27]. For the infrared measurement, the infrared thermometer and seven landmarks adapted from equine IRT were used. Although this study evaluated differences in the surface temperatures of donkeys of varying age groups under changing environmental conditions, no studies to date have compared the thermal images of donkeys and horses obtained in the same circumstances. The scarcity of works on the imaging of donkeys is a motivation to try to answer the question of whether there are significant differences in the thermal images of horses and donkeys. If the images of these animals were similar, it would suggest that intensively researched methods for analyzing equine images are applicable to donkeys.

In this study, the imaging of horses and donkeys was performed under the same environmental conditions. Following the methodology of previous equine researchers, body surface temperatures in healthy animals were evaluated. The normal thermal image was already described, for e.g., the coronary band [28], distal forelimb joints [26,29], the thoracolumbar region [30], the back, and pelvic regions [31] in the horse. It showed a high degree of symmetry between the left and right sides of the body [1,26] and reproducibility over hourly, daily, and weekly intervals up to 90% [30].

The thermal images were manually segmented into fifteen regions of interest (ROIs) corresponding to underlying large muscles. Since the phenomena observable in thermal images often includes more than one ROI, individual ROIs were combined into groups of ROIs (GORs), and the differences in their mean temperatures were examined. The differences, the occurrence of which was statistically confirmed, constituted thermal patterns, which were the basis for the comparison of both species and the analysis of special cases (outliers). This comparison was the main focus of the experiments in this study. The hypothesis of this study is that the thermal patterns of horses and donkeys are similar.

2. Materials and Methods

2.1. Animals

Eighteen donkeys (nine mares, seven geldings, and two stallions; mean age 7.78 ± 3.04 years, minimum age 2 years, maximum age 13 years; mean height 119.00 ± 11.72 cm) and sixteen horses (eight mares, six geldings, and two stallions; mean age 7.53 ± 2.83 years, minimum age 2 years, maximum age 11 years; mean height 137.40 ± 9.33 cm) participated in the study. All horses met the growth criteria for ponies, i.e., individual height at withers ≤148.00 cm, according to the standards of the International Federation for Equestrian Sport. However, to facilitate comparison between species, they are called horses throughout the manuscript. Most of the donkeys in the study were mixed breeds; however, the following pedigrees could be listed: two half-breed Romanian donkeys, one half-breed Martina Franca donkey, one half-breed Andalusian donkey, one half-breed Magyar Parlagi Szamér donkey, two quarter-breed Grigio Siciliano donkeys, one quarter-breed Andalusian donkey, five mixed breed donkeys with a quarter-blood Romanian donkey and a quarter-blood Andalusian donkey, two mix-breed donkeys with a quarter-blood Martina Franca donkey and a quarter-blood Andalusian donkey, and two local mix-breed donkeys. Furthermore, one donkey represented a pure bred Poitou donkey (the donkey *D.17* discussed as an outlier case). Horses, in general, represented two typical polish pony breeds (eight Polish Koniks and five Hucul ponies); however, two Haflinger ponies and one half-breed Connemara pony were also included. The donkeys and horses were privately owned and were housed in the same stable located in southern Poland in Lubachów. The owners of the animals consented to our research. The ethics approval was deemed unnecessary according to the regulations of the II Local Ethical Committee on Animal Testing in Warsaw and the National Ethical Committees on Animal Testing because all procedures in the study were non-invasive and did not cause distress and/or pain equal to or greater than a needlestick. The equids were fed three times a day with a dose of hay personalized to each animal to maintain an optimal, healthy condition and had daily access to a grassy paddock no shorter than 8 h per day. All horses received a BCS 3 (body condition score) [32], and all donkeys obtained an FNS 3 (fatty neck score) [33], both on a five-point scale. Both during the study and the month preceding the study, equids were not used in riding, nor were harnessed. Before the IRT imaging, physical examinations were conducted to ensure that the equids were free from preexisting inflammatory conditions. The general examination including the evaluation of internal temperature, heart rate, respiratory rate, mucous membranes, capillary refill time, and lymph nodes and was carried out following international veterinary standards. The detailed examination of the musculoskeletal system was performed following the guidelines for the lameness evaluation of athletic horses [34]. All donkeys and horses were clinically healthy, with no clinical signs of lameness, back problems, or musculoskeletal injury. They had normal species conformation and normal growth pattern. No horses were excluded due to the physical examination results. Two donkeys were excluded due to the properties of their hair coat: the first of them due to hair loss caused by abrasions during transport the week preceding the study and the second due to the significantly longer hair length (7.6 ± 1.2 cm) in comparison with the hair length of the other donkeys (3.4 ± 0.7 cm). Finally, sixteen donkeys were qualified for the formal analysis; however, an analysis of the two donkeys deviating from the accepted uniform appearance is included in the Section 4.2.

2.2. Data Collection

To ensure the best possible conditions for the comparison of the collected thermal images, the skin thickness, the hair coat length, and the constant thermal gradient between the skin surface and the environment were taken into account. The study was performed in mid-September, and all measurements were taken on the same day under the same circumstances (ambient temperature

20.2 °C; humidity 45%). A total of 68 images were taken in a closed space, protected from wind and sun radiation, to minimize the influence of external environmental conditions [35]. The imaging of donkeys and horses was carried out following equine international veterinary standards [36]. The imaged area was brushed, and dirt and mud were removed 15 min before imaging. The thermal images were acquired on the left and right sides at a 90° camera angle from a distance of approximately 2 m from the animal. During each imaging session, two images of each individual were taken. The images were focused on the center of the trunk. The animals were imaged on the side where the mane was less visible. The images were taken by the same researcher (M.M.) using an infrared radiation camera (FLIR Therma CAM E25, Brazil) with an emissivity (e)∼0.99. The temperature range was standardized in the professional software (FLIR Tools Professional, Brazil) during the preprocessing of the images at the 10–30 °C level.

After each IRT imaging, an ultrasonographic image was taken with an ultrasound scanner (SonoScape S9, SonoScape, Shenzhen, China) using a linear 5–12 MHz transducer (L752, SonoScape, Shenzhen, China). Ultrasound scans were performed with the transducer placed on the animal's back, over the third lumbar vertebra, perpendicular to the spine. All images were collected on the left side of the animal [37]. The hair was trimmed at the measurement place, and ultrasound gel (Aquasonic 100, Parker Laboratories Inc., Fairfield, NJ, USA) was used as a coupling medium. The real-time ultrasonographic examination was frozen, and the image was saved, as well as the subcutaneous fat (SF) plus skin thickness (SF-Skin) measurements were obtained. An example of an ultrasonographic image is presented in Figure 1. The hair coat samples were taken from the mid-neck approximately 5 cm below the base of the mane. The length of individual hairs was determined from a random sample of five pulled strands, including the roots [38].

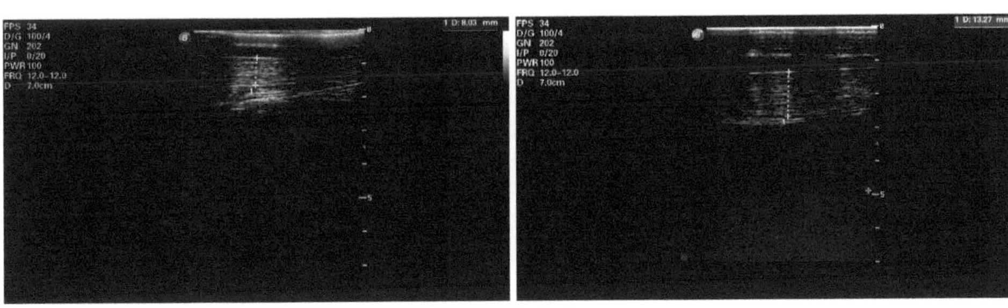

(**a**) SF-Skin of horse (**b**) SF-Skin of donkey

Figure 1. Example of an ultrasonographic image taken over the third lumbar vertebra: (**a**) the horse *H.1*; (**b**) the donkey *D.3*. The subcutaneous fat plus skin thickness (SF-Skin) is highlighted.

2.2.1. Dataset Preparation

Based on collected data, a dataset was prepared that was later used in the experiments. The dataset consisted of images from a thermal camera and the corresponding annotations in the form of class maps of the main muscle areas. Every thermal image was a table of 320 × 240 pixels. The value in each pixel was the measured temperature value. A corresponding class map was a table, where the value in every pixel was the ROI number, and a value of zero was used for pixels without annotation. An example class map is presented in Figure 2. The class maps were produced by hand annotating the fifteen identified regions of interest (ROIs) in each image. The following ROIs corresponding to the underlying large muscles were annotated:

1. ROI 1 *m. brachiocephalicus*—a parallelogram-shaped area from the lateral surface of the atlas, behind the angle of the mandible, to the regio supraspinatus of the scapula.
2. ROI 2 *mm. splenius capitis and cervicis*—a triangle-shaped area from the lateral surface of the axis to the regio supraspinatus of the scapula above ROI 3.

3. ROI 3 *m. trapezius pars cervicalis*—a triangle ranging from the middle of the neck to the regio cartilaginis of the scapula and along the regio supraspinatus of the scapula up to two-thirds of the length of the scapula.
4. ROI 4 *m. trapezius pars thoracica*—a triangle ranging from the regio cartilaginis of the scapula along the regio supraspinatus of the scapula up to one-thirds of the length of the scapula.
5. ROI 5 *m. latissimus dorsi*—a triangle-shaped area from the regio infraspinatus of the scapula up to two-thirds of the length of the scapula along the back to the tuber coxae.
6. ROI 6 *mm. glutei* (superficialis and medius)—an irregular area in the regio tuberis coxae.
7. ROI 7 *m. biceps femoris*—an oblong s-shaped area in the regio femoris cranially from the m. semitendinosus.
8. ROI 8 *m. semitendinosus*—an oblong s-shaped area in the regio femoris caudally from the m. biceps femoris.
9. ROI 9 *mm. in regio cruris*—a rectangular-shaped area in the regio cruris between articulatio genus and articulatio tarsi.
10. ROI 10 *m. tensor fasciae latae*—an irregular area between the regio tuberis coxae and the flank.
11. ROI 11 *m. obliquus externus abdominis*—a trapezoid-shaped area from the lower two-thirds of the regio infraspinatus of the scapula to the tuber coxae and the regio of processus xiphoideus sterni.
12. ROI 12 *m. pectoralis transversus*—a triangle-shaped area behind the regio olecranon to the regio processus xiphoideus sterni.
13. ROI 13 *mm. in regio antebrachii*—a rectangular-shaped area in the regio antebrachii between articulatio humeri and articulatio cubiti.
14. ROI 14 *m. pectoralis descendens*—an irregular area in the projection of the regio infraspinatus of the scapula.
15. ROI 15 *m. deltoideus*—an irregular area in the projection of the regio supraspinatus of the scapula.

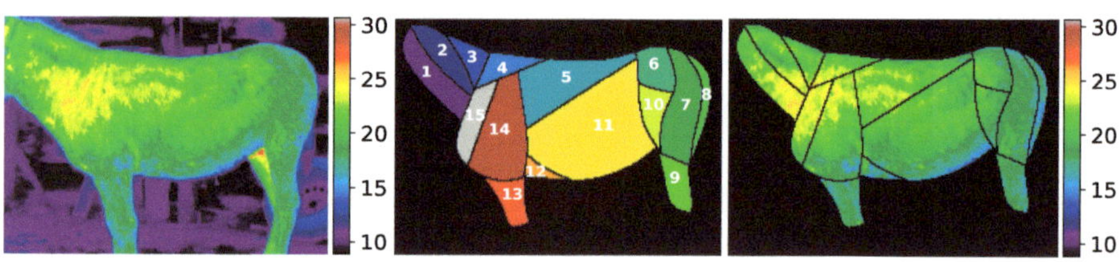

(**a**) Thermal map (**b**) ROIs (**c**) Extracted pixels

Figure 2. Visualization of a donkey *D.3*: (**a**) thermal data from the camera as a thermal map; (**b**) annotated classes corresponding to selected ROIs (see Section 2.2.1); (**c**) extracted pixels used in the experiments.

2.2.2. Dataset Availability

In order to facilitate the replication of the experiments presented in this work, the dataset [39] (dataset location: https://zenodo.org/record/4085075) and the experimental source code (source code location: https://github.com/iitis/thermal_patterns.git) are made available to the public under an open license.

2.3. Thermal Patterns in the IRT Images of Horses and Donkeys

The main goal of this study was to find patterns in the thermal images of both species and determine if these patterns share similarities. A thermal pattern was defined as a statistically significant difference between the mean temperatures in any two areas composed of groups of ROIs.

2.3.1. Testing the Statistical Significance of Temperature Differences

The statements in this work are usually associated with the comparison of temperatures between areas (subsets of pixels) in a thermal image or images, e.g., a statement "Animals A were warmer than Animals B in area C" means that based on the available sample, the surface temperatures of Animals A were on average higher in this region. Therefore, to test the statistical significance of these statements, the one-sided Mann–Whitney–Wilcoxon (MWW) test [40] was used. MWW is a non-parametric statistical hypothesis test that allows for the comparison of two related sequences of samples. A one-sided test was used because the direction of the difference was known, as it was the average temperature difference in the compared areas. A non-parametric test was used because the temperature distributions in ROIs were diverse and often non-Gaussian. The two sequences of samples were obtained by randomly, uniformly sampling the compared ROIs or groups of ROIs. The number of samples was the size of the smaller set; when comparing individual ROIs between animals, the average difference in the sample count was $(10.25 \pm 6.58)\%$ of the average ROI size for both species. Unless stated otherwise, the statistical significance of the p-value was set at $p < 0.001$.

2.3.2. Finding Thermal Patterns in Animal Species

The definition of thermal patterns in this paper followed the assumption that the surface temperature differences between different parts of the animal's body repeat within species. To identify these patterns, the following methodology was used:

Combining ROIs into Groups of ROIs

In the first step, based on the observation that visible patterns in the thermal images from our dataset were often located in several ROIs, ten groups of ROIs (GORs) were manually designated for analysis. The designated GORs, presented in Figure 3, were as follows:

1. GOR 1 *Neck*, ROIs $\{1,2,3\}$—represents an area of skin covering muscles located cranially from the cranial border of the scapula.
2. GOR 2 *Front quarter*, ROIs $\{1,2,3,4,14,15\}$—represents an area of skin covering muscles located cranially from the spinous processes of the scapula.
3. GOR 3 *Trunk*, ROIs $\{5,11\}$—represents an area of skin covering muscles between the caudal border of the scapula and the vertical line defined by tuber coxae, excluding the area of m. pectoralis transversus.
4. GOR 4 *Hindquarter*, ROIs $\{6,7,8,9,10\}$—represents an area of skin covering the examined muscles of the pelvic limbs laid caudally from the vertical line defined by tuber coxae.
5. GOR 5 *Rump*, ROIs $\{8,9\}$—represents an area of two ROIs of the pelvic limbs lying the most caudally.
6. GOR 6 *Dorsal aspect*, ROIs $\{3,4,5,6\}$—collects an area of skin covering muscles located above the horizontal line halfway up the trunk.
7. GOR 7 *Ventral aspect*, ROIs $\{9,10,11,12,13\}$—collects an area of skin covering muscles located below the horizontal line halfway up the trunk.
8. GOR 8 *Abdomen*, ROI $\{11\}$—represents an area of skin covering muscles lying between the caudal border of the scapula and the vertical line defined by tuber coxae, excluding the area of m. pectoralis transversus and m. latissimus dorsi.
9. GOR 9 *Groins* (Girth and Flank), ROIs $\{10,12\}$—represents two areas of skin mostly covering large muscles of thoracic and pelvic limbs represented by the girth area and flank area.
10. GOR 10 *Legs*, ROIs $\{9,13\}$—represents two areas of skin covering muscles of the proximal parts of limbs, both thoracic and pelvic.

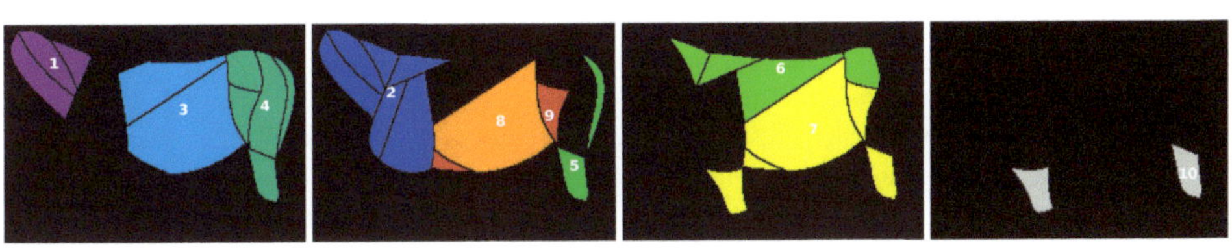

Figure 3. Visualization of a donkey *D.3*, divided into groups of ROIs (GORs).

Comparing GOR Temperatures

In order to compare the average temperatures between the designated groups of ROIs (GORs) and test whether the difference was statistically significant, the following methodology was used:

Let there be set of animals of a given species $\mathcal{A} = \{a_1, \ldots, a_{16}\}$ and a set of GORs defined in the previous paragraph $\mathcal{G} = \{g_1, \ldots, g_{10}\}$. Let a set $\mathcal{T}_g^a$ be a set of pixels (temperatures) of an animal $a \in \mathcal{A}$ from a group $g \in \mathcal{G}$, and let the mean value of pixels in a set be denoted by δ, e.g., $\delta(\mathcal{T}_g^a)$. For every pair of groups $(i, j) \in \mathcal{G} \times \mathcal{G}$, a difference in average values of temperatures in these groups for all animals was computed, i.e.,

$$\Delta_{(i,j)} = \delta\left(\bigcup_{k\in\mathcal{A}} \mathcal{T}_k^j\right) - \delta\left(\bigcup_{k\in\mathcal{A}} \mathcal{T}_k^j\right).$$

Values $\Delta_{(i,j)}$ form a matrix of differences $\boldsymbol{M}^{\Delta} \in \mathbb{R}^{|\mathcal{G}|\times|\mathcal{G}|}$. Due to the fact that values in the matrix $\boldsymbol{M}^{\Delta}$ represent temperature differences, the matrix is not symmetric.

In the next step, the MWW test described in Section 2.3.1 was applied to verify the statistical significance of the difference $\Delta_{(i,j)}$ for every pair of groups $(i, j) \in \mathcal{G} \times \mathcal{G}$. This was done in two ways:

1. Globally—for every pair $(i, j) \in \mathcal{G} \times \mathcal{G}$, the MWW test was applied to the whole population, i.e., the union of sets $\bigcup_{k\in\mathcal{A}} \mathcal{T}_i^k$ was compared with the union of sets $\bigcup_{k\in\mathcal{A}} \mathcal{T}_j^k$.
2. Locally—for every pair $(i, j) \in \mathcal{G} \times \mathcal{G}$, the MWW test was applied separately for every animal $a \in \mathcal{A}$, by comparing the set $\mathcal{T}_i^a$ with the set $\mathcal{T}_j^a$. For the dataset used in this study, this resulted in 16 tests for every pair.

The results of "local" tests formed a matrix $\boldsymbol{M}^{L} \in \mathbb{R}_+^{|\mathcal{G}|\times|\mathcal{G}|}$, where the value in every cell represented the number of animals for which the difference was significant.

Thermal Patterns

For a given animal species, a statistically significant difference $\Delta_{(i,j)}$ between two GORs $(i, j) \in \mathcal{G} \times \mathcal{G}$ was treated as a thermal pattern. A thermal pattern can thus be interpreted as a statement based on available data, e.g., the GOR 5 *Rump* was colder than the GOR 1 *Neck*. If the significance of the difference was confirmed by the "global" test, but not for every animal, by the "local" test, i.e., the value in the matrix $\boldsymbol{M}^{L}$ for this difference was less than 16, and this means that while the pattern emerged in a population, it was susceptible to individual differences of animals; thus, there were animals that did not show this pattern. If the pattern also appeared individually in all of the animals tested, it was considered to be stable.

2.4. Thermal Images' Visualization

In order to visualize the visible structures in IRT images from the dataset used in this study, the temperatures are presented in the form of a color map, modeled on the visible part of the electromagnetic spectrum, i.e., ranging from violet to red. To improve the clarity of the images, the zero values representing the areas outside the ROIs are shown in black. By manipulating the color map threshold values (assigned to its extreme colors), patterns common to all animals or patterns specific to a particular animal are highlighted.

Temperature distributions within a specific ROI are visualized using histograms where the y-axis is presented as a probability density, i.e., bin counts are divided by a total number of counts. Alternatively, boxplots where the box extends from the lower to the upper quartile values are used. The line in the boxplot denotes the median, the whiskers the range of $\{q_1 - 1.5 * (q_3 - q_1), q_3 - 1.5 * (q_3 - q_1)\}$ where q_1, q_3 denote the first and the third quartiles, and circles outliers.

Unless stated otherwise, in all thermal map visualizations, the presented color map temperature values t_c are limited to the common range of $t_c \in \langle 8.8, 30.65\rangle$ °C, which were extreme values in annotated ROIs for all animals included in the study (temperatures in ROIs: horses, $t_h \in \langle 10.64, 30.65\rangle$

°C, $E(t_h) = 22.72 \pm 2.46$ °C, donkeys, $t_d \in \langle 8.8, 29.56 \rangle$ °C, $E(t_h) = 18.88 \pm 2.30$ °C), not including the animals *D.17*, *D.18*, for reasons explained in Section 2.1. Alternatively, t_c values were selected as extreme temperature values in the ROIs for a given animal to highlight the features of visible thermal patterns; these special cases are clearly indicated.

2.4.1. Data Visualization

An individual animal in the dataset could be represented by a vector $v_i \in \mathbb{R}^d$ of d features corresponding, e.g., to the means or variances of temperatures in every ROI, which led to $d \geq 15$. The extraction and visualization of data structures in a high-dimensional space are often performed by using the principal component analysis [41] (PCA) and projecting data onto the first principal components. However, PCA uses a sample covariance matrix. Since the dataset contained a limited number of examples, the computation of a reliable covariance matrix was difficult. Therefore, the t-distributed stochastic neighbor embedding (t-SNE) [42] algorithm was used for data presentation. T-SNE visualizes data by giving each example a location in a two-dimensional map. An important feature of the t-SNE is that its output is non-deterministic, which results from the fact that the optimization problem solved by the technique has a cost function that is not convex. Since in this work, t-SNE was only used to visualize patterns emerging in the data, it was considered acceptable. The presented visualizations were selected as representative examples after several executions of t-SNE. When t-SNE was applied for data visualization, its perplexity parameter was set to a value of five.

Data features were extracted with common statistics such as the mean, standard deviation, kurtosis, and skewness. In addition, a scenario in which the data were normalized by subtracting the average global temperature of every animal from values of all pixels in this animal's image was tested.

2.5. Implementation

Experiments were implemented in Python 3.6.9 using the libraries: NumPy 1.16.4, SciPy 1.3.1, scikit-learn 0.22.1, Matplotlib 3.2.2, seaborn 0.11.0.

Experiments were conducted using a computer with Intel(R) Core i7-5820K CPU @ 330 GHz with 64 GB of RAM and with the Windows 10 Pro system. The running time of the experiments could be measured in seconds.

3. Results

A comparison of temperatures between ROIs is presented in Figure 4. The surface temperatures for horses were, on average, higher than for donkeys, which was confirmed as statistically significant for every ROI (MWW, $p < 0.001$). Considerable variances in ROIs' temperatures and the presence of many outliers were observed. Example histograms for two ROIs with the most extreme differences in mean temperatures of horses and donkeys are presented in Figure 5. The individual differences in animal surface temperatures resulted in multi-modal temperature distributions, as, e.g., in Figure 5a. Histograms of temperatures for all ROIs can be found in Figure A1 in Appendix A.

A comparison of hair coat length and SF-Skin values between donkeys and horses is presented in Table 1. The skin and the subcutaneous fat were thicker and the hair coat was longer in donkeys than in horses (MWW, $p < 0.0001$).

Table 1. Measured features (mean ± SD) of horses (*H.1–H.16*) and donkeys (*D.1–D.16*): the length of the hair coat and the thickness of the subcutaneous fat plus skin (SF-Skin).

Animals	Hair Coat (cm)	SF-Skin (mm)
Donkeys	3.39 ± 0.46^{a}	12.01 ± 0.83^{c}
Horses	1.78 ± 0.38^{b}	8.80 ± 0.87^{d}
p-value	< 0.0001	< 0.0001

Different superscript letters indicate significant differences between horses and donkeys for hair coat (*a*, *b*) and SF-Skin (*c*, *d*) respectively according to the Mann–Whitney–Wilcoxon (MWW) test.

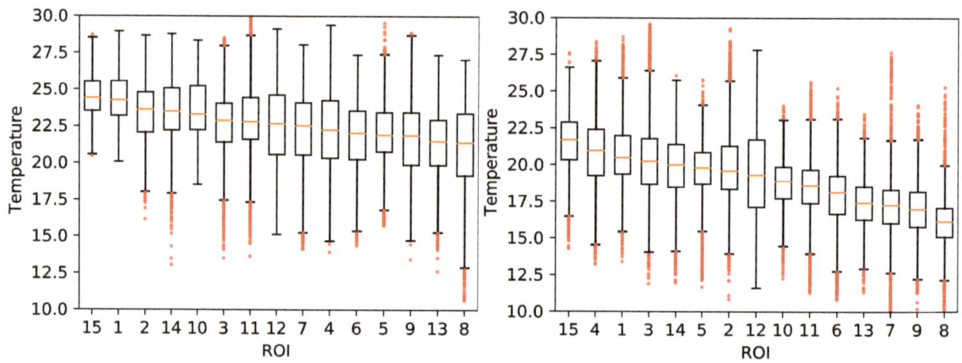

(**a**) Horses (**b**) Donkeys

Figure 4. Visualization of temperatures in ROIs (ordered by their medians) of all animals: (**a**) horses; (**b**) donkeys.

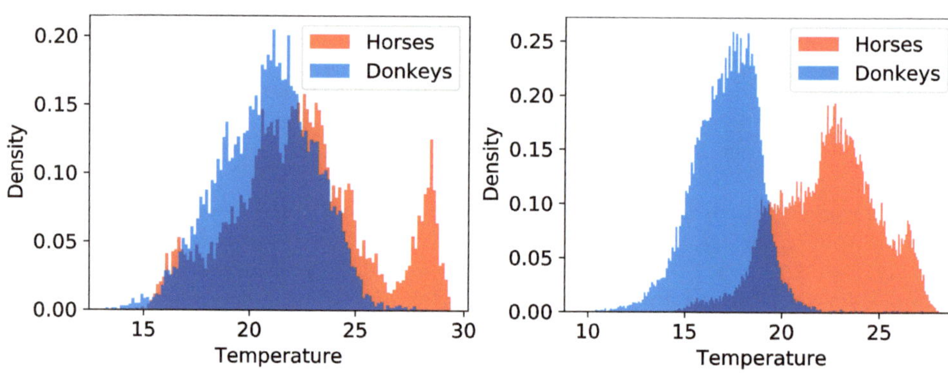

(**a**) ROI 4 (**b**) ROI 7

Figure 5. Histograms of temperatures for two ROIs where the difference Δ_t between mean values of temperatures for the two animal species is: (**a**) the smallest (ROI 4, $\Delta_t = 1.59$) and (**b**) the largest (ROI 7, $\Delta_t = 5.26$).

t-SNE data visualization is presented in Figure 6. When the examples were represented by mean temperature vectors in ROIs, the data formed distinct clusters, separated by low-density areas, as, e.g., in Figure 6a. However, the labels of classes within a cluster were mixed, suggesting that the mean temperature in ROIs alone was not a definite species descriptor. For features based on the standard deviation, clusters ceased to be clearly separable, as, e.g., in Figure 6b. For features based on skewness, kurtosis, or normalized temperatures, it was difficult to observe a consistent clustering of data.

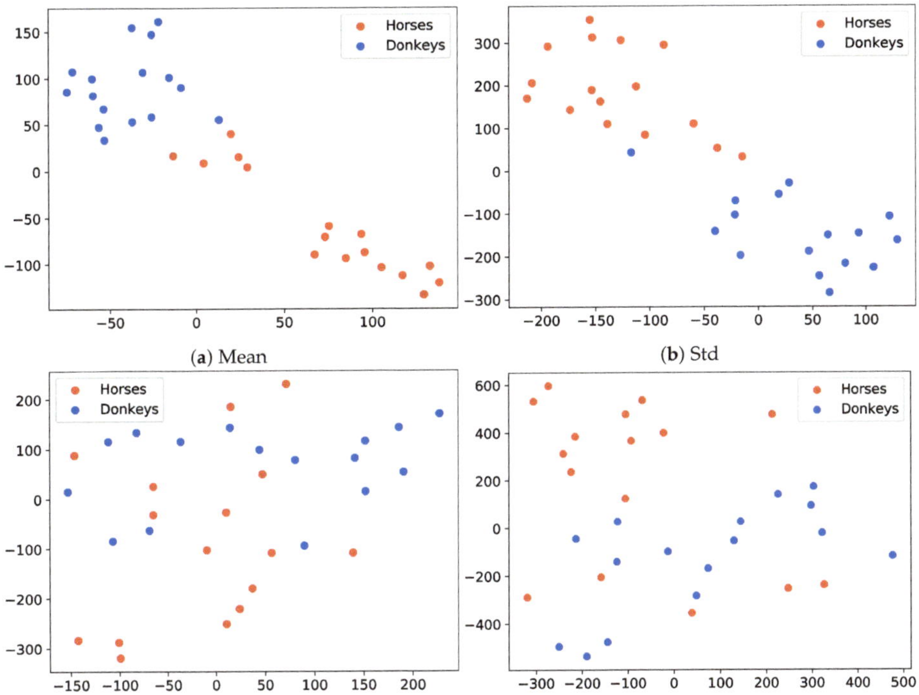

(**a**) Mean (**b**) Std (**c**) Kurtosis (**d**) Mean (normalized)

Figure 6. t-SNE visualization of the dataset. Every dot represents an animal described with features extracted from the pixels of its 15 ROIs. Plots present different feature extraction statistics: (**a**) the mean; (**b**) the standard deviation; (**c**) the kurtosis; (**d**) the mean, after removing the global mean temperature of an animal from all pixel values. Notice that the examples in Plot (**a**) form clusters that correspond to the species of the animal, although some examples are in the wrong cluster.

The visualization of thermal maps for horses in this study is presented in Figure 7 and for donkeys in Figure 8. As visual comparison of the images reveals that visible temperature patterns were more complex for horses, e.g.: average temperature values for horses *H.8, H.13* were globally higher; GOR 8 *Abdomen* was visibly warmer for horses *H.4, H.7, H.8, H.3*; and GOR 4 *Hindquarter* was visibly warmer for horses *H.4, H.7, H.8, H.10, H.13*.

The donkey temperatures were more uniform. Temperature values in GOR 5 *Rump* were visibly lower than in other GORs, while in GOR 2 *Front quarter*, warm areas were observed. A comparison of the histograms for four selected GORs is presented in Figure 9, where it can be observed that the overlap between histograms is greater for GOR 2 than for GOR 5.

To highlight the visible patterns, individual thermal maps for two example animals are presented in Figure 10. Visual examination of, e.g., the GOR 2 *Front quarter* in Figure 10c,d reveals how the characteristic patterns are usually associated with groups of ROIs rather than an individual ROI.

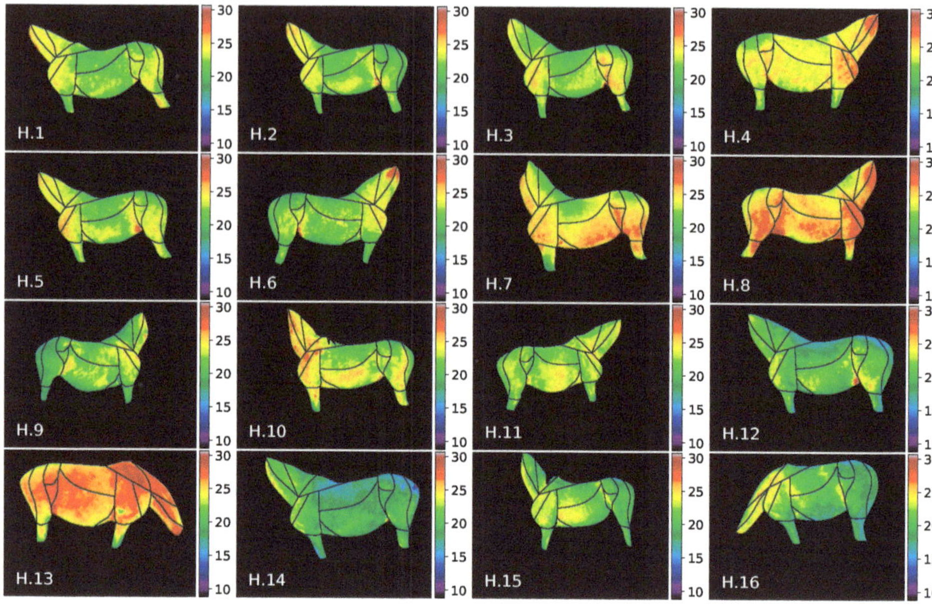

Figure 7. Thermal maps of annotated ROIs for horses in our dataset.

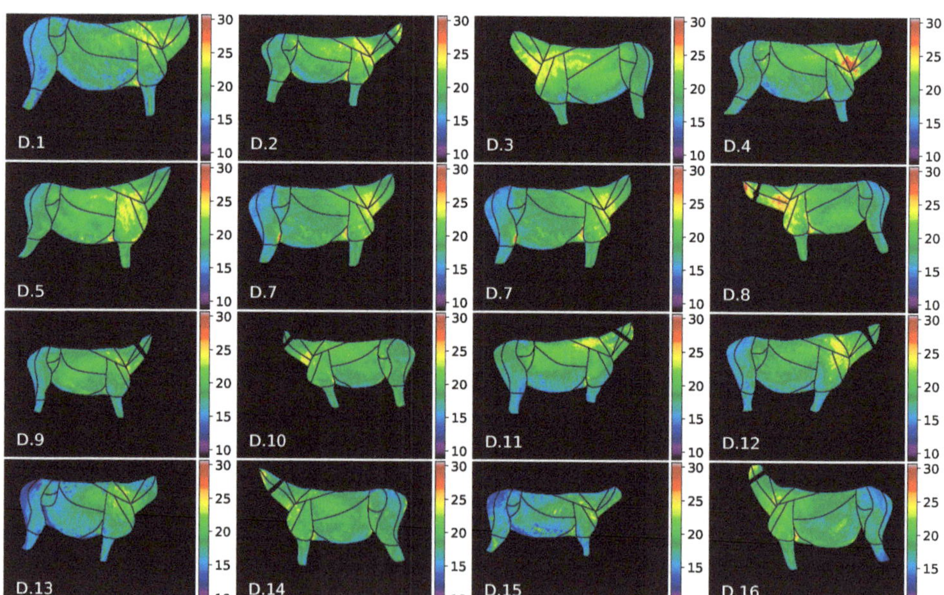

Figure 8. Thermal maps of annotated ROIs for donkeys in our dataset.

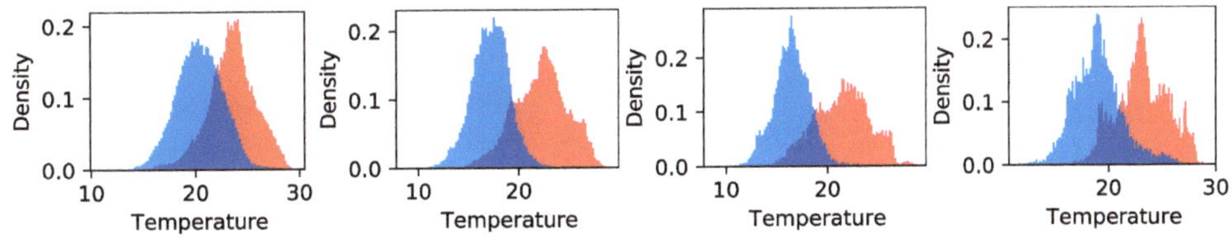

(**a**) *Front quarter* (**b**) *Hindquarter* (**c**) *Rump* (**d**) *Groins*

Figure 9. Comparison of temperature histograms between animal species in identified characteristic areas corresponding to selected groups of ROIs: (**a**) GOR 2 *Front quarter*; (**b**) GOR 4 *Hindquarter*; (**c**) GOR 5 *Rump*; (**d**) GOR 9 *Groins*. Horses are represented in red and donkeys in blue.

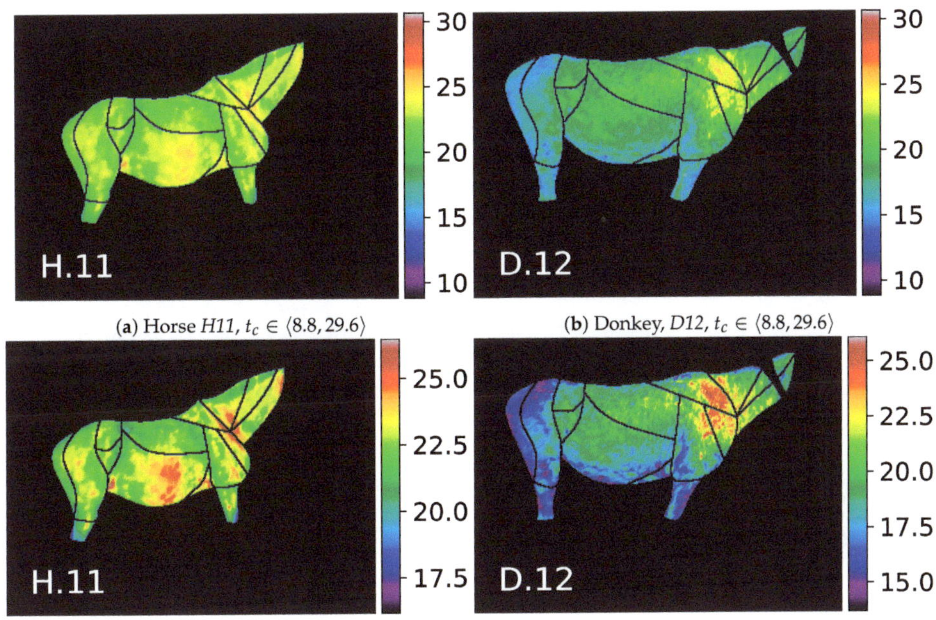

(**a**) Horse *H11*, $t_c \in \langle 8.8, 29.6 \rangle$ (**b**) Donkey, *D12*, $t_c \in \langle 8.8, 29.6 \rangle$

(**c**) Horse, *H11*, $t_c \in \langle 13.7, 26 \rangle$ (**d**) Donkey, *D12*, $t_c \in \langle 16.2, 26.4 \rangle$

Figure 10. Selected examples of two animals from the dataset. The color map values t_c for images in the upper row are scaled to the common range, which makes them easy to compare: (**a**) horses; (**b**) donkeys. Images in the bottom row are scaled to the minimal and maximal temperatures in the annotated ROIs of each animal, which highlights individual thermal patterns: (**c**) horses; (**d**) donkeys; e.g., warm horse's GORs *Abdomen* and *Neck*, cool donkey's GOR *Rump*, and warm donkey's GOR *Front quarter*.

Thermal patterns for both species, i.e., the differences in the temperatures between designated GORs, are presented in Figure 11. For both species, GORs *Rump* and *Legs* were consequently colder than the others, while GORs *Neck* and *Front quarter* were warmer. The majority of differences were globally significant ($p < 0.001$). For horses, there were five exceptions: *Neck/Front quarter*, *Trunk/Ventral aspect*, *Trunk/Abdomen*, *Ventral aspect/Abdomen*, and *Rump/Legs*. For donkeys, there were only two exceptions: *Ventral aspect/Abdomen* and *Trunk/Groins*.

However, as for the local significance of differences, for horses, there were only five patterns that consequently appeared for all animals: *Dorsal aspect/Front quarter*, *Groins/Trunk*, *Groins/Hindquarter*, *Legs/Neck*, *Legs/Front quarter*. On the contrary, for donkeys, there were 21 such patterns, which indicates that donkeys were individually more consistent with the global trend.

A summary of the pattern similarities between both species is presented in Figure 12. Figure 12a presents patterns that are similar for both species, e.g., the relation in temperatures between GORs *Rump* and *Neck* was the same for both species (the GOR 5 *Rump* was colder than the GOR 1 *Neck*), and this relation was globally, statistically significant ($p < 0.001$), which is indicated with the green color (the similar and globally statistically significant (SPS) class) in the image. Figure 12b presents the minimal number of animals in each species that shared the corresponding pattern, e.g., for the pair *Rump* and *Neck*, there existed at least 15 horses and 15 donkeys for which the pattern was also locally, statistically significant ($p < 0.001$). The majority of patterns fell under the SPS class, which supports the thesis about the similarities in the patterns for both species. In addition, the dissimilar patterns were most common in GOR 6 *Dorsal aspect* and GOR 3 *Trunk*.

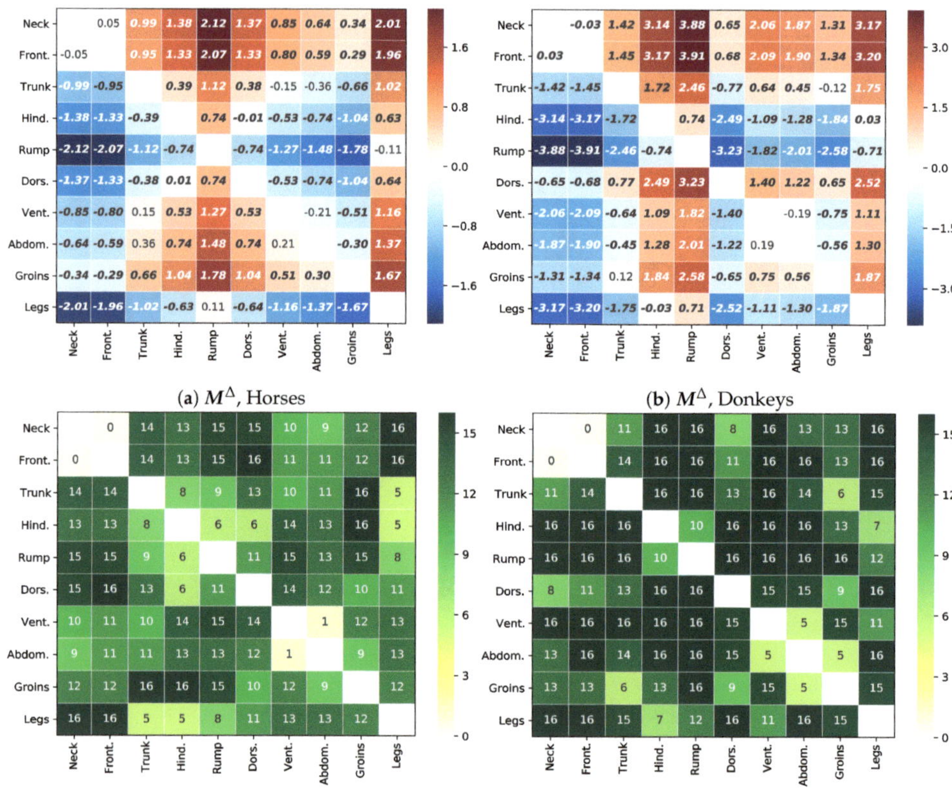

(**a**) M^{Δ}, Horses (**b**) M^{Δ}, Donkeys

(**c**) M^{L}, Horses (**d**) M^{L}, Donkeys

Figure 11. Thermal patterns and statistically significant differences between GORs. The upper panels present the matrix of differences within one species: (**a**) horses; (**b**) donkeys; e.g., the value $M^{\Delta}_{[4,0]} = -2.12$ in the cell $[4,0]$ in Panel (**a**) is the difference between the mean temperatures for the pair *Rump* and *Neck*, indicating that the *Rump* GOR is colder. Bold font indicates "global" statistical significance of this difference ($p < 0.001$). The bottom panels present tables for (**c**) horses and (**d**) donkeys, with the number of animals for which the corresponding temperature difference in the table above is statistically significant considering individual thermal pattern of this animal ($p < 0.001$); e.g., the value $M^{L}_{[4,0]} = 15$ in Panel (**c**), which indicates that the pattern *Rump* and *Neck* is locally significant for 15 horses. A stable pattern should be statistically significant simultaneously for all data combined and for each of the 16 animals of a given species.

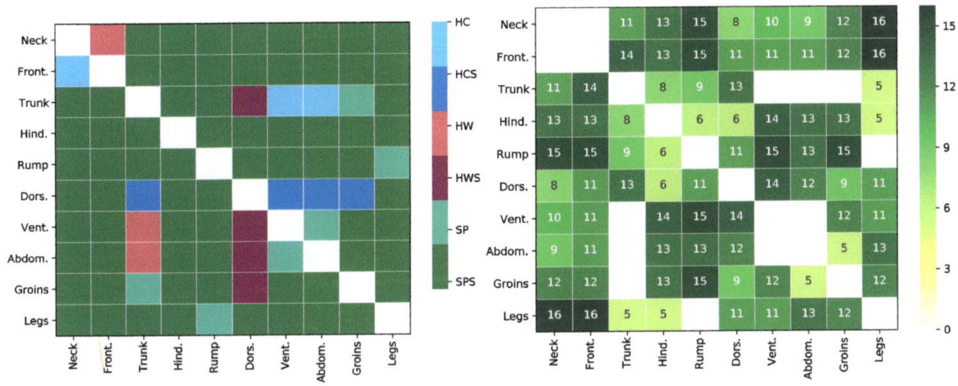

(**a**) Global similarity of patterns (**b**) Min. No. animals with a given pattern

Figure 12. Comparison of thermal patterns for both species: (**a**) division of thermal patterns into six classes: SPS denotes thermal patterns that are similar and globally statistically significant ($p < 0.001$) for both species; SP: similar patterns, but not significant; HWS: opposite patterns where horses are warmer (and donkeys colder), which are statistically significant; HW: same as HWS, but not significant; HCS: significant patterns where horses are colder (and donkeys warmer); HC: same as HCS, but not significant. Note that the SPS class is the most common, which suggests the global similarity of patterns. (**b**) The minimum number of animals that locally confirm the global trend for classes SPS, HWS, and HCS, i.e., for both species, at least this number of animals share a given pattern individually ($p < 0.001$). Note that the maximum value in the table is 16, which indicates a stable pattern.

4. Discussion

To the best of the authors' knowledge, this study is the first to describe the whole body surface thermal patterns of donkeys using infrared thermography. Since there are many applications of IRT in horses' veterinary diagnostic procedures, training monitoring, and welfare evaluations, the initial comparison between species is essential for further donkey IRT applications. Therefore, the motivation of this work is to contribute to a better understanding of the normal thermal pattern for rested donkeys.

The surface temperatures of the horses in our study were on average higher than those of the donkeys, and also, their individual temperatures varied more within the species. This was largely due to the differences in the thermal properties of the skin and the hair coat. The skin and the subcutaneous fat were thicker and the hair coat was longer in donkeys than in horses (see the values in Table 1), providing a better thermal insulation for donkeys. Recent results suggest that the hair coat properties of donkeys and horses are significantly different [38], even in animals with shorter hair than in this study. This difference might be due to the considerably large seasonal variation in hair weight and length typical for horses, but not for donkeys, or different breeds of horses participating in our research (Polish-native warmblooded horses/ponies) compared to [38] (U.K.-native coldblooded horses/ponies).

The use of a heterogeneous group of donkeys and horses in terms of their breed is one of the limitations of this research. It should be taken into account that the horse breed affects coat growth [43]. Therefore, there is a considerable variation in the thermal insulation of the coat between different breeds [44]. In this study, individuals in both groups could differ in skin thickness and/or hair coat, which could influence their thermal insulation and thermal patterns. This is well visible in the case of a long-haired donkey (pure-bred Poitou donkey *D.17*). On the other hand, the SD of each hair length measurement was less than one-quarter of the mean, which indicates an acceptable homogeneity. However, differences in the properties of the hair coat and skin within breeds, as well as other breed-related features could not be excluded as a factor that influenced the presented results. Further studies should consider a larger panel of populations, e.g., different breeds represented by a larger number of individuals. Similar studies were carried out on the warmblooded horses, and the comparison of the superficial body temperatures between thoroughbreds, Arabian, and Polish half-breed horses was reported in [45]. Thoroughbreds were reported as significantly warmer than Arabian and Polish half-breed horses at most ROIs located on a distal part of the limbs and back region. However, no differences in measured temperatures between Arabian and Polish half-breeds were observed [45]. The authors did not find a comparison of whole body temperature patterns between other breeds, especially Polish-native warmblooded ponies, in the available literature. In recent studies, the strong relationship between BCS (body condition score) and SF-Skin, for both donkeys and horses, was demonstrated [37,46]. The higher SF-Skin thickness in donkeys than in horses in this study may indicate greater adiposity of donkeys and thus better insulation. As a result, slight local changes in donkey surface body temperature may be difficult to observe. This makes the warm area visible in regio scapularis associated with the GOR 2 *Front quarter* particularly interesting. Additionally, it suggests the validity of animal temperature analysis through comparing the characteristics of different regions of a given animal.

4.1. Similarities in Thermal Patterns of Horses and Donkeys

It was determined in this study that patterns in IRT images were often visible in groups of ROIs, and a methodology of assessing these patterns based on the difference of temperatures in groups was proposed. Figure 12 shows that the thermal patterns for both species share similarities: 77.8% of patterns visible in Figure 12a were similar and statistically significant ($p < 0.001$); 8.9% of patterns were the opposite; and the rest of them could not be statistically confirmed. For 88.8% of globally significant patterns, half or more individual animals from every species shared this pattern. In the authors' opinion, this supports the thesis about similarities in IRT images of both species.

The results presented in Figure 11 indicate that donkeys were more "uniform" in their GORs, which resulted in larger maximum differences between GORs and the fact that more individual animals shared the global trend, when compared with horses.

The opposite thermal patterns were usually associated with the GORs *Dorsal aspect* and *Trunk*. Both GORs cover a relatively large area of the animal's body, which raises the question of whether a more granular segmentation of these areas would allow for the discovery of further similarities.

An important question is whether the trends observed in the studied dataset are characteristic of the entire population. As the number of cases was limited by practical considerations, we believe that our results should be treated as a significant indication of the existence of the relationships described. At the same time, we emphasize the need to further verify these conclusions in a larger population. To facilitate this, all research data related to this study are available to the public under open licenses. However, the authors speculate that many imaging approaches applied successfully in equine veterinary medicine [1,25] can also be used for donkey IRT imaging, subject to the visualization conditions presented here. The IRT may become a useful diagnosis tool in donkeys, particularly in the field of orthopedics [5–7,14] or effort assessment [21,22,24], but even in the field of animal welfare [2,18,33]. Given the increase in general interest in donkeys, as milk producers, companion animals [33], and working animals under a saddle and in a harness [47], the new noninvasive imaging approaches in donkeys are in demand.

4.2. Special Cases

The two animals identified as outliers, i.e., *D.17–18*, allowed for an interesting study of how general or specific the proposed thermal patterns were. The visualization of the differences between the patterns of these animals and the rest of the donkeys is presented in Figure 13. In Panels b and c, the green color indicates the compliance of the animal thermal pattern with the global trend, i.e., the difference for a given pair of GORs has the same sign, and it is statistically significant ($p < 0.001$) for the animal. Thermal patterns for the donkey *D.17* with a thick hair coat were usually in line with the global pattern, except for four pairs of GORs, where the statistical significance of the local animal pattern could not be confirmed. The patterns of the donkey *D.18* were much less in line with the global pattern. This was an expected result, as the patchy hair losses visibly affected its thermal characteristics in the image. This also suggests that the proposed thermal patterns may be the basis for creating temperature indexes, as, e.g., in [2], or features for detecting anomalies. The temperature difference matrices for these two animals are provided in Figure A2 in Appendix A.

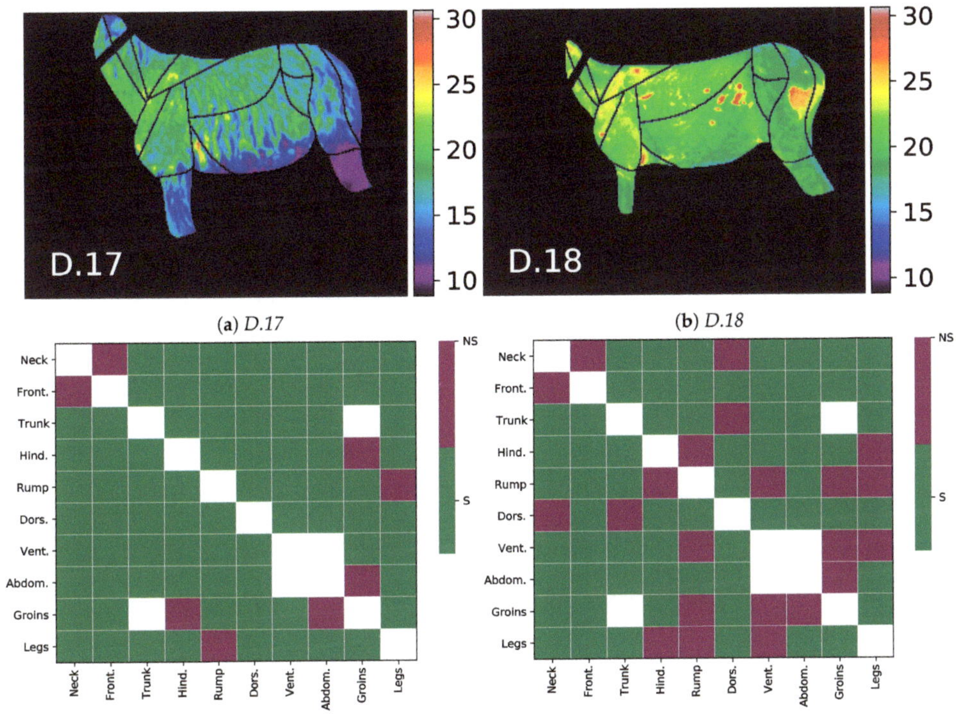

(**a**) *D.17* (**b**) *D.18*

(**c**) *D.17* and Donkeys (**d**) *D.18* and Donkeys

Figure 13. Differences between donkeys *D.17* and *D.18*, outlier cases, and the rest of the donkeys, i.e., animals *D.1-16*. The upper panels present thermal maps of the two cases: (**a**) donkey *D.17*; (**b**) donkey *D.18*. Donkey *D.17* was colder than other animals due to its long hair length. Donkey *D.18* had an unusual pattern of warm areas resulting from patchy hair loss. Bottom plots show differences in the thermal patterns of these donkeys compared to the global pattern of other donkeys: (**c**) donkey *D.17* compared to other donkeys; (**d**) donkey *D.18* compared to other donkeys. The *S* class (green) indicates that the individual animal pattern was in line with the global trend, and class *NS* (red) indicates the opposite.

5. Conclusions

The results of this study indicate that the characteristic thermal patterns of both horses and donkeys are mostly associated with groups of ROIs (GORs) rather than an individual ROI. Based on this observation, the thermal pattern is defined as a statistically significant difference between the mean temperatures of the designated GORs for a given animal species. The thorough verification of the significance (both globally for all data and locally for individual animals) reveals the similarity for the majority of proposed thermal patterns in both studied species. It is worth noting that the thermal patterns of the donkeys are more uniform than those of the horses, and the donkeys are individually more consistent with the global trend. The average surface temperatures compared within the proposed thermal patterns are higher for the studied horses than for the donkeys, which may be related to different thermal properties of their skin and hair coat.

Author Contributions: Conceptualization, M.D. and T.J.; methodology, M.D., M.R.; software, M.R.; validation, M.D. and M.R.; formal analysis, M.R.; investigation, M.D., M.R., M.M. and T.J.; resources, M.D., M.R., M.M. and T.J.; data curation, M.D. and M.R.; writing, original draft preparation, M.D. and M.R.; writing, review and editing, M.D. and M.R.; visualization, M.D. and M.R.; supervision, M.D.; project administration, M.D. All authors read and agreed to the published version of the manuscript.

Funding: This research received no external funding.

Acknowledgments: This work was conducted at the Veterinary Research Centre WULS (WCB) and the Center for Biomedical Research (CBB) supported by EFRR RPO WM 2007–2013. The authors are grateful to the Słupski, owners of the Mariaż agritourism farm in Lubachów, for letting them observe donkeys and horses and for their help on site. The authors would like to thank Zbigniew Puchała, Przemysław Głomb, and Michał Cholewa for their remarks on the methodology and Hanna Christoffersson, Victoria Czarnota, and Hannah York for their help with language editing. The authors would also like to thank the reviewers for their valuable comments and suggestions.

Conflicts of Interest: The authors declare no conflict of interest.

Abbreviations

The following abbreviations are used in this manuscript:

BCS	Body condition score
FNS	Fatty neck score
ROI	Region of interest
GOR	Group of ROIs
IRT	Infrared thermography
SF-Skin	Subcutaneous fat (SF) plus skin thickness
t-SNE	t-distributed stochastic neighbor embedding

Appendix A

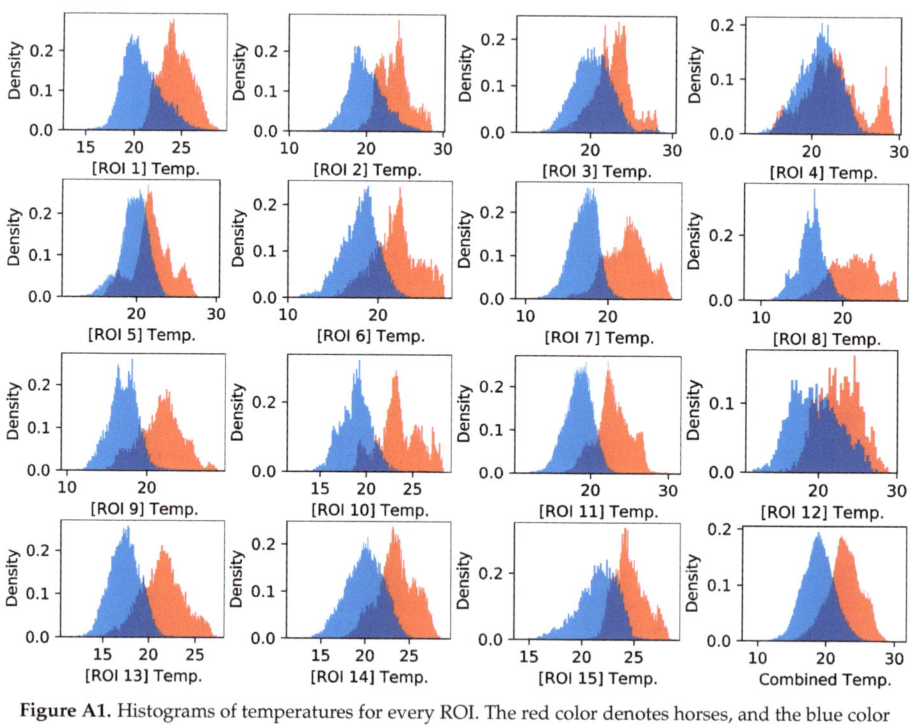

Figure A1. Histograms of temperatures for every ROI. The red color denotes horses, and the blue color denotes donkeys. The last plot presents the combined histogram for all ROIs.

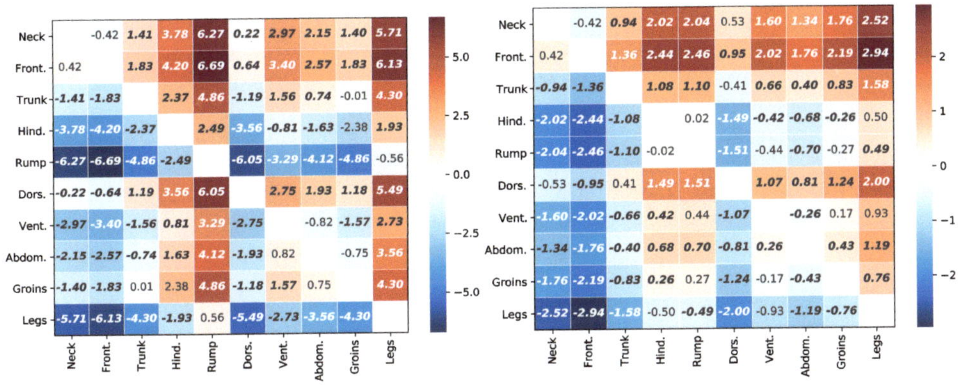

(**a**) M^{Δ}, *D.17* (**b**) M^{Δ}, *D.18*

Figure A2. Thermal patterns, i.e., statistically significant ($p < 0.001$) differences between the mean temperatures in GORs for the two donkeys identified as outliers (see Section 2.1): (**a**) donkey *D.17*; (**b**) donkey *D.18*. Bold font indicates the statistical significance of the difference for the given pattern ($p < 0.001$).

References

1. Soroko, M.; Howell, K. Infrared thermography: Current applications in equine medicine. *J. Equine Vet. Sci.* **2018**, *60*, 90–96. [CrossRef]
2. Soroko, M.; Zaborski, D.; Dudek, K.; Yarnell, K.; Górniak, W.; Vardasca, R. Evaluation of thermal pattern distributions in racehorse saddles using infrared thermography. *PLoS ONE* **2019**, *14*, e0221622. [CrossRef] [PubMed]
3. Masko, M.; Zdrojkowski, L.; Domino, M.; Jasinski, T.; Gajewski, Z. The Pattern of Superficial Body Temperatures in Leisure Horses Lunged with Commonly Used Lunging Aids. *Animals* **2019**, *9*, 1095. [CrossRef] [PubMed]
4. Smith, W. Applications of thermography in veterinary medicine. *Ann. N. Y. Acad. Sci.* **1964**, *121*, 248. [CrossRef]
5. Eddy, A.; Van Hoogmoed, L.; Snyder, J. The role of thermography in the management of equine lameness. *Vet. J.* **2001**, *162*, 172–181. [CrossRef]
6. Cetinkaya, M.; Demirutku, A. Thermography in the assessment of equine lameness. *Turk. J. Vet. Anim. Sci.* **2012**, *36*, 43–48.
7. Ciutacu, O.; Tanase, A.; Miclaus, I. Digital infrared thermography in assessing soft tissues injuries on sport equines. *Bull. Univ. Agric. Sci. Vet. Med. Cluj Napoca. Vet. Med.* **2006**, *63*, 228–233.
8. Kastberger, G.; Stachl, R. Infrared imaging technology and biological applications. *Behav. Res. Methods Instruments Comput.* **2003**, *35*, 429–439. [CrossRef]
9. Westermann, S.; Buchner, H.; Schramel, J.; Tichy, A.; Stanek, C. Effects of infrared camera angle and distance on measurement and reproducibility of thermographically determined temperatures of the distolateral aspects of the forelimbs in horses. *J. Am. Vet. Med. Assoc.* **2013**, *242*, 388–395. [CrossRef]
10. Gómez Álvarez, C.; Rhodin, M.; Byström, A.; Back, W.; Van Weeren, P. Back kinematics of healthy trotting horses during treadmill versus over ground locomotion. *Equine Vet. J.* **2009**, *41*, 297–300. [CrossRef]
11. Bächi, B.; Wiestner, T.; Stoll, A.; Waldern, N.; Imboden, I.; Weishaupt, M. Changes of ground reaction force and timing variables in the course of habituation of horses to the treadmill. *J. Equine Vet. Sci.* **2018**, *63*, 13–23. [CrossRef]
12. Seeherman, H. The use of high-speed treadmills for lameness and hoof balance evaluations in the horse. *Vet. Clin. N. Am. Equine Pract.* **1991**, *7*, 271–309. [CrossRef]
13. Tabor, G.; Nankervis, K.; Fernandes, J.; Williams, J. Generation of Domains for the Equine Musculoskeletal Rehabilitation Outcome Score: Development by Expert Consensus. *Animals* **2020**, *10*, 203. [CrossRef] [PubMed]

14. Fonseca, B.; Alves, A.; Nicoletti, J.; Thomassian, A.; Hussni, C.; Mikail, S. Thermography and ultrasonography in back pain diagnosis of equine athletes. *J. Equine Vet. Sci.* **2006**, *26*, 507–516. [CrossRef]
15. Soroko, M.; Jodkowska, E.; Zabłocka, M. The use of thermography to evaluate back musculoskeletal responses of young racehorses to training. *Thermol. Int.* **2012**, *22*, 114.
16. Arruda, T.; Brass, K.; Flavio, D. Thermographic assessment of saddles used on jumping horses. *J. Equine Vet. Sci.* **2011**, *31*, 625–629. [CrossRef]
17. Becker-Birck, M.; Schmidt, A.; Wulf, M.; Aurich, J.; Von der Wense, A.; Möstl, E.; Berz, R.; Aurich, C. Cortisol release, heart rate and heart rate variability, and superficial body temperature, in horses lunged either with hyperflexion of the neck or with an extended head and neck position. *J. Anim. Physiol. Anim. Nutr.* **2013**, *97*, 322–330. [CrossRef] [PubMed]
18. Redaelli, V.; Luzi, F.; Mazzola, S.; Bariffi, G.; Zappaterra, M.; Nanni Costa, L.; Padalino, B. The Use of Infrared Thermography (IRT) as Stress Indicator in Horses Trained for Endurance: A Pilot Study. *Animals* **2019**, *9*, 84. [CrossRef]
19. Valera, M.; Bartolomé, E.; Sánchez, M.J.; Molina, A.; Cook, N.; Schaefer, A. Changes in eye temperature and stress assessment in horses during show jumping competitions. *J. Equine Vet. Sci.* **2012**, *32*, 827–830. [CrossRef]
20. Bartolomé, E.; Sánchez, M.; Molina, A.; Schaefer, A.; Cervantes, I.; Valera, M. Using eye temperature and heart rate for stress assessment in young horses competing in jumping competitions and its possible influence on sport performance. *Anim. Int. J. Anim. Biosci.* **2013**, *7*, 2044. [CrossRef]
21. Soroko, M.; Śpitalniak-Bajerska, K.; Zaborski, D.; Poźniak, B.; Dudek, K.; Janczarek, I. Exercise-induced changes in skin temperature and blood parameters in horses. *Arch. Anim. Breed.* **2019**, *62*, 205. [CrossRef] [PubMed]
22. Witkowska-Piłaszewicz, O.; Maśko, M.; Domino, M.; Winnicka, A. Infrared thermography correlates with lactate concentration in blood during race training in horses. *Animals* **2020**, *11*, 2072. [CrossRef] [PubMed]
23. Hinchcliff, K.; Kaneps, A.; Geor, R. *Equine Sports Medicine and Surgery: Basic and Clinical Sciences of the Equine Athlete*; Elsevier: London, UK, 2013.
24. Simon, E.; Gaughan, E.; Epp, T.; Spire, M. Influence of exercise on thermographically determined surface temperatures of thoracic and pelvic limbs in horses. *J. Am. Vet. Med Assoc.* **2006**, *229*, 1940–1944. [CrossRef] [PubMed]
25. Turner, T. Diagnostic thermography. *Vet. Clin. N. Am. Equine Pract.* **2001**, *17*, 95–114. [CrossRef]
26. Soroko, M.; Howell, K.; Dudek, K. The effect of ambient temperature on infrared thermographic images of joints in the distal forelimbs of healthy racehorses. *J. Therm. Biol.* **2017**, *66*, 63–67. [CrossRef]
27. Zakari, F.; Ayo, J.; Rekwot, P.; Kawu, M.; Minka, N. Daily rhythms of rectal and body surface temperatures in donkeys during the cold-dry (harmattan) and hot-dry seasons in a tropical savannah. *Int. J. Biometeorol.* **2018**, *62*, 2231–2243. [CrossRef]
28. Rosenmeier, J.; Strathe, A.; Andersen, P. Evaluation of coronary band temperatures in healthy horses. *Am. J. Vet. Res.* **2012**, *73*, 719–723. [CrossRef]
29. Machado, L.F.S.; Dittrich, R.L.; Pavelski, M.; Afonso, A.M.C. da F.; Deconto, I.; Dornbusch, P.T.; Standardization of thermographic examination in joints of horses in training. *Arch. Vet. Sci.* **2013**, *18*, 40–45.
30. Tunley, B.; Henson, F. Reliability and repeatability of thermographic examination and the normal thermographic image of the thoracolumbar region in the horse. *Equine Vet. J.* **2004**, *36*, 306–312. [CrossRef]
31. Pavelski, M.; Basten, M.; Busato, E.; Dornbusch, P. Infrared thermography evaluation from the back region of healthy horses in controlled temperature room. *Ciênc. Rural.* **2015**, *45*, 1274–1279. [CrossRef]
32. Rowland, M.; Coombs, T.; Connor, M. A study of traveller horse owners' attitudes to horse care and welfare using an equine Body Condition Scoring system. *Animals* **2019**, *9*, 162. [CrossRef] [PubMed]
33. Valle, E.; Raspa, F.; Giribaldi, M.; Barbero, R.; Bergagna, S.; Antoniazzi, S.; Mc Lean, A.; Minero, M.; Cavallarin, L. A functional approach to the body condition assessment of lactating donkeys as a tool for welfare evaluation. *PeerJ* **2017**, *5*, e3001. [CrossRef] [PubMed]
34. Davidson, E. Lameness evaluation of the athletic horse. *Vet. Clin. Equine Pract.* **2018**, *34*, 181–191. [CrossRef] [PubMed]

35. Satchell, G.; McGrath, M.; Dixon, J.; Pfau, T.; Weller, R. Effects of Time of Day, Ambient Temperature and Relative Humidity on the Repeatability of Infrared Thermographic Imaging in Horses. *Equine Vet. J.* **2015**, *47*, 13–14. [CrossRef] [PubMed]
36. Purohit, R. Standards for thermal imaging in veterinary medicine. In Proceedings of the 11th European Congress of Thermology, Mannheim, Germany, 18–20 September 2009; Volume 19, p. 99.
37. Silva, S.; Payan-Carreira, R.; Quaresma, M.; Guedes, C.; Santos, A. Relationships between body condition score and ultrasound skin-associated subcutaneous fat depth in equids. *Acta Vet. Scand.* **2016**, *58*, 37–42. [CrossRef] [PubMed]
38. Osthaus, B.; Proops, L.; Long, S.; Bell, N.; Hayday, K.; Burden, F. Hair coat properties of donkeys, mules and horses in a temperate climate. *Equine Vet. J.* **2018**, *50*, 339–342. [CrossRef]
39. Domino, M.; Romaszewski, M.; Jasiński, T.; Maśko, M. Comparison of surface thermal patterns of horses and donkeys in IRT images. *arXiv* **2020**, arXiv:2010.09302, 2020.
40. Derrick, B.; White, P. Comparing two samples from an individual Likert question. *Int. J. Math. Stat.* **2017**, *18*, 55084474.
41. Hotelling, H. Analysis of a complex of statistical variables into principal components. *J. Educ. Psychol.* **1933**, *24*, 417. [CrossRef]
42. Van der Maaten, L.; Hinton, G. Visualizing data using t-SNE. *J. Mach. Learn. Res.* **2008**, *9*, 2579–2605.
43. Cymbaluk, N.; Christison, G. Environmental effects on thermoregulation and nutrition of horses. *Vet. Clin. North Am. Equine Pract.* **1990**, *6*, 355–372. [CrossRef]
44. Ousey, J.; McArthur, A.; Murgatroyd, P.; Stewart, J.; Rossdale, P. Thermoregulation and total body insulation in the neonatal foal. *J. Therm. Biol.* **1992**, *17*, 1–10. [CrossRef]
45. Soroko, M.; Howell, K.; Dudek, K.; Henklewski, R.; Zielińska, P. The influence of breed, age, gender, training level and ambient temperature on forelimb and back temperature in racehorses. *Anim. Sci. J.* **2017**, *88*, 347–355. [CrossRef] [PubMed]
46. Quaresma, M.; Payan-Carreira, R.; Silva, S. Relationship between ultrasound measurements of body fat reserves and body condition score in female donkeys. *Vet. J.* **2013**, *197*, 329–334. [CrossRef] [PubMed]
47. McLean, A.; Navas González, F.; Canisso, I. Donkey and Mule Behavior. *Vet. Clin. N. Am. Equine Pract.* **2019**, *35*, 575–588. [CrossRef]

animals

Article

Stage-Dependent Expression of Protein Gene Product 9.5 in Donkey Testes

Yeonju Choi [1], Youngwook Jung [1], Seongmin Kim [1], Junyoung Kim [1], Heejun Jung [1] and Minjung Yoon [1,2,*]

1 Department of Animal Science and Biotechnology, Kyungpook National University, Sangju 37224, Korea; yjchoi2031@gmail.com (Y.C.); dyddnr02070@naver.com (Y.J.); kimhandles@gmail.com (S.K.); cpm_911@naver.com (J.K.); hjjung1019@gmail.com (H.J.)

2 Department of Horse, Companion, and Wild Animal Science, Kyungpook National University, Sangju 37224, Korea

* Correspondence: mjyoonemail@gmail.com; Tel.: +82-05-4530-1233; Fax: +82-05-4530-1959

Received: 30 September 2020; Accepted: 17 November 2020; Published: 20 November 2020

Simple Summary: Spermatogenesis and steroidogenesis are key functions of the testes. Molecular markers that identify each stage of germ cells and Leydig cells can identify and isolate specific germ or Leydig cells. Protein gene product (PGP)9.5 is observed in neuroendocrine cells and tumors; it is also used for the immunohistochemical detection of spermatogonial stem cells (SSCs) in various species of animals. It was found that the immunolabeling of PGP9.5 in testicular tissue was not observed in the seminiferous tubules in the pre-pubertal stage. However, in the post-pubertal stage, spermatogonia were immunolabeled with PGP9.5. Interestingly, some Leydig cells were immunolabeled with PGP9.5 in both pre- and post-pubertal stages. This study reflects that the PGP9.5 antibody can be used as a tool to identify and isolate spermatogonia from seminiferous tubules in the post-pubertal stage of donkey testes.

Abstract: Molecular markers can be used to identify and isolate specific developmental stages of germ cells and Leydig cells. Protein gene product (PGP)9.5 expression in spermatogonia and Leydig cells has been reported in several species. The stages of spermatogonia and Leydig cells expressing PGP9.5 vary depending on the species and reproductive stages. Thus, the objectives of this study were (1) to identify the localization of PGP9.5 in donkey testicular cells, and (2) to compare the expression patterns of PGP9.5 in donkey testicular cells between pre- and post-pubertal stages. Testes samples were collected following the routine field castration of six donkeys. Western blotting was performed to verify the cross-reactivity of the rabbit anti-human PGP9.5 antibody to donkey testes. Immunofluorescence was performed to investigate the expression pattern of PGP9.5 in testicular tissues at different reproductive stages. In Western blotting, the protein band of the PGP9.5 antibody appeared at approximately 27 kDa, whereas the band was not observed in the negative control treated with normal mouse IgG. In the pre-pubertal stage, the expression of deleted in azoospermia-like (DAZL) was found in some spermatogonia in pre-pubertal testicular tissues. However, the immunolabeling of PGP9.5 in testicular tissue was not observed in the seminiferous tubules. In stages 1 and 2, spermatogonia were immunolabeled with either PGP9.5 or DAZL. In contrast, PGP9.5 and DAZL were co-immunolabeled in some of the spermatogonia in stages 3 to 8. Interestingly, some Leydig cells were immunolabeled with PGP9.5 in both pre- and post-pubertal stages. In conclusion, the PGP9.5 antibody can be used as a tool to identify and isolate spermatogonia from seminiferous tubules.

Keywords: PGP9.5; donkey; testes; spermatogenesis

1. Introduction

Spermatogenesis and steroidogenesis are key functions of the testes. Molecular markers that identify each stage of germ cells and Leydig cells can identify and isolate specific germ or Leydig cells. Using tools such as molecular markers, biological processes for the proliferation and differentiation of germ and Leydig cells can be explored. These markers can also be applied to develop advanced assisted reproductive techniques such as spermatogonial stem cell (SSC) or Leydig cell transplantation and diagnose subfertility or infertility in donkeys.

Previous studies have examined molecular markers for germ cells. Glial cell-derived neurotrophic factor family receptor alpha-1, promyelocytic leukemia zinc finger, and colony-stimulating factor 1 receptor were found to be markers for SSCs [1]. In a previous study, our laboratory reported that a co-immunolabeling system with undifferentiated embryonic cell transcription factor 1 (UTF1) and deleted in azoospermia-like (DAZL) could be used to identify undifferentiated (UTF1 only), differentiating (UTF1 and DAZL), or differentiated spermatogonia (DAZL only) [2]. Thus, the ability to more efficiently detect more molecular markers that can identify various stages of germ and Leydig cells would be beneficial.

Protein gene product (PGP)9.5, also known as ubiquitin C-terminal hydrolase L1 (UCHL1), was initially found to be expressed in neuroendocrine cells and tumors [3]. The function of PGP9.5 is to increase the available pool of ubiquitin that can attach to the proteins which are about to be degraded by the proteasome [4]. The expression of PGP9.5 was also reported in reproductive systems such as the testes in several animals, which include mice [5], cattle [6], pigs [7], sheep [8], and goats [9]. The expression of PGP9.5 was detected in spermatogonia and Leydig cells, but the immunolabeling pattern varies depending on the species. In addition, the PGP9.5 was immunolabeled in a reproductive stage-dependent manner. Thus, to confirm the localization of PGP9.5 in testes, the expression patterns of the marker must be tested in each species at different reproductive stages.

We recently reported that PGP9.5 was expressed in the germ cells adjacent to the seminiferous tubule membrane and cytoplasmic area of Leydig cells in stallions [10]. Based on the results of our study and other previous studies, we hypothesized that PGP9.5 is expressed in donkey testes, and it can be used as a marker for specific stages of germ cells and Leydig cells.

Thus, the objectives of this study were (1) to identify the localization of PGP9.5 in donkey testicular cells, and (2) to compare the expression patterns of PGP9.5 in donkey testicular cells between pre- and post-pubertal stages.

2. Materials and Methods

2.1. Testicular Sample Preparation

Testes were collected following the routine field castration of six donkeys from a private farm in Icheon, South Korea. The reproductive stages of donkeys were categorized based on the morphological characteristics of the cross sections of seminiferous tubules within donkey testes. Following castration, the testes were directly transported to the laboratory in a 4 °C icebox. The testes samples were then cut into small pieces (approximately 1 cm^3), and the tissues were treated with paraformaldehyde for at least 24 h at room temperature. During the dehydration process, the sample tissues were immersed in phased ethanol concentrations of 25%, 50%, 70%, 80%, 90%, and 100%. Then, the tissues were embedded in paraffin blocks, and the tissue pieces (1 mg each) were snap-frozen in liquid nitrogen and stored at −80 °C for further Western blotting.

2.2. Western Blotting

Western blotting was performed to verify the cross-reactivity of rabbit anti-human PGP9.5 (7863-2004, Bio-Rad, Hercules, CA, USA) to the PGP9.5 present in donkey testes according to the previously reported protocol with minor modifications [2,11]. The testicular samples, which had been stored at −80 °C, were thawed at room temperature and homogenized using a Polytron PT

1200 CL homogenizer (Kinematica AG, Littau, Lucerne, Switzerland). Sample preparation buffer (0.5 M Tris-HCL (pH 6.8), 0.1% glycerol, 10% sodium dodecyl sulfate (SDS), 0.05% 2-β-mercaptoethanol, and bromophenol blue in distilled water) was used to dilute the concentration of quantified protein to 2 mg/mL. After being heated in a boiling water bath for 15 min, 15-μL samples were loaded onto a 10% SDS-polyacrylamide gel and separated using a Mini-Protean system (Bio-Rad, Hercules, CA, USA). The samples were electrotransferred to a polyvinylidene difluoride membrane (Millipore) and blocked with skim milk (BD Biosciences, San Jose, CA, USA). The membrane was incubated with PGP9.5 antibody diluted to 1:500 in skim milk overnight at 4 °C. For the negative control, the membrane was treated with normal mouse immunoglobulin G (IgG) at the same concentration of primary antibody. Horseradish peroxidase-conjugated anti-mouse IgG diluted in skim milk (1:10,000) was used as a secondary antibody.

2.3. *Immunofluorescence*

Immunofluorescence was performed using a previously described protocol [12]. The testicular tissues (5 μm), which were attached to the slides, were then deparaffinized using xylene solution. The tissues were dehydrated with a series of ethanol baths (100%, 95%, 80%, 50%, and 25%), treated in a citrate buffer for 30 min at 97.5 °C for antigen retrieval, and cooled via a cooling process for 30 min to room temperature. The slides were blocked with 5% donkey serum (Sigma, St. Louis, MO, USA) diluted in phosphate-buffered saline. The slides were co-stained with mouse anti-human PGP9.5 antibody and goat anti-human DAZL antibody diluted in blocking buffer at the ratio of 1:1000 and 1:100, respectively. Donkey anti-mouse IgG (Alexa Fluor 488, Thermo Fisher Scientific, Waltham, MA, USA) and donkey anti-goat IgG (Alexa Fluor 594, Thermo Fisher Scientific, Waltham, MA, USA) were used as secondary antibodies.

2.4. *Imaging*

Immunolabeling of the sample tissues was observed using a Leica DM2500 fluorescent microscope (Leica, Wetzlar, Germany). The microscope was equipped with an EL 6000 external light source (Leica). Green and red fluorescent expressions were observed using a dual-emission fluorescein-5-isothiocyanate (FITC)/ tetramethyl rhodamine isothiocyanate (TRITC) filter. Cells showing green or red fluorescence were considered positive for either PGP9.5 or DAZL, whereas cells with no fluorescence were considered negative. The immunolabeled images were captured using a Leica DFC 450 C digital camera.

3. Results

3.1. *Cross-Reactivity of the PGP9.5 Antibody in Donkey Testes*

The cross-reactivity of the PGP9.5 antibody in testicular tissues of pre-pubertal and post-pubertal donkeys was evaluated using Western blot analysis. The protein band appeared in the PGP9.5 antibody at approximately 27 kDa (Figure 1). However, the band was not observed in the negative control lane that was treated with normal mouse IgG with the same concentration as the primary antibody. The β-actin band, which was used for positive control, was also detected at an approximate molecular weight of 45 kDa in donkey testes. This result proves that the mouse anti-human PGP9.5 antibody used in this study suggests cross-reactivity for the PGP9.5 protein of donkey testes.

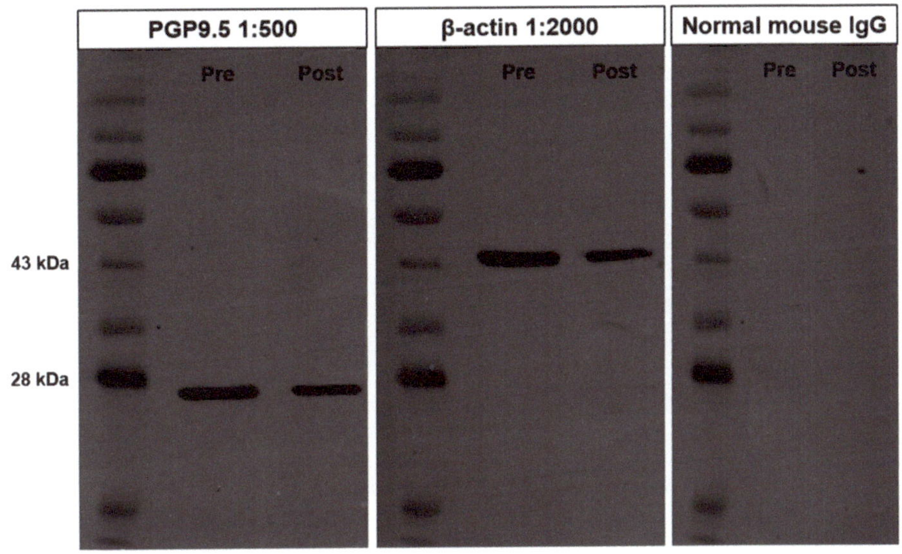

Figure 1. Cross-reactivity of protein gene product (PGP)9.5 in donkey testes. The PGP9.5 protein band appeared at the molecular size of approximately 27 kDa. The β-actin band was also detected at approximately 45 kDa in donkey testes. The negative control lane was treated with normal mouse IgG with the same concentration as the primary antibody and showed no bands.

3.2. Identification of Reproductive Stageof Donkey Testes

The reproductive stage of donkeys is characterized according to the seminiferous tubular morphology. The testes sample of pre-pubertal donkeys had relatively undeveloped seminiferous tubules with lumen closing (Figure 2a,b). In contrast, the seminiferous tubules of post-pubertal donkeys had all developmental stage of germ cells with lumen opening (Figure 2c,d).

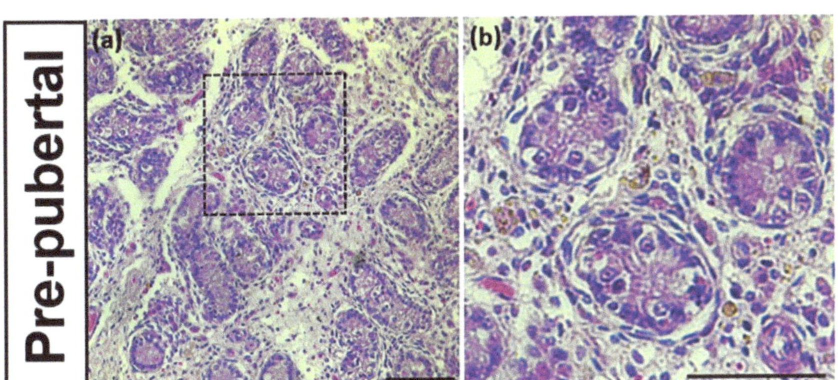

Figure 2. *Cont.*

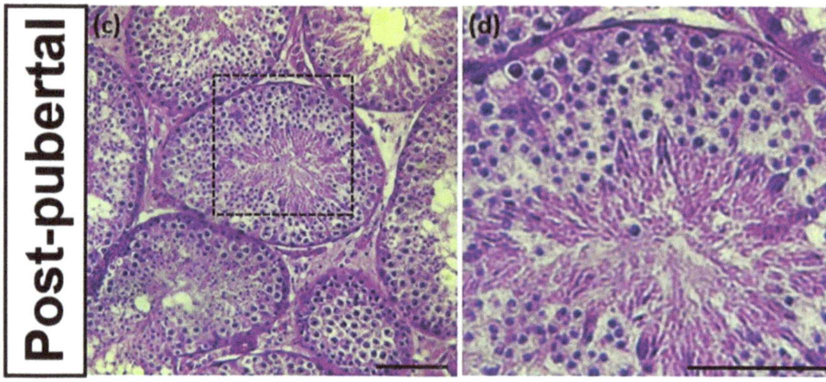

Figure 2. Morphological characteristics of seminiferous tubules of pre- and post-pubertal donkey with H&E staining. Pre-pubertal donkey testes had undeveloped seminiferous tubules with lumen closing (**a**,**b**). In contrast, post-pubertal donkeys had all developmental stage of germ cells with lumen opening gin the seminiferous tubules (**c**,**d**). The size bar = 50 μm.

3.3. PGP9.5 and DAZL Expression in the Testes of Pre-Pubertal Donkeys

The deleted in azoospermia-like (DAZL) is a molecular marker for undifferentiated spermatogonia of pre-pubertal donkeys [10]. In this study, the expression of DAZL was detected in the spermatogonia in pre-pubertal testicular tissues (Figure 3c,d,e). The immunolabeling of PGP9.5 in testicular tissue was not observed in germ cells in the seminiferous tubules. However, the cytoplasms of myoid cells located basal lamina in seminiferous tubules were immunostained with PGP9.5 (Figure 3b,d,e).

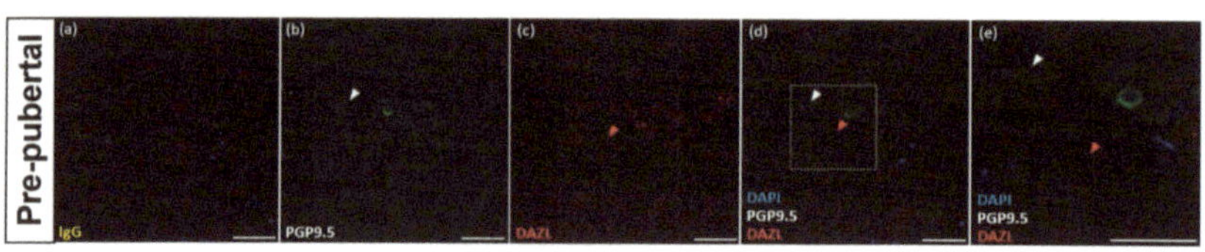

Figure 3. Immunostaining of PGP9.5 and deleted in azoospermia-like (DAZL) in the seminiferous tubules of pre-pubertal donkeys. Testicular tissue stained with IgG showed no immunolabeling (**a**). PGP9.5 immunolabeling was detected in the cytoplasms of myoid cells (**b**,**d**,**e**). The DAZL immunolabeling was detected in the cytoplasm of germ cells (**c**,**d**,**e**). The white arrowhead indicates myoid cells immunolabeled with PGP9.5. The red arrowhead indicates germ cells immunolabeled with DAZL. The size bar = 50 μm.

3.4. PGP9.5 and DAZL Expressions in Testes of Post-Pubertal Donkeys in Different Seminiferous Epithelium Cycle Stages

The stages of the seminiferous epithelium cycle of donkey testes were categorized with eight stages following the results of previous studies [13,14] to determine the germ cell types immunostained with PGP9.5. The DAZL is a molecular marker of undifferentiated spermatogonia and primary spermatocytes in the post-pubertal stage of donkeys [10]. Most spermatogonia and primary spermatocytes near the basal membrane of seminiferous tubules were immunostained with DAZL antibody in all reproductive stages (Figure 4a(3–5), b(3–5), c(3–5), d(3–5), e(3–5), f(3–5), g(3–5), and h(3–5)). In addition, PGP9.5 displayed positive expression in spermatogonia in all eight different stages of the seminiferous epithelium cycle (Figure 4a(4–5), b(4–5), c(4–5), d(4–5), e(4–5), f(4–5), g(4–5), and h(4–5)). The PGP9.5 was co-immunolabeled with DAZL from stages 3 to 8 (c(4–5), d(4–5), e(4–5), f(4–5), g(4–5), and h(4–5)), whereas PGP9.5 and DAZL were immunolabeled, respectively, in spermatogonia in stages 1 and 2 (Figure 4a(4–5) and b(4–5)).

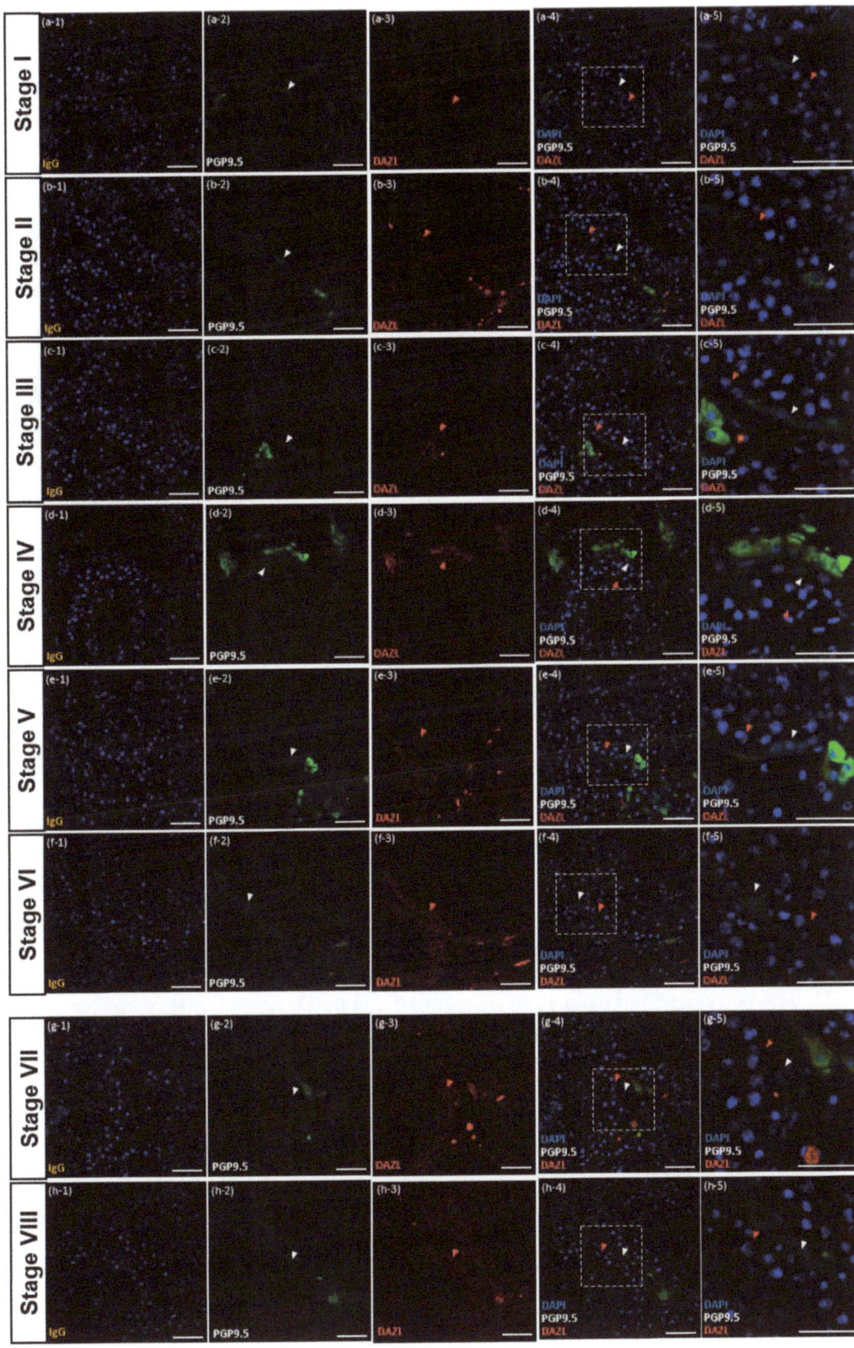

Figure 4. Reproductive stage-dependent expression of PGP9.5 and DAZL in the spermatogonia of a post-pubertal donkey. PGP9.5 immunolabeling was discovered in the cytoplasm of spermatogonia in the post-pubertal stage co-stained with DAZL (**a–h**). The white arrowhead indicates germ cells immunolabeled with PGP9.5. The red arrowhead indicates germ cells immunolabeled with DAZL. The size bar = 50 μm.

3.5. Expression of PGP9.5 in Leydig Cells of the Testes of Donkeys

Leydig cells were immunolabeled with the PGP9.5 antibody at both the pre-pubertal and post-pubertal stages (Figure 5a–h). At both reproductive stages, PGP9.5 expression was localized in the cytoplasm of Leydig cells.

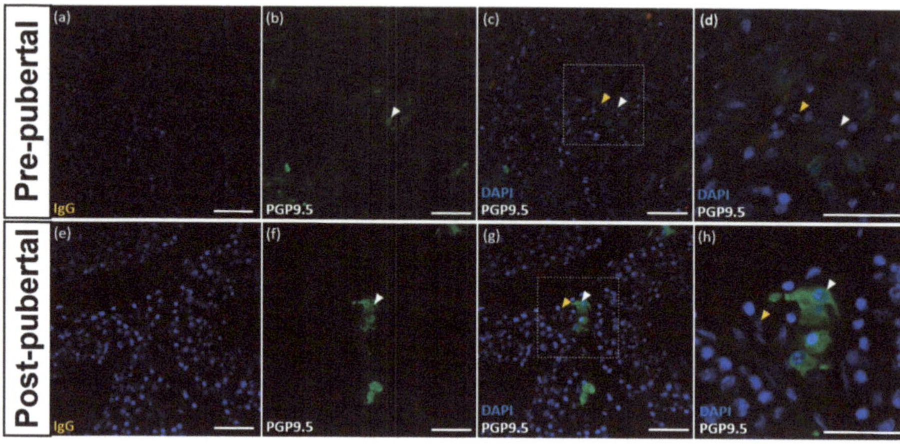

Figure 5. Expression of PGP9.5 in Leydig cells of pre- and post-pubertal donkeys. Testicular tissue stained with IgG showed no immunolabeling (**a**,**e**). The expression of PGP9.5 was observed in the testes of pre-pubertal (**b**–**d**) and post-pubertal (**f**–**h**) donkeys. The white arrowheads indicate Leydig cells stained with PGP9.5. The yellow arrowheads indicate Leydig cells without PGP9.5 staining. The white bar = 50 μm.

4. Discussion

This study was conducted to find molecular markers to specify germ cell developmental stages in donkey testes. First, we demonstrated the cross-reactivity of each antibody to its respective protein in the testicular tissues of donkeys using Western blot analysis. The PGP9.5 protein band appeared at the adequate molecular size (~27 kDa). The antibodies also had cross-reactivity with PGP9.5 expressed in donkey testicular tissues, similar to a result seen in horses [15]. UCHL1 mRNA of horses is 99.44% identical with donkey mRNA based on the sequence analysis. The similar immunolabeling pattern of UCHL1 between horse and donkey testes appears to be due to high rate of sequence similarity. In addition, in the study of chromosomal polymorphisms in the genus *Equus*, authors suggested that the chromosome arms are homologous in UCHL1 [16]. Such results indicate that the structural motifs recognized by the PGP9.5 antibody are identical in the two species.

PGP9.5 is observed in neuroendocrine cells and tumors; it is also used for the immunohistochemical detection of SSCs in several animals such as mice [17], bovines [18], goats [9], and porcine [19]. Since SSCs exist in the testes of either pre-mature or mature males, we expected that the expression of PGP9.5 is present in both the pre- and post-pubertal tissues of donkeys. However, in the pre-pubertal stage, PGP9.5 was not immunolabeled in germ cells, while post-pubertal stage spermatogonia were immunolabeled with the PGP9.5 antibody. This result indicates that PGP9.5 expression in spermatogonia is reproductive stage-dependent in the donkey.

PGP9.5 is an essential modulator for germ cell apoptosis in testes [20–22]. The apoptosis of germ cells regulates the number of germ cells by modulating dynamic cell differentiation of spermatogenesis. In addition, apoptosis can eliminate defective germ cells during active spermatogenesis to maintain testicular homeostasis [23,24]. During the pre-pubertal period, the gonocytes are gradually replaced with spermatogonia, which is reserved, rather than undergoing active differentiation [25,26]. The static condition of spermatogenesis during the pre-pubertal stage may be why PGP9.5 expression was not expressed in the seminiferous tubules of pre-pubertal donkeys.

In post-pubertal tissues, seminiferous tubules were categorized with eight different cyclic stages based on the stage of the seminiferous epithelium cycle of the donkey [13,14]. In stages 1 and 2, spermatogonia were immunolabeled with either PGP9.5 or DAZL. In contrast, the PGP9.5 and DAZL were co-immunolabeled in some spermatogonia in stages 3 to 8. The distinct morphological difference of spermatogonia immunolabeled with PGP9.5 only compared to immunolabeled with PGP9.5 and DAZL was the absence of cytoplasmic bridges. Type A SSCs do not have cytoplasmic bridges with which progenitor cells are connected [27]. Thus, spermatogonia immunolabeled with PGP9.5 only should be SSCs in donkeys. We previously reported that DAZL might be a marker for differentiating and differentiated spermatogonia and primary spermatocytes in donkeys [10]. Thus, germ cells stained with DAZL only in stages 1 and 2 appear to be differentiating spermatogonia. In the stages of 3 to 8, several PGP9.5-positive spermatogonia were connected with cytoplasmic bridges and placed adjacent to the membrane of seminiferous tubules. This result also indicates that PGP9.5 is expressed in the cytoplasmic area of spermatogonia A and B.

Interestingly, Leydig cells of donkeys were also immunolabeled with PGP9.5 at both pre- and post-pubertal stages in this study. Our finding is consistent with previous research in other species such as horses [15] and humans [28], showing the strong intensity of immunolabeling in the cytoplasmic area of Leydig cells. However, this result is contrary to the results found in domestic dogs [29] and mice [17]. The results of those studies suggest that PGP9.5 expression in Leydig cells is species-dependent. In horses, all types of Leydig cells were immunolabeled with PGP9.5 [15]. In contrast, in donkey testes, some Leydig cells were immunolabeled with PGP9.5 in both pre- and post-pubertal stages. Based on the developmental stages, different stages of Leydig cells can be identified as stem, progenitor, immature, adult, or aged Leydig cells [30]. In this study, we could not identify the stages of Leydig cells immunolabeled with PGP9.5. Interestingly, in the pre-pubertal stages, various Leydig cells were immunolabeled with PGP9.5, whereas PGP9.5-positive Leydig cells had a round nucleus and wide cytoplasmic area, assuming that Leydig PGP9.5-positive cells are adult Leydig cells. However, further study should be conducted to confirm the developmental stages of Leydig cells expressing PGP9.5 at different reproductive stages. The results of this study indicate that PGP9.5 is a critical molecular marker for the study of donkey Leydig cell development.

5. Conclusions

In conclusion, PGP9.5 expresses in germ cells and Leydig cells differently depending on the reproductive stage. PGP9.5 antibody can be used as a marker to identify undifferentiated, differentiating, and differentiated donkey spermatogonia. This antibody can also be used as a marker to study the development of Leydig cells in donkey testes.

Author Contributions: Conceptualization, M.Y.; methodology, M.Y., Y.C., and Y.J.; validation, Y.C., Y.J., and S.K.; formal analysis, M.Y., Y.C., Y.J., and S.K.; investigation, J.K.; writing—original draft preparation, Y.C.; writing—review and editing, H.J.; supervision, M.Y.; project administration, Y.C. and Y.J. All authors have read and agreed to the published version of the manuscript.

Funding: This research was funded by National Research Foundation of Korea, grant number MSIT No. 2019R1A2C1003701.

Conflicts of Interest: The authors declare no conflict of interest.

References

1. Costa, G.M.; Avelar, G.F.; Rezende-Neto, J.V.; Campos-Junior, P.H.; Lacerda, S.M.; Andrade, B.S.; Thomé, R.G.; Hofmann, M.C.; Franca, L.R. Spermatogonial stem cell markers and niche in equids. *PLoS ONE* **2012**, *7*, e44091. [CrossRef]
2. Jung, H.; Roser, J.F.; Yoon, M. UTF1, a putative marker for spermatogonial stem cells in stallions. *PLoS ONE* **2014**, *9*, e108825. [CrossRef]

3. Thompson, R.J.; Doran, J.F.; Jackson, P.; Dhillon, A.P.; Rode, J. PGP 9.5—A new marker for vertebrate neurons and neuroendocrine cells. *Brain Res.* **1983**, *278*, 224–228. [CrossRef]
4. Gong, B.; Cao, Z.; Zheng, P.; Vitolo, O.V.; Liu, S.; Staniszewski, A.; Moolman, D.; Zhang, H.; Shelanski, M.; Arancio, O.; et al. Ubiquitin hydrolase Uch-L1 rescues beta-amyloid-induced decreases in synaptic function and contextual memory. *Cell* **2006**, *126*, 775–788. [CrossRef]
5. Kwon, J.; Wang, Y.L.; Setsuie, R.; Sekiguchi, S.; Sakurai, M.; Sato, Y.; Lee, W.W.; Ishii, Y.; Kyuwa, S.; Noda, M.; et al. Developmental regulation of ubiquitin C-terminal hydrolase isozyme expression during spermatogenesis in mice. *Biol. Reprod.* **2004**, *71*, 515–521. [CrossRef]
6. Wrobel, K.H. Prespermatogenesis and spermatogoniogenesis in the bovine testis. *Anat. Embryol. (Berl.)* **2000**, *202*, 209–222. [CrossRef]
7. Luo, J.; Megee, S.; Rathi, R.; Dobrinski, I. Protein gene product 9.5 is a spermatogonia-specific marker in the pig testis: Application to enrichment and culture of porcine spermatogonia. *Mol. Reprod. Dev.* **2006**, *73*, 1531–1540. [CrossRef]
8. Rodriguez-Sosa, J.R.; Dobson, H.; Hahnel, A. Isolation and transplantation of spermatogonia in sheep. *Theriogenology* **2006**, *66*, 2091–2103. [CrossRef]
9. Heidari, B.; Rahmati-Ahmadabadi, M.; Akhondi, M.M.; Zarnani, A.H.; Jeddi-Tehrani, M.; Shirazi, A.; Naderi, M.M.; Behzadi, B. Isolation, identification, and culture of goat spermatogonial stem cells using c-kit and PGP9.5 markers. *J. Assist. Reprod. Genet.* **2012**, *29*, 1029–1038. [CrossRef]
10. Lee, Y.S.; Jung, H.J.; Yoon, M.J. Undifferentiated embryonic cell transcription factor 1 (UTF1) and deleted in azoospermia-like (DAZL) expression in the testes of donkeys. *Reprod. Domest. Anim.* **2017**, *52*, 264–269. [CrossRef]
11. Jung, H.; Song, H.; Yoon, M. Stage-dependent DAZL localization in stallion germ cells. *Anim. Reprod. Sci.* **2014**, *147*, 32–38. [CrossRef]
12. Lee, G.; Jung, H.; Yoon, M. The Lin28 expression in stallion testes. *PLoS ONE* **2016**, *11*, e0165011. [CrossRef]
13. Heninger, N.L.; Staub, C.; Blanchard, T.L.; Johnson, L.; Varner, D.D.; Forrest, D.W. Germ cell apoptosis in the testes of normal stallions. *Theriogenology* **2004**, *62*, 283–297. [CrossRef]
14. Neves, E.S.; Chiarini-Garcia, H.; França, L.R. Comparative testis morphometry and seminiferous epithelium cycle length in donkeys and mules. *Biol. Reprod.* **2002**, *67*, 247–255. [CrossRef]
15. Jung, H.; Yoon, M. Protein Gene Product 9.5 Expression in Stallion Testes. *J. Equine Vet. Sci.* **2016**, *45*, 1–7. [CrossRef]
16. Myka, J.L.; Lear, T.L.; Houck, M.L.; Ryder, O.A.; Bailey, E. Homologous fission event(s) implicated for chromosomal polymorphisms among five species in the genus Equus. *Cytogenet. Genome Res.* **2003**, *102*, 217–221. [CrossRef]
17. Kon, Y.; Endoh, D.; Iwanaga, T. Expression of protein gene product 9.5, a neuronal ubiquitin C-terminal hydrolase, and its developing change in Sertoli cells of mouse testis. *Mol. Reprod. Dev. Inc. Gamete Res.* **1999**, *54*, 333–341. [CrossRef]
18. Herrid, M.; Davey, R.J.; Hill, J.R. Characterization of germ cells from pre-pubertal bull calves in preparation for germ cell transplantation. *Cell Tissue Res.* **2007**, *330*, 321–329. [CrossRef]
19. Lee, W.-Y.; Lee, K.-H.; Heo, Y.-T.; Kim, N.-H.; Kim, J.-H.; Kim, J.-H.; Moon, S.-H.; Chung, H.-J.; Yoon, M.-J.; Song, H. Transcriptional coactivator undifferentiated embryonic cell transcription factor 1 expressed in spermatogonial stem cells: A putative marker of boar spermatogonia. *Anim. Reprod. Sci.* **2014**, *150*, 115–124. [CrossRef]
20. Kwon, J.; Mochida, K.; Wang, Y.-L.; Sekiguchi, S.; Sankai, T.; Aoki, S.; Ogura, A.; Yoshikawa, Y.; Wada, K. Ubiquitin C-terminal hydrolase L-1 is essential for the early apoptotic wave of germinal cells and for sperm quality control during spermatogenesis. *Biol. Reprod.* **2005**, *73*, 29–35. [CrossRef]
21. Kwon, J.; Wang, Y.-L.; Setsuie, R.; Sekiguchi, S.; Sato, Y.; Sakurai, M.; Noda, M.; Aoki, S.; Yoshikawa, Y.; Wada, K. Two closely related ubiquitin C-terminal hydrolase isozymes function as reciprocal modulators of germ cell apoptosis in cryptorchid testis. *Am. J. Pathol.* **2004**, *165*, 1367–1374. [CrossRef]
22. Kwon, J. The new function of two ubiquitin C-terminal hydrolase isozymes as reciprocal modulators of germ cell apoptosis. *Exp. Anim.* **2007**, *56*, 71–77. [CrossRef]
23. Matsui, Y. Regulation of germ cell death in mammalian gonads. *APMIS* **1998**, *106*, 142–148. [CrossRef]
24. Print, C.G.; Loveland, K.L. Germ cell suicide: New insights into apoptosis during spermatogenesis. *Bioessays* **2000**, *22*, 423–430. [CrossRef]

25. Culty, M. Gonocytes, the forgotten cells of the germ cell lineage. *Birth Defects Res.* **2009**, *87*, 1–26. [CrossRef]
26. Paniagua, R.; Nistal, M. Morphological and histometric study of human spermatogonia from birth to the onset of puberty. *J. Anat.* **1984**, *139*, 535.
27. De Kretser, D.M.; Loveland, K.L.; Meinhardt, A.; Simorangkir, D.; Wreford, N. Spermatogenesis. *Hum. Reprod.* **1998**, *13*, 1–8. [CrossRef]
28. Wilson, P.; Barber, P.; Hamid, Q.; Power, B.; Dhillon, A.; Rode, J.; Day, I.; Thompson, R.; Polak, J. The immunolocalization of protein gene product 9.5 using rabbit polyclonal and mouse monoclonal antibodies. *Br. J. Exp. Pathol.* **1988**, *69*, 91.
29. Lee, K.H.; Lee, R.; Lee, W.Y.; Kim, D.H.; Chung, H.J.; Kim, J.H.; Kim, N.H.; Choi, S.H.; Kim, J.H.; Song, H. Identification and in vitro derivation of spermatogonia in beagle testis. *PLoS ONE* **2014**, *9*, e109963. [CrossRef]
30. Chen, H.; Ge, R.-S.; Zirkin, B. Leydig cells: From stem cells to aging. *Mol. Cell. Endocrinol.* **2009**, *306*, 9–16. [CrossRef]

animals

MDPI

Case Report

Management of Thermal Injuries in Donkeys: A Case Report

Jorge Lohse [1,*], Pierpaolo Pietrantoni [2] and Christian Tummers [3]

1 Equine Sport Medicine Advisor, Equinos Chile, Chorrillos, Viña del Mar 399, Chile
2 School of Agricultural Sciences and Veterinary Medicine, Viña del Mar University, Agua Santa, Viña del Mar 7055, Chile; ppietrantoni@uvm.cl
3 Natural Resources Faculty, Department of Veterinary Medicine and Public Health, Temuco Catholic University, Manuel Montt, Temuco 056, Chile; ctuemmers@uct.cl
* Correspondence: lohse.j@gmail.com; Tel.: +56-991396797

Received: 25 September 2020; Accepted: 10 November 2020; Published: 17 November 2020

Simple Summary: Reports or descriptions of medical management of thermal injuries in donkeys is lacking. Four donkeys sustained burn injuries during the wildland–urban interface fire in Valparaiso, Chile, in 2014. The donkeys received first aid care at the scene of the fire, and then were hospitalized and treated for three months for thermal injuries, of various degrees of severity, in several body parts. The clinical findings and treatment of four of the donkeys are described in this paper. All donkeys recovered and were retired to an animal shelter.

Abstract: Burn injuries are uncommon in large animals and there are no reports of these injuries in donkeys. Burns cause local and systemic effects. Extensive thermal injuries can be challenging to manage and the extent of the burn surface affected will directly impact the severity of the illness and the prognosis. Burns are classified according to the depth of injury into four categories, from first-degree burns, and the least affect to fourth-degree burns, which are the more severely affected patients. This case report describes the medical management of four donkeys that sustained various degrees of external burn injuries during the wildland–urban interface fire in Valparaiso, Chile. The donkeys were treated topically for several weeks and closely monitor for inadequate nutritional intake. Water based topical medications are preferred in burn cases because they can be easily applied and removed without interfering with wound healing. Of note, the caloric demands of these cases can be achieved by increasing the amount of grain, adding fat (i.e., vegetable oil), and free-choice alfalfa hay. All donkeys recovered and were retired to an animal shelter.

Keywords: donkeys; thermal injurie; burns; fire

1. Introduction

Burns are damage to body tissues caused by the action of high temperature or hot solutions, lightning, friction, abrasions, or certain chemicals (alkalis, acids, salts of heavy metals) [1]. Burns are uncommon in horses and there are no reports of thermal injuries in donkeys in scientific literature. Emergency treatment guidelines are extrapolated from human and equine medical literature [2–6].

From 12 to 14 April 2014, a wildland-urban interface (WUI) fire occurred at a catastrophic scale. Over 2827 acres were burned, 11,000 people affected, and 2900 homes destroyed in Valparaiso, Chile. WUI fires occur more frequently during periods of drought and climate change. The wildfires also significantly affect land used for agriculture and wildlife habitats, injuring and killing large numbers of wildlife and farm animals. Valparaiso has a peculiar topography characterized by 45 hills, and the risk for wildfires exist in most parts of rural areas every summer. The city of Valparaiso was declared a World

Heritage Site by United Nations Educational, Scientific and Cultural Organization (UNESCO) in 2003. The estimated number of affected domestic animals associated with the WUI fire was 146 (70 horses, 33 donkeys, 29 goats, 11 pigs, 2 cows, and 1 sheep), of which many of them died. Many of those animals were examined and treated by veterinarians for burns and/or smoke inhalation. Four donkeys suffered burns, two of them severe, requiring long-term hospitalization, and costly medical treatment, therefore, the owner donated the donkeys to the Veterinary Clinic Service of Viña del Mar University.

During fires, thermal injuries are caused by direct exposition to the flames and/or inhalation of toxic gases [7]. Inhalation injury is a common sequel of closed-space fires and develops through three mechanisms: direct thermal injury, carbon monoxide poisoning, and chemical insult [4]. Horses exposed can suffer respiratory injury of varying degrees, ranging from mild irritation to severe smoke inhalation-induced airway or lung damage [8]. The extent of the burns can vary in terms of the surface area affected and the depth of tissue injury. Thermal injury can be life-threatening due to subsequent denaturation of cellular metabolic processes and apoptosis, leading to necrosis [9].

Classification of the burn injuries is considered one of the more important determinants of outcome. First-degree burns affect only superficial layers of the epidermis, causing erythema, edema, and desquamation, and they heal without complication. Second-degree burns involve the superficial and/or deep portion of the epidermis. Superficial second-degree burns affect the stratum corneum, stratum granulosum, and a few cells of the basal layer; however, deeper burns involve the epidermis, including the basal layers. Such burns are characterized by erythema and edema at the epidermal-dermal junction, necrosis of the epidermis, accumulation of leukocytes at the basal layer of the burn, eschar formation, and minimal pain [10]. These may heal within 3 to 4 weeks if further dermal ischemia is prevented. Third-degree burns affect all layers of the epidermis and dermis, including the adnexa. These burns heal by contraction and epithelialization from the wound margins and secondary infections are common complications. Fourth-degree burns affect the skin and underlying muscle, bone, ligaments, fat, and fascia [10–12].

In addition to the depth of the burn, the extent or total body surface area (TBSA) affected by thermal injuries also needs to be assess. The "Wallace Rule of Nines" system used in human medicine estimates the prognosis according to the extent of the burn. This system divides the body surface area into regions that represent multiples of nine and allocates body regions as follows: each limb, head, neck, thorax, and abdomen represent 9% [13]. Specific guidelines of the TBSA do not exist in large animals [10–12]. Other evaluation methods are described [12].

Donkeys are particularly susceptible to hyperlipemia in stressful circumstances. It can progress rapidly and is often life-threatening. Prompt diagnosis and treatment is required to improve the outcome. If managed correctly, with an appropriate diet and regular (routine) preventive healthcare, this risk can be minimized [14].

This report describes the medical management of veterinary care of thermal injuries in four donkeys.

2. Materials and Methods

History and Case Presentation

Case 1 (D1): a 7-year-old male donkey, weighing 150 kg was examined and treated within 2 h after being burned. The donkey was kept in a small pen (10 × 10 m) with three other donkeys. Upon physical examination, the body condition score (BCS) was 3/5, tachycardic 64 beats/min (reference range 36–52), tachypneic 30 breath/min (reference range 12–28), furthermore, the donkey had congested mucous membranes. There were burn marks over the lips, face, ears, dorsum, and both front limbs. All serous exudate was observed over the coronary bands in front limbs, but had a normal gait. Immediate treatment included flunixin meglumine 1.1 mg/kg intravenously (IV) (83 mg/mL, Febrectal, Drag Pharma Laboratory Chile Invetec S.A., Santiago, Chile); hosing with tap water (13 o 15°C) all over the body for 30 min to protect the tissue, and bandage with sterilized gauze and liquid paraffin gauze dressing over the neck and back prior to transportation to the clinic 2 h later (Figure 1A).

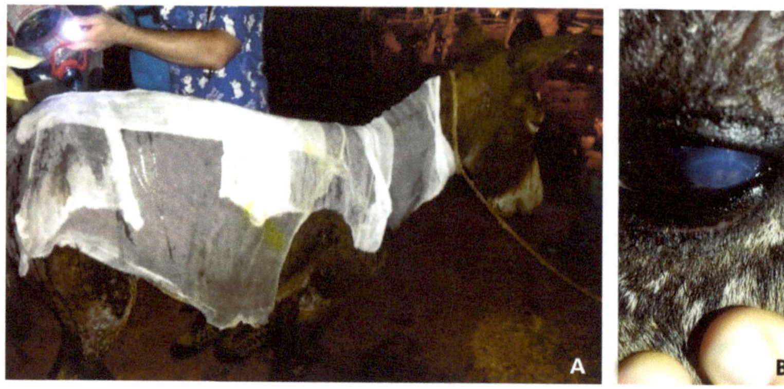

Figure 1. Donkey 1 received emergency medical attention for thermal injuries. (**A**) A large gauze dressing was applied over his body and soaked with tap water. (**B**): corneal ulcer.

Approximately 12 h after the incident, the estimated TBSA was approximately 40–50%. Most of the affected areas were classified as deep second-degree burns with some small areas of third-degree burns on the distal area of the front limbs and face. There was also facial, abdominal, and limb edema. The donkey was eating and drinking with difficulty (dysphagia—possibly caused by edema and pain). He was placed on IV fluids with Hartmann's solution 10 mL/kg bwt/h (Lactated Ringer's solution, Baxter, Toongabbie, New South Wales, Australia) for 3 h until he was stable and started to eat and drink. Both eyes had bilateral corneal ulcers and uveitis (Figure 1B), with signs, such as epiphora, blepharospasm, corneal edema, and episcleral congestion. Treatment with ophthalmic antibiotic ointment containing zinc bacitracin 400 iu/g, neomycin sulfate 3.5 mg/g, and polymyxin B sulfate 6000 IU/g q 12 h (Oftabiotico, Saval S.A. Laboratory, Santiago, Chile), ophthalmic atropine sulfate drops q 12 h × 2 days (10 mg /mL, Atropina 1%, Saval S.A. Laboratory, Santiago, Chile), and dextran-70 lubricant drops 1 mg with hidroxipropilmetilcelulosa 3 mg (Tears Naturale II, Alcon Laboratory Chile Ltd.a., Santiago, Chile) was initiated several times a day. The eyes were treated with triple antibiotic ointment for six weeks, and natural tear drops for four weeks.

A topical cream of sulfanilamide with zinc oxide was applied to the coronary bands (sulfanilamide 5 g, zinc oxide 5 g, vitamin A 250,000 IU, vitamin D2 90,000 IU/100 g, Pomada Sulfavitaminada, Drag Pharma Laboratory Chile Invetec S.A., Santiago, Chile).

Respiratory distress was observed, characterized by tachypnea, nostril flaring, and cough—a suspected result from smoke inhalation; therefore, treatment with corticosteroid, dexamethasone sodium phosphate 0.2 mg/kg IV (5,26 mg/mL, Hasyun, Drag Pharma Laboratory Chile Invetec S.A., Santiago, Chile) was administered. A parasympatholytic drug and beta-2 agonist inhalation therapy by aerochamber (Aerofacidose Aerocamara A.E., Chile Laboratory S.A. Teva, Santiago, Chile), using ipratropium bromide 0.8 μg/kg q 6 h (250 μg/mL, Atrovent, Boehringer Ingelheim International GmbH, Rhein, Germany) and albuterol 2 μg/kg q 4 h (100 μg/doses, Sinasmal SF Salbutamol, Chile Laboratory S.A. Teva, Santiago, Chile) were also administered. The aerochamber was placed in one nostril and the opposite side was occluded (Figure 2). Airway endoscopy was performed the second day (1800PVS9150, PortaScope, FL, USA), and mild to moderate irritation with hyperemia of the mucous membrane from a common nasal meatus, and the first part of a ventral nasal meatus of both nostrils, as well as mild hyperemia in the pharynx and trachea, were observed. The cough resolved within 48 h of treatment; therefore, the bronchodilators were discontinued after three days.

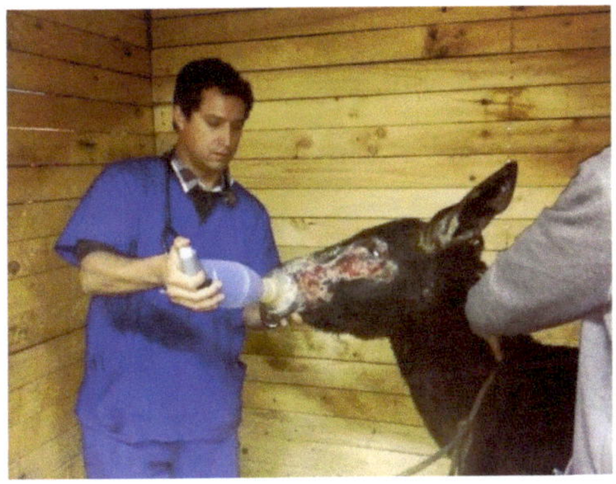

Figure 2. Use of the aerochamber for inhalator therapy in donkey 1, placed in the left nostril with the opposite side occluded.

Complete blood cell count and serum biochemistry profile were performed on the second and eight days post rescue (Biosystems BA400, Biosystems S.A., Barcelona, Spain). The first analysis, done at day two, revealed: neutrophilia, increased alkaline phosphatase, muscle enzymes creatine kinase, aspartate aminotransferase and triglycerides, decreased total protein and albumin, increased glucose, and decreased total calcium and chloride concentration. Slightly cloudy serum was observed in the clotted blood sample (Table 1).

Table 1. Serial complete blood count and serum biochemistry profile of two burned donkeys (D1–D2).

Parameters	D1	D2	D1	D2	D1	D2	RI LRL-URL
	Day 2		Day 8		Day 15		
RBC (×106/μL)	4.6	5.5	6.6	6.6	6.8	7.0	4.4–7.1
Hemoglobin (g/dL)	11.1	9.3	11.4	11.4	12.2	13.9	8.9–14.7
Hematocrit (%)	30.0	29.0	32.0	36.0	32.0	38.0	27–42
WBC (×109/L)	15.1	15.8	14.8	14.2	11.1	13.4	6.2–15.0
S. neutrophils (×109/L)	8.4	7.1	6.2	6.0	6.0	5.5	2.4–6.3
Lymphocytes (×109/L)	4.1	1.5	1.9	3.0	2.6	2.7	2.9–9.6
Eosinophils (×109/L)	0.3	0.1	0	0,7	0.1	0.1	0.1–0.9
Monocytes (×109/L)	0.2	0	0.1	0	0	0	0–0.75
Thrombocyte (×109/L)	210	146	167	117	133	158	95–384
Urea (mmol/L)	4.0	3.0	4.2	3.6	1.9	2.5	1.5–5.2
Creat (μmol/L)	55	72	81	87	70	82	53–118
ALP (U/L)	328	293	291	270	123	201	98–252
AST (U/L)	829	712	621	589	399	512	238–536
GGT (U/L)	26	20	36	41	29	36	14–69
CK (U/L)	1024	619	319	375	326	208	128–525
Trig (mmol/L)	4.2	3.2	2.5	2.1	2.2	1.8	0.6–2.8
Total Proteins (g/L)	55	48	58	62	71	58	58–76
Albumin (g/L)	21	19	23	22	26	28	22–32
Globulin (g/L)	33	46	39	41	35	47	32–48
Tbil (μmol/L)	0.9	0.3	0.9	1.0	2.1	2.2	0.1–3.7
Chol (mmol/L)	2.9	3.0	2.5	2.8	1.9	2.3	1.4–2.9
Glucose (mmol/L)	5.1	4.0	4.5	4.6	4.1	4.5	3.9–4.7

Table 1. *Cont.*

Parameters	D1	D2	D1	D2	D1	D2	RI LRL-URL
	Day 2		Day 8		Day 15		
Calcium (mmol/L)	2.1	1.9	2.5	2.9	3.0	3.2	2.2–3.4
Na (mmol/L)	130	131	129	135	129	133	128–138
K (mmol/L)	3.5	5.0	3.9	4.1	4.9	3.9	3.2–5.1
Cl (mmol/L)	95	100	98	104	99	104	96–106

D1: case donkey 1; D2: case donkey 2; RBC: Red Blood Cells; ALP: Alkaline phosphatase; AST: Aspartate aminotransferase; GGT: Gamma-glutamyl transpeptidase; Creat: Creatinine, BUN: blood urea nitrogen, CK: Creatine phosphokinase, Tri: Triglycerides, Tbil: Total bilirubin, Cho: Cholesterol, Na: Sodium, K: Potassium, Cl: Chloride. RI = Reference interval. LRL = Low RI. URL = Upper RL [14].

The second analysis at day 8 showed increased alkaline phosphatase, aspartate aminotransferase. Blood analysis was repeated on day 15 of the hospitalization and all of the parameters were within normal limits (Table 1). Antibiotics therapy was initiated with procaine g penicillin 20,000 IU/kg + dihydrostreptomycin 15 mg/kg q 24 h intramuscular (IM)/7 days (200,000 IU procaine G penicillin + 250 mg dihydrostreptomycin sulfate/mL, (Pentril, Drag Pharma Laboratory Chile Invetec S.A., Santiago, Chile).

During the hospitalization time period (90 days), the burn surfaces, appetite, and pain level of the donkey were monitored. The donkey appeared more comfortable, with good appetite, and showed interest in socializing with other donkeys and his environment; these responses became evident since day 7 in the clinic. The edema (described) was observed since the first day of the rescue and decreased after 6 days, the skin became dry and sensitive to touch, and he experienced hair loss on his legs, neck, and back.

Flunixin was given daily for pain control for 10 days since the rescue (3 days, 1.1 mg/kg q12 h, then 7 days, 1.1 mg/kg q 24 h IV), and continued with a natural product (devil's claw 726 mg + *Yucca schidigera* extract 954 mg + vitamin B12 10 mcg/ 10 mL, Bute-Less Paste, Absorbine, Massachusetts, MA, USA), 5 mL oraly q 24 h/15 days, supplement to discomfort and pain relief. Omeprazole 2 mg/kg per os q 24 h/4 weeks (20 mg tablets, Omeprazole, Chile Laboratory S.A. Teva, Santiago, Chile) was given for treatment and/or prevention of gastric ulceration.

During the following 2 weeks, the eyelids, ears, and back became dry, hairless, and painful, and there was a loss of elasticity (Figure 3). Topical treatment with aloe vera gel was initiated twice daily to exposed areas before washing and debridement. The pinna of both ears developed ulcers and contracture of the cartilage.

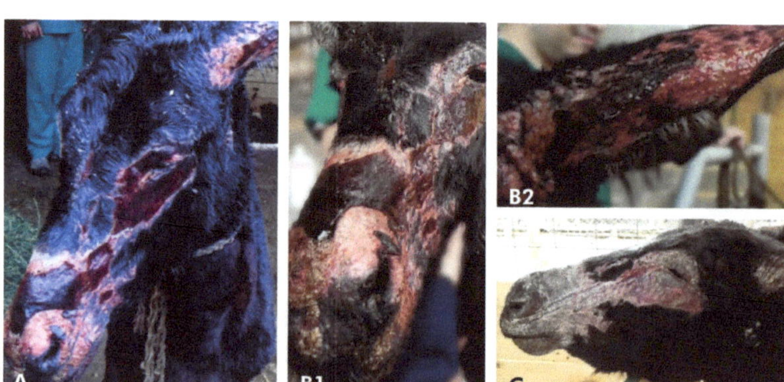

Figure 3. Donkey case 1, showing changes of the burned skin overtime. Appearance of the face after (**A**) after 5 days; (**B1**) 12 days; (**B2**) 12 days, external pinna area of the left ear; and (**C**) 20 days, respectively.

Daily skin care initially required sedation xylazine 0.5 mg/kg IV, with butorphanol tartrate 0.01 mg/kg IV (10 mg/ mL, Torbugesic, Zoetis, NJ, USA) as potent analgesic for the first six days in the clinic. On day seven, sedation and potent analgesic for treatment was no longer required. Local anesthetic instillation with lidocaine hydrochloride was used in some affected areas on the back and face (20 mg/ 1 mL, Lidocalm, Drag Pharma Laboratory Chile Invetec S.A., Santiago, Chile).

Wounds were lavaged every 24–36 h, for debridement of devitalized tissue, and dead skin was removed with sterile scissors or a hypodermic needle. The blisters were left intact, however, gauze dressing with impregnated paraffin (Jelonet) was used in some areas where blisters opened. Local antibacterial drugs were used every 24 h. The wounds were lavaged with chlorhexidine gluconate solution (2 g/100 mL, Inveclor, Drag Pharma Laboratory Chile Invetec S.A., Santiago, Chile) and diluted in water with sterilized gauze using gloves and triclosan (0.3%/100 g, Sanigermin, Sanitas S.A. Laboratory, Santiago, Chile). A variety of topical products were used in conjunction with the dressings for the following weeks: cream of *Buddleja globosa* (Matihorse, Drag Pharma Laboratory Chile Invetec S.A., Santiago, Chile), raw honey, aloe vera, and nitrofurazone cream (0.2 g/100 g, Furasep Drag Pharma Laboratory Chile Invetec S.A., Santiago, Chile). Nitrofurazone was used in areas with exudation and erythema and, locally, raw honey on areas with less exudation and erythema. Dry skin areas were treated with Jelonet, and then it was changed to *Buddleja globosa* cream. Fly repellent was used around wounds (dichlorvos 1250 mg + triclosan, 500 mg/ 100 g, Moskation larvicida spray, Drag Pharma Laboratory Chile Invetec S.A., Santiago, Chile).

Three weeks later, the donkey developed pruritus, which was treated with chlorpheniramine maleate 0.5 mg/kg IV (50 mg/mL, Histamil, Troy Laboratories Pty. Limited, Glendenning, Australia) as needed.

After 2 months, a thin layer of skin started to regrow. Exuberant granulation tissue on some areas was periodically de-bulked. After 3 months, all treatments were discontinued, and most of the burned areas had healed, but the area of the face and front limbs were without hair growth. At this time, the only product used on his legs and body was aloe vera gel. His coronary bands from the front limbs did not separate significantly, but a horizontal line of damage was evident around the hoof wall (1 cm below the coronary band), walk, and trot sound.

Case 2 (D2): a 15-year-old female donkey, 120 kg, was examined with D1. The jenny appeared anxious and painful. Upon physical examination, her BCS was 3/5, she was tachycardic (72 beats/min), tachypneic (30 breath/min), and had congested mucous membranes and muscle tremors. Burns over her face, neck, back, perineal area, vulva, and coronary band swelling of the left hind limb were observed. She was transported to the clinic with the other affected animals and was treated, as per case 1.

Her burns were classified as second-degree, with some areas with third-degree (perineal area); estimated TBSA approximately 40–50%. She had bilateral corneal ulcers and uveitis, and her eyes were treated with the same antibiotic ointments and regime as case 1. The left coronary band was inflamed and a serious exudate was present; therefore, topic cream with sulfanilamide + zinc oxide was applied. She was five-months in gestation, which was closely monitored at the clinic. Intravenous fluids with Hartmann's solution were administered at the same rate as described above, as well as flunixin and omeprazole. Fluids were discontinued on the second day at the clinic, when the laboratory report indicated adequate hydration. Moreover, she was eating hay with a good appetite at this time.

Complete blood cell count and serum biochemistry profile were performed in the clinic similar to D1. The first blood analysis, taken at day two, revealed: neutrophilia, increased alkaline phosphatase, muscle enzymes creatine kinase, aspartate aminotransferase, triglycerides and cholesterol, decreased total protein and albumin, increased glucose, and decreased mineral calcium and chloride concentration. Slightly cloudy serum was observed in the clotted blood sample (Table 1).

Antibiotic therapy started with procaine penicillin + dihydrostreptomycin IM, and for the skin, aloe vera cream, at the same doses rates described in D1. The wounds from the skin were periodically lavaged and cleaned using disinfectant, and a similar topic cream, as described in D1, removing the devitalized tissue when necessary (sedation and potent analgesic, described in D1, were previously used). The pinna of the left ear also developed ulcers and permanent cartilage contracture, as D1. A second blood analysis performed eight days post rescue showed increases in alkaline phosphatase and aspartate aminotransferase. Concerning the third blood analysis performed on the 15th day of hospitalization, all parameters were within normal limits (Table 1). Pruritus was evident on day 10; therefore, the antihistamine product used was the same as D1. Over the next 2 weeks, her wounds continued to improve and the burns re-epithelialized rapidly (Figure 4).

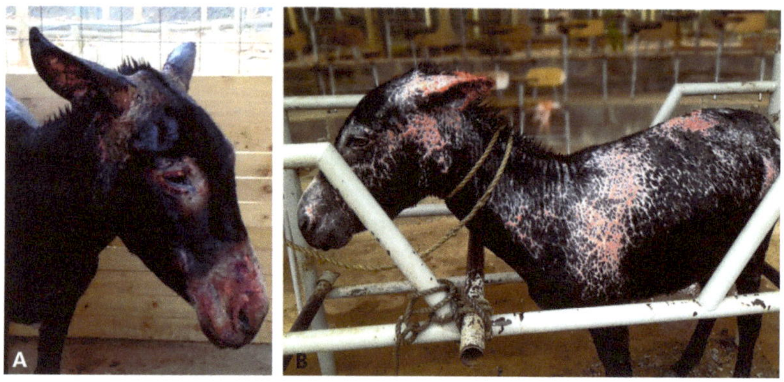

Figure 4. Donkey 2, showing changes of the burned skin over time. Appearance of her face and body at (**A**) 10 days; and (**B**) 20 days, respectively.

The left coronary band lesion, increasing in size and lameness, was graded 2/5. Radiographs were taken, which revealed no sign of distal phalangeal rotation. Flunixin meglumine (1.1 mg/kg) for pain was continued, daily, for 12 days (3 days 1.1 mg/kg q12 h, then 9 days 1.1 mg/kg q 24 h IV), and the supplement for pain relief was the same dose rates as D1. After six weeks, she could canter when turned out into a paddock in the evenings, or early mornings, when the outdoor temperatures were cooler. His left coronary band from the hind limb did not separate, but a horizontal line of damage was evident around the hoof wall (1 cm below the coronary band).The jenny was discharged after 3 months of hospitalization. She was doing well on her 3- and 8-month pregnancy checks. She foaled a healthy donkey foal after 13 months of gestation (range of 11–14.5 months) [11].

Cases 3 and 4 (D3–D4): the next two cases were a 12-year-old female donkey, 160 kg, and a 5-year-old male, 110 kg. Both were examined at the site of the fire, as described for both previous cases (D1 and D2). They both had a BCS of 2/5, were tachycardic, and small burn areas with hair loss on their heads (noses and cheeks), as well as necks, were observed. Their TBSAs were approximately 5% and the burn areas classified as first degree. These two donkeys were in the same paddock with the others; however, they were less affected. Cases D1–D2 were tied up to a fence, close to the burning vegetation until the rope was burned by fire, while cases 3 and 4 were free. This most likely explains the severity of the injuries in the first two cases.

D3 and D4 were also hosed with tap water all over their bodies, and moved to the same clinic. They were also treated with flunixin meglumine for 3 days and the burn areas were treated with aloe vera as the previous two cases. Superficial burns healed without complication. The areas affected became dry between 5 and 8 days. The wounds from the skin were periodically lavaged and cleaned using disinfectant, and similar topic cream, as described in D1, was used. It was necessary to use xylazine and butorphanol tartrate for wound management for 2 days after they arrived at the clinic. After 3 weeks, most of the burned areas healed (Figure 5).

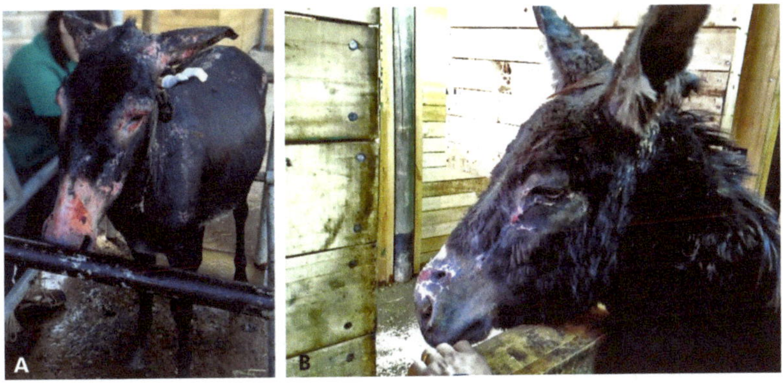

Figure 5. Appearance of burned skin in donkeys 3 (**A**) and 4 (**B**), 5 days after the fire.

The hematologic and biochemical analyses of D3–D4 performed two days after the recue showed no abnormalities (Table 2). They showed no signs of discomfort and had good appetite. They were treated for 5 days and then discharged from the clinic to a free ranged pasture.

Table 2. Complete blood count and serum biochemistry profile of two burned donkeys (D3–D4).

Parameters	D1	D2	RI LRL-URL
RBC ($\times10^6/\mu$L)	4.9	6.3	4.4–7.1
Hemoglobin (g/dL)	8.7	9.8	8.9–14.7
Hematocrit (%)	31	38	27–42
WBC ($\times10^9$/L)	7.4	14.2	6.2–15.0
S. neutrophils ($\times10^9$/L)	3.1	5.8	2.4–6.3
Lymphocytes ($\times10^9$/L)	3.5	6.9	2.9–9.6
Eosinophils (×109/L)	0.1	0.3	0.1–0.9
Monocytes (×109/L)	0.2	0	0–0.75
Thrombocyte (×109/L)	128	305	95–384
Urea (mmol/L)	4.0	3.6	1.5–5.2
Creat (μmol/L)	53	92	53–118
ALP (U/L)	104	211	98–252
AST (U/L)	257	322	238–536
GGT (U/L)	26	49	14–69
CK (U/L)	130	513	128–525
Tri (mmol/L)	0.7	2.1	0.6–2.8
Total Proteins (g/L)	63	70	58–76
Albumin (g/L)	26	23	22–32
Globulin (g/L)	33	38	32–48
Cho (mmol/L)	2.2	2.7	1.4–2.9
Tbil (μmol/L)	0.4	1.5	0.1–3.7
Glucose (mmol/L)	3.9	4.1	3.9–4.7
Calcium (mmol/L)	2.3	2.6	2.2–3.4
Na (mmol/L)	130	134	128–138
K (mmol/L)	3.5	4.7	3.2–5.1
Cl (mmol/L)	98	104	96–106

D1: case donkey 1; D2: case donkey 2; RBC: Red Blood Cells; ALP: Alkaline phosphatase; AST: Aspartate aminotransferase; GGT: Gamma-glutamyl transpeptidase; Creat: Creatinine, BUN: blood urea nitrogen, CK: Creatine phosphokinase, Tri: triglycerides, Tbil: Total bilirubin, Cho: cholesterol, Na: Sodium, K: Potassium, Cl: Chloride. RI = Reference interval. LRL = Low RI. URL = Upper RL [14].

All of the donkeys were fed alfalfa hay and grass (*Medicago sativa* and *Lolium perenne*), 1.7% daily of their bwt in dry matter (DM) intake, which is equivalent to 2.6 kg D1, 2.1 kg D2, 2.7 kg D3, 1.9 kg D4 per day (Figure 6). What was also given daily included: vegetable oil, 30 mL, and 20 mL of multivitamin and mineral oral supplements for horses containing: copper 150 mg, cobalt 6 mg, potassium 330 mg, magnesium 70 mg, manganese 130 mg, zinc 370 mg, iron 1000 mg, calcium pantothenate 170 mg, folic acid 35 mg, biotin 0.08 mg, vitamin: A 80,000 IU, B1 200 mg, B2 100 mg, B6 35 mg, B12 400 µg, D2 12,000 IU, E 150 IU, sucrose 50 G, choline chloride 675 mg, and selenium 3.1 mg/ 100 mL (Equifort, Drag Pharma Laboratory Chile Invetec S.A., Santiago, Chile). All were orally de-wormed with fenbendazole 7.5 mg/ mL (10 g/100 mL, Lombrimic, Drag Pharma Laboratory Chile Invetec S.A., Santiago, Chile).

Figure 6. Donkeys D1–4 fully recovered, eating alfalfa hay and grass (*Medicago sativa* and *Lolium perenne*) 90 days after the fire. Left to right is D4, D2, D3, and D1.

3. Results and Discussion

To the best of our knowledge, this is the first case report on the medical management of severely burned donkeys. There is extensive literature regarding thermal injuries in humans and horses, and the lack of reports of thermal injuries in donkeys may be due to the costs associated with treatments, which are usually prolonged and expensive. The market value of donkeys is usually lower than horses, and owners cannot afford to cover the medical treatment. Therefore, euthanasia may be their only option. In these cases, the owner donated the donkeys to the attending veterinary clinic for treatments, and then, rehomed them to a shelter.

After first aid assessment and treatment at the site of the fire, the donkeys were transferred and fully evaluated at the clinic. The day after the fire, D1–D2 injuries were worse than initially assessed, and the extent of burns was greater than the predicted survival rate reported in the literature [15].

Furthermore, estimates of the extent and depth of the burns made prior to admission to a burn center, as well as experts in this field, have consistently shown to be inaccurate, despite standardization attempts and availability of tools, such as the Rule of Nines [9,16]. Given the unreliability of burn size and depth assessment by clinicians who are not burn experts, developing a scale for burn equid patients, we believe, is essential. A new approach from human medicine includes the use of computer-assisted programs to improve burn size estimation [17,18]. The treatment of donkeys with extensive burns was a major challenge. After the initial fluid therapy, however, the most severely affected donkeys (D1 and D2) improved and started to eat and drink. At this stage, however, prognosis for survival remained guarded.

Earlier intervention of burn patients is associated with favorable outcomes, lower complication, and mortality rates. The immediate emergency recommendation for severe burns is the application of copious amounts of cold water for a minimum of 20 min, within the first three hours of injury [19,20]. Water with temperature between 2 °C and 15 °C is preferred [20]. All donkeys described in this report, were managed using the considerations described (2 h after bring burned). Overcooling, resulting in hypothermia, was reported to be associated with adverse outcomes, including clotting disorders and increased mortality [19,20]. Animal studies showed that the immediate application of cold water was associated with faster re-epithelialization and reduced scarring [21].

D1 showed respiratory signs from smoke inhalation, which resolved in 3 days with the established treatment. The area where the donkeys were rescued from the fire that day was windy, which helped blow away smoke and ash into the atmosphere—this could be why D2 to D4 did not present problems in their respiratory systems. A case report of horses injured in open range fires described one horse with problems from smoke inhalation, which resolved in 24 h [5]. Barn fires are unfortunately too common and, each year, hundreds of horses die or are severely injured in these incidents. Gimenez et al. describe a review of strategies to prevent and respond to barn fires in the horse industry [22].

All of the parameters of complete blood count and serum biochemistry profile were normal, since the second analysis.

In some areas of deep second-degree and third-degree burns, skin grafting may be considered [4]; however, in these donkeys this intervention was unnecessary, since the wounds healed as per the reported time, according to the burn depth classification [4].

There are several methods to treat burn wounds in a horse, and the choice depends on the extent and location of the injury. Full-thickness burns can be managed by occlusive dressings (closed technique), continuous wet dressings (semi-open technique), eschar formation (exposed technique), or excision and grafting [4]. In this case report, bandaging the burned limbs was not performed as mentioned in burned horses [5], and was managed as open wounds with no complications. The use of the compression garments or compression suits post-burn injury is described as an important component of a human patient's rehabilitation program, and reported for horses [5,23,24].

Cleaning with mild soap and water was used to facilitate softening and removal of dead skin. Aloe vera appeared to be helpful for managing dry skin, which appeared to prevent the skin from sloughing, and allegedly has anti-inflammatory effects and promotes wound healing [25,26]. This product had been reported to cause pruritus in some horses [5], but such adverse side effects were not observed in D1 and D2. The precise mechanism of post-burn pruritus has not been elucidated in humans or animals, but it appears to have pruritogenic and neuropathic aspects [27]. The occasional pruritus presented in D1 and D2 were managed by the use of chlorphenamine. Many burned equine patients are pruritic, and measures must be taken to prevent self-mutilation of the wound [4]. Reserpine had been used in burned horses for this complication, decreasing the urge to scratch by successfully breaking the itch–scratch cycle [4,5]. One author mentioned that the use of air conditioning fans directed to the stall may help [5]. The same author mentioned that topical corticosteroid creams were ineffective at controlling the itching in burned horses [5].

Honey has been used in burn patients for many years. The antibacterial activity of honey, its low pH, high viscosity, hygroscopic effect, and hydrogen peroxide content, may play a combined role in its effectiveness for the treatment of burns, and may also provide a moist environment for optimum healing conditions [28,29]. There are several types of honey—raw honey was used in these donkeys, and appeared to be useful in superficial and moderate burns, as reported in horses [5].

The donkeys with corneal ulcers (D1 and D2) responded well to treatment and healed with no complications. Human studies report 7.7%–18% of the incidence of chemical and thermal injuries to the eye of all ocular trauma, with corneal ulcers and blepharospasm [30]. Burned horse may have blepharospasm, epiphora, or both, which signify corneal damage [4]. Ocular emergency treatment by thermal injuries in humans is described [30]. Case reports in horses injured in open range fires that reported corneal ulcers were successfully treated with twice-daily treatments of topical antibiotics (zinc bacitracin, neomycin sulfate and polymyxin B sulfate), as well as daily cloxacillin benzathine eye ointment [5].

Long-term use of nonsteroidal anti-inflammatory drugs (NSAIDs) did not result in adverse effects during hospitalization, but other NSAIDs could have been used described for donkeys [14,31]. Gastrointestinal motility and appetite were normal in the first two days for donkeys 1 and 2. The prophylactic use of omeprazole may have been beneficial in preventing gastric ulceration with long-term use of NSAIDs. Omeprazole intended for use in humans was prescribed to these cases by the economic possibility of that moment, but there is no evidence about their efficacy in preventing ulcers in these donkeys. The authors recommend the use of an approved omeprazole product manufactured for horses, using the standard equine dosage appropriate for donkeys [14].

The application of sulfanilamide with zinc oxide around the coronary band was an effective topical treatment in these cases. Sulfanilamide acted with a differential bactericidal activity, and mild astringent, and decongestive effect, in addition to the absorbent and insulating properties conferred by zinc oxide [32,33]. Coronary band effusion was noted on the first day on D1 and D2, but separation was also noted. Abnormal hoof growth was observed—moved down over time until the old hoof was replaced—as was noted 8 months after the fire. The thickened hoof wall distal to the coronary band protected the majority of the laminae from thermal injury [5].

Preventive antibiotic therapy was used in D1 and D2, but the literature mentioned that the role of prophylactic antibiotics for severe burns is controversial, both in humans and horses [34]. Therefore, their use has not been advocated in recent guidelines or recommendations owing to a lack of evidence for efficacy and induction of antibiotic resistance. Systemic antibiotics do not favorably influence wound healing, fever, or mortality, and can encourage the emergence of resistant microorganisms. Additionally, circulation to the burned areas is often compromised, making it highly unlikely that parenteral administration of antibiotics can achieve therapeutic levels to the wound [4]. However, topical antimicrobials have been the mainstay of nonsurgical burn treatment [16]. In human reports, the use of prophylactic antibiotics may result in improved 28-day in-hospital mortality in mechanically ventilated patients with severe burns, but not in those who do not receive mechanical ventilation [35]. The use of antibiotics had been recommended in equine burn cases because of puncture wounds that occurred while attempting to escape the fire [5]. We do not find evidence that systemic therapy was useful, and we will not recommend it in future cases.

A hallmark of burn injury is a hypermetabolic response that results in significant pathological alterations in a number of tissues [36]. Studies in large animal models of burn hypermetabolism showed that 25% TBSA burns can generate a hypermetabolic response greater than smaller animals and closer to that seen in human patients [37,38]. It is recommended to gradually increase the grain, adding fat in the form of vegetable oil, and offering free-choice alfalfa hay increase caloric intake as nutritional needs for burned horses [4]. An anabolic steroid may be used to help restore a positive nitrogen balance [4]. Donkeys are highly efficient at digesting poor nutritional quality fiber and have lower energy requirements than horses and ponies of similar size (it is even lower in sick donkeys) [14,39,40]. Equine feedstuff based upon cereals or containing high levels of molasses should be avoided. It has been shown that they are risk factors for the development of gastric ulcers, laminitis, hyperlipidemia, and fatty liver disease in donkeys [14,40]. Early intervention to restore a positive energy balance, even before triglyceride values, are known to greatly increase the chance of survival. The restoration of a positive energy balance will stimulate endogenous insulin secretion and switch off lipolysis. Soya bean meal or alfalfa are excellent sources of digestible protein for convalescent donkeys [14,40]. The primary goal of this response is to provide adequate energy levels to maintain organ function and whole-body homeostasis. In addition to the food offered, multivitamin and mineral supplements, and vegetable oil to increase the energy intake, were added.

In veterinary emergency medicine and critical care, teamwork to provide adequate patient care is essential. The entire team consisted of two veterinarians, a technician, and eight graduate students to provide continues care. The care and devotion provided to each animal by the veterinarian staff highlights the concept of the animal-human bond and the dedication veterinarians provide to their patients.

4. Conclusions

Domestic and wild animals can suffer major thermal injuries during wildfires, and reports of these injuries in donkeys is lacking. This manuscript describes the clinical findings, medical treatments, and evolution of four donkeys that suffered thermal burns. All of the donkeys recovered and were sent to an animal shelter after their recovery.

Author Contributions: Preparation of this submission J.L., P.P., C.T., J.L and P.P. treated the donkeys, C.T. gave advice during the treatment process. J.L. wrote the manuscript; it was reviewed and edited by J.L., P.P., and C.T. All authors have read and agreed to the published version of the manuscript.

Funding: This research received no external funding.

Acknowledgments: We are grateful to the people who helped with the pharmaceutical donations for the donkeys at Drag Pharma Laboratory; Tia Clafira Shelter and Raul Cabello for the information concerning the number of animals affected in the fire; Jorge Oyarce for helping us convert the photograph formats; and Luis Arroyo for reviewing the paper.

Conflicts of Interest: The authors declare no conflict of interest.

References

1. Efimova, O.I.; Dimitriev, A.I.; Nesterova, O.P.; Aldyakov, A.V.; Obukhova, A.V.; Ivanova, T.N. Methods for the effective treatment of animal burns. In *IOP Conference Series: Earth and Environmental Science, Proceedings of the 5th International Conference on Agricultural and Biological Sciences (ABS), Macau, China, 21–24 July 2019*; IOP Publishing: Bristol, UK, 2019; Volume 346, p. 012057. [CrossRef]
2. Litt, J.S. Evaluation and Management of the Burn Patient: A Case Study and Review. *Mo. Med.* **2018**, *115*, 443–446. [PubMed]
3. Geiser, D.R.; Walker, R.D. Management of thermal injuries in large animals. *Vet. Clin. N. Am. Large. Anim. Prac.* **1984**, *6*, 91–105. [CrossRef]
4. Hanson, R.R. Management of burn injuries in the horse. *Vet. Clin. N. Am. Equine Pract.* **2005**, *21*, 105–123. [CrossRef] [PubMed]
5. Herbert, E.W. Findings and strategies for treating horses injured in open range fires. *Equine. Vet. Educ.* **2018**, *30*, 177–186. [CrossRef]
6. Silva, S.; Ríspoli, V.; Graner, C.; Sá, L.R.; Belli, C.; De Zoppa, A.L. Using tilapia skin (*Oreochromis niloticus*) as an occlusive biological curative in equine wounds: Short communication. *Braz. J. Vet. Res. Anim. Sci.* **2019**, *56*, e154079. [CrossRef]
7. Wohlsein, P.; Peters, M.; Schulze, C.; Baumgärtner, W. Thermal Injuries in Veterinary Forensic Pathology. *Vet. Pathol.* **2016**, *53*, 1001–1017. [CrossRef]
8. Magdesian, G.; Watson, J. What Horse Owners Can Do To Monitor Horses Evacuated from Fire Areas. Guidelines UC Davis VMTH Large Animal Clinic. 2020. Available online: https://www.vetmed.ucdavis.edu/sites/g/files/dgvnsk491/files/inline-files/smoke-inhalation-quick-reference-guide.pdf (accessed on 21 October 2020).
9. Church, D.; Elsayed, S.; Reid, O.; Winston, B.; Lindsay, R. Burn Wound Infections. *Clin. Microbiol. Rev.* **2006**, *19*, 403–434. [CrossRef]
10. Toussaint, J.; Singer, A.J. The evaluation and management of thermal injuries: 2014 update. *Clin. Exp. Emerg. Med.* **2014**, *1*, 8–18. [CrossRef]
11. Hanson, R.R.; Munsterman, A.S. Treatment of Burn Injuries, Gunshot Wounds, and Dog-Bite Wound. In *Equine Wound Management*, 3rd ed.; Theoret, C., Schumacher, J., Eds.; John Wiley & Sons: Ames, IA, USA, 2017; Volume 21, pp. 476–490.
12. Ministerio de Salud Gobierno de Chile. *Gran Quemado. Guías Clínicas AUGE*; Santiago, Chile, 2016; pp. 1–109. Available online: http://www.bibliotecaminsal.cl/wp/wp-content/uploads/2016/04/GPC-GRAN-QUEMADO-FINAL-18-MARZO-2016_DIAGRAMADA.pdf (accessed on 18 May 2020).
13. Warden, G.D. Outpatient care of thermal injuries. *Surg. Clin. N. Am.* **1987**, *67*, 147–157. [CrossRef]
14. The Donkey Sactuary. *The Clinical Companion of the Donkey*, 1st ed.; Matador, Troubador Ltd.: Leicestershire, UK, 2018; p. 267.
15. Pavoni, P.; Gianesello, L.; Paparella, L.; Buoninsegni, L.T.; Barboni, E. Outcome predictors and quality of life of severe burn patients admitted to intensive care unit. *Scand. J. Trauma Resusc. Emerg. Med.* **2010**, *18*, 24. [CrossRef]

16. Jeschke, M.G.; van Baar, M.E.; Choudhry, M.A.; Chung, K.K.; Gibran, N.S.; Logsetty, S. Burn injury. *Nat. Rev. Dis. Primers.* **2020**, *6*, 11. [CrossRef] [PubMed]
17. Benjamin, N.C.; Lee, J.O.; Norbury, W.B.; Branski, L.K.; Wurzer, P.; Jimenez, C.J.; Benjamin, D.A.; Herndon, D.N. Accuracy of Currently Used Paper Burn Diagram vs. a Three-Dimensional Computerized Model. *J. Burn Care Res.* **2017**, *38*, e254–e260. [CrossRef] [PubMed]
18. Williams, J.F.; King, B.T.; Aden, J.K.; Serio-Melvin, M.; Chung, K.K.; Fenrich, C.A.; Salinas, J.; Renz, E.M.; Wolf, S.E.; Blackbourne, L.H.; et al. Comparison of traditional burn wound mapping with a computerized program. *J. Burn Care Res.* **2013**, *34*, e29–e35. [CrossRef] [PubMed]
19. Wood, F.M.; Phillips, M.; Jovic, T.; Cassidy, J.T.; Cameron, P.; Edgar, D.W. Steering Committee of the Burn Registry of Australia and New Zealand (BRANZ) Water First Aid Is Beneficial In Humans Post-Burn: Evidence from a Bi-National Cohort Study. *PLoS ONE* **2016**, *11*, e0147259. [CrossRef] [PubMed]
20. Hammett, E. The latest advice on burns. *BDJ Team* **2019**, *6*, 16–18. [CrossRef]
21. Cuttle, L.; Pearn, J.; McMillan, J.R.; Kimble, R.M. A review of first aid treatments for burn injuries. *Burns* **2009**, *35*, 768–775. [CrossRef]
22. Gimenez, R.M.; Woods, J.A.; Dwyer, R.M.; Gimenez, T. A Review of Strategies to Prevent and Respond to Barn Fires Affecting the Horse Industry. *AAEP Proc.* **2008**, *54*, 160–179.
23. Kim, J.Y.; Willard, J.J.; Supp, D.M.; Roy, S.; Gordillo, G.M.; Sen, C.K.; Powell, H.M. Burn Scar Biomechanics after Pressure Garment Therapy. *Plast. Reconstr. Surg.* **2015**, *136*, 572–581. [CrossRef]
24. Coghlan, N.; Copley, J.; Aplin, T.; Strong, J. The experience of wearing compression garments after burn injury: "On the inside it is still me". *Burns* **2019**, *45*, 1438–1446. [CrossRef]
25. Somboonwong, J.; Thanamittramanee, S.; Jariyapongskul, A.; Patumraj, S. Therapeutic Effects of Aloe Vera on Cutaneous Microcirculation and Wound Healing in Second Degree Burn Model in Rats. *J. Med. Assoc. Thai.* **2000**, *83*, 417–425.
26. Maenthaisong, R.; Chaiyakunapruk, N.; Niruntraporn, S.; Kongkaew, C. The Efficacy of Aloe Vera Used for Burn Wound Healing: A Systematic Review. *Burns* **2007**, *33*, 713–718. [CrossRef] [PubMed]
27. Chung, B.Y.; Kim, H.B.; Jung, M.J.; Kang, S.Y.; Kwak, I.S.; Park, C.W.; Kim, H.O. Post-Burn Pruritus. *Int. J. Mol. Sci.* **2020**, *21*, 3880. [CrossRef] [PubMed]
28. Subrahmanyam, M. Topical Application of Honey for Burn Wound Treatment-an Overview. *Ann. Burns Fire Disasters* **2007**, *20*, 137–139. [PubMed]
29. Zbuchea, A. Up-to-date use of honey for burns treatment. *Ann. Burns Fire Disasters* **2014**, *27*, 22–30.
30. Kuckelkorn, R.; Schrage, N.; Keller, G.; Redbrake, C. Emergency treatment of chemical and thermal eye burns. *Acta Ophthalmol. Scand.* **2002**, *80*, 4–10. [CrossRef]
31. The Brook. The Working Equid Veterinary Manual. Whittet Books. 2013, p. 528. Available online: https://www.thebrooke.org/our-work/working-equid-veterinary-manual (accessed on 16 October 2020).
32. Dai, T.; Huang, Y.Y.; Sharma, S.K.; Hashmi, J.T.; Kurup, D.B.; Hamblin, M.R. Topical Antimicrobials for Burn Wound Infections. *Recent. Pat. Antiinfect. Drug Discov.* **2010**, *5*, 124–151. [CrossRef]
33. Heyneman, A.; Hoeksema, H.; Vandekerckhove, D.; Pirayesh, A.; Monstrey, S. The Role of Silver Sulphadiazine in the Conservative Treatment of Partial Thickness Burn Wounds: A Systematic Review. *Burns* **2016**, *42*, 1377–1386. [CrossRef]
34. Putra, O.N.; Saputro, I.D.; Nurrahman, N.D.; Herawati, E.D.; Dewi, L.K. Effects of empirical antibiotic administration on the level of C-Reactive protein and inflammatory markers in severe burn patients. *Ann. Burns Fire Disasters* **2020**, *33*, 20–26.
35. Takashi, T.; Hiroki, M.; Kiyohide, F.; Hideo, Y. Prophylactic Antibiotics May Improve Outcome in Patients with Severe Burns Requiring Mechanical Ventilation: Propensity Score Analysis of a Japanese Nationwide Database. *Clin. Infect. Dis.* **2016**, *62*, 60–66. [CrossRef]
36. Abdullahi, A.; Amini-Nik, S.; Jeschke, M.G. Animal Models in Burn Research. *Cell. Mol. Life Sci.* **2014**, *71*, 3241–3255. [CrossRef]
37. Hu, R.H.; Yu, Y.M.; Costa, D.; Young, V.R.; Ryan, C.M.; Burke, J.F.; Tompkins, R.G. A rabbit model for metabolic studies after burn injury. *J. Surg. Res.* **1998**, *75*, 153–160. [CrossRef] [PubMed]
38. Fernandez, M.L.; Volek, J.S. Guinea pigs: A suitable animal model to study lipoprotein metabolism, atherosclerosis and inflammation. *Nutr. Metab.* **2006**, *3*, 17. [CrossRef] [PubMed]

39. Mendoza, F.J.; Toribio, R.E.; Perez-Ecijaa, A. Donkey Internal Medicine—Part I: Metabolic, Endocrine, and Alimentary Tract Disturbances. *J. Equine Vet. Sci.* **2018**, *65*, 66–74. [CrossRef]
40. Burden, F.; Thiemann, A. Donkeys Are Different. *J. Equine Vet. Sci.* **2015**, *35*, 376–382. [CrossRef]

Publisher's Note: MDPI stays neutral with regard to jurisdictional claims in published maps and institutional affiliations.

animals

MDPI

Article

Single Layer Centrifugation Improves the Quality of Fresh Donkey Semen and Modifies the Sperm Ability to Interact with Polymorphonuclear Neutrophils

Marion Papas [1], Jaime Catalán [1], Sandra Recuero [2,3], Jane M. Morrell [4], Marc Yeste [2,3,*,†] and Jordi Miró [1,*,†]

1 Equine Reproduction Service, Department of Animal Medicine and Surgery, Faculty of Veterinary Sciences, Autonomous University of Barcelona, E-08193 Bellaterra (Cerdanyola del Vallès), Spain; papas.marion@gmail.com (M.P.); dr.jcatalan@gmail.com (J.C.)

2 Biotechnology of Animal and Human Reproduction (TechnoSperm), Institute of Food and Agricultural Technology, University of Girona, E-17003 Girona, Spain; sandra.recuero@udg.edu

3 Unit of Cell Biology, Department of Biology, Faculty of Sciences, University of Girona, E-17003 Girona, Spain

4 Clinical Sciences, Swedish University of Agricultural Sciences, Box 7054, SE-75007 Uppsala, Sweden; jane.morrell@slu.se

* Correspondence: marc.yeste@udg.edu (M.Y.); jordi.miro@uab.cat (J.M.); Tel.: +34-972-419514 (M.Y.); +34-93-5814293 (J.M.)

† These authors contribute equally.

Received: 26 September 2020; Accepted: 13 November 2020; Published: 16 November 2020

Simple Summary: Donkey Artificial Insemination (AI) with frozen/thawed semen results in poor fertility outcomes. Jennies show a significant post-AI endometrial reaction, with a large amount of defense cells—polymorphonuclear neutrophils (PMN)—migrating to the uterine lumen. Seminal plasma (SP) has a detrimental effect on sperm conservation and its removal is a necessary step in the semen freezing protocol. However, several SP proteins seem to control sperm-PMN binding. Single layer centrifugation (SLC) with colloids, which has been used to select spermatozoa and improve reproductive performance in different species, is known to remove SP proteins attached to the sperm membrane. In this study, two experiments were performed. The first one compared the quality of SLC-selected and non-selected fresh donkey spermatozoa. In the second experiment, PMN obtained from the peripheral blood were co-incubated with selected and unselected spermatozoa, and the interaction between PMN and spermatozoa was analyzed. In conclusion, SLC of fresh donkey semen increases the proportion of functionally intact spermatozoa and appears to remove the SP proteins that inhibit sperm-PMN binding, thus increasing sperm phagocytosis by PMN.

Abstract: This study sought to determine whether single layer centrifugation (SLC) of fresh donkey semen with Equicoll has any impact on sperm quality parameters and on the modulation of endometrial reaction following semen deposition using an in vitro model. Seventeen ejaculates from five jackasses were obtained using an artificial vagina and diluted in a skim-milk extender. Samples were either selected through SLC (Equicoll) or non-treated (control). Two experiments were performed. The first one consisted of incubating selected or non-selected spermatozoa at 38 °C for 180 min. Integrity and lipid disorder of sperm plasma membrane, mitochondrial membrane potential, and intracellular levels of calcium and reactive oxygen species were evaluated at 0, 60, 120, and 180 min. In the second experiment, polymorphonuclear neutrophils (PMN) isolated from jennies blood were mixed with selected and unselected spermatozoa. Interaction between spermatozoa and PMN was evaluated after 0, 60, 120, and 180 min of co-incubation at 38 °C. SLC-selection increased the proportions of spermatozoa with an intact plasma membrane and low lipid disorder, of spermatozoa with high mitochondrial membrane potential and with high calcium levels, and of progressively motile spermatozoa. In addition, selection through SLC augmented the proportion of phagocytosed spermatozoa, which supported the modulating role of seminal plasma proteins on sperm-PMN interaction. In conclusion, SLC of fresh donkey semen increases the proportions of functionally intact and motile spermatozoa, and appears to remove the seminal plasma proteins that inhibit sperm-PMN binding.

Keywords: sperm; fresh semen; donkey; single layer centrifugation; sperm–PMN binding

1. Introduction

Proper management of breeding and genetic diversity in equine species requires the use of assisted reproduction technologies (ART), which includes artificial insemination (AI), embryo transfer (ET), intracytoplasmic sperm injection (ICSI), and cryopreservation [1,2]. It has been reported that when sperm are selected, fertility outcomes following ART increase [3,4]. This selection is usually based on sperm motility, morphology, and the integrities of plasma membrane, acrosome, and DNA. In addition, sperm selection allows for the removal of seminal plasma, non-viable sperm, pathogens, and debris particles [5]. While seminal plasma has been described to have a detrimental effect on sperm motility and viability during storage, the presence of non-viable, morphologically abnormal spermatozoa may be a source of reactive oxygen species (ROS), which can also be detrimental to sperm survival during storage [6].

Several sperm preparation techniques have been designed to separate male gametes from seminal plasma and to select spermatozoa with better quality. Because simple sperm washing does not remove all seminal plasma components [7], the other two alternatives, i.e., sperm migration and colloid centrifugation, are the most used in equine practice [8]. Sperm migration through swim-up is based on sperm motility and yields low recovery rates [7]. In contrast, centrifugation with colloids uses silane-coated silica particles in a species-specific formulation that allows the separation of heterogeneous sperm into sub-populations according to their density. Not only does this technique select sperm on the basis of their motility, but also on their normal morphology, and plasma membrane and chromatin integrities [8].

Single layer centrifugation (SLC) is a simplification of the density gradient centrifugation as only one layer of colloid is employed [9]. Previous studies have demonstrated that, following SLC, spermatozoa show higher motility, less DNA fragmentation, and higher pregnancy rates in subfertile stallions [10,11]. With respect to jackasses, SLC using Equicoll improves sperm quality parameters, such as motility, viability, and morphology, after 24 h of cooled storage [12], and after thawing of cryopreserved doses [13]. To the best of our knowledge, however, no study with fresh donkey semen has been performed nor has any evaluated the effects of SLC with Equicoll on mitochondrial activity, and intracellular levels of ROS and calcium. Other studies in rams [14], pigs [15], bulls [16], and brown bears [17,18] have demonstrated that centrifugation with SLC increases the proportions of spermatozoa with an intact plasma membrane and acrosome, high mitochondrial activity, an intact DNA, and progressive motility, and has also been found to increase sperm cryotolerance when performed before cryopreservation [19,20].

While elimination of seminal plasma is beneficial for sperm cryopreservation [21] and improves the stability of plasma membrane in fresh and cooled-stored stallion spermatozoa [22], one should bear in mind that several seminal plasma factors are involved in the regulation of the inflammatory response within the female reproductive tract. In addition, studies conducted in other species, such as the pig, showed that SLC removes seminal plasma components, including porcine spermadhesins PSP-I and PSP-II, and some cholesterol molecules from sperm plasma membrane [23]. Remarkably, the presence of seminal plasma has been shown to reduce chemotaxis, sperm-neutrophil binding, phagocytosis, and the formation of DNA-based neutrophil extracellular traps (NETs) [24]. In the donkey, in vitro models suggest that seminal plasma can suppress sperm-PMN attachment [25,26] and that, when added to frozen/thawed sperm, downregulates the expression of *COX2* in endometrial cells, which leads PMN chemotaxis to reduce [27].

Against this background, we hypothesized that SLC of jackass semen with Equicoll could increase the proportions of donkey spermatozoa exhibiting high motility, an intact and functional plasma membrane and high mitochondrial activity, and modify sperm-PMN binding. The effects of SLC on the proportions of spermatozoa with high levels of intracellular calcium and reactive oxygen species were also investigated. In addition, we tested the effects of SLC with Equicoll on donkey sperm not only after centrifugation, but also after 60, 120, and 180 min of incubation at 38 °C. The rationale of testing SLC-effects over this incubation period is related to the time that sperm are deemed to reside within the uterus (3 h).

2. Materials and Methods

2.1. Animals

Semen samples (N = 17) were collected from five Catalonian jackasses, whose age ranged from 3 to 11 years, that were in good health and shown to be fertile. Jackasses underwent a regular semen collection, every other day during the week, and were housed in individual paddocks at the Equine Reproduction Service, Autonomous University of Barcelona (Bellaterra, Cerdanyola del Vallès, Spain), which is an EU-approved equine semen collection center (ES09RS01E).

Polymorphonuclear neutrophils were isolated from the peripheral blood of three Catalonian jennies as described previously [26]. These animals, which were also in good health and of proven fertility, were housed in a big paddock at the Equine Reproduction Service, Autonomous University of Barcelona.

Animals were handled following the European Union Directive 2010/63/EU for animal experiments and the Animal Welfare Law from the Regional Government of Catalonia (Spain). In addition, this study was approved by the Ethics Committee, Autonomous University of Barcelona (Code: CEEAH 1424).

2.2. Experimental Design

This study consisted of two separate experiments. The goal of the first experiment was to determine the effects of SLC with Equicoll on sperm function parameters, such as motility, plasma membrane integrity, membrane lipid disorder, intracellular calcium levels, intracellular ROS levels, and mitochondrial activity. The effects of SLC-selection on donkey sperm subpopulations were also considered. Although previous works evaluated the effects of SLC with Equicoll on donkey semen, they used cooled-stored and cryopreserved sperm [12,13]; in contrast, the current study used fresh semen. Ejaculates (N = 10) were collected through a Hannover artificial vagina (Minitüb GmbH, Tiefenbach, Germany) with an in-line nylon mesh filter to remove gel and debris. Immediately after collection, gel-free semen was extended at a ratio of five volumes of skim milk extender (4.9% glucose, 2.4% skim milk, 100 mL double distilled water), previously warmed at 38 °C, and one volume of semen. After assessing sperm concentration, viability, morphology, and motility, each ejaculate was divided into two fractions. The first one was utilized as a control and was directly adjusted to a final concentration of 25×10^6 sperm/mL. The second fraction was centrifuged through a single layer of a silane-coated silica-based colloid formulation (Equicoll, SLU, Sweden; formerly known as Androcoll-E) according to the protocol for small centrifuge tubes described by Morrell et al. [28]. Briefly, 1.5 mL of extended semen containing up to 100×10^6 sperm/mL was pipetted on top of 4 mL Equicoll in a 15-mL tube. After centrifugation at 300× *g* and room temperature for 20 min, the supernatant and most of the colloid was discarded and the sperm pellet was transferred to a clean centrifuge tube containing 0.5 mL of the same skim milk extender. Sperm concentration was evaluated and subsequently adjusted to a final concentration of 25×10^6 sperm/mL. Aliquots were incubated at 38 °C under aerobic conditions for 3 h. Sperm motility and other functional parameters evaluated with flow cytometry were determined immediately after SLC-selection (0 min), and after 60 min, 120 min, and 180 min of incubation at 38 °C.

The second experiment aimed to determine the effects of SLC with Equicoll on the interaction of donkey sperm with polymorphonuclear neutrophils (PMN). With this purpose, each ejaculate (N = 7) was split into two fractions. One fraction was used as a control and its sperm concentration was adjusted to 500 × 10^6 sperm/mL. The other fraction was centrifuged through a single layer of a silane-coated silica-based colloid formulation (Equicoll) as described for Experiment 1. After centrifugation, sperm concentration was evaluated and subsequently adjusted to a final concentration of 500 × 10^6 sperm/mL. Treatments consisted of in vitro co-incubation of sperm cells with PMN-rich samples (1:1; v:v) in a water bath at 38 °C for 3 h, as previously described by Miró et al. [26]. Briefly, peripheral blood neutrophils were isolated from healthy jennies. Following incubation in a water bath at 38 °C for 30 min, heparinized blood was subjected to centrifugation at 400× *g* and 4 °C for 5 min prior to removing plasma. The buffy coat was mixed with an isotonic saline solution (PBS) and again centrifuged at 400× *g* and 4 °C for 5 min. The PMN-rich buffy coat was collected and re-suspended in PBS to a final concentration of 100 × 10^6 PMN/mL. All samples (selected and non-selected) were examined for sperm phagocytosis immediately after SLC-selection (0 min), and after 60 min, 120 min, and 180 min of incubation at 38 °C.

2.3. Flow Cytometry

Sperm quality parameters were determined using a Cell Laboratory Quanta SC™ flow cytometer (Beckman Coulter; Fullerton, CA, USA). The following sperm parameters were evaluated: plasma membrane integrity (viability), membrane lipid disorder, acrosome integrity, mitochondrial activity (mitochondrial membrane potential), and intracellular levels of calcium (Ca^{2+}), superoxides ($O_2^{-\bullet}$) and peroxides (H_2O_2). Prior to any staining, each sample was diluted with HEPES buffered saline solution (10 mmol/L HEPES (4-(2-hydroxyethyl)-1-piperazineethanesulfonic acid); 150 mmol/L NaCl, 10% BSA; pH = 7.4) to a final concentration of 1 × 10^6 spermatozoa/mL. Samples were subsequently incubated with the corresponding fluorochromes under dark conditions. After fluorochrome staining, an argon ion laser was used to excite samples at 488 nm (22 mW). Spermatozoa were selected on the basis of electronic volume (EV) and side scatter (SS) and the sheath flow rate was set at 4.17 µL/min. Based on EV/SS dot-plots, sperm cells were selected and the other events, including subcellular debris (particle diameter <7 µm) and cell aggregates (particle diameter >12 µm), were gated out. The flow cytometer was calibrated every day using 10-µm flow-check fluorospheres (Beckman Coulter), positioning the bead size (10 µm) on the EV channel (EV = 200).

Up to 10,000 events were analyzed per replicate and a total of three independent replicates per sample were evaluated. FL1 detector (BP, band pass: 525 nm; DRLP, dichroic long pass: 550 nm) was used for monitoring the green fluorescence emitted by SYBR14, YO-PRO-1 (1-(4-[3-methyl-2,3-dihydro-(benzo-1,3-oxazole)-2-methylidene]-quinolinium)-3trimethylammonium propane diodide), $JC1_{mon}$ (5,5′,6,6′-tetrachloro-1,1′3,3′tetraethyl-benzimidazolylcarbocyanine iodide monomers), H_2DFCDA (2′,7′-dichlorodihydrofluorescein diacetate), and Fluo3-AM (acetoxymethyl, ester form) fluorochromes. FL2 detector (BP: 575 nm; DRLP: 600 nm) was used for monitoring the orange fluorescence emitted by $JC1_{agg}$. FL3 detector (LP, long pass: 670 nm) was used for monitoring the red fluorescence emitted by HE (hydroethidine), PI (propidium iodide), M540 (merocyanine 540), and Rhod5.

Post-acquisition analyses were conducted using the Cell Lab Quanta® SC MPL Analysis Software (version 1.0; Beckman Coulter; Fullerton, CA, USA). Data were corrected by subtracting the percentage of debris particles found in SYBR14/PI staining from the percentages of particles appearing at the lower left quadrant in the other tests [29]. The proportions of the other sperm populations were normalized.

2.3.1. Evaluation of Sperm Membrane Integrity (SYBR14/PI)

Sperm viability was assessed using a combination of two fluorochromes (SYBR14 and propidium iodide, PI) available from a commercial kit (Live/Dead Sperm Viability kit, Molecular Probes, ThermoFisher Scientific; Waltham, MA, USA). After adjusting cell concentration to 1×10^6 sperm/mL, spermatozoa were stained with SYBR14 (final concentration: 100 nM) at 38 °C for 10 min. Following this, PI (final concentration: 12 μM) was added and samples were incubated at 38 °C for 5 min. The following three categories of spermatozoa were identified: (i) spermatozoa with an intact plasma membrane, stained in green (viable spermatozoa; $SYBR14^+/PI^-$); (ii) spermatozoa with an altered plasma membrane, stained in red (non-viable spermatozoa; $SYBR14^-/PI^+$); and (iii) spermatozoa with an altered plasma membrane, stained in both red and green (non-viable spermatozoa; $SYBR14^+/PI^+$). Non-DNA containing particles (debris particles; $SYBR14^-/PI^-$), which appeared at the lower left quadrant, were subtracted from the total number of events and, as aforementioned, were used to correct the other tests. The protocol included compensation of FL1-spill over into the FL3-channel (2.45%).

2.3.2. Evaluation of Sperm Membrane Lipid Disorder (M540/YO-PRO-1)

Lipid disorder of plasma membrane was evaluated following the protocol set by Rathi et al. [30] with minor modifications [31], which is based on M540 and YO-PRO-1 fluorochromes (Molecular Probes, ThermoFisher Scientific). Briefly, samples were incubated with M540 (final concentration: 2.6 μM) and YO-PRO-1 (final concentration: 25 nM) at 38 °C for 10 min. A total of four sperm populations was identified in flow cytometry dot-plots: (1) viable spermatozoa with low membrane lipid disorder ($M540^-/YO\text{-}PRO\text{-}1^-$); (2) viable spermatozoa with high membrane lipid disorder ($M540^+/YO\text{-}PRO\text{-}1^-$); (3) non-viable spermatozoa with low membrane lipid disorder ($M540^-/YO\text{-}PRO\text{-}1^+$); and (4) non-viable spermatozoa with high membrane lipid disorder ($M540^+/YO\text{-}PRO\text{-}1^+$). Correction was performed following the procedure described previously and data were not compensated.

2.3.3. Evaluation of Mitochondrial Membrane Potential (JC1)

Assessment of mitochondrial membrane potential (MMP) was conducted using JC1-staining (Molecular Probes, ThermoFisher Scientific), following the protocol set by Ortega-Ferrusola et al. [32]. With this purpose, samples were incubated with JC1 (final concentration: 0.3 μM) at 38 °C for 30 min. When MMP is low, JC1 remains as monomers ($JC1_{mon}$); when MMP is high, JC1 forms aggregates ($JC1_{agg}$). Three different sperm populations were identified: (1) spermatozoa with green-stained mitochondria, which had low MMP ($JC1_{mon}$); (2) spermatozoa with orange-stained mitochondria; and (3) spermatozoa with green- and orange-stained mitochondria. Populations 2 and 3 were considered as having high MMP ($JC1_{agg}$). The protocol included compensation of FL1-spill over into the FL3-channel (68.50%).

2.3.4. Evaluation of Intracellular Calcium Levels (Fluo3/PI and Rhod5/YO-PRO-1)

Two different co-staining protocols were used to determine intracellular calcium levels. Fluo3-AM is able to penetrate cell membranes and displays better affinity for the calcium present in the sperm mid-piece [33]. This test, which is combined with PI, was performed as described by Harrison et al. [34] and Kadirvel et al. [35]. Briefly, spermatozoa were incubated with Fluo3-AM (final concentration: 1 μM) and PI (final concentration: 12 μM) at 38 °C for 10 min. Four different sperm populations were identified: (1) viable spermatozoa with low levels of intracellular calcium ($Fluo3^-/PI^-$); (2) viable spermatozoa with high levels of intracellular calcium ($Fluo3^+/PI^-$); (3) non-viable spermatozoa with low levels of intracellular calcium ($Fluo3^-/PI^+$); and (4) non-viable spermatozoa with high levels of intracellular calcium ($Fluo3^+/PI^+$). The protocol included compensation of FL1-spill over into the FL3-channel (28.72%), and of FL3-spill over into the FL1-channel (2.45%).

Rhod5 has more affinity for the calcium residing in the sperm head than for that found in the mid-piece. This test was performed following the protocol set by Yeste et al. [33]. In brief, spermatozoa were incubated with Rhod-5N (final concentration: 5 μM) and YO-PRO-1 (final concentration: 25 nM) at 38 °C for 10 min. Four sperm populations were distinguished in flow cytometry dot-plots: (1) viable spermatozoa with low levels of intracellular calcium ($Rhod5^-/YO\text{-}PRO\text{-}1^-$); (2) viable spermatozoa with high levels of intracellular calcium ($Rhod5^+/YO\text{-}PRO\text{-}1^-$); (3) non-viable spermatozoa with low levels of intracellular calcium ($Rhod5^-/YO\text{-}PRO\text{-}1^+$); and (4) non-viable spermatozoa with high levels of intracellular calcium ($Rhod5^+/YO\text{-}PRO\text{-}1^+$). The protocol included compensation of FL3-spill over into the FL1-channel (3.16%).

2.3.5. Evaluation of Intracellular Reactive Oxygen Species (ROS) Levels: H_2O_2 and $O_2^{-\bullet}$ (H_2DCFDA/PI and HE/YO-PRO-1)

The evaluation of intracellular levels of hydrogen peroxides (H_2O_2) was performed following the protocol set by Morrell et al. [36]. In brief, sperm samples were stained with H_2DCFDA (final concentration: 140 μM) and PI (final concentration: 12 μM) and incubated at 25 °C for 30 min. H_2DCFDA consists of a non-fluorescent probe that is able to penetrate plasma membranes. Upon intracellular oxidation, H_2DCFDA is de-esterified and converted into highly green fluorescent 2′,7′-dichlorofluorescein (DCF^+), which is detected through FL1. Flow cytometry dot-plots depicted four sperm populations: (1) viable spermatozoa with high levels of intracellular peroxides (DCF^+/PI^-); (2) viable spermatozoa with low levels of intracellular peroxides (DCF^-/PI^-); (3) non-viable spermatozoa with high levels of intracellular peroxides (DCF^+/PI^+); and (4) non-viable spermatozoa with low levels of intracellular peroxides (DCF^-/PI^+). The protocol included compensation of FL1-spill over into the FL3-channel (2.45%).

The determination of intracellular levels of superoxide ($O_2^{-\bullet}$) radicals was performed through co-staining with hydroethidine (HE) and YO-PRO-1, following the protocol set by Guthrie and Welch [37]. Briefly, sperm samples were mixed with HE (final concentration: 4 μM) and YO-PRO-1 (final concentration: 25 nM) and incubated at 25 °C for 30 min. In the presence of $O_2^{-\bullet}$, HE, which is known to pass across plasma membranes, is oxidized to ethidium (E^+; detected by FL3). A total of four sperm populations were identified: (1) viable spermatozoa with high levels of intracellular superoxides ($E^+/YO\text{-}PRO\text{-}1^-$); (2) viable spermatozoa with low levels of intracellular superoxides ($E^-/YO\text{-}PRO\text{-}1^-$); (3) non-viable spermatozoa with high levels of intracellular superoxides ($E^+/YO\text{-}PRO\text{-}1^+$); and (4) non-viable spermatozoa with low levels of intracellular superoxides ($E^-/YO\text{-}PRO\text{-}1^+$). The protocol included compensation of FL1-spill over into the FL3-channel (5.06%).

2.4. Evaluation of Sperm Motility

Sperm motility and kinetic parameters were assessed using a computer-assisted sperm motility analysis (CASA) system (ISAS v. 1.0; Proiser S.L., Valencia, Spain) and an Olympus BX41 microscope (Olympus 20x 0.30 PLAN objective; Olympus Europe, Hamburg, Germany). With this purpose, 5 μL of each sample was placed into a Neubauer Chamber and then observed using a phase contrast microscope with a pre-warmed stage (38 °C). For each sample, three fields per drop and at least 1000 spermatozoa were assessed. The following variables were analyzed: percentages of total motile spermatozoa (TMOT), percentages of progressively motile spermatozoa (PMOT), straight line velocity (VSL), curvilinear velocity (VCL), average path velocity (VAP), linearity index (LIN), straightness index (STR), oscillation index (WOB), amplitude of lateral head displacement (ALH), and beat cross frequency (BCF) [38]. Cut-off values for a given sperm cell to be considered as motile or progressively motile were VAP ≥ 10 μm/s and STR ≥ 75%, respectively.

2.5. Evaluation of Sperm–PMN Binding

Smears were prepared on microscope slides for each treatment and stained with a modified Wright's dye (Diff Quick®, Quimica Clinica Aplicada, Amposta, Spain). Phagocytosis was determined at 1000× magnification (Olympus Europe, Hamburg, Germany) under immersion oil, and was expressed as the percentage of PMN that ingested at least one spermatozoon [26]. The percentages of spermatozoa in contact with PMN were also counted. Two hundred PMN per slide were counted, and three technical replicates were made.

2.6. Statistical Analysis

Data obtained from the analysis of all sperm parameters were first tested for normality (Shapiro–Wilk test) and homogeneity of variances (Levene test). Following this, all variables except those related to the effects of SLC-selection upon sperm motile populations (i.e., % $SYBR14^+/PI^-$ spermatozoa, % $M540^-$/YO-PRO-1^- spermatozoa, % spermatozoa with high MMP, $JC1_{agg}/JC1_{mon}$ ratios, % $Fluo3^+$/viable spermatozoa, % $Rhod5^+$/viable spermatozoa, % DCF^+/viable spermatozoa, % E^+/viable spermatozoa, % TMOT spermatozoa, % PMOT spermatozoa, VCL, VSL, VAP, LIN, STR, WOB, ALH, and BCF) were evaluated through a linear mixed model (intra-subjects factor: time of incubation; inter-subjects factor: control vs. SLC-Equicoll), followed by Sidak post-hoc test.

In order to classify spermatozoa into separate motile subpopulations, the procedure described by Luna et al. [39] was conducted with minor modifications; this procedure combines Principal Component (PCA) and cluster analyses. First, a PCA with the individual kinematic parameters (VSL, VCL, VAP, LIN, STR, WOB, ALH, and BCF) obtained for each spermatozoon (CASA assessment) was run; data were sorted into PCA components and the matrix was rotated using the Varimax approach and Kaiser standardization. Thereafter, cluster analyses were conducted through a two-way procedure based on Schwarz's Bayesian Criterion and log-likelihood distance using the individual regression scores of each PCA component. Four separate motile sperm subpopulations were identified, and proportions of spermatozoa belonging to each subpopulation (SP1, SP2, SP3, or SP4) were subsequently calculated. The effects of SLC-selection upon sperm motile populations were evaluated through a Scheirer–Ray–Hare ranked ANOVA followed by Mann–Whitney test for pair-wise comparisons. Proportions of SP1, SP2, SP3, and SP4 spermatozoa were considered as dependent variables.

Results are shown as mean ± standard error of the mean (SEM); the minimal level of significance was set at $p \leq 0.05$ in all cases.

3. Results

3.1. Integrity and Lipid Disorder of Sperm Plasma Membrane

Figure 1a shows (mean ± SEM) the proportions of viable spermatozoa (i.e., spermatozoa with an intact plasma membrane; $SYBR14^+/PI^-$) in the control and samples selected through SLC with Equicoll. Incubation led to a significant ($p < 0.05$) decrease in the proportions of viable spermatozoa in both the control and SLC-selected samples. However, proportions of viable spermatozoa were significantly ($p < 0.05$) higher in samples selected through SLC than in their control counterparts, immediately after selection (0 min) and after 60 min and 120 min of incubation at 38 °C. At 180 min of incubation, no significant differences between treatments were observed.

Proportions of viable spermatozoa with low membrane lipid disorder ($M540^-$/YO-PRO-1^-; mean ± SEM) also decreased ($p < 0.05$) over incubation time (Figure 1b). Again, these percentages were significantly ($p < 0.05$) higher in SLC-selected samples than in the control immediately after centrifugation (0 min) and after 60 min of incubation at 38 °C. No significant differences between the control and SLC-selected samples were observed after 120 min and 180 min of incubation.

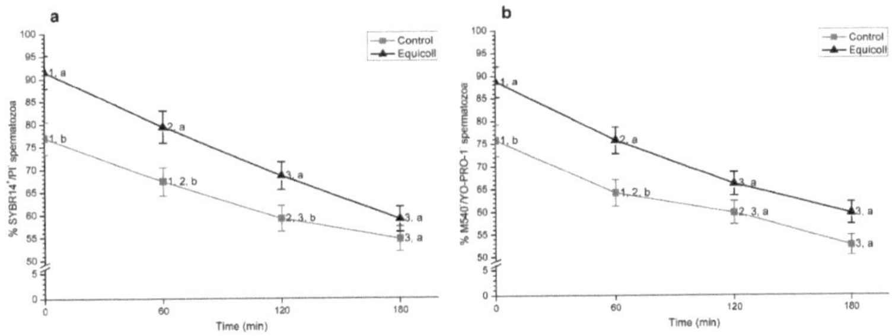

Figure 1. Percentages of (**a**) membrane-intact spermatozoa (SYBR14$^+$/PI$^-$) and (**b**) viable spermatozoa with low membrane lipid disorder (M540$^-$/YO-PRO-1$^-$) in control (non-selected samples) and SLC-selected samples (Equicoll). Different superscript letters (a, b) indicate significant differences ($p \leq 0.05$) between treatments within a given time point, and different superscript numbers (1–3) mean significant differences between time points within the control or samples selected with Equicoll. Data are shown as mean ± SEM for 10 separate experiments. SLC: single layer centrifugation.

3.2. Mitochondrial Membrane Potential

Incubation at 38 °C led to a significant ($p < 0.05$) decrease in the proportions of spermatozoa with high MMP (Figure 2a; mean ± SEM). Upon selection (0 min) and after 60 min of incubation at 38 °C, proportions of spermatozoa with high MMP were significantly ($p < 0.05$) higher in samples selected through SLC than in their control counterparts. No significant differences between the control and SLC-selected samples were observed after 120 min and 180 min of incubation at 38 °C. In addition, $JC1_{agg}/JC1_{mon}$ ratios in the sperm population with high MMP were significantly ($p < 0.05$) lower in the control than in SLC-selected samples at 0 min, and after 60 and 120 min of incubation at 38 °C (Figure 2b; mean ± SEM).

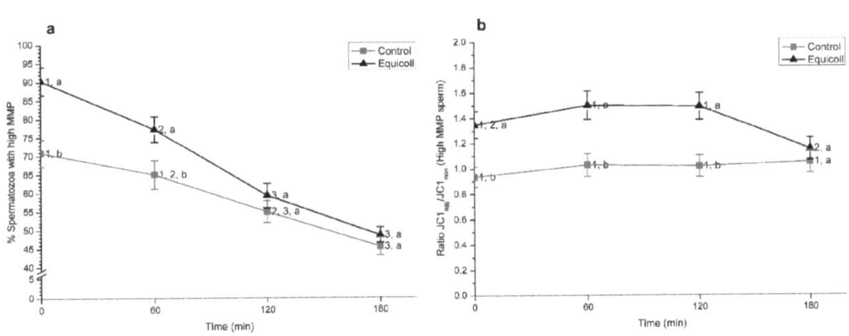

Figure 2. (**a**) Percentages of spermatozoa with high mitochondrial membrane potential (MMP) and (**b**) ratios between $JC1_{agg}$ and $JC1_{mon}$ of the sperm population with high MMP in control (non-selected samples) and SLC-selected samples (Equicoll). Different superscript letters (a, b) indicate significant differences ($p \leq 0.05$) between treatments within a given time point, and different superscript numbers (1–3) mean significant differences between time points within the control or samples selected with Equicoll. Data are shown as mean ± SEM for 10 separate experiments. SLC: single layer centrifugation.

3.3. Intracellular Calcium Levels

As shown in Figure 3a (mean ± SEM), proportions of spermatozoa with high levels of intracellular calcium calculated over the total number of viable spermatozoa and evaluated through both Fluo3 and Rhod5 fluorochromes were significantly ($p < 0.05$) higher in SLC-samples than in the control, not only immediately after selection (0 min) but also after 60 min, 120 min, and 180 min of incubation at 38 °C. While proportions of spermatozoa with high levels of intracellular calcium calculated over the total number of viable spermatozoa and evaluated through Fluo3/PI staining significantly ($p < 0.05$) decreased in SLC-selected samples after 180 min of incubation at 38 °C, this effect was not observed when the Rhod5/YO-PRO-1 test was used (mean ± SEM; Figure 3b).

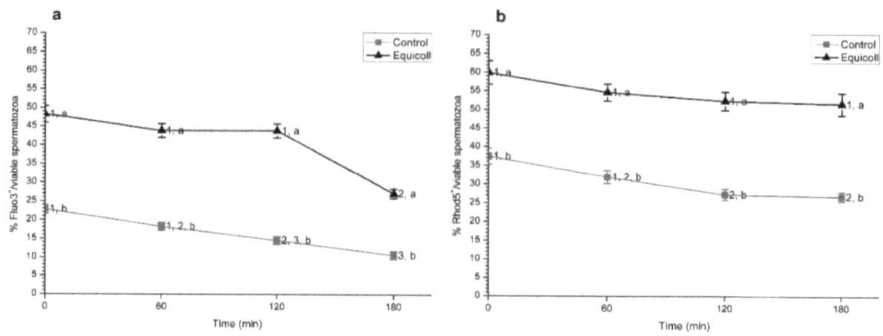

Figure 3. Proportions of viable spermatozoa with high intracellular calcium levels evaluated with two separate fluorochromes: Fluo3 (**a**) and Rhod5 (**b**) in control (non-selected samples) and SLC-selected samples (Equicoll). Data are given considering the viable sperm population. Different superscript letters (a, b) indicate significant differences ($p \leq 0.05$) between treatments within a given time point, and different superscript numbers (1–3) indicate significant differences between time points within the control or samples selected with Equicoll. Results are shown as mean ± SEM for 10 separate experiments. SLC: single layer centrifugation.

3.4. Intracellular Levels of Peroxides and Superoxides

Figure 4 shows, as mean ± SEM, the proportions of spermatozoa with high levels of peroxides and superoxides over the number of viable spermatozoa. While incubation at 38 °C significantly ($p < 0.05$) decreased the proportions of spermatozoa with high levels of intracellular peroxides over the total number of viable spermatozoa (DCF^+/viable spermatozoa), no significant differences between the control and SLC-selected samples were observed (Figure 4a). In contrast, not only did incubation at 38 °C induce a significant ($p < 0.05$) decrease in the proportions of spermatozoa with high levels of superoxides over the total number of spermatozoa (E^+/viable spermatozoa), but SLC-selected samples showed significantly ($p < 0.05$) higher values of this parameter than their control counterparts (Figure 4b).

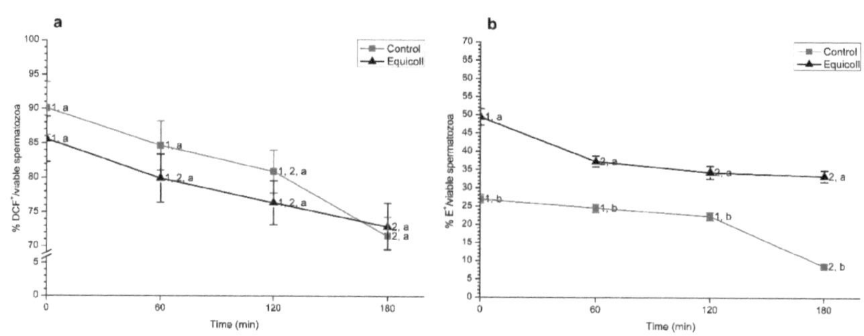

Figure 4. Proportions of viable spermatozoa with high intracellular levels of peroxides (DCF^+; **a**) and superoxides (E^+; **b**) in control (non-selected samples) and SLC-selected samples (Equicoll). Data are given considering the viable sperm population. Different superscript letters (a, b) indicate significant differences ($p \leq 0.05$) between treatments within a given time point, and different superscript numbers (1–3) mean significant differences between time points within the control or samples selected with Equicoll. Results are shown as mean ± SEM for 10 separate experiments. SLC: single layer centrifugation.

3.5. *Sperm Motility*

Proportions of total and progressively motile spermatozoa after SLC-selection and incubation at 38 °C are shown, as mean ± SEM, in Figure 5. Proportions of total and progressively motile spermatozoa were significantly ($p < 0.05$) higher in SLC-samples than in the control at 0 and 60 min. After 120 min of incubation at 38 °C, progressive but not total motility was significantly ($p < 0.05$) higher in SLC-selected samples than in the control.

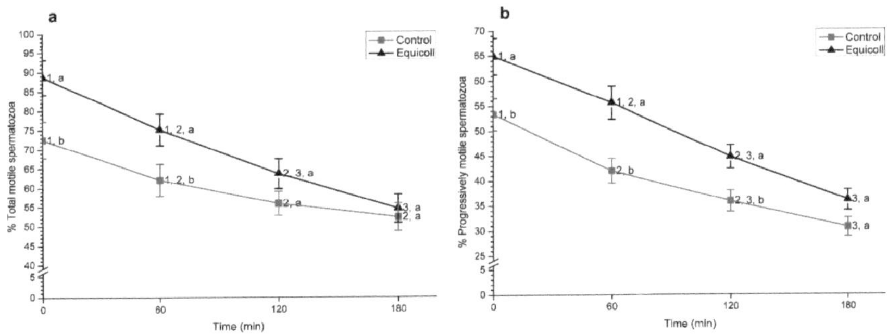

Figure 5. Proportions of (**a**) total and (**b**) progressively motile spermatozoa in control (non-selected samples) and SLC-selected samples (Equicoll). Different superscript letters (a, b) indicate significant differences ($p \leq 0.05$) between the control and SLC-selected samples, and different superscript numbers (1–3) mean significant ($p \leq 0.05$) differences between time points within the control or samples selected with Equicoll. Data are shown as mean ± SEM for 10 separate experiments. SLC: single layer centrifugation.

Sperm kinematics parameters are shown, as mean ± SEM, in Table 1. After 120 min of incubation at 38 °C, samples selected with SLC showed significantly ($p < 0.01$) lower values of VCL than the control. After 60 min, 120 min, and 180 min of incubation, VSL and VAP were significantly ($p < 0.01$) lower in SLC-selected samples than in the control. At 180 min, significantly ($p < 0.05$) lower values of WOB were observed in samples selected through SLC than in their control counterparts.

Table 1. Sperm kinematic parameters (as mean ± SEM) in control (non-selected samples) and SLC-selected samples (Equicoll).

Kinematic Parameters	0 min		60 min		120 min		180 min	
	Control	Equicoll	Control	Equicoll	Control	Equicoll	Control	Equicoll
VCL	155.0 ± 7.8 ^{a}	138.4 ± 9.3 ^{a}	143.4 ± 12.2 ^{a}	102.5 ± 11.2 ^{a}	136.3 ± 14.6 ^{a}	84.8 ± 5.4 ^{b}	108.5 ± 10.7 ^{a}	67.7 ± 4.7 ^{a}
VSL	115.3 ± 6.0 ^{a}	96.8 ± 6.9 ^{a}	112.4 ± 5.3 ^{a}	80.6 ± 6.0 ^{b}	109.1 ± 9.0 ^{a}	68.5 ± 4.2 ^{b}	106.9 ± 3.8 ^{a}	53.9 ± 4.2 ^{b}
VAP	141.6 ± 5.7 ^{a}	121.9 ± 7.6 ^{a}	130.0 ± 9.5 ^{a}	91.9 ± 9.2 ^{b}	124.7 ± 12.0 ^{a}	76.3 ± 5.2 ^{b}	100.1 ± 9.1 ^{a}	58.9 ± 4.5 ^{b}
LIN	74.7 ± 3.4 ^{a}	70.2 ± 2.3 ^{a}	79.6 ± 3.8 ^{a}	79.8 ± 3.1 ^{a}	81.6 ± 4.1 ^{a}	81.0 ± 1.2 ^{a}	84.5 ± 2.3 ^{a}	79.5 ± 1.2 ^{a}
STR	81.4 ± 2.6 ^{a}	79.4 ± 2.5 ^{a}	87.2 ± 3.0 ^{a}	88.6 ± 2.5 ^{a}	88.4 ± 3.0 ^{a}	90.1 ± 1.9 ^{a}	91.3 ± 1.6 ^{a}	91.5 ± 0.8 ^{a}
WOP	91.6 ± 1.7 ^{a}	88.2 ± 1.0 ^{a}	91.1 ± 1.6 ^{a}	90.0 ± 1.1 ^{a}	92.2 ± 1.6 ^{a}	89.8 ± 0.7 ^{a}	92.4 ± 1.4 ^{a}	86.8 ± 1.2 ^{b}
ALH	3.2 ± 0.4 ^{a}	3.1 ± 0.3 ^{a}	3.0 ± 0.4 ^{a}	2.4 ± 0.2 ^{a}	3.0 ± 0.4 ^{a}	2.1 ± 0.1 ^{a}	2.5 ± 0.3 ^{a}	2.1 ± 0.1 ^{a}
BCF	9.5 ± 0.8 ^{a}	10.1 ± 1.1 ^{a}	10.0 ± 10.9 ^{a}	9.9 ± 0.6 ^{a}	10.1 ± 1.0 ^{a}	9.1 ± 0.2 ^{a}	8.7 ± 0.1 ^{a}	9.7 ± 0.3 ^{a}

Different superscript letters (a, b) indicate significant differences ($p \leq 0.05$) between treatments at 0, 60, 120, and 180 min. Data are shown as mean ± SEM for 10 separate experiments. VCL: sperm curvilinear velocity (µm/s); VSL: sperm linear velocity (µm/s); VAP: mean velocity (µm/s); LIN: linear coefficient (%); STR: straightness coefficient (%); WOB: wobble coefficient (%); ALH: mean lateral head displacement (µm); BCF: frequency of head displacement (Hz). SLC: single layer centrifugation.

3.6. *Motile Sperm Subpopulations*

A two-step clustering procedure based on log-likelihood distance and the Schwarz Bayesian criterion was run with 24,399 motile spermatozoa. Four sperm subpopulations (SP1, SP2, SP3, and SP4) were identified. SP1 and SP2 exhibited the highest average path velocity (VAP), whereas SP3 was characterized by moderate VAP, and SP4 showed the lowest VAP (Table 2; mean ± SEM).

Table 2. Kinematic parameters (mean ± SEM) of each motile sperm population.

Kinematic Parameters	SP1	SP2	SP3	SP4
n	8422	8133	6152	1692
VCL	165.8 ± 0.2	103.8 ± 0.3	120.1 ± 0.1	0.6 ± 0.1
VSL	132.8 ± 0.3	90.7 ± 0.3	46.4 ± 0.3	2.0 ± 0.1
VAP	150.5 ± 0.2	97.5 ± 0.3	94.4 ± 0.6	0.3 ± 0.0
LIN	80.2 ± 0.1	87.3 ± 0.1	40.6 ± 0.2	0.3 ± 0.1
STR	88.2 ± 0.1	93.2 ± 0.1	53.2 ± 0.3	0.67 ± 0.1
WOB	90.6 ± 0.1	93.4 ± 0.1	767.0 ± 0.1	1.0 ± 0.2
ALH	3.6 ± 0.0	2.1 ± 0.0	3.6 ± 0.0	0.1 ± 0.0
BCF	11.0 ± 0.0	8.4 ± 0.0	7.8 ± 0.0	0.1 ± 0.0

Abbreviations: SP1: subpopulation 1; SP2: subpopulation 2; SP3: subpopulation 3; SP4: subpopulation 4; VCL: sperm curvilinear velocity (μm/s); VSL: sperm linear velocity (μm/s); VAP: mean velocity (μm/s); LIN: linear coefficient (%); STR: straightness coefficient (%); WOB: wobble coefficient (%); ALH: mean lateral head displacement (μm); BCF: frequency of head displacement (Hz).

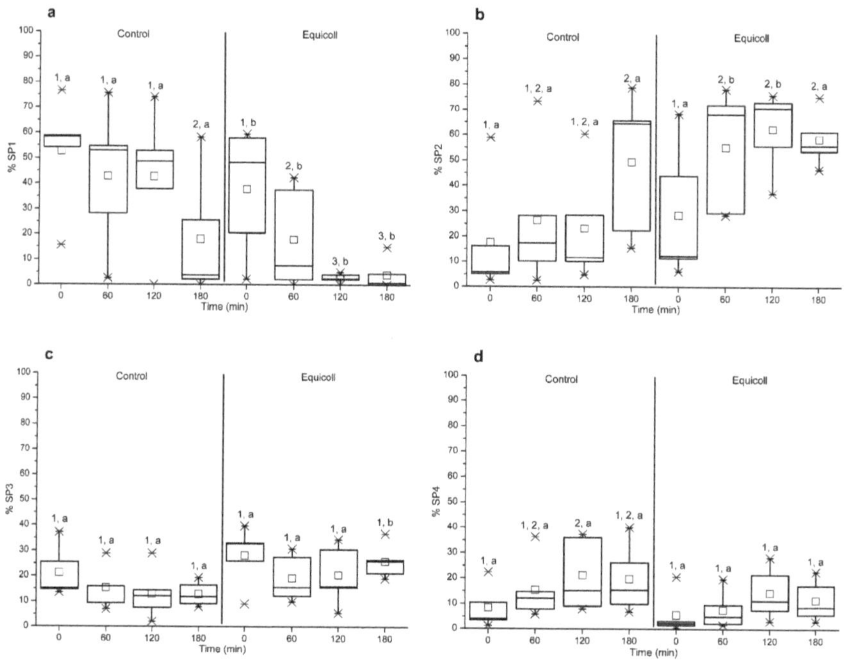

Figure 6. Proportions of spermatozoa belonging to one of each sperm subpopulations ((**a**), SP1; (**b**), SP2; (**c**), SP3; (**d**), SP4) in control (non-selected samples) and SLC-selected samples (Equicoll). Different superscript letters (a, b) indicate significant differences ($p \leq 0.05$) between treatments within a given time point, and different superscript numbers (1–3) mean significant differences between time points within the control or samples selected with Equicoll. Data are shown as box-whisker plots for 10 separate experiments. SLC: single layer centrifugation. SP1: subpopulation 1; SP2: subpopulation 2; SP3: subpopulation 3; SP4: subpopulation 4.

Immediately after centrifugation with Equicoll (0 min) and after 60, 120, and 180 min of incubation at 38 °C, proportions of motile spermatozoa belonging to SP1 were significantly ($p < 0.05$) lower in SLC-samples than in the control (Figure 6a). The proportions of motile spermatozoa belonging to SP2 were significantly ($p < 0.05$) lower in the control than in SLC-selected samples (Figure 6b). With regard to SP3, the proportions of spermatozoa belonging to that population were significantly ($p < 0.05$) higher in SLC-selected samples than in the control after 180 min of incubation (Figure 6c).

Finally, no significant differences in the proportions of sperm belonging to SP4 were observed between the control and SLC-selected samples either at 0 min or over the incubation period at 38 °C (60, 120, and 180 min; Figure 6d).

3.7. Sperm-PMN Interaction

At 0 min, proportions of sperm bound to PMN were significantly ($p < 0.05$) higher in SLC-selected samples than in the control. In contrast, no significant differences between control and SLC-selected samples were observed after 60, 120, or 180 min of incubation (Figure 7a; mean ± SEM). Moreover, and as shown in Figure 7b (mean ± SEM), percentages of spermatozoa phagocytosed by PMN were significantly ($p < 0.05$) higher in SLC-selected samples than in the control after 60, 120, and 180 min of incubation at 38 °C.

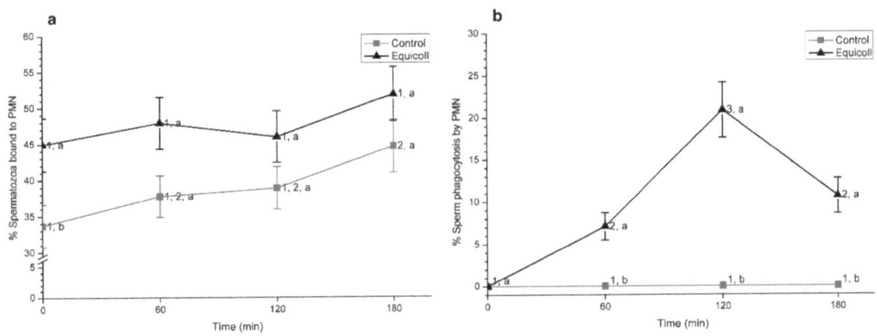

Figure 7. Percentages of (**a**) spermatozoa bound to PMN and (**b**) sperm phagocytosed by PMN in control (non-selected samples) and SLC-selected samples (Equicoll). Different superscript letters (a, b) indicate significant differences ($p \leq 0.05$) between treatments within a given time point, and different superscript numbers (1, 2) mean significant differences between time points within the control or samples selected with Equicoll. Data are shown as mean ± SEM for 7 separate experiments. SLC: single layer centrifugation. PMN: Polymorphonuclear neutrophils.

4. Discussion

Pregnancy rates after AI are affected by sperm quality and by the ability of the female reproductive tract to support an embryo. Consequently, selection of high-quality sperm represents a challenge in equine assisted reproduction. Previous studies have already shown that SLC with cooled-stored and frozen/thawed semen significantly improves sperm quality parameters in both stallions [11] and jackasses [12]. On the other hand, one should note that using a species-specific formulated colloid is highly advisable, as the specific particularities of each species require the design of concrete formulations. In the current study, we observed that fresh donkey semen samples previously centrifuged with SLC (Equicoll) showed higher proportions of spermatozoa with an intact plasma membrane (viable spermatozoa), low membrane lipid disorder, high intracellular calcium (which was evaluated with two separate fluorochromes [33,34]), high superoxide levels, and high MMP, and of progressively and total motile spermatozoa than the control. In contrast, after centrifugation with SLC, spermatozoa showed lower VSL, VCL, and VAP, and the proportions of spermatozoa belonging to SP1, which was the motile sperm subpopulation that exhibited the highest VAP, were also significantly lower than the control. Furthermore, the removal of seminal plasma components due to centrifugation with SLC led to an increase in the percentages of sperm phagocytosed by PMN after 60, 120, and 180 min of incubation, and in the percentages of spermatozoa bound to PMN, although in this case significant differences were only observed at 0 min.

Overall, our data support that SLC selects a sperm functional population, which exhibits high plasma membrane integrity. This increase in plasma membrane integrity is not only apparent from the analysis of SYBR14/PI staining but also from that of M540/YO-PRO-1. However, the higher plasma membrane integrity and lower membrane lipid disorder of SLC-selected spermatozoa compared to the control were statistically significant immediately after centrifugation and after 60 min of incubation at 38 °C but not later, which indicates that further incubation at this temperature abolishes these differences. In this context, it is worth mentioning that the 3-h period was set to mimic the time deemed for sperm to reside within the uterus. However, this post-selection incubation time was not included in previous studies, which made our results difficult to compare. Moreover, most of the previous works investigating how selection with SLC affects sperm quality were focused on stored semen (cooled and frozen). In donkeys, Ortiz et al. obtained significantly higher values of sperm quality parameters with colloid centrifugation using Equicoll after 24 h of cooled storage [12] and after freezing/thawing [13]. It is worth highlighting that sperm samples included in this study were initially within the standard values for donkey semen [40]. In this regard, even if SLC-selected spermatozoa are the ones exhibiting the highest quality, it is difficult to observe much apparent differences between the control and SLC-selected samples when the raw semen is of good quality. Under this perspective, our results are in agreement with those obtained by Ortiz et al. who observed that the use of SLC-selection to improve post-thaw sperm quality yielded better results when ejaculates with poor freezability were used [41]. Be that as it may, further research comparing the effects of SLC-selection on groups of fresh ejaculates with high and low sperm quality is warranted.

As aforementioned, in this work, we found a significant improvement in different sperm function parameters. In agreement with our data, centrifugation with species-specific designed colloids gives similar results. In horses, centrifugation of cooled-stored semen with Equicoll has been reported to increase the proportions of the most functional spermatozoa [42,43]. Despite not being evaluated in our study, the aforementioned research also demonstrated that SLC with Equicoll removes much of the bacterial load present in their ejaculates [42,43] and increases pregnancy rates after AI [44,45]. In rams, Šterbenc et al. [14] observed that centrifugation of frozen/thawed semen with a species-specific colloid (Androcoll-O) increases the proportions of spermatozoa with an intact plasma membrane and DNA, and of motile spermatozoa. In pigs, SLC with Porcicoll (formerly known as Androcoll-P) has been found to enhance the quality of frozen/thawed sperm, in terms of viability, mitochondrial activity and intracellular levels of ROS [15]. Also in pigs, centrifugation of semen with SLC with Porcicoll prior to cryopreservation increases the cryotolerance and fertilizing ability of frozen/thawed sperm [19,20]. Even in brown bears, centrifugation with SLC has also been reported to improve the proportion of spermatozoa with intact plasma membrane and acrosome [17,18]. In bulls, centrifugation through SLC with a species-specific colloid (Androcoll-B/Bovicoll) also results in an increase in the proportions of spermatozoa with high mitochondrial membrane potential and high relative levels of peroxides [16], and in some kinematic parameters, but has no impact upon total and progressive motility of cooled-stored semen [46].

Another novel and interesting finding of this study is related to the evaluation of the percentages of sperm bound to PMN and those of sperm phagocytosed by PMN. We observed that, immediately after SLC-selection (0 min), the percentages of spermatozoa bound to PMN were higher in samples washed through SLC than in the control. In addition, the percentages of sperm phagocytozed by PMN were significantly higher in washed SLC-samples than in the control after 60, 120, and 180 min of incubation. As aforementioned, not only is centrifugation through SLC expected to select the sperm population with higher sperm quality and fertilizing ability, but it also removes seminal plasma components [23,47]. This means that donkey spermatozoa, especially when seminal plasma components are removed, are very susceptible to bind to PMN and, in some cases, be phagocytozed. In this context, it is worth mentioning that while sperm separation methods seek to mimic in vitro the natural selection occurring within the female reproductive tract [48], they also entail the removal of seminal plasma, which plays a crucial role in the response of the endometrium [26]. Upon ejaculation, specific components of seminal plasma, notably proteins such as spermadhesins, are adsorbed onto the sperm surface [49]. These sperm coating components confer important properties to the plasma membrane and prevent premature sperm capacitation [50]. Furthermore, previous studies have described that seminal plasma components are able to regulate the endometrial inflammatory reaction [25,51]. In effect, in the horse, the seminal protein CRISP3 has been found to inhibit the binding mechanism between viable spermatozoa and PMN involved in phagocytosis [52], improving fertility in mares [53]. In jennies, suppression of sperm-PMN binding by the presence of seminal plasma has also been identified [18], which would also be in agreement with the observed effect of seminal plasma in this study. We can thus hypothesize that sperm selection with Equicoll leads to the removal of seminal plasma components involved in the sperm protection from phagocytosis, since a previous study reported that the percentages of phagocytosis are low in the presence of seminal plasma [26]. In addition, one should also take into consideration that donkey semen has been reported to activate PMN and that PMN can create complexes known as neutrophil extracellular traps (NETosis) in the presence of spermatozoa [54–56]. NETosis is another type of programmed cell death that differs from apoptosis and is triggered by neutrophils (although monocytes and macrophages also show that ability, which is known as METosis). During NETosis, PMN undergo morphological changes, which include the disintegration of nuclear and granule membranes and the combination of nuclear, granular, and cytoplasmic components; the final result is the formation of traps made up of DNA and antimicrobial proteins. These extracellular traps allow capturing microbial agents and spermatozoa [57]. Therefore, further studies should elucidate whether the modifications of the sperm surface due to selection through SLC modify the ability of PMN to trigger NETosis, and determine the impact upon sperm fertilizing ability and endometrial reaction in the jenny.

5. Conclusions

We can conclude that previous centrifugation of donkey spermatozoa with SLC (Equicoll) likely selects the most functional spermatozoa, which are the ones that exhibit intact membranes, high mitochondrial membrane potential, and high intracellular calcium levels. In addition, selection with SLC increases the proportion of progressively and total motile spermatozoa. Finally, since selection of donkey sperm through SLC augments phagocytosis by PMN, further studies analyzing the effect of SLC-selection on sperm–PMN interaction, particularly NETosis, are much warranted.

Author Contributions: Conceptualization, J.M.M., J.M., and M.Y.; methodology, M.P., J.C., S.R., M.Y., and J.M.; software, M.P.; validation, J.M. and M.Y.; formal analysis, M.P.; investigation, M.P., J.C., S.R., M.Y., and J.M.; resources, J.M.M., M.Y., and J.M.; data curation, M.P.; writing—original draft preparation, M.Y.; writing—review and editing, M.P. and J.M.; visualization, M.Y. and J.M.; supervision, M.Y. and J.M.; project administration, M.Y. and J.M.; funding acquisition, M.Y. and J.M. All authors have read and agreed to the published version of the manuscript.

Funding: This research was supported by the Ministry of Science, Innovation and Universities, Spain (Grants: RYC-2014-15581 and AGL2017-88329-R), and the Regional Government of Catalonia (2017-SGR-1229). J.C. was funded by the National Agency for Research and Development (ANID), Ministry of Education, Chile (Scheme: Becas Chile Doctorado en el Extranjero, PFCHA; Grant: 2017/72180128).

Acknowledgments: The authors would like to thank the support from Sebastián Bonilla-Correal and Sabrina Gazem from Autonomous University of Barcelona, Spain, and from Marc Llavanera, Ariadna Delgado-Bermúdez, Beatriz Fernandez-Fuertes, Sergi Bonet, and Isabel Barranco from University of Girona for their technical support.

Conflicts of Interest: J.M.M. is the inventor and one of the patent holders of the colloid formulation used in this study. The other authors declare no conflicts of interest.

References

1. Hinrichs, K. Assisted reproductive techniques in mares. *Reprod. Domest. Anim.* **2018**, *53*, 4–13. [CrossRef]
2. Alvarenga, M.A.; Papa, F.O.; Neto, C.R. Advances in Stallion Semen Cryopreservation. *Vet. Clin. Equine Pract.* **2016**, *32*, 521–530. [CrossRef]
3. Morrell, J.M.; Wallgren, M. Colloid centrifugation of boar semen. *Reprod. Domest. Anim.* **2011**, *46*, 18–22. [CrossRef]
4. Oseguera-López, I.; Ruiz-Díaz, S.; Ramos-Ibeas, P.; Pérez-Cerezales, S. Novel Techniques of Sperm Selection for Improving IVF and ICSI Outcomes. *Front. Cell Dev. Biol.* **2019**, *7*. [CrossRef] [PubMed]

5. Morrell, J.M. Stallion Sperm Selection: Past, Present, and Future Trends. *J. Equine Vet. Sci.* **2012**, *32*, 436–440. [CrossRef]
6. Roca, J.; Parrilla, I.; Gil, M.A.; Cuello, C.; Martinez, E.A.; Rodriguez-Martinez, H. Non-viable sperm in the ejaculate: Lethal escorts for contemporary viable sperm. *Anim. Reprod. Sci.* **2016**, *169*, 24–31. [CrossRef] [PubMed]
7. Morrell, J.M.; Kumaresan, A.; Johannisson, A. Practical implications of sperm selection techniques for improving reproduction. *Anim. Reprod.* **2017**, *14*, 572–580. [CrossRef]
8. Morrell, J.; Alsina, M.S.; Abraham, M.C.; Sjunnesson, Y. Practical applications of sperm selection techniques for improving reproductive efficiency. *Anim. Reprod.* **2016**, 340–345. [CrossRef]
9. Morrell, J.M.; Rodriguez-Martinez, H. Biomimetic Techniques for Improving Sperm Quality in Animal Breeding: A Review. *Open Androl. J.* **2009**, *1*, 1–9. [CrossRef]
10. Morrell, J.; Dalin, A.M.; Rodriguez-Martinez, H. Comparison of density gradient and single layer centrifugation of stallion spermatozoa: Yield, motility and survival. *Equine Vet. J.* **2009**, *41*, 53–58. [CrossRef]
11. Morrell, J.; Garcia, B.M.; Penã, F.J.; Johannisson, A. Processing stored stallion semen doses by Single Layer Centrifugation. *Theriogenology* **2011**, *76*, 1424–1432. [CrossRef]
12. Ortiz, I.; Dorado, J.; Ramírez, L.; Morrell, J.; Acha, D.; Urbano, M.; Gálvez, M.J.; Carrasco, J.J.; Hidalgo, M. Effect of single layer centrifugation using Androcoll-E-Large on the sperm quality parameters of cooled-stored donkey semen doses. *Animal* **2014**, 308–315. [CrossRef]
13. Ortiz, I.; Dorado, J.; Morrell, J.M.; Crespo, F.; Gosálvez, J.; Gálvez, M.J.; Acha, D.; Hidalgo, M. Effect of single-layer centrifugation or washing on frozen–thawed donkey semen quality: Do they have the same effect regardless of the quality of the sample? *Theriogenology* **2015**, *84*, 294–300. [CrossRef] [PubMed]
14. Šterbenc, N.; Morrell, J.M.; Kosec, M.; Rath, D.; Klein, S.; Klinc, P. Single layer colloid centrifugation technique improves motility, viability and chromatin integrity of ram spermatozoa after thawing. *Cryobiology* **2019**, *86*, 77–83. [CrossRef] [PubMed]
15. Crespo-Félez, I.; Castañeda-Sampedro, A.; Sánchez, D.I.; Fernández-Alegre, E.; Álvarez-Rodríguez, M.; Domínguez, J.C.; Morrell, J.M.; Martínez-Pastor, F. Effect of Single Layer Centrifugation Porcicoll (70%, 80% and 90%) or supplementation with reduced glutathione, seminal plasma and bovine serum albumin on frozen-thawed boar sperm. *Anim. Reprod. Sci.* **2017**, *187*, 167–173. [CrossRef] [PubMed]
16. Goodla, L.; Morrell, J.M.; Yusnizar, Y.; Stålhammar, H.; Johannisson, A. Quality of bull spermatozoa after preparation by single-layer centrifugation. *J. Dairy Sci.* **2014**, *97*, 2204–2212. [CrossRef] [PubMed]
17. Álvarez-Rodríguez, M.; Álvarez, M.; Anel-López, L.; López-Urueña, E.; Manrique, P.; Borragán, S.; Morrell, J.M.; de Paz, P.; Anel, L. Effect of colloid (Androcoll-Bear, Percoll, and PureSperm) selection on the freezability of brown bear (Ursus arctos) sperm. *Theriogenology* **2016**, *85*, 1097–1105. [CrossRef]
18. Anel-Lopez, L.; Ortega-Ferrusola, C.; Álvarez, M.; Borragán, S.; Chamorro, C.; Peña, F.J.; Morrell, J.; Anel, L.; de Paz, P. Improving sperm banking efficiency in endangered species through the use of a sperm selection method in brown bear (Ursus arctos) thawed sperm. *BMC Vet. Res.* **2017**, *13*, 200. [CrossRef]
19. Martinez-Alborcia, M.J.; Morrell, J.M.; Parrilla, I.; Barranco, I.; Vázquez, J.M.; Martinez, E.A.; Roca, J. Improvement of boar sperm cryosurvival by using single-layer colloid centrifugation prior freezing. *Theriogenology* **2012**, *78*, 1117–1125. [CrossRef]
20. Martinez-Alborcia, M.J.; Morrell, J.M.; Gil, M.A.; Barranco, I.; Maside, C.; Alkmin, D.V.; Parrilla, I.; Martinez, E.A.; Roca, J. Suitability and effectiveness of single layer centrifugation using Androcoll-P in the cryopreservation protocol for boar spermatozoa. *Anim. Reprod. Sci.* **2013**, *140*, 173–179. [CrossRef]
21. Moore, A.I.; Squires, E.L.; Graham, J.K. Effect of seminal plasma on the cryopreservation of equine spermatozoa. *Theriogenology* **2005**, *63*, 2372–2381. [CrossRef] [PubMed]
22. Barrier-Battut, I.; Bonnet, C.; Giraudo, A.; Dubois, C.; Caillaud, M.; Vidament, M. Removal of Seminal Plasma Enhances Membrane Stability on Fresh and Cooled Stallion Spermatozoa. *Reprod. Domest. Anim.* **2013**, *71*, 64–71. [CrossRef]
23. Kruse, R.; Dutta, P.C.; Morrell, J. Colloid centrifugation removes seminal plasma and cholesterol from boar spermatozoa. *Reprod. Fertil. Dev.* **2011**, 858–865. [CrossRef] [PubMed]
24. Alghamdi, A.S.; Lovaas, B.J.; Bird, S.L.; Lamb, G.C.; Rendahl, A.K.; Taube, P.C.; Foster, D.N. Species-specific interaction of seminal plasma on sperm-neutrophil binding. *Anim. Reprod. Sci.* **2009**, *114*, 331–344. [CrossRef] [PubMed]

25. Miró, J.; Vilés, K.; García, W.; Jordana, J.; Yeste, M. Effect of donkey seminal plasma on sperm movement and sperm-polymorphonuclear neutrophils attachment in vitro. *Anim. Reprod. Sci.* **2013**, *140*, 164–172. [CrossRef]
26. Miró, J.; Marín, H.; Catalán, J.; Papas, M.; Gacem, S.; Yeste, M. Seminal plasma, sperm concentration, and sperm-PMN interaction in the donkey: An in vitro model to study endometrial inflammation at post-insemination. *Int. J. Mol. Sci.* **2020**, *21*, 3478. [CrossRef]
27. Vilés, K.; Rabanal, R.; Rodríguez-Prado, M.; Miró, J. Influence of seminal plasma on leucocyte migration and amount of COX-2 protein in the jenny endometrium after insemination with frozen-thawed semen. *Anim. Reprod. Sci.* **2013**, *143*, 57–63. [CrossRef]
28. Morrell, J.; Van Wienen, M.; Wallgren, M. Single Layer Centrifugation Can Be Scaled-Up Further to Process up to 150 mL Semen. *ISRN Vet. Sci.* **2011**, *2011*. [CrossRef]
29. Petrunkina, A.M.; Waberski, D.; Bollwein, H.; Sieme, H. Identifying non-sperm particles during flow cytometric physiological assessment: A simple approach. *Theriogenology* **2010**, *73*, 995–1000. [CrossRef]
30. Rathi, R.; Colenbrander, B.; Bevers, M.M.; Gadella, B.M. Evaluation of in vitro capacitation of stallion spermatozoa. *Biol. Reprod.* **2001**, *65*, 462–470. [CrossRef]
31. Yeste, M.; Estrada, E.; Rivera Del Álamo, M.M.; Bonet, S.; Rigau, T.; Rodríguez-Gil, J.E. The increase in phosphorylation levels of serine residues of protein HSP70 during holding time at 17 °C is concomitant with a higher cryotolerance of boar spermatozoa. *PLoS ONE* **2014**, *9*, e90887. [CrossRef] [PubMed]
32. Ortega-Ferrusola, C.; Sotillo-Galan, Y.; Varela-Fernandez, E.; Gallardo-Bolanos, J.M.; Muriel, A.; Gonzalez-Fernandez, L.; Tapia, J.A.; Pena, F.J. Detection of 'Apoptosis-Like' Changes During the Cryopreservation Process in Equine Sperm. *J. Androl.* **2007**, *29*, 213–221. [CrossRef] [PubMed]
33. Yeste, M.; Fernández-Novell, J.M.; Ramió-Lluch, L.; Estrada, E.; Rocha, L.G.; Cebrián-Pérez, J.A.; Muiño-Blanco, T.; Concha, I.I.; Ramírez, A.; Rodríguez-Gil, J.E. Intracellular calcium movements of boar spermatozoa during "in vitro" capacitation and subsequent acrosome exocytosis follow a multiple-storage place, extracellular calcium-dependent model. *Andrology* **2015**, *3*, 729–747. [CrossRef] [PubMed]
34. Harrison, R.A.P.; Mairet, B.; Miller, N.G.A. Flow cytometric studies of bicarbonate-mediated Ca^{2+} influx in boar sperm populations. *Mol. Reprod. Dev.* **1993**, *35*, 197–208. [CrossRef]
35. Kadirvel, G.; Kumar, S.; Kumaresan, A.; Kathiravan, P. Capacitation status of fresh and frozen-thawed buffalo spermatozoa in relation to cholesterol level, membrane fluidity and intracellular calcium. *Anim. Reprod. Sci.* **2009**, *116*, 244–253. [CrossRef]
36. Morrell, J.M.; Winblad, C.; Georgakas, A.; Stuhtmann, G.; Humblot, P.; Johannisson, A. Reactive oxygen species in stallion semen can be affected by season and colloid centrifugation. *Anim. Reprod. Sci.* **2013**, *140*, 62–69. [CrossRef]
37. Guthrie, H.D.; Welch, G.R. Determination of intracellular reactive oxygen species and high mitochondrial membrane potential in Percoll-treated viable boar sperm using fluorescence-activated flow cytometry1. *J. Anim. Sci.* **2006**, *84*, 2089–2100. [CrossRef]
38. Yeste, M.; Bonet, S.; Rodríguez-Gil, J.E.; Rivera Del Álamo, M.M. Evaluation of sperm motility with CASA-Mot: Which factors may influence our measurements? *Reprod. Fertil. Dev.* **2018**, *30*, 789–798. [CrossRef]
39. Luna, C.; Yeste, M.; Rivera Del Alamo, M.M.; Domingo, J.; Casao, A.; Rodriguez-Gil, J.E.; Pérez-Pé, R.; Cebrián-Pérez, J.A.; Muiño-Blanco, T. Effect of seminal plasma proteins on the motile sperm subpopulations in ram ejaculates. *Reprod. Fertil. Dev.* **2017**, *29*, 394–405. [CrossRef]
40. Miró, J.; Lobo, V.; Quintero-Moreno, A.; Medrano, A.; Peña, A.; Rigau, T. Sperm motility patterns and metabolism in Catalonian donkey semen. *Theriogenology* **2005**, *63*, 1706–1716. [CrossRef]
41. Ortiz, I.; Dorado, J.; Acha, D.; Gálvez, M.J.; Urbano, M.A.; Hidalgo, M. Colloid single-layer centrifugation improves post-thaw donkey (*Equus asinus*) sperm quality and is related to ejaculate freezability. *Reprod. Fertil. Dev.* **2015**, *27*, 332–340. [CrossRef] [PubMed]
42. Guimarães, T.; Lopes, G.; Pinto, M.; Silva, E.; Miranda, C.; Correia, M.J.; Damásio, L.; Thompson, G.; Rocha, A. Colloid centrifugation of fresh stallion semen before cryopreservation decreased microorganism load of frozen-thawed semen without affecting seminal kinetics. *Theriogenology* **2015**, *83*, 186–191. [CrossRef] [PubMed]

43. Varela, E.; Rey, J.; Plaza, E.; Muñoz de Propios, P.; Ortiz-Rodríguez, J.M.; Álvarez, M.; Anel-López, L.; Anel, L.; De Paz, P.; Gil, M.C.; et al. How does the microbial load affect the quality of equine cool-stored semen? *Theriogenology* **2018**, *114*, 212–220. [CrossRef] [PubMed]
44. Morrell, J.M.; Stuhtmann, G.; Meurling, S.; Lundgren, A.; Winblad, C.; Macias Garcia, B.; Johannisson, A. Sperm yield after single layer centrifugation with Androcoll-E is related to the potential fertility of the original ejaculate. *Theriogenology* **2014**, *81*, 1005–1011. [CrossRef] [PubMed]
45. Morrell, J.M.; Richter, J.; Martinsson, G.; Stuhtmann, G.; Hoogewijs, M.; Roels, K.; Dalin, A.M. Pregnancy rates after artificial insemination with cooled stallion spermatozoa either with or without Single Layer Centrifugation. *Theriogenology* **2014**, *82*, 1102–1105. [CrossRef] [PubMed]
46. Yulnawati, Y.; Abraham, M.; Laskowski, D.; Johannisson, A.; Morrell, J. Changes in Bull Sperm Kinematics after Single Layer Centrifugation. *Reprod. Domest. Anim.* **2014**, *49*, 954–956. [CrossRef]
47. Morrell, J.M. Effect of colloid centrifugation on boar sperm quality during storage and function in in vitro fertilization. *Theriogenology* **2019**, *137*, 122–126. [CrossRef]
48. Beydola, T.; Sharma, R.; Agarwal, A. Sperm preparation and selection techniques. In *Medical and Surgical Management of Male Infertility*, 1st ed.; Rizk, B., Aziz, N., Agarwal, A., Sabanegh, E., Eds.; Jaypee Brothers Medical Publishers: New Delhi, India, 2014; pp. 244–251.
49. Björkgren, I.; Sipilä, P. The impact of epididymal proteins on sperm function. *Reproduction* **2019**, *158*, R155–R167. [CrossRef]
50. Recuero, S.; Fernandez-Fuertes, B.; Bonet, S.; Barranco, I.; Yeste, M. Potential of seminal plasma to improve the fertility of frozen-thawed boar spermatozoa. *Theriogenology* **2019**, *137*, 36–42. [CrossRef]
51. Troedsson, M.H.T. Breeding-Induced Endometritis in Mares. *Vet. Clin. North. Am. Equine Pract.* **2006**, *22*, 705–712. [CrossRef]
52. Doty, A.; Buhi, W.C.; Benson, S.; Scoggin, K.E.; Pozor, M.; Macpherson, M.; Mutz, M.; Troedsson, M.H.T. Equine CRISP3 modulates interaction between spermatozoa and polymorphonuclear neutrophils. *Biol. Reprod.* **2011**, *85*, 157–164. [CrossRef] [PubMed]
53. Alghamdi, A.S.; Foster, D.N.; Troedsson, M.H.T. Equine seminal plasma reduces sperm binding to polymorphonuclear neutrophils (PMNs) and improves the fertility of fresh semen inseminated into inflamed uteri. *Reproduction* **2004**. [CrossRef] [PubMed]
54. Alghamdi, A.S.; Foster, D.N. Seminal DNase Frees Spermatozoa Entangled in Neutrophil Extracellular Traps. *Biol. Reprod.* **2005**, *73*, 1174–1181. [CrossRef]
55. Alghamdi, A.S.; Funnell, B.J.; Bird, S.L.; Lamb, G.C.; Rendahl, A.K.; Taube, P.C.; Foster, D.N. Comparative studies on bull and stallion seminal DNase activity and interaction with semen extender and spermatozoa. *Anim. Reprod. Sci.* **2010**, *121*, 249–258. [CrossRef] [PubMed]
56. Fichtner, T.; Kotarski, F.; Gärtner, U.; Conejeros, I.; Hermosilla, C.; Wrenzycki, C.; Taubert, A. Bovine sperm samples induce different NET phenotypes in a NADPH oxidase-, PAD4-, and Ca-dependent process. *Biol. Reprod.* **2020**, *102*, 902–914. [CrossRef] [PubMed]
57. Thiam, H.R.; Wong, S.L.; Wagner, D.D.; Waterman, C.M. Cellular Mechanisms of NETosis. *Annu. Rev. Cell Dev. Biol.* **2020**, *36*, 191–218. [CrossRef]

animals

MDPI

Article

Optimization of CASA-Mot Analysis of Donkey Sperm: Optimum Frame Rate and Values of Kinematic Variables for Different Counting Chamber and Fields

Sabrina Gacem [1], Jaime Catalán [1], Anthony Valverde [2], Carles Soler [3,*] and Jordi Miró [1,*]

1 Equine Reproduction Service, Department of Animal Medicine and Surgery, Faculty of Veterinary Sciences, Autonomous University of Barcelona, 08193 Bellaterra, Spain; swp.sabrina.gacem@gmail.com (S.G.); dr.jcatalan@gmail.com (J.C.)

2 Costa Rica Institute of Technology, School of Agronomy, San Carlos Campus, 223-21001 Alajuela, Costa Rica; anvalverde@itcr.ac.cr

3 Departamento de Biología Celular, Biología Funcional y Antropología Física, Universitat de València, 46100 Burjassot, Valencia, Spain

* Correspondence: carles.soler@uv.edu (C.S.); jordi.miro@uab.cat (J.M.)

Received: 30 September 2020; Accepted: 26 October 2020; Published: 29 October 2020

check for updates

Simple Summary: A reliable sperm motility exam is important for semen analysis and breeding soundness examination. Different parameters can affect the Computer Assisted Sperm Analysis (CASA) motility results. Today, new high-resolution cameras and different chambers are introduced to CASA systems, and protocol optimization is required to render the estimation results for donkey sperm. The objective of this study is the optimization of the conditions used for donkey semen motility analysis with CASA-Mot by defining the optimum frame rate for different chamber types. Additionally, to study the effect of different chamber types, chamber field and sperm dilution on the sperm kinematic parameters with higher frame rates are examined.

Abstract: In order to optimize the donkey sperm motility analysis by the CASA (Computer Assisted Sperm Analysis)-Mot system, twelve ejaculates were collected from six jackasses. Capillary loaded chamber (CLC), ISAS®D4C depths 10 and 20 μm, ISAS®D4C Leja 20 and drop displacement chamber (DDC), Spermtrack® (Spk) depths 10 and 20 μm were used. Sperm kinematic variables were evaluated using each chamber and a high-resolution camera capable of capturing a maximum of 500 frames/second (fps). The optimum frame rate (OFR) (defined according to curvilinear velocity—VCL) was dependent on chamber type. The highest OFR obtained was 278.46 fps by Spk20. Values for VCL, straight-line velocity (VSL), straightness (STR), amplitude of lateral head displacement (ALH) and beat cross frequency (BCF) were high in DDC and 10 μm depth. In both DDC 10 and 20 μm, the sperm velocities (VCL, VSL, VAP) and ALH values decreased significantly from the centre to the edges, while Wobble and BCF increased. No defined behavior was observed along the CLC. However, all the kinematic variables had a higher value in a highly concentrated sample, in both chamber types. In conclusion, analyzing a minimum of nine fields at 250 fps from the centre to the edges in Spk10 chamber using a dilution of 30×10^6 sperm/mL offers the best choice for donkey computerised sperm motility analysis.

Keywords: frame rate; drop displacement chambers; capillary loaded chambers; chamber depth; field; sperm dilution

1. Introduction

The domestic donkey (*Equus asinus*) is one of the two domestic species of the genus Equus along with the horse (*Equus caballus*) [1]. In developed zones, donkeys have suffered a significant decrease due to industrialization and mechanization of agriculture. However, in recent years, there has been an increase in donkey product interest: milk, meat or skin [2]. Donkey farming is expanding and the research interest about donkey production and reproduction optimization is increasing [3,4]. This implies a more complete knowledge of their semen quality and general reproductive characteristics especially for achieving a productive-assisted reproduction [5].

Semen quality can be defined upon certain criteria performing certain tests, such as motility, concentration and morphology [5–7]. Nonetheless, from all those tests, sperm motility is commonly considered the most significant parameter for breeding soundness examination [8,9]. Usually, motility is analysed subjectively looking for just the total and progressive motility, but in the last few decades, an objective method—CASA (Computer Assisted Sperm Analysis)—was introduced and is now available and used widely by veterinarians and laboratories [10]. CASA-Mot technology is based on the computational reconstitution of sperm trajectory from image sequences; the last ones are captured by a video camera mounted on a microscope. Then, the sperm sequences are automatically analysed by the computer in a concise time [11]. Moreover, CASA-Mot systems provide a battery of kinematic quantitative parameters that define the sperm cell motility rather than the progressivity [12].

A critical review of the literature revealed that no standard practices have been embraced or recommended by professional societies in the case of donkey samples and thus, no defined protocols are followed within or across CASA-Mot instruments [13]. In fact, in other species, it has been proved that the accuracy and the sensitivity of the measurements obtained with CASA-Mot systems can be affected by different factors, such as the mathematical algorithms, suspending medium, sample concentration, frame rate, chamber type and depth, hardware, and instrument settings [14]. Usually, for the donkey semen assessment within CASA, we use the same setting and protocol as horses. However, comparative studies in both species have identified differences in reproductive strategy as well as in the sperm form and function [15,16]. Effectively, the donkey testis is bigger and has been proved to have more efficient spermatogenesis than a stallion [17–20]. In effect, donkey sperm heads are smaller with a larger mid-piece than the stallion resulting in differences in motility patterns [21]. It was reported that donkey spermatozoon is faster than the horse when using the same CASA set up [22]. However, we need to consider those differences in the definition of the most adequate setting when analysing donkey sperm.

On the other hand, in recent years, the development of high-resolution cameras, the improvement of informatic systems and the development of specific chambers to analyze sperm motility have changed the published data completely on sperm motility patterns in different species by CASA-Mot systems.

The objective of this study was to standardise the method of sperm motility assessment in the donkey, by first defining the optimal frame rate (OFR) based on VCL (curvilinear velocity) data for different counting chambers for the use in CASA-Mot system. Secondly, we analyse the kinematic variables at OFR in different chamber types and depths (including the considered counting area). Finally, an investigation of the effect of semen dilution on donkey sperm motility parameters is carried out.

2. Materials and Methods

2.1. Animals Used and Ethics Statement

The experiment was carried out on six Catalan donkeys with two ejaculates for each one. All males were aged between 3 and 20 years old and they are known for their successful fertility. Animals involved in the study were housed at the Equine Reproduction Service, Autonomous University of Barcelona (Bellaterra, Cerdanyola del Valles, Spain) at a Europe-approved equine semen collection centre (authorization number: ES09RS01E). The centre operates under strict protocols of animal welfare and health control. All jackasses were semen donors, which were housed in an individual paddock at the centre. Semen has been collected under CEE health conditions (free of Equine Arteritis, Infectious Anaemia and Contagious Metritis). It is important to note that the service runs under the Catalonia Regional Government's approval (located in Spain) and no manipulations to the animals other than semen collection were carried out. The Ethics committee of this institution indicated that no further ethical approval was required. Additionally, all the animals received a standard diet (with mixed hay and basic concentrate) and were provided with water ad libitum. Three times a week, donkeys underwent regular semen collection once a day under the same conditions and samples were collected throughout the year.

2.2. Semen Preparation

The semen was collected manually using an artificial Hannover vagina (Minitüb GmbH, Tiefenbach, Germany) with an in-liner nylon filter to eliminate the gel fraction. Once semen collected, the volume was recorded, then was immediately diluted 1:5 (*v*/*v*) in a skim-milk-based semen extender (Kenney) [23] and allowed in 50 mL conical tubes. Morphological abnormalities and viability were determined by bright field microscopy (mag. ×1000) when examining 200 cells after smear staining with Eosin–Nigrosin. Total sperm concentrations were determined using a Neubauer Chamber (Paul Marienfeld GmbH and Co. KG, Lauda-Königshofen, Germany); counting was performed in triplicate using a phase-contrast microscope (mag. ×20).

2.3. Semen Dilution

The sperm concentration of semen diluted previously was reevaluated and adjusted to obtain two groups: the first one with a high concentration (80×10^6 spz/mL) and the second, low concentrated, (30×10^6 spz/mL) in order to analyze the effect of concentration on sperm motility.

2.4. Counting Chambers and Loading Technique

Five commercial counting chambers (all from Proiser R + D S.L., Paterna, Spain) were used: (1) three disposable, capillary loaded chambers ISAS®D4C having a fixed cover-slide attached by glue, ISAS®D4C10, ISAS®D4C20 (hereafter D4C10 and 20) of 10 and 20 µm depth (Figure 1a) and ISAS®D4C20L (hereafter D4CL20) of 20 µm depth (Figure 1b). (2) two reusable drops displacement chambers Spermtrack® (Proiser R + D, Paterna, Spain) having a separate cover slide, Spermtrack®10 and Spermtrack®20 (hereafter Spk10 and 20) of 10 and 20 µm depth (Figure 1c).

The samples were well homogenized just before being charged in the chambers. All chambers were loaded with the adequate technique and quantity (Figure 1) of semen as recommended by the manufacturer. When using the Spermtrack® chambers, the covers were rapidly but gently put in place to achieve a homogenous distribution of the sample. The other chambers were loaded by depositing the sperm sample in the loading area. Then, the sperm travels by capillarity into the different areas of the chamber. Chambers were maintained on a thermo-plate for 15 s at 37 °C to prevent heat shock and allowing the fluid to cease permitting correct observation. For the analysis, the order of chambers used was randomized to avoid the effect of incubation time.

To study the effect of the field location on sperm motility, 7 fields were identified longitudinally in D4C10 and 20 (Figure 1c). However, in D4C20L two captures were made in each zone A, B and C from the proximal position to the distal position as presented in Figure 1b. Finally, in Spermtrack® chamber different fields were captured from the center to the edges as shown in Figure 1c. An average of 500 sperm was captured per field.

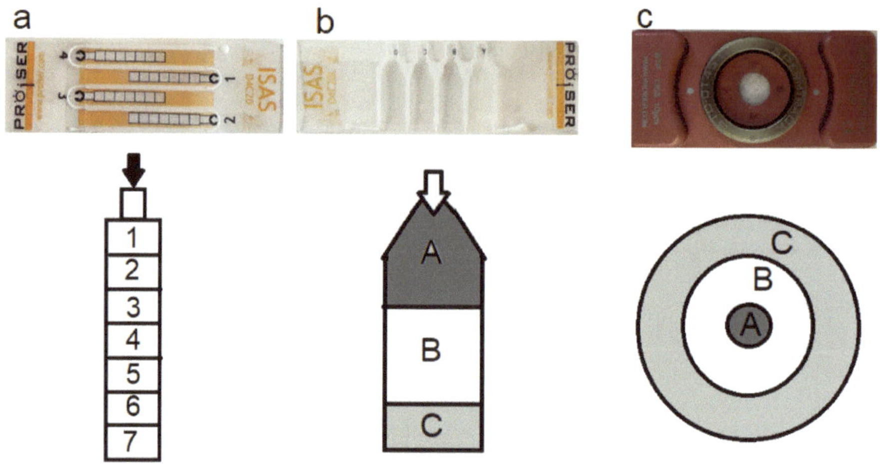

Figure 1. The different chambers types and forms used and the filed analyzed. (**a**): Disposable chamber D4C10, D4C20. (**b**): Disposable chamber D4C20L and (**c**): Reusable chambers SpK10, SpK 20. The arrow demonstrates the place of drop deposition. Numbers (1–7) and characters (A, B, C) inside the chambers demonstrate different capture fields.

2.5. CASA-Mot Analysis

Objective motility assessment was performed using ISAS® v1.2 CASA-Mot system (Proiser R + D, Paterna, Spain) combined with a UOP200i microscope (Proiser R + D) equipped with a negative phase contrast 10× objective (AN 0.25) and an MQ003MG- CM digital camera (Proiser R + D S.L.) which was capable of capturing a maximum frame rate of 500 per second. The final resolution was 0.48 μm/pixel in both x- and y-axis. The system was set with particle area between 4 and 70 μm^2 and connectivity set value of 6 μm. All samples analysis was performed by the same technician to avoid errors related and biases.

Sequences were captured at 500 fps and recorded during 3 s in different fields. For the study, these original videos were later segmented into 25, 50, 100, 150, 200 and 250 fps working videos, using the following command: [echo off: set fps = 25, 50, 75, 100, 150: for %%i in (.Ä*.avi) do (set fname = %%~ni) & call: encodeVideo; goto eof: encodeVideo: ffmpeg.exe -i%fname%.avi -r %fps% -clibx264 -preset slow -qp 0"%fname%_(%fps%fps).avi"; goto eof].

The following kinematic parameters were considered for this study: the sum of the distances between each measured sperm position divided by the analysis time (VCL, μm/s), the straight-line distance between the first and last sperm position divided by the analysis time (VSL, μm/s), the average path velocity is the time-averaged velocity of a sperm head along its average path (VAP, μm/s), the wobble is a measure of oscillation of the actual path about the average path (WOB = VAP/VCL, dimensionless), the straightness is a measure of the linearity of the average path (STR = VSL/VAP, dimensionless), the linearity of forward progression, the linearity of the curvilinear path (LIN = VSL/VCL, dimensionless), the average distance of the sperm head from the average sperm-swimming path where the average path (ALH, μm) and the beat cross frequency the number of lateral oscillatory movements of the sperm head around the mean trajectory (BCF, Hz).

2.6. Calculating the Optimum Frame Rate

Optimal frame rates were determined from the VCL of the sperms using point-to-point reconstructions of their trajectories at each tested FR. For this, the results were subjected to exponential regression analysis:

$$y = \beta.\alpha \exp(-\beta/x) \quad (1)$$

where y is the VCL, x is the FR, α is the asymptotic level, β is the rate of increase to the asymptote and exp based on the natural logarithm. The biological meaning of the formulae is that the asymptotic level (α) represents the maximum achievable when the FR is above the threshold value. The threshold level is conventionally calculated as the FR needed to obtain 95% of the maximum value. The rate of the approach to the asymptote represents the dependence on the curve on the FR; a higher value of β indicates high growth of the VCL as FR increases and vice versa. There is no substantial increase in the VCL with the increase of FR which represents α the asymptotic value (at least 95% of the maximum VCL has been achieved). The rate of approach to the asymptote represents the dependence of the curve on the FR, thus, a high β value indicates an increase in VCL with increasing FR and vice versa.

Although, the maximum image captured by the camera was 500 f/s. There is no software that can analyse the huge amount of data generated by capture frequency higher than 250 f/s. Therefore, given the OFR for all the studied chambers was either virtually identical to 250 f/s or >250 f/s. The following experiments were performed at 250 f/s for every chamber.

2.7. Statistical Analysis

The data obtained for the analysis of all sperm variables were first assessed for normality and homoscedasticity by using Shapiro–Wilk and Levene tests respectively. A normal probability plot was used to assess normal distribution. In trying to obtain a normal distribution, data were transformed using acrsine square root (acrsin$\sqrt{x}$) before repeated-measures ANOVA was run. An ANOVA was applied to evaluate statistical differences in the distributions of observation (individual spermatozoa) within disposable and reusable counting chambers and then a generalized linear model (GLM) procedure was used to determine the effects on the mean kinematic values defining the different fields of disposable counting chambers ISAS®D4C depth. Differences between means were analyzed by a Bonferroni test. The statistical model used was:

$$Xijk = \mu + Ai + Bj + AB\ (ij) + \varepsilon ijk \tag{2}$$

where Xijk = the measured sperm motility variable, μ = the overall mean of variable x, Ai = the effect of depth, Bj = the effect of the counting chamber; AB (ij) = the effect of the interaction depth-counting chamber; and εijk = the residual.

Results are presented as mean ± standard error of the mean (SEM) Statistical significance was considered at $p < 0.05$. All calculations were performed using the IBM SPSS V.23.0 package for Windows (IBM Inc., Chicago, IL, USA).

3. Results

3.1. Optimum Frame Rate in Different Chambers

The OFR was calculated for each chamber type (Figure 2). The higher OFR found was obtained by means of reusable chambers. However, OFR was higher in the chamber of 20 µm in depth compared to 10 µm in depth whatever the chamber type. Therefore, the highest value of the OFR was in Spk20 (278.4 fps) while the lowest was in D4C20L (225.3 fps) as presented in Table 1.

The OFR of all the chambers was between 225 f/s and 278 f/s. Thus, all the subsequent experiments were performed at 250 f/s for every chamber assuring a reliable measurement for sperm kinematic parameters in donkey semen.

Sperm trajectory showed a different form as the FR increase. At a higher FR, the trajectory showed a much higher oscillation that was not possible to appreciate at lower FR as shown in Figure 3.

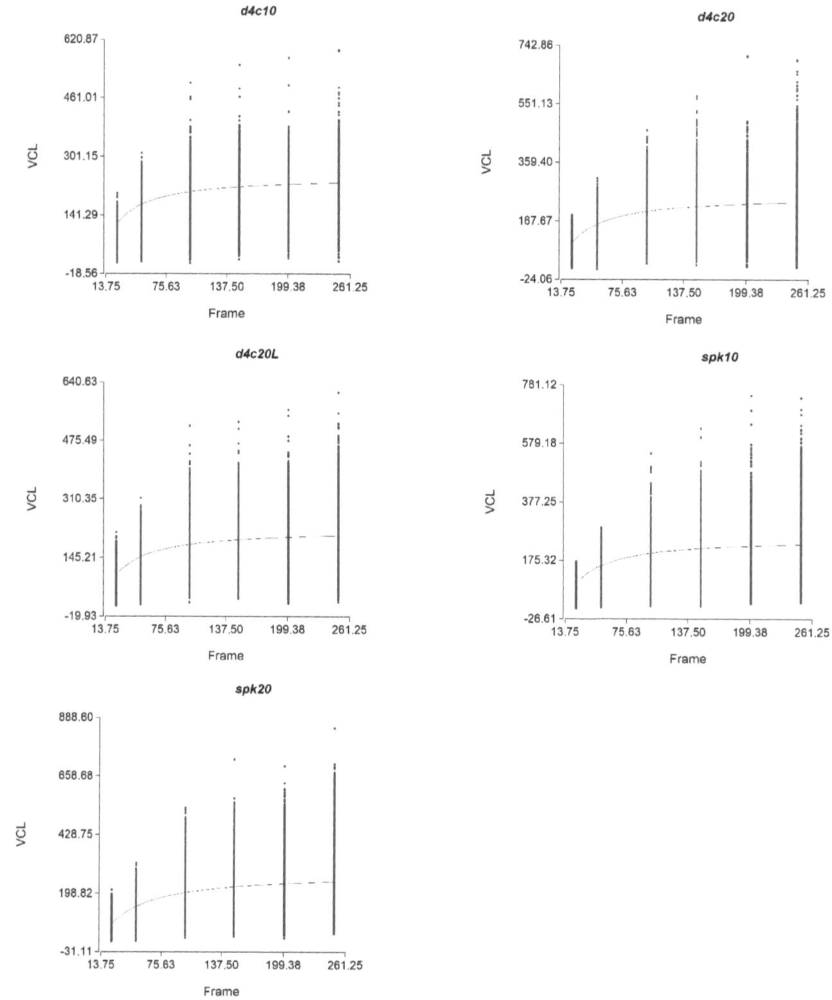

Figure 2. Sperm curvilinear velocity (VCL. μm/s) obtained at different frame rates in the different counting chambers.

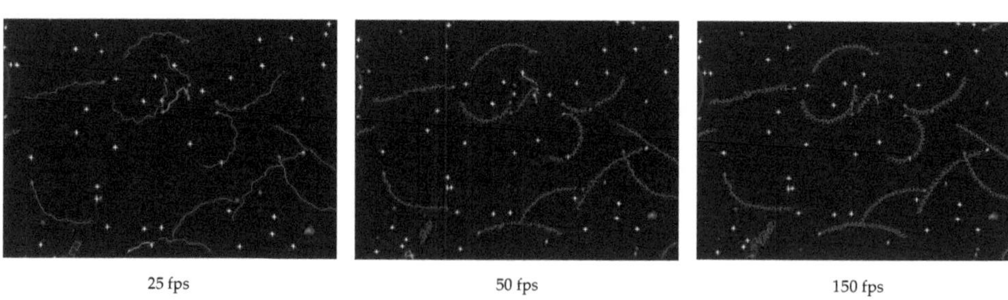

Figure 3. Donkey sperm motility tracks at different frames (25, 50 and 150), exhibiting 4 groups of spermatozoa velocity: rapid (red), medium (green), slow (blue) and static (yellow) in D4C10 chamber, fps = frames per second.

Table 1. Optimum frame rate calculated to have the threshold level for different chambers types and depths. Curvilineal velocity of donkey sperm for the optimal frame rate (VCLα) and with different frames rates (25–250) calculated based on α and β values.

Chamber Type	Chamber	*n*	α	SEMα	β	SEMβ	VCLα	VCL25	VCL50	VCL100	VCL150	VCL200	VCL250
Disposable													
	D4C10	14,451	247.6	0.97	18.43	0.31	229.79	118.46	171.26	205.92	218.97	225.8	230.00
	D4C20	64,732	252.96	0.62	23.64	0.25	230.30	98.26	157.65	199.7	216.07	224.75	230.13
	D4C20L	76,812	225.71	0.49	21.18	0.18	207.72	96.74	147.76	182.62	195.98	203.02	207.37
Reusable													
	Spk10	105,679	255.01	0.55	24.23	0.19	231.89	96.74	157.07	200.13	216.97	225.91	231.45
	Spk20	172,918	278.46	0.46	31.80	0.15	248.36	78.04	147.41	202.60	225.26	237.52	245.20

n: number of spermatozoa analyzed; α = asymptote of curvilinear velocity; β = rate of increase; SEM = standard error of the mean; VCL = curvilinear velocity (μm/s). D4C10, D4C20: Disposable chambers 10 μm and 20 μm depth. D4C20L: Leja disposable chamber 20 μm depth. Spk10, Spk20: Reusable chambers (Spermtrack®) 10 μm and 20 μm depth.

3.2. Effect of Chamber Type and Depth on Sperm Kinematic Parameters

The analysis of sperm kinematic parameters (VCL, VSL, VAP, LIN, STR, ALH) was generally characterized by a significantly higher value in reusable chambers compared to disposable chambers ($p < 0.05$). On the contrary, WOB and BCF were lower in reusable chamber Spk.

Then, when comparing the motility parameters in different depths for the same type and design chamber, a disposable D4C chamber of 10 µm depth reveals a higher value for the VAP, LIN, WOB and BCF rather than the D4C 20 µm depth chamber. Nevertheless, ISAS®D4C20L reusable chamber had the lowest value for all kinematic parameters except for WOB, STR and LIN which showed a higher value compared to D4C 20 µm (Table 2).

Table 2. Motility parameters (mean ± SEM) of donkey spermatozoa determined by CASA system using disposable and reusable chambers obtained at 250 frames.

Chambers	Disposable			Reusable	
	D4C	D4C	D4CL	Spk	Spk
Depth	10 µm	20 µm	20 µm	10 µm	20 µm
VCL	227.91 ± 1.19 ^{d}	238.16 ± 0.83 ^{c}	223.77 ± 0.77 ^{d}	260.47 ± 0.78 ^{b}	268.07 ± 0.54 ^{a}
VSL	70.19 ± 0.67 ^{c,d}	71.16 ± 0.45 ^{c}	68.93 ± 0.42 ^{d}	83.57 ± 0.45 ^{a}	80.46 ± 0.3 ^{b}
VAP	181.57 ± 0.8 ^{a}	177.18 ± 0.52 ^{b}	168.15 ± 0.5 ^{c}	183.8 ± 0.48 ^{a}	181.66 ± 0.33 ^{a}
LIN	29.82 ± 0.27 ^{a,b}	28.3 ± 0.18 ^{c}	29.32 ± 0.19 ^{b}	30.23 ± 0.18 ^{a}	28.46 ± 0.12 ^{a,b}
STR	37.67 ± 0.38 ^{d}	38.09 ± 0.25 ^{d}	39.06 ± 0.24 ^{c}	42.61 ± 0.23 ^{a}	41.47 ± 0.16 ^{b}
WOB	79.1 ± 0.19 ^{a}	73.72 ± 0.13 ^{c}	74.39 ± 0.15 ^{b}	70.19 ± 0.13 ^{d}	67.5 ± 0.09 ^{e}
ALH	1.17 ± 0.01 ^{e}	1.26 ± 0.01 ^{c}	1.22 ± 0.01 ^{d}	1.39 ± 0.01 ^{b}	1.44 ± 0.0049 ^{a}
BCF	42.58 ± 0.2 ^{a}	40.35 ± 0.14 ^{c}	38 ± 0.14 ^{d}	40.79 ± 0.12 ^{b}	39.96 ± 0.08 ^{c}

SEM: Standard Error of the Mean; VCL (µm/s) curvilinear velocity; VSL (µm/s) straight line velocity; VAP (µm/s) average path velocity; LIN (%) linearity; STR (%) straightness; WOB (%) wobble; ALH (µm) amplitude of lateral head displacement; BCF (Hz) beat-cross frequency. Disposable chambers D4C 10 µm and 20 µm depth. Reusable chambers. Spk 10 µm and 20 µm depth. ^{a,b,c,d} Within columns, rates with different superscripts differed ($p < 0.05$).

3.3. Effect of the Capture Field Inside the Counting Chamber

In D4C10, there was no clear tendency observed in the kinematic parameters as the spermatozoids travel from the point of deposition to the last field. However, the velocities (VCL, VSL, VAP) and BCF were higher in the last field while LIN increased along the counting way (form field 1 to 7) (Table 3).

The same results were found in the D4C20 chamber; we can observe an oscillatory change along the counting way. However, we observed a reduction in the velocity value (VCL, VSL, VAP) and the linearity in the last field (Table 3).

Regarding the D4C20L chamber, the highest values for all kinematic parameters (exception from the WOB) were observed closer to the point of deposition. However, the sperm had the lowest kinematic values in the middle of the chamber. The only value that showed no changes in the three counting zones was the BCF (Table 4).

Finally, in the reusable chambers, Spk spermatozoa showed similar behavior for both depths. In both Spk10 and 20 the sperm velocities (VCL, VSL, VAP) and ALH, values decreased significantly from the centre to the edges, while WOB and BCF increased ($p < 0.05$). Yet, VAP and LIN showed the highest value in the second counting ring in Skp20 (Table 5).

Table 3. CASA motility parameters (mean ± SEM) in different fields of disposable chamber ISAS®D4C10 and 20 µm depth.

Chamber	VCL	VSL	VAP	LIN	STR	WOB	ALH	BCF
D4C10								
1	237.0 ± 3.3 ^{a}	71.96 ± 1.69 ^{b}	186.6 ± 2.71 ^{a}	28.94 ± 0.56 ^{b,c}	37.37 ± 0.71 ^{b,c}	77.73 ± 0.39 ^{b}	1.20 ± 0.01 ^{a}	41.26 ± 0.67 ^{a}
2	226.9 ± 3.1 ^{a,b,c}	66.70 ± 1.59 ^{b}	180.39 ± 2.56 ^{a,b}	28.03 ± 0.53 ^{c}	35.51 ± 0.67 ^{c}	78.79 ± 0.37 ^{a}	1.18 ± 0.01 ^{a}	41.62 ± 0.63 ^{a}
3	218.56 ± 4.22 ^{c,d}	66.62 ± 2.16 ^{b}	177.59 ± 3.47 ^{a,b,c}	29.63 ± 0.72 ^{b,c}	36.63 ± 0.91 ^{b,c}	80.53 ± 0.5 ^{a}	1.14 ± 0.01 ^{a}	42.02 ± 0.86 ^{a}
4	233.54 ± 3.11 ^{a,b}	71.18 ± 1.6b	183.96 ± 2.56 ^{a,b}	29.66 ± 0.53 ^{b}	37.63 ± 0.67 ^{b}	78.58 ± 0.37 ^{a,b}	1.2 ± 0.01 ^{a}	41.99 ± 0.63 ^{a}
5	218.26 ± 4.22 ^{c,d}	67.45 ± 2.16 ^{b}	177.22 ± 3.47 ^{a,b,c}	30.46 ± 0.72 ^{a,b}	37.76 ± 0.91 ^{b}	80.63 ± 0.5 ^{a}	1.14 ± 0.01 ^{a}	42.34 ± 0.86 ^{a}
6	212.65 ± 4.09 ^{d}	69.66 ± 2.1 ^{b}	170.19 ± 3.37 ^{c}	31.97 ± 0.7 ^{a}	40.27 ± 0.88 ^{a}	79.41 ± 0.49 ^{a}	1.12 ± 0.01 ^{a}	40.91 ± 0.83 ^{a}
7	252.96 ± 16.8 ^{a}	93.72 ± 8.61 ^{a}	198.38 ± 13.83 ^{a}	35.64 ± 2.87 ^{a}	45.5 ± 3.61 ^{a}	78.06 ± 1.99 ^{a,b}	1.23 ± 0.06 ^{a}	43.78 ± 3.42 ^{a}
D4C20								
1	238.94 ± 2.64 ^{b}	68.96 ± 1.32 ^{c}	171.01 ± 2.04 ^{b}	26.99 ± 0.4 ^{c}	37.94 ± 0.51 ^{b}	70.54 ± 0.27 ^{c}	1.29 ± 0.01 ^{b}	38.41 ± 0.48 ^{c}
2	215.94 ± 2.58 ^{c}	65.06 ± 1.29 ^{d}	162.94 ± 1.99 ^{c}	28.2 ± 0.39 ^{b}	37.66 ± 0.5 ^{b}	74.28 ± 0.26 ^{a}	1.17 ± 0.01 ^{d}	36.79 ± 0.46 ^{d}
3	241.81 ± 2.51 ^{b}	70.26 ± 1.26 ^{b,c}	180.81±1.94 ^{a}	27.7 ± 0.38 ^{b,c}	37 ± 0.48 ^{b}	74.31 ± 0.26 ^{a}	1.27 ± 0.01 ^{b}	40.52 ± 0.45 ^{b}
4	243.05 ± 2.38 ^{b}	72.97 ± 1.19 ^{b}	182.02±1.84 ^{a}	28.35 ± 0.36 ^{b}	37.88 ± 0.46 ^{b}	74.47 ± 0.24 ^{a}	1.27 ± 0.01 ^{b}	40.88 ± 0.43 ^{b}
5	244.19 ± 2.71 ^{b}	73.77 ± 1.36 ^{b}	182.1±2.09 ^{a}	28.7 ± 0.4 ^{b}	38.39± 0.52 ^{b}	73.99 ± 0.28 ^{a}	1.28 ± 0.01 ^{b}	39.7 ± 0.49 ^{b,c}
6	253.55 ± 3.51 ^{a}	78.47 ± 1.76 ^{a}	184.84±2.72 ^{a}	30.13 ± 0.53 ^{a}	41.1 ± 0.68 ^{a}	72.75 ± 0.36 ^{b}	1.33 ± 0.01 ^{a}	41.51 ± 0.63 ^{a}
7	222.69 ± 3.91 ^{c}	63.82 ± 1.96 ^{d}	169.34 ± 3.02 ^{b,c}	26.61 ± 0.59 ^{c}	34.95 ± 0.75 ^{c}	74.77 ± 0.4 ^{a}	1.22 ± 0.01 ^{c}	38.54 ± 0.71 ^{c}

SEM: Standard error of the mean; VCL (µm/s) curvilinear velocity; VSL (µm/s) straight line velocity; VAP (µm/s) average path velocity; LIN (%) linearity; STR (%) straightness; WOB (%) wobble; ALH (µm) amplitude of lateral head displacement; BCF (Hz) beat-cross frequency. Disposable chambers D4C 10 µm and 20 µm depth. Reusable chambers. Different letters of the alphabet within columns indicate significant differences between different fields ($p < 0.05$).

Table 4. CASA motility parameters (mean ± SEM) in different fields of disposable chamber ISAS®D4C20L.

Capture Fields	VCL	VSL	VAP	LIN	STR	WOB	ALH	BCF
A	227.53 ± 1.89 ^{a}	70.55 ± 0.95 ^{a}	169.65 ± 1.49 ^{a}	29.54 ± 0.3 ^{a}	39.71 ± 0.38 ^{a}	74.03 ± 0.19 ^{b}	1.24 ± 0.01 ^{a}	38.6 ± 0.37 ^{a}
B	219.63 ± 1.65 ^{b}	65.43 ± 0.83 ^{b}	164.74 ± 1.3 ^{b}	28.28 ± 0.26 ^{b}	37.83 ± 0.33 ^{b}	74.12 ± 0.17 ^{b}	1.21 ± 0.01 ^{b}	37.68 ± 0.32 ^{a}
C	222.51 ± 1.5 ^{b}	69.13 ± 0.76 ^{a}	167.8 ± 1.19 ^{a,b}	29.52 ± 0.24 ^{a}	39.1 ± 0.3 ^{a}	74.58 ± 0.16 ^{a}	1.21 ± 0.01 ^{b}	37.91 ± 0.29 ^{a}

SEM: Standard error of the mean; VCL (µm/s) curvilinear velocity; VSL (µm/s) straight line velocity; VAP (µm/s) average path velocity; LIN (%) linearity; STR (%) straightness; WOB (%) wobble; ALH (µm) amplitude of lateral head displacement; BCF (Hz) beat-cross frequency. Different letters of the alphabet within columns indicate significant differences between different fields ($p < 0.05$). Capture fields for semen analysis proximal (A), medium (B), distal (C) as described in Figure 1.

Table 5. Mean (±SEM) of sperm kinetic parameters (VCL, VSL, VAP, LIN, STR, WOB, ALH and BCF) in donkey semen in reusable chamber Spermtrack® (SpK) 10 and 20 µm depth.

Chamber and Capture Field	VCL	VSL	VAP	LIN	STR	WOB	ALH	BCF
SpK10								
A	274.11 ± 2.31 [a]	88.37 ± 1.28 [a]	188.94 ± 1.69 [a]	30.33 ± 0.36 [a,b]	43.8 ± 0.48 [a]	68.53 ± 0.22 [b]	1.45 ± 0.01 [a]	41.35 ± 0.39 [a]
B	257.82 ± 1.24 [b]	81.41 ± 0.69 [b]	182.23 ± 0.91 [b]	29.6 ± 0.19 [b]	41.61 ± 0.26 [b]	70.3 ± 0.12 [a]	1.38 ± 0.0045 [b]	39.84 ± 0.21 [b]
C	257.6 ± 1.63 [b]	83.48 ± 0.91 [b]	182.8 ± 1.2 [b]	30.69 ± 0.25 [a]	43.07 ± 0.34 [a]	70.53 ± 0.15 [a]	1.37 ± 0.01 [b]	40.3 ± 0.28 [b]
SpK20								
A	271.45 ± 1.76 [a]	82.66 ± 0.93 [a]	177.73 ± 1.25 [b]	28.08 ± 0.26 [b]	42.47 ± 0.34 [a]	64.84 ± 0.15 [c]	1.47 ± 0.01 [a]	38.54 ± 0.29 [b]
B	271.84 ± 1.11 [a]	81.69 ± 0.59 [a]	183.81 ± 0.79 [a]	28.71 ± 0.16 [a]	41.89 ± 0.22 [a]	67.51 ± 0.1 [b]	1.45 ± 0.0041 [b]	39.42 ± 0.18 [a]
C	263.56 ± 1.05 [b]	77.72 ± 0.56 [b]	180.56 ± 0.74 [b]	27.94 ± 0.15 [b]	40.27 ± 0.21 [b]	68.18 ± 0.09 [a]	1.42 ± 0.0038 [c]	39.54 ± 0.17 [a]

SEM: Standard error of the mean; VCL (µm/s) curvilinear velocity; VSL (µm/s) straight line velocity; VAP (µm/s) average path velocity; LIN (%) linearity; STR (%) straightness; WOB (%) wobble; ALH (µm) amplitude of lateral head displacement; BCF (Hz) beat-cross frequency. Different letters of the alphabet within columns indicate significant differences between different fields ($p < 0.05$). Capture fields for semen analysis central field (A), medium field (B), border field (C) as described in Figure 1.

4. Discussion

In the present study, we attempted to standardize the technique used for the motility assessment of donkey sperm using high-resolution cameras and specific chambers.

The first step was to define the OFR for sperm kinematic analysis. Our work evidenced a direct positive relationship between capture frequency and VCL. Thus, capture frequency is the number of captures that permit the tracking of the sperm from one dot to another, hence we can reconstruct the sperm trajectory while VCL is the speed of the sperm (sperm trajectory/time) and dependent on sperm trajectory. As demonstrated in this study using non-linear regression, we could calculate the maximum frames needed, defined as OFR representing the threshold (α) where more captures will lead to the same sperm path. The OFR for the donkey sperm was different between chambers with a higher value in Spermtrack® (278 fps) having a VCL of 248.36 μm/s. The lower frames were needed in the D4C20L chamber (225.7 fps). It has been noted in other studies in horses that the calculated OFR (309 fps) with a VCL is remarkably superior compared to the donkey (not published). Contrarily to what was found with lower frames (25 fps), the VCL was higher in donkeys compared to the horse [22]. This can be explained by the loss of information with a lower frame. It is believed that donkey sperm is faster than a stallion, but the current study showed less VCL with higher VSL and a straighter trajectory in donkey than the horse. Thus, this explains the lower OFR found in the horse compared to the donkey. Additionally, in recent studies, while using a high-performance camera and the same study design, depending on each species that the sperm shows non-linear trajectory will need more captures to define the correct track. This explains the differences obtained in the calculation of the OFR in the bull (256 fps) [24], the boar (212 fps) [25], and the salmon (250 fps) [26].

It was also found that the OFR is dependent on the chamber type and depth. In fact, the sperm motion or the flagellar beat is affected by the space where it is placed. In a similar study designed for bulls [24], this was also found in a different frame for different chambers.

The second step was to analyse the sperm behavior in different chamber types. As we can observe, the kinematic parameters were high in the reusable chamber compared to disposable chambers, except that the BCF and WOB values were higher in the disposable chamber. In fact, the same results were observed for VC, VSL and VAP on different species like horses, bulls, boars and bucks. Yet, the other parameters were variable depending on the species [24,27–32]. Those differences could be explained by the loading technique that could generate physical forces on the spermatozoa affecting its motility. As well, it was showed the existence of a certain interaction between sperm and ions of the glass mounted on the chamber, which would be toxic in certain species [14]. Furthermore, the chamber's designs can exercise a certain force and affect sperm movement as seen in the D4C20L chamber where donkey sperm showed the lowest values for all kinematic parameters compared to other chambers which could be due to turbulence created inside this chamber [14]. The same results were found for the bull [33]. All these effects, as we can see, are specific to each species. So, it is essential to take these parameters into account for each species when choosing the chamber for sperm motility analysis.

The results of our study additionally demonstrated that the depth of the counting chamber influenced kinematic values. In fact, comparing the chambers of the same type for a different depth, we can observe that donkey sperm had greater VAP, VSL, LIN, STR, WOB and BCF in a narrow chamber of 10 μm rather than 20 μm chamber while VCL and ALH were greater in 20um chamber. The same result was found in other species such as the goat [27]. In the Belgian Blue bull and the Limousine bull breed [24], D4C10, 20 μm depths were used. This kind of linear motility movement is defined by low lateral amplitude and a high straight-line velocity. In actuality, it was found that this movement was essential for the sperm migration from the cervix to the uterus and then to the oviduct [34,35]. Moreover, this movement was also observed in the seminal plasma and the uterine fluids, with a low concentration of glucose. However, a study on the boar [36], the Holstein bull breed [24] and the stallion [29] determined that the sperm was faster (high VCL, VSL and VAP) with higher index values in the deepest chamber 20 μm. Yet, it has also been suggested that sperm kinematics are unaffected by the chamber depth [13]. Looking at the studies referred to above, it was observed that the frame rate setting and the chambers used to achieve different depths were not identical in all other respects. Therefore, differences in sperm kinematics could not be attributed solely to a chamber depth. A study conducted by [36], using lensless microscopy, showed that the kinetic parameters of boar semen in the deepest chambers of 100 μm increase significantly. Considering all these parameters discussed previously, it was demonstrated that the sperm behave differently depending on the chamber's depth and the species. It is important to contemplate that there is a possibility that in the OFR with a higher chamber depth (more than the size of the sperm), the sperm will show different behavior.

To the best of our knowledge, this is the first work that studies the effect of the chamber field in the analysis of donkey sperm. The semen analysis demonstrated that the zone analysis in the chamber had a significant effect on the kinematic outcomes. The analysis in D4C10 rectangular chamber showed that the sperm increased in velocity as the drop moved to the last field. However, it was observed that the sperm lost its linearity. On the contrary, it was demonstrated that in the D4C20 chamber, the sperm showed opposite comportment for the velocity parameters. The same outcome was found in the fox samples [37] using the same chambers. It was suggested that the Segre–Silderberg effect altered the sperm movement as a consequence of the hydrodynamic drive of the fluid within the capillary-loaded chambers [38]. In this case, it affected the sperm tail [30,32] and the vitality [33]. In fact, we can observe that the results were emphasised in a narrow chamber when observing the sperm of the donkey. In the circle form chamber, Spermtrack®, the drop is moved by the force of the coverslip. In our study, it was found that when the drop moves from the centre to the edges, the motility decreased in both chambers' depth, but the linearity increased. This result could be explained by the force of the fluid in the centre, which is generated by the cover that moves the sperm forward increasing the linearity but affecting the velocity. A similar result was found in the ram using a slide-coverslip [27,32]. However, [39] when using a Makler® for the bull sperm, no differences between the centre and the edges could be determined [39]. As we can see, this could implicate the biology or metabolism of each species which behave differently [40].

Finally, it is important to also consider the dilution rate which had a direct effect on sperm motility parameters as observed in our study. The sperm velocity (VAP and VCL) and the progressiveness (BCF, STR and LIN) values were higher in the high-density sample (80×10^6 sperm/mL) compared to a low density (30×10^6 sperm/mL) sample for all the chambers. As previously reported in many studies, the semen concentration could affect sperm motility parameter values recorded by CASA [41]. The same results were found for different studies in dogs [42] and bulls [43] with a higher initial sperm concentration resulted in an increase in sperm velocity (VAP, VSL and VCL) and progressiveness. This difference can be related to the "dilution effect" referring to the detrimental effect on sperm quality, motility and resistance to a cold shock when adding a high volume of diluent of raw semen. The dilution effect was observed in a low sperm concentration (i.e., $<20 \times 10^6$ sperm/mL) and demonstrated in various species [44–46]. Nevertheless, a study on horses [45,47,48] and humans [49] using a lower sperm density (2.5×10^6 sperm/mL) revealed higher motility parameters due to the minor effect that the "dilution effect" had on the sperm motility comparatively to other species. We propose a concentration of 80×10^6 sperm/mL for further experiments in the study of the donkey when using a narrow chamber of 10 μm and especially in a disposable chamber. Usually, at high concentrations, the sperm aggregates affecting the CASA system results. This was reported for a concentration higher than 100×10^6 sperm/mL [32]. Yet, in the present study, with a concentration of 80×10^6 sperm/mL, the videos were clear, and the trajectories were defined correctly. On the other hand, in reusable chambers, we recommend using a lower concentration since the concentration of 80×10^6 sperm/mL was too dense. Nonetheless, when using the playback facility for sperm tracking, the result showed a wrong trajectory reconstruction due to erroneous head detection in following frames, collision, and cross-tracks. Essentially, a different sperm concentration was proposed for CASA evaluation for different species [43,50,51] such as the horse where the concentration used is between (25×10^6 and 50×10^6 sperm/mL) [46,52]. The differences in the sperm concentration suggested for the analysis, in different studies, that it is likely due to the different usage of CASA devices and chambers and it is directly related to the species-specificities.

In the present study, a classical skim-milk extender (Kenney extender) was used. Rota et al. [53], by means of a previous CASA system, observed an effect of extender on sperm motility patterns of Amiata donkey spermatozoa. Further studies are needed to better understand the changes in motility patterns of donkey spermatozoa caused by semen extenders, with different composition and fluidity and using new CASA devices and chambers.

5. Conclusions

Each species, including horse and donkey, has its own sperm motility patterns; as a result, each species needs its own CASA system analysis conditions. These conditions are changing according to the constant improvement of related technologies. Current and new high-resolution video-cameras, informatic software and hardware-increased capacities (but limited) or new specific chambers to analyse sperm motility are changing previously defined sperm motility patterns. Then, to define the OFR, the kinds and depths of analysis chambers or the sample dilution are very important to more accurately describe the species-specific sperm motility. Thus, when examining the motility of donkey sperm, the 250 fps in Spk10 chamber and analyzing a minimum of nine fields considering all the capture area (centre and edges) and means determined at a concentration 30×10^6 sperm/mL represents an excellent choice.

Author Contributions: Conceptualization. C.S. and J.M.; methodology. S.G., J.C. and J.M.; software. C.S.; validation. C.S. and J.M.; formal analysis. A.V.; investigation. S.G., J.C. and J.M.; data curation. S.G., C.S.; writing—original draft preparation. S.G.; writing—review and editing. S.G., C.S., J.M.; visualization. C.S. and J.M.; supervision. C.S. and J.M. All authors have read and agreed to the published version of the manuscript.

Funding: This work was funded by the Equine Reproduction Service, Department of Animal Medicine and Surgery, Faculty of Veterinary Medicine, Autonomous University of Barcelona. Spain.

Acknowledgments: We are grateful to Peggy Le Bleguet for English correction of the manuscript.

Conflicts of Interest: The authors declare no conflict of interest. Some of the authors (C.S. and D.B.) are employees for the company that manufacture the CASA-Mot system and the counting chambers used or collaborate through a collaborative agreement between the València University and the company that manufacture some of the components used here. In any case, it does not imply any conflict of interest, because no publicity (direct or indirect) is done in the manuscript. These components were just used in the same way we could use from other brands and the elements studied are not related with the brand but with physic and biological topics.

References

1. Carneiro, G.F.; Lucena, J.E.C.; Barros, L.D.O. The Current Situation and Trend of the Donkey Industry in South America. *J. Equine Vet. Sci.* **2018**, *65*, 106–110. [CrossRef]
2. Faccia, M.; D'Alessandro, A.G.; Summer, A.; Hailu, Y. Milk products from minor dairy species: A review. *Animals* **2020**, *10*, 1260. [CrossRef] [PubMed]
3. Camillo, F.; Rota, A.; Biagini, L.; Tesi, M.; Fanelli, D.; Panzani, D. The Current Situation and Trend of Donkey Industry in Europe. *J. Equine Vet. Sci.* **2018**, *65*, 44–49. [CrossRef]
4. Yılmaz, O.; Boztepe, S.; Ertuğrul, M. The domesticated donkey: III-economic importance, uncommon usages, reproduction traits, genetics, nutrition and health care. *Can. J. Appl. Sci.* **2012**, *3*, 320–338.
5. Varner, D.D.D. Developments in stallion semen evaluation. *Theriogenology* **2008**, *70*, 448–462. [CrossRef]
6. Giaretta, E.; Munerato, M.; Yeste, M.; Galeati, G.; Spinaci, M.; Tamanini, C.; Mari, G.; Bucci, D. Implementing an open-access CASA software for the assessment of stallion sperm motility: Relationship with other sperm quality parameters. *Anim. Reprod. Sci.* **2017**, *176*, 11–19. [CrossRef]
7. Colenbrander, B.; Gadella, B.; Stout, T. The Predictive Value of Semen Analysis in the Evaluation of Stallion Fertility. *Reprod. Domest. Anim.* **2003**, *38*, 305–311. [CrossRef]
8. McCue, P.M. Breeding Soundness Evaluation of the Stallion. In *Equine Reproductive Procedures*; John Wiley & Sons, Inc: Hoboken, NJ, USA, 2014; pp. 319–324.

9. Whitesell, K.; Stefanovski, D.; McDonnell, S.; Turner, R. Evaluation of the effect of laboratory methods on semen analysis and breeding soundness examination (BSE) classification in stallions. *Theriogenology* **2020**, *142*, 67–76. [CrossRef]
10. Broekhuijse, M.L.W.J.; Šostarić, E.; Feitsma, H.; Gadella, B.M. Additional value of computer assisted semen analysis (CASA) compared to conventional motility assessments in pig artificial insemination. *Theriogenology* **2011**, *76*, 1473–1486. [CrossRef]
11. Holt, W.V.; Cummins, J.M.; Soler, C. Computer-assisted sperm analysis and reproductive science; a gift for understanding gamete biology from multidisciplinary perspectives. *Reprod. Fertil. Dev.* **2018**, *30*, 3–5. [CrossRef]
12. Yániz, J.L.; Silvestre, M.A.; Santolaria, P.; Soler, C. CASA-Mot in mammals: An update. *Reprod. Fertil. Dev.* **2018**, *30*, 799–809. [CrossRef]
13. Canisso, I.F.; Panzani, D.; Miró, J.; Ellerbrock, R.E. Key Aspects of Donkey and Mule Reproduction. *Vet. Clin. N. Am. Equine Pract.* **2019**, *35*, 607–642. [CrossRef]
14. Bompart, D.; García-Molina, A.; Valverde, A.; Caldeira, C.; Yániz, J.; De Murga, M.N.; Soler, C. CASA-Mot technology: How results are affected by the frame rate and counting chamber. *Reprod. Fertil. Dev.* **2018**, *30*, 810–819. [CrossRef]
15. Contri, A.; Gloria, A.; Robbe, D.; De Amicis, I.; Carluccio, A. Characteristics of donkey spermatozoa along the length of the epididymis. *Theriogenology* **2012**, *77*, 166–173. [CrossRef]
16. Quartuccio, M.; Marino, G.; Zanghì, A.; Garufi, G.; Cristarella, S. Testicular Volume and Daily Sperm Output in Ragusano Donkeys. *J. Equine Vet. Sci.* **2011**, *31*, 143–146. [CrossRef]
17. Carluccio, A.; Panzani, S.; Contri, A.; Bronzo, V.; Robbe, D.; Veronesi, M.C. Influence of season on testicular morphometry and semen characteristics in Martina Franca jackasses. *Theriogenology* **2013**, *79*, 502–507. [CrossRef]
18. Gacem, S.; Papas, M.; Catalan, J.; Miró, J. Examination of jackass (*Equus asinus*) accessory sex glands by B-mode ultrasound and of testicular artery blood flow by colour pulsed-wave Doppler ultrasound: Correlations with semen production. *Reprod. Domest. Anim.* **2020**, *55*, 181–188. [CrossRef] [PubMed]
19. Moustafa, M.N.K.; Sayed, R.; Zayed, A.E.; Abdel-Hafeez, H.H. Morphological and Morphometric Study of the Development of Seminiferous Epithelium of Donkey (*Equus asinus*) from Birth to Maturity. *J. Cytol. Histol.* **2015**, *6*, 1.
20. Rota, A.; Puddu, B.; Sabatini, C.; Panzani, D.; Lainé, A.L.; Camillo, F. Reproductive parameters of donkey jacks undergoing puberty. *Anim. Reprod. Sci.* **2018**, *192*, 119–125. [CrossRef]
21. Miró, J.; Flotats, A.; Rivera, M.; Ocaña, M.; Taberner, E.; Peña, A.; Rigau, T. OC3 Morphometry Characterisation of Catalan Donkey Spermatozoa and Identification of Sperm Morphometric Subpopulations. *Reprod. Domest. Anim.* **2006**, *41*, 103. [CrossRef]
22. Miró, J.; Lobo, V.; Quintero-Moreno, A.; Medrano, A.; Peña, A.; Rigau, T. Sperm motility patterns and metabolism in Catalonian donkey semen. *Theriogenology* **2005**, *63*, 1706–1716. [CrossRef] [PubMed]
23. Kenney, R.M. Minimal contamination techniques for breeding mares: Techniques and priliminary findings. *Proc. Am. Assoc. Equine Pract.* **1975**, *21*, 327–336.
24. Bompart, D.; Vázquez, R.F.; Gómez, R.; Valverde, A.; Roldán, E.R.S.; García-Molina, A.; Soler, C. Combined effects of type and depth of counting chamber, and rate of image frame capture, on bull sperm motility and kinematics. *Anim. Reprod. Sci.* **2019**, *209*, 106169. [CrossRef]
25. Valverde, A.; Madrigal, M.; Caldeira, C.; Bompart, D.; de Murga, J.N.; Arnau, S.; Soler, C. Effect of frame rate capture frequency on sperm kinematic parameters and subpopulation structure definition in boars, analysed with a CASA-Mot system. *Reprod. Domest. Anim.* **2019**, *54*, 167–175. [CrossRef]
26. Caldeira, C.; Hernández-Ibáñez, S.; Valverde, A.; Martin, P.; Herranz-Jusdado, J.G.; Gallego, V.; Asturiano, J.F.; Dzyuba, B.; Pšenička, M.; Soler, C. Standardization of sperm motility analysis by using CASA-Mot for Atlantic salmon (*Salmo salar*), European eel (*Anguilla anguilla*) and Siberian sturgeon (*Acipenser baerii*). *Aquaculture* **2019**, *502*, 223–231. [CrossRef]
27. Del Gallego, R.; Sadeghi, S.; Blasco, E.; Soler, C.; Yániz, J.L.L.; Silvestre, M.A.A. Effect of chamber characteristics, loading and analysis time on motility and kinetic variables analysed with the CASA-mot system in goat sperm. *Anim. Reprod. Sci.* **2017**, *177*, 97–104. [CrossRef] [PubMed]

28. Basioura, A.; Tsousis, G.; Boscos, C.; Lymberopoulos, A.; Tsakmakidis, I. Method agreement between three different chambers for comparative boar semen computer-assisted sperm analysis. *Reprod. Domest. Anim.* **2019**, *54*, 41–45. [PubMed]
29. Hoogewijs, M.K.; De Vliegher, S.P.; Govaere, J.L.; De Schauwer, C.; De Kruif, A.; Van Soom, A. Influence of counting chamber type on CASA outcomes of equine semen analysis. *Equine Vet. J.* **2012**, *44*, 542–549. [CrossRef]
30. Lenz, R.W.; Kjelland, M.E.; Vonderhaar, K.; Swannack, T.M.; Moreno, J.F. A comparison of bovine seminal quality assessments using different viewing chambers with a computer-assisted semen analyzer. *J. Anim. Sci.* **2011**, *89*, 383–388. [CrossRef]
31. Valverde, A.; Arnau, S.; García-Molina, A.; Bompart, D.; Campos, M.; Roldán, E.; Soler, C. Dog sperm swimming parameters analysed by computer-assisted semen analysis of motility reveal major breed differences. *Reprod. Domest. Anim.* **2019**, *54*, 795–803. [CrossRef]
32. Palacín, I.; Vicente-Fiel, S.; Santolaria, P.; Yániz, J.L. Standardization of CASA sperm motility assessment in the ram. *Small Rumin. Res.* **2013**, *112*, 128–135. [CrossRef]
33. Gloria, A.; Carluccio, A.; Contri, A.; Wegher, L.; Valorz, C.; Robbe, D. The effect of the chamber on kinetic results in cryopreserved bull spermatozoa. *Andrology* **2013**, *1*, 879–885. [CrossRef]
34. Suarez, S.S.; Pacey, A.A. Sperm transport in the female reproductive tract. *Hum. Reprod. Update* **2006**, *12*, 23–37. [CrossRef] [PubMed]
35. Shalgi, R.; Smith, T.T.; Yanagimachi, R. A Quantitative Comparison of the Passage of Capacitated and Uncapacitated Hamster Spermatozoa through the Uterotubal Junction. *Biol. Reprod.* **1992**, *46*, 419–424. [CrossRef]
36. Soler, C.; Picazo-Bueno, J.; Micó, V.; Valverde, A.; Bompart, D.; Blasco, F.J.; Álvarez, J.G.; García-Molina, A. Effect of counting chamber depth on the accuracy of lensless microscopy for the assessment of boar sperm motility. *Reprod. Fertil. Dev.* **2018**, *30*, 924–934. [CrossRef]
37. Soler, C.; García, A.; Contell, J.; Segervall, J.; Sancho, M. Kinematics and subpopulations' structure definition of blue fox (*Alopex lagopus*) sperm motility using the ISAS®V1 CASA system. *Reprod. Domest. Anim.* **2014**, *49*, 560–567. [CrossRef] [PubMed]
38. Douglas-Hamilton, D.H.; Smith, N.G.; Kuster, C.E.; Vermeiden, J.P.W.; Althouse, G.C. Capillary-loaded particle fluid dynamics: Effect on estimation of sperm concentration. *J. Androl.* **2005**, *26*, 115–122.
39. Valverde, A.; Areán, H.; Fernández, A.; Bompart, D.; García-Molina, A.; López-Viana, J.; Soler, C. Combined effect of type and capture area of counting chamber and diluent on Holstein bull sperm kinematics. *Andrologia* **2018**, *51*, 1–10. [CrossRef]
40. Dresdner, R.D.; Katz, D.F. Relationships of Mammalian Sperm Motility and Morphology to Hydrodynamic Aspects of Cell Function1. *Biol. Reprod.* **1981**, *25*, 920–930. [CrossRef]
41. Spiropoulos, J. Computerized semen analysis (CASA): Effect of semen concentration and chamber depth on measurements. *Arch. Androl.* **2001**, *46*, 37–42. [CrossRef]
42. Rijsselaere, T.; Van Soom, A.; Maes, D. Effect of technical settings on canine semen motility parameters measured by the Hamilton-Thorne analyzer. *Theriogenology* **2003**, *60*, 1553–1568. [CrossRef]
43. Contri, A.; Valorz, C.; Faustini, M.; Wegher, L.; Carluccio, A. Effect of semen preparation on casa motility results in cryopreserved bull spermatozoa. *Theriogenology* **2010**, *74*, 424–435. [CrossRef] [PubMed]
44. Makler, A. Sealed mini-chamber of variable depth for direct observation and extended evaluation of sperm motility under the influence of various gases. *Hum. Reprod.* **1991**, *6*, 1275–1278. [CrossRef]
45. Buss, T.; Aurich, J.; Aurich, C. Evaluation of a portable device for assessment of motility in stallion semen. *Reprod. Domest. Anim.* **2019**, *54*, 514–519. [CrossRef]
46. Iguer-ouada, M.; Verstegen, J.P. Evaluation of the "Hamilton thorn computer-based automated system" for dog semen analysis. *Theriogenology* **2019**, *55*, 733–749. [CrossRef]
47. Hayden, S.S.; Blanchard, T.L.; Brinsko, S.P.; Varner, D.D.; Hinrichs, K.; Love, C.C. Theriogenology The "dilution effect" in stallion sperm. *Theriogenology* **2015**, *83*, 772–777. [CrossRef] [PubMed]
48. Varner, D.D.; Blanchard, T.L.; Love, C.L.; Garcia, M.C.; Kenney, R.M. Effects of semen fractionation and dilution ratio on equine spermatozoal motility parameters. *Theriogenology* **1987**, *28*, 709–723. [CrossRef]
49. Makler, A.; Deutch, M.; Vilensky, A.; Palti, Y. Factors affecting sperm motility VIII. Velocity and survival of human spermatozoa as related to temperatures above zero. *Int. J. Androl.* **1981**, *4*, 559–569. [CrossRef]

50. Mortimer, D.; Goel, N.; Shu, M.A. Evaluation of the CellSoft automated semen analysis system in a routine laboratory setting. *Fertil. Steril.* **1988**, *50*, 960–968. [CrossRef]
51. Neuwinger, J.; Knuth, U.A.; Nieschlag, E. Evaluation of the Hamilton–Thorn 2030 motility analyser for routine semen analysis in an infertility clinic. *Int. J. Androl.* **1990**, *13*, 100–109. [CrossRef]
52. Verstegen, J.; Iguer-Ouada, M.; Onclin, K. Computer assisted semen analyzers in andrology research and veterinary practice. *Theriogenology* **2002**, *57*, 149–179. [CrossRef]
53. Rota, A.; Magelli, C.; Panzani, D.; Camillo, F. Effect of extender, centrifugation and removal of seminal plasma on cooled-preserved Amiata donkey spermatozoa. *Theriogenology* **2008**, *69*, 176–185. [CrossRef] [PubMed]

animals

MDPI

Article

Factors Affecting Embryo Recovery Rate, Quality, and Diameter in Andalusian Donkey Jennies

J. Dorado [1,*], M. Bottrel [1], I. Ortiz [1], M. Díaz-Jiménez [1], B. Pereira [1], C. Consuegra [1], J. J. Carrasco [2], V. Gómez-Arrones [2], A. Domingo [2] and M. Hidalgo [1]

[1] Veterinary Reproduction Group, Department of Medicine and Animal Surgery, Faculty of Veterinary Medicine, University of Cordoba, 14071 Cordoba, Spain; ma.bottrel@gmail.com (M.B.); isabel.ortiz.vet@gmail.com (I.O.); mariadijim@gmail.com (M.D.-J.); blasypereiraaguilar@gmail.com (B.P.); mtc93vet@gmail.com (C.C.); mhidalgo@uco.es (M.H.)

[2] Equine Reproduction Center, Centro de Selección y Reproducción Animal, (CENSYRA-Extremadura Government), 06007 Badajoz, Spain; juanjesus.carrasco@juntaex.es (J.J.C.); vanearrones@gmail.com (V.G.-A.); adomingomontes@gmail.com (A.D.)

* Correspondence: jdorado@uco.es; Tel.: +34-957-212-136

Received: 23 September 2020; Accepted: 25 October 2020; Published: 26 October 2020

check for updates

Simple Summary: Embryo transfer has been successfully used for the conservation of equine endangered species, but a number of factors may affect the outcome of these techniques in mares. However, only a few studies have evaluated these factors in donkeys. The present study was conducted to determine which factors affect the recovery rate, morphological quality, and diameter in embryos from Andalusian donkey jennies. According to our results, the factors affecting embryo recovery rate were donor jenny, donor age, successive cycle within donor, number of flushings, and jack. Day of flushing and number of flushings had an effect on embryo diameter, whereas donor jenny and day of flushing had an effect on embryo quality. The knowledge of these factors is crucial to achieve a higher efficiency of embryo transfer in endangered donkey breeds.

Abstract: Embryo transfer and the vitrification of embryos could be used for the conservation and recovery of endangered donkey breeds. It is important to develop techniques that optimize recovery rates and the cryotolerance of donkey embryos. This study evaluates factors affecting the recovery rate, quality, and diameter of embryos obtained from donor jennies as a starting point for the use of vitrification and embryo transfer in the conservation of the Andalusian donkey. A total of 100 embryos were recovered out of 124 estrous cycles (80.6%). The donor jenny affected the rates of positive flushings (PFR; $p = 0.040$) and embryo recovery (ERR; $p < 0.05$) as well as embryo quality ($p = 0.004$). ERR was also affected by the number of flushings ($p < 0.001$), donor age ($p < 0.05$), successive cycle within donor ($p < 0.001$), and jacks ($p < 0.05$). Number of flushings ($p < 0.001$) and jack ($p < 0.05$) had a significant effect on PFR, whereas the day of flushing influenced the developmental stage ($p < 0.001$), embryo quality ($p < 0.05$), and diameter of embryos ($p < 0.001$). The number of flushings significantly influenced the diameter ($p = 0.038$) and embryo developmental stage ($p = 0.001$), whereas the developmental stage was statistically different between herds ($p = 0.020$). The factors influencing the success of this assisted reproductive technique were donor jenny, donor age, successive cycle within donor, day of flushing, number of flushings, and jack. The identification of these key points is crucial to achieve a higher efficiency of embryo transfer and vitrification processes, before considering their application in the conservation of endangered donkey breeds.

Keywords: donkey; embryo transfer; embryo recovery rate; embryo quality

1. Introduction

In the past, domestic donkeys (*Equus africanus asinus*) were used as pack animals in agricultural activities, commerce, and militia [1,2], mainly due to their easy care, their resistance to diseases, and their physical resistance [3]. However, the mechanization of agriculture in Europe together with the consequent sharp decrease in mule breeding caused a drastic reduction of the donkey population [4–7]. Currently, all Spanish donkey breeds (Andaluza, Catalana, Balear, Majorera, Asno de las Encartaciones, and Zamorano-Leonés) are considered endangered (Real Decreto 2129/2008, regulation of the National Catalogue of Endangered Species). Although the population size of the Andalusian donkey has increased to 839 animals in 2018, only 14 females were pure breed. Moreover, the number of herds across Spain has decreased (163 herds), and the average herd size is five heads [8], thereby increasing the possibility of mating of related animals. Considering the contribution of donkeys to biodiversity [9], milk and meat products production [3,10], or pet therapy [2], strategies for the preservation of the genetic pool of donkey breeds and for the maintenance of the genetic heterozygosis of equine endangered species is highly advisable.

The conservation of endangered species is an excellent opportunity for applying assisted reproductive technologies such as embryo transfer, embryo cryopreservation, and germplasm cryobanking. Embryo transfer (ET) has been successfully used for the conservation of equine endangered species such as Przewalski´s horses (*Equus przewalskii*) [11], and numerous studies have been conducted in the past decades to investigate the suitability and efficiency of equine ET [12]. Together with this technique, the cryopreservation of embryos and their storage in embryo banks offer several advantages to the preservation and management of equine endangered species [13,14]. However, in donkeys, the studies on both procedures are scarce and recent [15–21].

It is known that some factors may affect the embryo recovery rate and embryo diameter and morphological quality in mares, including the day of flushing, number of ovulations, age of the donor, and quality of semen [22], and that morphological embryo quality has a major effect on pregnancy rates [13]. Other factors such as the size and age of embryos and storage of embryos may also affect pregnancy rates after ET in horses [23,24]. In addition, previous studies have demonstrated that early-stage horse embryos (<300 μm) show better survival rates after cryopreservation than large embryos collected at a later day [25,26]. In donkeys, only a few studies have been conducted, and the results could not prove the influence of embryo quality and age on embryo recovery rate [16,17]. Similarly, no effect on embryo recovery rate and quality was observed by Pérez-Marín et al. [20]

The aim of the present study was to determine which factors affect the recovery rate, morphological quality, and diameter in embryos from Andalusian donkey jennies as a prerequisite to improve the success of both embryo transfer and cryopreservation in this endangered donkey breed.

2. Materials and Methods

2.1. Experimental Animals and Study Location

All animal procedures were approved by the Ethical Committee for Animal Experimentation of the University of Cordoba (no. 31/08/2017/105) and are in accordance with Spanish laws for animal welfare and experimentation (Real Decreto 53/2013).

From February to December of three consecutive years (2015–2017), a total of twenty-six healthy Andalusian jennies (3–13 years old), of known fertility, served as embryo donors, and eight Andalusian jacks (6–9 years old) known to be fertile were used to mate the donors. To assess the effect of the age, the donor jennies were divided into three categories: ≤3 (n = 5), 4–9 (n = 17), and ≥10 years old (n = 5).

General health and reproductive history were recorded, and jennies were submitted to a general and reproductive physical examination [23]. Donors were housed, monitored, mated, and flushed in three different herds: the Equine Center for Assisted Reproduction of the Centro de Selección y Reproducción Animal (CENSYRA, Badajoz, Spain), the Centro Rural Malpica (Palma del Río, Cordoba, Spain) or the Centro de Medicina Deportiva Equina (CEMEDE, Cordoba, Spain). The jennies were housed in paddocks, the jacks were housed in stalls, and they were fed with hay, barley, and water ad libitum.

2.2. Oestrus Synchronization and Mating

Ovarian activity was evaluated by transrectal ultrasonography (Aloka SSD 500, ALOKA Co. Ltd., Tokyo, Japan) on a biweekly schedule during diestrus and daily during oestrus until ovulation (Day 0 = day of ovulation). Estrus was induced with one intramuscular injection of 5.25 mg luprostiol (Prosolvin®, Virbac, Barcelona, Spain) in the presence of corpus luteum. Donor jennys received human chorionic gonadotropin (hCG; 1500 IU, intramuscularly; Veterin Corion®, Divasa-Farmavic S.A., Barcelona, Spain) to induce ovulation when a follicle of 35–40 mm was detected. Next day, donor jennies were bred by live cover every other day until ovulation.

2.3. Embryo Recovery and Evaluation

Six to nine days after ovulation, donor jennies were flushed 3 times with a total of 3 L of Lactated Ringer´s solution (B. Braun VetCare S.A., Rubí, Spain), as described by Camillo et al. [17] for donkeys. After the flushing, luprostiol was administered to donors to induce luteolysis. Recovered embryos were evaluated for developmental stage (morula, early blastocyst, blastocyst, or expanded blastocyst) and morphological quality, and they were graded on a scale of 1–4 [27], 1 being excellent, 2 being good, 3 being fair, and 4 being poor, degenerate, or dead (Figure 1). After the quality evaluation, the embryos were washed ten times in Syngro® holding (Bioniche Animal Health, Washington, DC, USA), as previously described [16]. The diameter of the embryos was measured under bright field conditions (SZ51 Olympus optical, Tokyo, Japan) using an ocular micrometer (scale of 1 mm/100), as previously described [28].

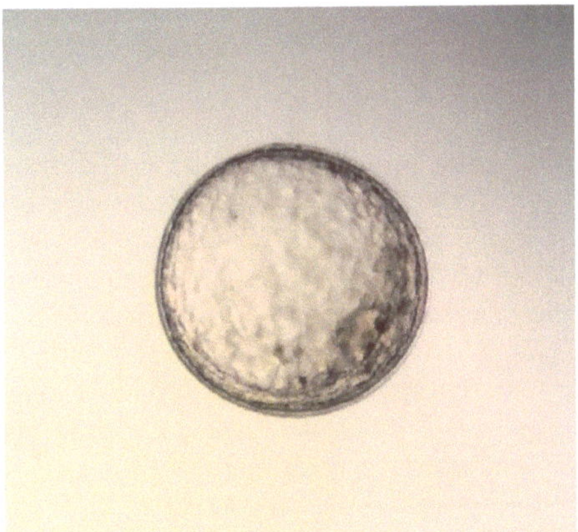

(a)

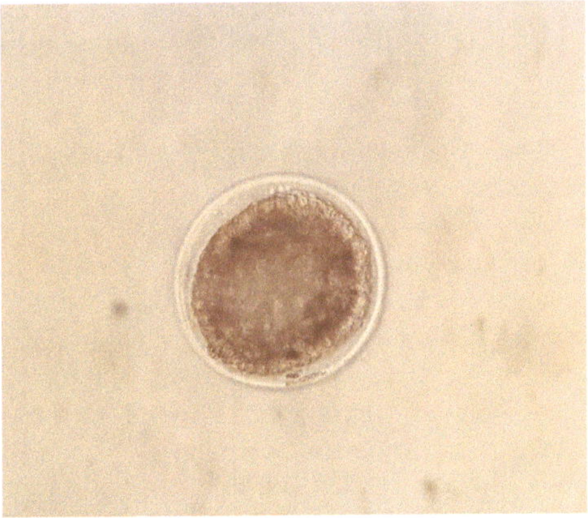

(b)

Figure 1. *Cont.*

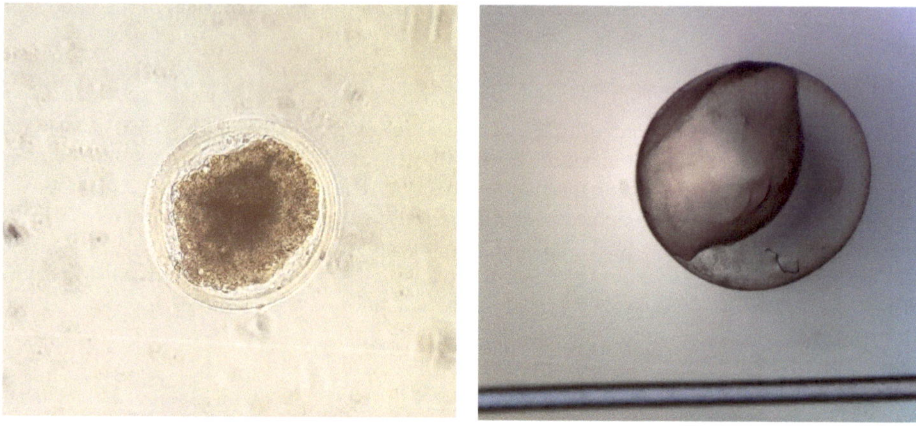

(c) (d)

Figure 1. Donkey embryos of various developmental stages and quality grades. (**a**) Expanded blastocyst stage embryo, Grade 1. Note blastocoele cavity and distinct inner cell mass. The zona pellucida has been shed, and the capsule is surrounding the embryo. No morphologic abnormalities are present in this embryo; (**b**) Blastocyst stage embryo, Grade 2. Note the distinct blastomere cells around the edge of the embryo and the capsule. A blastocoele cavity is just beginning to form within the center of the embryo. Note minor imperfections, such as a few extruded cells, occasional discolored cells, and slight shrinkage of trophoblast from zona pellucida; (**c**) Early blastocyst stage embryo, Grade 3. Note thick zona pellucida and capsule. Note moderate level of imperfections, such as a high proportion of extruded cells, discoloration of remaining cell mass, and moderate shrinkage of trophoblast from zona pellucida; (**d**) Expanded blastocyst stage embryo, Grade 4. Note complete collapse of the blastocoele.

2.4. Statistical Analysis

Descriptive statistical analysis was performed, presenting the qualitative variables as frequencies and percentages, and quantitative as a mean ± standard error of the mean (SEM). The effects of the year in which the study was performed (1–3), season (winter: December 22–March 20; spring: March 21–June 20; summer: June 21–September 22; autumn: September 23–December 21), photoperiod (positive: March–October; negative: November–February), herd (1–3), days of flushing for embryo recovery (6–9), number of flushings (1–3), donor (25 jennies), donor age (≤3; 4–9; ≥10 years old), parity (nulliparous vs. multiparous), successive cycle within donor (1–5; 6–14; 15–23), number of ovulations per cycle (single vs. double), and jack (8 donkeys) on positive uterine flushing rates (PFR; flushing where at least one embryo was recovered), embryo recovery rate (ERR; embryos recovered per cycle), and on ovulation rate (OR; number of ovulations per cycles) were analyzed by the Chi-square test and by the Kruskal–Wallis one-way ANOVA, respectively. When the effect was statistically significant, post-hoc multiple comparisons were made using Chi-square tests for categorical variables and Mann–Whitney U tests for continuous variables.

To evaluate the effects of single factors (year, season, photoperiod, herd, day of flushing, number of flushings, donor, donor age, parity, successive cycle within donor, number of ovulations, and jack) on embryo quality (Grade 1–4), diameter (μm), and developmental stage (morula, early blastocyst, blastocyst, expanded blastocyst) the Kruskal–Wallis one-way ANOVA was performed. Mean values were compared by Duncan's test.

All analyses were performed using the statistical package SPSS v15.0 (IBM Spain, Madrid, Spain). Differences were considered statistically significant when $p < 0.05$.

3. Results

The average ovulation rate in donor jennies is shown in Table 1. The average rate per jenny was 1.26 ± 0.04 and varied significantly among jennies ($p = 0.01$). Results of statistical analysis also showed differences among herds ($p = 0.031$), while no differences ($p > 0.05$) were observed among donor age categories, years, photoperiods, and seasons (Table 1). The single ovulation rate was 59.2% (93/157), while for double ovulation, the rate was 40.8% (64/157). Single ovulation occurred with equal frequency ($p > 0.05$) on both ovaries (left ovary: 42.5%; right ovary: 32.5%). However, the incidence of bilateral double ovulation (18.3%) was significantly higher ($p < 0.01$) than that of ipsilateral double ovulation (6.7%).

Table 1. Ovulation rate (mean ± SEM) in 124 cycles of 26 donor jennies according to donor age categories (≤3, 4–9 or ≥10 years old), parity (nulliparous or multiparous), year of the study (first, second or third), season of the year (spring, summer, autumn, or winter), photoperiod (positive or negative), and herd (1, 2, or 3).

Variable	No. of Cycles	Ovulation Rate	p-Values
Donor age			
≤3 years	34	1.24 ± 0.07	0.433
4–9 years	65	1.23 ± 0.05	
≥10 years	25	1.36 ± 0.10	
Parity			
Nulliparous	35	1.26 ± 0.08	0.988
Multiparous	89	1.26 ± 0.05	
Year			
First	33	1.21 ± 0.07	0.073
Second	42	1.17 ± 0.06	
Third	49	1.37 ± 0.07	
Season			
Spring	44	1.18 ± 0.06	0.095
Summer	12	1.08 ± 0.08	
Autumn	52	1.37 ± 0.07	
Winter	16	1.25 ± 0.11	
Photoperiod			
Positive	66	1.20 ± 0.05	0.099
Negative	58	1.33 ± 0.06	
Herd			
1	6	1.33 ± 0.21 ^{ab}	0.031
2	71	1.17 ± 0.05 ^{b}	
3	47	1.38 ± 0.07 ^{a}	
Total	124	1.26 ± 0.04	

^{a,b} Values with different superscript differ significantly.

A total of 124 uterine flushings were carried out during the study, of which 92 were positive (PFR: 74.2%; 92/124), and 100 embryos were recovered out of 124 estrous cycles (ERR: 80.6%; 100/124) and 157 ovulations (embryo recovery per ovulation: 63.7%; 100/157).

The embryo diameter and morphological quality score of donkey embryos are shown in Table 2. Overall, 77 of 100 embryos (77%) were classified as Grade 1 (excellent), 17 (17%) were classified as Grade 2 (good) and 6 (6%) were classified as Grade 3 (fair). The most frequent stages of development observed were early blastocyst (37%, 37/100) and expanded blastocyst (36%, 36/100), which were followed by morula (20%, 20/100) and blastocyst (7%, 7/100) stages. The embryo quality score significantly ($p < 0.05$) varied according to developmental stage and day of recovery, being lower for blastocysts or when flushed at day 8 after ovulation (Table 2). As expected, the embryo diameter was also affected ($p < 0.001$) by the developmental stage and day of flushing. The mean diameter of embryos was 179.39 ± 9.61 µm (range: 150–300 µm) for morulae, 210.81 ± 7.90 µm (range: 150–325 µm) for early blastocysts, 425.00

± 45.32 μm (range: 275–600 μm) for blastocysts, and 1022.92 ± 125.22 μm (range: 250–3300 μm) for expanded blastocysts. Moreover, it was observed that embryos recovered at 6 days after ovulation had a diameter of 187.50 ± 15.23 μm (range: 150–300 μm); those collected on day 7 measured 236.48 ± 13.83 μm (range: 150–600 μm); while the mean diameter at 8 and 9 were 806.25 ± 83.31 μm (range: 275–2400 μm) and 2275 ± 300.14 μm (range: 1525–3300 μm), respectively (Table 2).

Table 2. Diameters (mean ± SEM) and morphological quality score (1–4) of donkey embryos collected at Days 6 to 9 after ovulation according to their developmental stage (morula, early blastocyst, blastocyst, or expanded blastocyst) and day of recovery (6, 7, 8, or 9).

Variable	Grade	Diameter (μm)	No. (%)	Embryo Quality at Collection		
				Grade 1	Grade 2	Grade 3
Developmental stages						
Morula	1.50 ± 0.15 ^{b}	179.39 ± 9.61 ^{a}	20 (20%)	12 (60%) ^{a}	6 (30%)	2 (10%)
Early blastocyst	1.38 ± 0.11 ^{ab}	210.81 ± 7.90 ^{a}	37 (37%)	26 (70.3%) ^{a}	8 (21.6%)	3 (8.1%)
Blastocyst	1.00 ± 0.00 ^{a}	425.00 ± 45.32 ^{a}	7 (7%)	7 (100%) ^{b}	0 (0%)	0 (0%)
Expanded blastocyst	1.14 ± 0.07 ^{ab}	1022.92 ± 125.22 ^{b}	36 (36%)	32 (88.9%) ^{b}	3 (8.3%)	1 (2.8%)
Day of recovery						
6	1.46 ± 0.22 ^{b}	187.50 ± 15.23 ^{a}	13 (13%)	9 (69.2%) ^{a}	2 (15.4%) ^{b}	2 (15.4%) ^{b}
7	1.38 ± 0.08 ^{ab}	236.48 ± 13.83 ^{a}	53 (53%)	36 (67.9%) ^{a}	14 (26.4%) ^{b}	3 (5.7%) ^{a}
8	1.04 ± 0.04 ^{a}	806.25 ± 83.31 ^{b}	28 (28%)	27 (96.4%) ^{b}	1 (3.6%) ^{b}	0 (0%) ^{a}
9	1.33 ± 0.33 ^{ab}	2275.00 ± 300.14 ^{c}	6 (6%)	5 (83.3%) ^{a}	0 (0%) ^{a}	1 (16.7%) ^{b}
Total	1.29 ± 0.06	515.24 ± 59.95	100 (100%)	77 (77%)	17 (17%)	6 (6%)

$^{a–c}$ Values with different superscript differ significantly ($p < 0.05$).

Presented in Table 3 are the developmental stage and embryo size for Day 6–9 embryos. At Day 6 after ovulation, embryos were mostly at the morula stage (12/13, 92.3%); meanwhile, 67.9% (36/53) of embryos recovered at Day 7 were early blastocyst stage embryos. Expanded blastocysts were recovered on Day 8 (25/28, 89.3%) and 9 (6/6, 100%). At Day 6–7, most of the embryos recovered were small (<200 μm) or medium (200–300 μm) embryos; meanwhile, large embryos (>300 μm) were recovered at Days 8 (27/28, 96.4%) and 9 after ovulation (6/6, 100%; Table 3).

Table 3. Developmental stage (morula, early blastocyst, blastocyst, or expanded blastocyst) and embryo size (<200 μm, 200–300 μm or >300 μm) of donkey embryos collected at Day 6 to 9 after ovulation.

Day of Recovery	No. (%)	Developmental Stage				Embryo Diameter		
		Morula	Early Blastocyst	Blastocyst	Expanded Blastocyst	<200 μm	200–300 μm	>300 μm
6	13 (13%)	12 (92.3%) ^{a}	1 (7.7%) ^{b}	0 (0%) ^{b}	0 (0%) ^{b}	9 (69.2%) ^{a}	4 (30.8%) ^{b}	0 (0%) ^{c}
7	53 (53%)	8 (15.1%) ^{b}	36 (67.9%) ^{a}	4 (7.5%) ^{b}	5 (9.4%) ^{b}	24 (45.3%) ^{a}	19 (35.8%) ^{a}	10 (18.9%) ^{b}
8	28 (28%)	0 (0%) ^{b}	0 (0%) ^{b}	3 (10.7%) ^{b}	25 (89.3%) ^{a}	0 (0%) ^{b}	1 (3.6%) ^{b}	27 (96.4%) ^{a}
9	6 (6%)	0 (0%) ^{b}	0 (0%) ^{b}	0 (0%) ^{b}	6 (100%) ^{a}	0 (0%) ^{b}	0 (0%) ^{b}	6 (100%) ^{a}
Total	100 (100%)	20 (20%)	37 (37%)	7 (7%)	36 (36%)	43 (43%)	14 (14%)	43 (43%)

$^{a–c}$ Values with different superscript differ significantly ($p < 0.05$).

Table 4 shows extrinsic factors that affect the rate of positive flushings (PFR) and the embryo recovery rate (ERR). None of the factors studied affected ($p > 0.05$) PFR or ERR; however, there was an effect of the number of flushings (1, 2, or 3) on both rates, which were significantly ($p < 0.001$) reduced in the third flushing (PFR: 15.8%; ERR: 0.16).

Table 4. Extrinsic factors affecting the rate of positive flushings and embryo recovery rate in Andalusian donkeys.

Factor	Positive Flushings	*p*-Value	Embryo Recovery Rate	*p*-Value
Year				
First	21/33 (63.6%)	NS	24/33 (0.73)	NS
Second	33/42 (78.6%)		36/42 (0.86)	
Third	38/49 (77.6%)		40/49 (0.82)	
Season				
Spring	33/44 (75.0%)	NS	34/44 (0.77)	NS
Summer	10/12 (83.3%)		10/12 (0.83)	
Autumn	36/52 (69.2%)		42/53 (0.79)	
Winter	13/16 (81.3%)		14/15 (0.93)	
Photoperiod				
Positive	50/66 (75.8%)	NS	53/66 (0.80)	NS
Negative	42/58 (72.4%)		47/58 (0.81)	
Herd				
1	4/6 (66.7%)	NS	5/6 (0.83)	NS
2	51/71 (71.8%)		56/71 (0.79)	
3	37/47 (78.7%)		39/47 (0.83)	
Day of flushing				
6	12/16 (75.0%)	NS	13/16 (0.81)	NS
7	49/68 (72.1%)		53/68 (0.80)	
8	25/31 (80.6%)		28/31 (0.90)	
9	6/9 (66.7%)		6/9 (0.67)	
No. of flushings				
1	66/66 (100%) [a]	0.001	73/66 (1.11) [a]	0.001
2	20/20 (100%) [a]		21/20 (1.05) [a]	
3	6/38 (15.8%) [b]		6/38 (0.16) [b]	
Total	92/124 (74.2%)		100/124 (0.81)	

[a,b] Values with different superscript differ significantly; NS, not significant.

The intrinsic factors that affect PFR and ERR are shown in Table 5. PFR did not vary ($p > 0.05$) with any of the studied variables except the donor ($p = 0.040$). No differences between parity ($p = 0.2610$) and number of ovulations ($p = 0.0971$) were detected for ERR (Table 5). In contrast, ERR not only varies among donors ($p < 0.05$) but also among donor age groups ($p < 0.05$) and successive cycles within the donor ($p < 0.001$). ERR was higher ($p < 0.05$) in jennies of 4–9 years of age (0.94; 51/54) with respect to the other groups (≤3 years: 0.77 (26/34); ≥10 years: 0.64 (23/36)). With regard to the number of ovulations in the same donor, ERR was significantly higher ($p < 0.001$) in the first group (1–5 cycles: 1.44; 49/34) with respect to the second (6–14 cycles: 0.78; 39/50) and third (15–23 cycles: 0.30; 12/40) groups.

As shown in Table 6, PFR and ERR varied ($p > 0.05$) among jacks. Four jacks (numbers 169, 192, 2895, and 4457) that were used in 79.7% of cycles (94/118) showed good results for both rates (PFR: 70.8–100%; ERR: 0.77–1.00). The other three jacks (numbers 95, 148, and 9025) that were used in 18.6% of cycles (22/118) yield a lower PFR (50–58.3%) and ERR (0.50–0.67) than the previous group, but this was not significant statistically ($p > 0.05$). No embryos were obtained with jack number 232, although he was used only two times.

Developmental stage, embryo quality, and diameter of the embryos recovered in this study are shown in Tables 7–9. None of the extrinsic factors studied significantly influenced ($p < 0.05$) these three variables, except for the day of flushing (6–9), which significantly influenced the developmental stage ($p < 0.001$), embryo quality ($p < 0.05$), and diameter of embryos ($p < 0.001$; Table 7). Similarly, the number of flushings (1–3) significantly influenced the diameter ($p = 0.038$) and embryo developmental stage ($p = 0.001$), whereas the developmental stage was statistically different among herds ($p = 0.020$; Table 7).

Table 5. Intrinsic factors affecting the rate of positive flushings and embryo recovery rate in Andalusian donkeys.

Factor	Positive Flushings	*p*–Value	Embryo Recovery Rate	*p*–Value
Donor				
60	0/1 (0%) [b]	0.040	0/1 (0.00) [b]	0.011
64	1/1 (100%) [a]		1/1 (1.00) [a]	
142	9/10 (90%) [a]		9/10 (0.90) [a]	
167	10/17(58.8%) [ab]		11/17 (0.65) [a]	
193	4/5(80%) [a]		4/5(0.80) [a]	
1159	1/2 (50%) [ab]		1/2 (0.50) [ab]	
1161	14/18 (77.8%) [a]		15/18 (0.83) [a]	
1220	1/1 (100%) [a]		1/1 (1.00) [a]	
1372	1/1 (100%) [a]		1/1 (1.00) [a]	
1826	0/1 (0%) [b]		0/1 (0.00) [b]	
2089	1/1 (100%) [a]		2/1 (2.00) [a]	
2339	1/1 (100%) [a]		1/1 (1.00) [a]	
2629	4/8 (50%) [ab]		4/8 (0.50) [ab]	
3223	6/6 (100%) [a]		6/6 (1.00) [a]	
3977	1/1 (100%) [a]		1/1 (1.00) [a]	
4103	12/13 (92.3%) [a]		12/13 (0.92) [a]	
5372	0/1 (0%) [b]		0/1 (0.00) [b]	
6069	1/1 (100%) [a]		1/1 (1.00) [a]	
6517	2/2 (100%) [a]		2/2 (1.00) [a]	
7148	12/15 (80%) [a]		14/15 (0.93) [a]	
7171	2/6 (33.3%) [ab]		2/6 (0.33) [b]	
7590	0/2 (0%) [b]		0/2 (0.00) [b]	
8310	2/2 (100%) [a]		3/2 (1.50) [a]	
8311	6/6 (100%) [a]		8/6 (1.33) [a]	
9695	1/2 (50%) [ab]		1/2 (0.50) [ab]	
Donor age				
≤3 years	25/34 (73.5%)	NS	26/34 (0.77) [b]	0.013
4–9 years	48/65 (73.8%)		51/54 (0.94) [a]	
≥10 years	19/25 (76%)		23/36 (0.64) [b]	
Parity				
Nulliparous	25/35 (71.4%)	NS	26/35 (0.74)	NS
Multiparous	67/89 (75.3%)		74/89 (0.83)	
Successive cycle within donor				
1–5	44/63 (69.8%)	NS	49/34 (1.44) [a]	0.001
6–14	36/45 (80%)		39/50 (0.78) [b]	
15–23	12/16 (75%)		12/40 (0.30) [c]	
No. of ovulations				
Single	70/92 (76.1%)	NS	71/92 (0.77)	NS
Double	22/32 (68.8%)		29/32 (0.91)	
Total	92/124 (74.2%)		100/124 (0.81)	

[a–c] Values with different superscript differ significantly; NS, not significant.

Table 6. Variation in the rate of positive flushings and embryo recovery rate between Andalusian jacks.

Jack	Mated Donors	Positive Flushings	*p*-Value	Embryo Recovery Rate	*p*-Value
95	12	7/12 (58.3%) [ab]	0.0253	8/12 (0.67) [ab]	0.0253
169	36	27/36 (75%) [a]		29/36 (0.81) [a]	
192	3	3/3 (100%) [a]		3/3 (1.00) [a]	
232	2	0/2 (0%) [b]		0/2 (0.00) [b]	
1481	2	1/2 (50%) [ab]		1/2 (0.50) [ab]	
2895	7	6/7 (85.7%) [a]		6/7 (0.86) [a]	
4457	48	34/48 (70.8%) [a]		37/48 (0.77) [a]	
9025	8	4/8 (50%) [ab]		4/8 (0.50) [ab]	
Total	118 *	82/118 (69.5%)		88/118 (0.75)	

[a,b] Values with different superscript differ significantly; * Missing data (n = 6).

Table 7. Extrinsic factors affecting the developmental stage, embryo quality, and diameter of embryos in Andalusian donkeys.

Factor	No.	Developmental Stage	*p*-Value	Embryo Grade	*p*-Value	Embryo Diameter	*p*-Value
Year							
First	24	2.25 ± 0.24	NS	1.25 ± 0.11	NS	548.91 ± 140.10	NS
Second	36	2.69 ± 0.20		1.28 ± 0.10		637.50 ± 124.22	
Third	40	2.70 ± 0.18		1.33 ± 0.09		385.83 ± 52.24	
Season							
Spring	34	2.74 ± 0.20	NS	1.26 ± 0.09	NS	660.29 ± 142.21	NS
Summer	10	3.20 ± 0.33		1.10 ± 0.10		602.50 ± 176.48	
Autumn	42	2.38 ± 0.18		1.40 ± 0.10		410.91 ± 62.66	
Winter	14	2.43 ± 0.33		1.14 ± 0.14		405.77 ± 87.90	
Photoperiod							
Positive	53	2.66 ± 0.16	NS	1.25 ± 0.07	NS	619.87 ± 100.27	NS
Negative	47	2.51 ± 0.17		1.34 ± 0.09		399.47 ± 56.92	
Herd							
1	5	1.20 ± 0.20 ^{a}	0.020	1.60 ± 0.40	NS	270.83 ± 30.90	NS
2	56	2.61 ± 0.16 ^{b}		1.25 ± 0.07		619.20 ± 97.36	
3	39	2.74 ± 0.12 ^{b}		1.31 ± 0.09		391.03 ± 53.33	
Day of flushing							
6	13	1.08 ± 0.08 ^{a}	0.000	1.46 ± 0.22 ^{b}	0.045	187.50 ± 15.23 ^{a}	0.000
7	53	2.11 ± 0.11 ^{b}		1.38 ± 0.08 ^{b}		236.48 ± 13.83 ^{a}	
8	28	3.89 ± 0.06 ^{c}		1.04 ± 0.04 ^{a}		806.25 ± 83.31 ^{b}	
9	6	4.00 ± 0.00 ^{c}		1.33 ± 0.33 ^{ab}		2275.00 ± 300.14 ^{c}	
No. of flushings							
1	73	2.84 ± 0.14 ^{c}	0.001	1.23 ± 0.06	NS	604.79 ± 78.25 ^{b}	0.038
2	21	2.05 ± 0.22 ^{ab}		1.48 ± 0.15		290.42 ± 32.19 ^{a}	
3	6	1.50 ± 0.22 ^{a}		1.33 ± 0.21		175.00 ± 9.13 ^{a}	
Total	100	2.59 ± 0.14		1.29 ± 0.06		515.24 ± 59.95	

$^{a–c}$ Values with different superscript differ significantly; NS, not significant.

Table 8. Intrinsic factors affecting the developmental stage, embryo quality, and diameter of embryos in Andalusian donkeys.

Factor	No.	Developmental Stage	*p*-Value	Embryo Grade	*p*-Value	Embryo Diameter	*p*-Value
Donor							
60	*	-	NS	-	0.004	-	NS
64	1	1.00 ± -		1.00 ± -		-	
142	9	2.89 ± 0.39		1.11 ± 0.11 ^{a}		631.11 ± 244.42	
167	11	2.36 ± 0.34		1.36 ± 0.15 ^{ab}		279.55 ± 55.17	
193	4	3.00 ± 0.58		1.25 ± 0.25 ^{ab}		368.75 ± 110.10	
1159	1	2.00 ± -		1.00 ± -		200.00 ± -	
1161	15	1.50 ± 0.22		1.20 ± 0.15 ^{ab}		405.00 ± 101.07	
1220	1	1.00 ± -		1.00 ± -		175.00 ± -	
1372	1	1.00 ± -		2.00 ± -		175.00 ± -	
1826	*	-		-		-	
2089	2	1.00 ± 0.00		2.00 ± 1.00 ^{b}		287.50 ± 12.50	
2339	1	1.00 ± -		1.00 ± -		175.00 ± -	
2629	4	3.50 ± 0.50		1.00 ± 0.00 ^{a}		1143.75 ± 722.51	
3223	6	2.67 ± 0.56		1.50 ± 0.34 ^{ab}		954.17 ± 440.29	
3977	1	4.00 ± -		1.00 ± -		1625.00 ± -	
4103	12	3.50 ± 0.26		1.17 ± 0.11 ^{a}		662.50 ± 181.36	
5372	*	-		-		-	
6069	1	1.00 ± -		2.00 ± -		183.33 ± -	
6517	2	2.00 ± 0.00		3.00 ± 0.00 ^{c}		325.00 ± 0.00	
7148	14	2.21 ± 0.33		1.43 ± 0.17 ^{ab}		360.71 ± 92.94	
7171	2	2.00 ± 1.00		1.00 ± 0.00 ^{a}		262.50 ± 112.50	
7590	*	-		-		-	
8310	3	3.00 ± 1.00		1.00 ± 0.00 ^{a}		725.00 ± 291.91	
8311	8	2.63 ± 0.42		1.13 ± 0.13 ^{a}		571.88 ± 222.88	
9695	1	2.00 ± -		1.00 ± -		325.00 ± -	
Donor age							
≤3 years	26	2.58 ± 0.22	NS	1.27 ± 0.09	NS	414.74 ± 91.94	NS
4–9 years	51	2.67 ± 0.17		1.29 ± 0.09		579.50 ± 96.80	
≥10 years	23	2.43 ± 0.26		1.30 ± 0.12		489.13 ± 108.25	
Parity							
Nulliparous	26	2.58 ± 0.22	NS	1.27 ± 0.09	NS	414.74 ± 91.94	NS
Multiparous	74	2.59 ± 0.14		1.30 ± 0.07		551.03 ± 74.30	

Table 8. *Cont.*

Factor	No.	Developmental Stage	*p*-Value	Embryo Grade	*p*-Value	Embryo Diameter	*p*-Value
Successive cycle within donor							
1–5	49	2.41 ± 0.17	NS	1.35 ± 0.09	NS	556.42 ± 93.53	NS
6–14	39	2.72 ± 0.19		1.26 ± 0.09		534.62 ± 97.97	
15–23	12	2.92 ± 0.29		1.17 ± 0.11		287.50 ± 37.75	
No. of ovulations							
Single	71	2.61 ± 0.14	NS	1.25 ± 0.06	NS	538.33 ± 79.70	NS
Double	29	2.55 ± 0.23		1.38 ± 0.13		459.48 ± 70.70	
Total	100	2.59 ± 0.14		1.29 ± 0.06		515.24 ± 59.95	

a–c Values with different superscript differ significantly; NS, not significant; * *Missing data.*

Table 9. Variation in developmental stage, embryo quality, and diameter of embryos between Andalusian jacks.

Jack	Mated Donors	Developmental Stage	*p*-Value	Embryo Grade	*p*-Value	Embryo Diameter	*p*-Value
95	8	3.00 ± 0.38	NS	1.25 ± 0.16	NS	475.00 ± 140.71	NS
169	29	3.03 ± 0.21		1.24 ± 0.08		447.41 ± 67.59	
192	3	2.33 ± 0.33		1.67 ± 0.67		258.33 ± 44.10	
1481	1	1.00 ± -		2.00 ± -		183.33 ± -	
2895	6	1.83 ± 0.48		1.50 ± 0.22		366.67 ± 177.09	
4457	37	2.51 ± 0.19		1.27 ± 0.10		629.05 ± 136.72	
9025	8	2.25 ± 0.75		1.00 ± 0.00		681.25 ± 412.36	
Total	88 *	2.65 ± 0.12		1.28 ± 0.06		521.97 ± 66.52	

NS, not significant; * Missing data ($n = 12$).

Table 8 shows that no influence was detected ($p > 0.05$) of the intrinsic factors on developmental stage, embryo quality, and diameter of embryos, except for the donor, which affected the embryo quality ($p = 0.004$). Moreover, no differences ($p > 0.05$) were detected among jacks for these three variables (Table 9).

4. Discussion

Due to the similarities in the reproductive physiology between horses and donkeys, several assisted reproductive technologies (ARTs) routinely used in horses have been applied directly to donkeys. Hence, previous studies have demonstrated the suitability of mare ET techniques for collecting embryos in jennies [16]. In this line, numerous studies have been conducted to examine the factors that affect embryo recovery, quality, and diameter in mares [13,22,29]. However, in donkeys, these studies have been very scarce [17,20], and they are often performed on a limited number of animals, cycles, or embryos. Therefore, more studies are needed to optimize embryo recovery rates and maximize the success of future ET programs in donkeys.

In this study, in which 26 donor jennies and 124 cycles were used, the average ovulation rate per jenny was 1.26 ± 0.04. This finding was slightly lower than the reported average in spontaneous (1.57 ± 0.06) and prostaglandin F2 alpha (PGF2α)-induced (1.56 ± 0.10) estrus of Andalusian jennies [30]. These small differences could be explained by other factors (such as feeding management, donor age, reproductive status, season of the year, and the use of drugs to induce ovulation) that can affect the incidence of multiple ovulations, as reported in mares [22,31–33] and donkeys [30,34].

In jennies, the incidence of multiple ovulation was reported to range between 5.3% and 61% [35–37]. In our study, the single ovulation rate was 59.2%, while the double ovulation rate was 40.8%. Double ovulation in jennies was similar to that reported in Catalonian jennies (42.45%) [34] and in the Asinina de Miranda jennies (36.36%) [38]. However, the incidence of double ovulation in this study was lower than that reported in spontaneous (51.7%) and PGF2α-induced cycles (56.5%) in Andalusian jennies [30]. It is interesting to note that single ovulation occurred with equal frequency on both ovaries, as also reported by other authors [30,34]. Similar to that reported by Taberner et al. [34], a minimally greater frequency of ovulation for the left ovary was found (42.5% vs. 32.5%), but the difference was not significant. On the other hand, the incidence of bilateral double ovulation was significantly higher than that of ipsilateral double ovulation, which agrees with findings for mares [31] and jennies [30].

No influence of the age of the donor on the ovulation rate was observed in this study; however, the ovulation rate was numerically (1.36 ± 0.10%), but not significantly higher, in the older jennies (≥10 years old). These results are consistent with previous findings in jennies [20,30] and mares [31]. Although the reason for this fact remains still unclear, it has been suggested that increased ovarian stimulation or enhanced ovarian receptivity to that stimulation may be involved. Thus, multiovulation would be a natural strategy to ensure gestation in older females, which have a reduced ability to become pregnant [32].

The ovulation rate did not vary among years, seasons, and photoperiods, but it tended to be lower in summer ($p = 0.095$). Similar findings have been reported in previous studies [20,30]. Ginther et al. [39], in a study using different breeds and geographical latitude than our study of donkeys, observed that the incidence of multiple ovulations was not affected by the season of the year. The statistical analysis revealed significant differences among herds ($p = 0.031$), which could be explained by the individuals comprising each herd. In fact, this study pointed out the existence of significant differences among jennies ($p = 0.01$).

In our study, in which 100 embryos were used, the overall ERR following non-surgical flushing on Days 6–9 was 80.6%, which was higher than those previously reported in different breeds of donkeys: 53.3% in jennies of unknown breed [40], 63.6% in Poitou jennies [41], 75.9% in Pantesca jennies [17], 50% in Amiata jennies [16], 52.3% in Pega jennies [42], and 40.7% in Andalusian and Zamorano-Leones jennies [20]. The EER obtained in our experiment was also higher than the rates reported in the literature for fertile mares in commercial ET programs [43]—60–77% for fresh, 44% for chilled, and 46% for frozen semen—but similar to that obtained in young fertile mares inseminated with fresh semen, 87% [44]. These results are likely due to the age of the jennies used in the study and the physical and reproductive assessment performed before, including donors in the experimental group.

It is known that the major factor affecting embryo recovery is the donors´ reproductive history. Hence, embryo recovery for old sub-fertile mares can be as low as 30–40% per cycle [45]. Other factors that affect embryo recovery include semen quality and semen type (fresh, cooled, or frozen) [45]. In our study, all donors were selected carefully, based on their reproductive history and clinical examination, and they were mated naturally with jacks of proven fertility. Although the PFR and ERR varied among jacks, seven out of eight jacks showed moderate to good results for both rates (≥50% and ≥0.50, respectively). Only one jack had low fertility (zero out of two positive flushes), but he was used only twice, which could mitigate its negative effect on average PFR and ERR. In addition, hCG was used as the ovulation inductor. Previous studies have clearly demonstrated that ovulation induction can enhance the efficiency of ARTs in domestic animal species, including the donkey [46].

The embryo morphology score is the most common method used to evaluate embryo quality [27]. In line with previous findings [16,17,20,42], 94% of the recovered embryos had a quality grade of excellent (Grade 1) or good (Grade 2).

In our experiment, none of the extrinsic factors analyzed (year of the study, season of the year, photoperiod, herd or day, of flushing) affected significantly PFR or ERR. Consistent with previous findings [17], the ERR obtained in the first year (0.73) was numerically lower than that obtained in the second (0.86) and third (0.82) year, which could be explained by the inexperience with this technology (i.e., ET) in donkeys. The absence of a photoperiod influence on embryo recovery has been previously described in donkeys [17,20]. The study carried out in Pantesca donkeys [17] also noted that the time of the year did not affect PFR and ERR. Considering these results, we could state that seasonality has little impact on reproductive performance of Andalusian jennies, which can get pregnant naturally all year round, as previously reported for other donkey breeds [17]. This fact brings the possibility of applying ET in Andalusian donkeys along the year.

In donkeys, the influence of the day of flushing on ERR has not been well established in the literature [17,20]. Under our experimental conditions, PFR and ERR were not different among Days 6, 7, 8, and 9. Similar findings have been reported in mares [22,47]. However, embryo diameter and developmental stage varied widely depending on the day of recovery, which is consistent with previous studies [16,41]. In line with these findings, poor-quality embryos were collected at Day 8, and ERR tended to be lower ($p = 0.08$), flushing the uterus 9 days after ovulation. Taken all together, our results emphasize the importance of collecting Day 6 to 7 donkey embryos, smaller than 300 μm in diameter and with good morphological score, to ensure vitrification success, as has been previously suggested [19].

From a practical point of view, another interesting observation was the effect of the number of flushings (1, 2, or 3) on both rates (PFR and ERR), which were significantly reduced in the third flushing (15.8% and 0.16, respectively), indicating that the majority of embryo collections require one or two maximum flushings. In addition, embryo diameter and developmental stage varied among flushings, with larger and older embryos in the first flushing. Thus, our results could suggest that the larger diameter of the older embryos could facilitate their recovery [17].

The embryos recovered in herd 1 were in earlier developmental stages than in the other herds. These results were probably due to the fact that all the embryo rerecovered in this herd (5 embryos) were flushed 6–7 days after ovulation, while in herds 2 (56 embryos) and 3 (39 embryos), flushes were carried out from Day 6 to 9 after ovulation, thus increasing the average diameter and developmental stage of the recovered embryos.

Regarding the intrinsic factors, PFR did not vary with any of the studied variables (i.e., donor age, parity, successive cycle within donor, and number of ovulations) except for the donor jenny. In contrast, ERR not only varied among donors but also among donor age categories, showing higher values in jennies of 4–9 years of age (0.94) compared with all the other categories of age. Moreover, both PFR and ERR were not different between younger (≤3 years old) and older (≥10 years old) jennies. Our findings are in contrast with the results of previous studies [17,20], which reported no effect of donor age and donor jenny on the aforementioned rates. However, the effect on ERR of donor age and donor mare have been reported by many authors [13,22,48,49], in which old age (>15 years old) and a history of sub-fertility were related to a lower ERR. Our results could suggest that embryo donors between 4 and 9 years are the best to be used in an ET program.

The effect of repeated uterine flushings has been previously described in mares [50], which was associated with increased chronic inflammation of the uterus. Although previous studies failed to observe this negative effect in donkeys [17], in our experiment, the ERR on successive cycles from 1 to 5 was higher (1.44) than in attempts from 6 to 14 (0.78) and from 15 to 23 (0.30). The differences observed between studies may be explained by different experimental conditions. Therefore, in our study, young (≤3 years), mature (4–9 years), and old (≥10 years) donor jennies were employed during the entire period of the study, while only young jennies (2–5 years old) were used in this previous work [17]. Moreover, a higher number of donors (10, 8, and 6, respectively) and cycles (63, 45, and 16, respectively) were used in each group.

It has been previously described that the occurrence of multiple ovulations enhances ERR in mares [22,51] and donkeys [17,20], but this effect was not shown in our study. However, despite the absence of statistical significance, ERR after single ovulation tended to be lower than that obtained after double ovulation (0.77 vs. 0.91; $p = 0.0971$). It has been also reported that ipsilateral double ovulations resulted in a lower ERR than bilateral double ovulations [52], which could be caused by interference between two or more simultaneous ovulations in the limited space of the ovulation fossa [23,53]. In the present study, the incidence of ipsilateral double ovulations was only 6.7%, and no significant difference in ERR was observed between bilateral and ipsilateral ovulations (1.62% vs. 1.88%; $p = 0.196$).

Conversely, the number of ovulations in the same donor influenced ERR, being significantly higher in the first group (1–5 cycles: 1.44; 49/34), but no effect on PFR was observed. These results partially agree with the findings of Camillo et al. [17], who observed a significant effect on both rates. Finally, we observed that the parity of the donors did not have an effect on PFR and ERR. In cattle, the parity of recipients does not affect pregnancy rates following the transfer of fresh and frozen embryos [54]. However, to our knowledge, no data are available for mare and jenny donors.

5. Conclusions

Based on our results, we can conclude that the donor jenny was the main factor that affects the rate of positive flushings and recovery rates as well as the embryo quality. Other factors that affected embryo recovery rate were the number of flushings, donor age, successive cycle within donor, and jack. Meanwhile, the rate of positive flushings was affected by the number of flushings and the jack. From a practical point of view, these findings could indicate that the majority of embryo collections require one or two maximum flushings per cycle. Moreover, the negative effect of repeated uterine flushings on embryo recovery rate was proven, being lower after six consecutive cycles. On the other hand, the day of flushing had a significant effect on embryo quality and diameter, which emphasizes the importance of collecting Day 6 to 7 donkey embryos, with good morphological score and smaller than 300 μm in diameter, if embryos are going to be cryopreserved.

Author Contributions: Conceptualization, M.H. and J.D.; methodology, M.B., I.O., M.D.-J., B.P., C.C., J.J.C., V.G.-A., A.D. and J.D.; software, M.B. and J.D.; validation, M.H., M.B. and J.D.; formal analysis, M.B., I.O., M.D.-J., B.P. and C.C.; investigation, M.B., I.O., M.D.-J., B.P. and C.C.; resources, M.H. and J.D.; data curation, M.B., M.H. and J.D.; writing—original draft preparation, I.O., M.H. and J.D.; writing—review and editing, I.O., M.H. and J.D.; visualization, M.H. and J.D.; supervision, M.H. and J.D.; project administration, M.H. and J.D.; funding acquisition, M.H. and J.D. All authors have read and agreed to the published version of the manuscript.

Funding: This work has been supported by Grant AGL2013-42726-R (Secretaría de Estado de Investigación, Desarrollo e Innovación, Ministerio de Economía y Competitividad, Spain). I.O. was supported by a Ph.D. fellowship from the ceiA3 (Andalusia, Spain) with funding provided by Banco Santander through its Global Division, Santander Universidades. M.D.-J. and C.C. were supported by a FPU fellowship from Spanish Ministry of Education, Culture, and Sports.

Acknowledgments: We are extremely grateful to the Centro Rural Malpica (Palma del Río, Cordoba, Spain) for providing the animals and facilities.

Conflicts of Interest: The authors declare no conflict of interest. The funders had no role in the design of the study; in the collection, analyses, or interpretation of data; in the writing of the manuscript, or in the decision to publish the results.

References

1. Gandini, G.C.; Villa, E. Analysis of the cultural value of local livestock breeds: A methodology. *J. Anim. Breed Genet.* **2003**, *120*, 1–11. [CrossRef]
2. Miragaya, M.H.; Neild, D.M.; Alonso, A.E. A Review of Reproductive Biology and Biotechnologies in Donkeys. *J. Equine Vet Sci.* **2018**, *65*, 55–61. [CrossRef]
3. Kugler, W.; Grunenfelder, H.P.; Broxham, E. *Donkey Breeds in Europe: Inventory, Description, Need for Action, Conservation*; Report 2007–2008; Monitoring Institute for Rare Breeds and Seeds in Europe: St. Gallen, Switzerland, 2008; p. 124.
4. Quaresma, M.; Martins, A.M.F.; Rodrigues, J.B.; Colaço, J.; Payan-Carreira, R. Viability analyses of an endangered donkey breed: The case of the Asinina de Miranda (*Equus asinus*). *Anim. Prod. Sci.* **2015**, *55*, 1184–1191. [CrossRef]
5. Colli, L.; Perrotta, G.; Negrini, R.; Bomba, L.; Bigi, D.; Zambonelli, P.; Verini Supplizi, A.; Liotta, L.; Ajmone-Marsan, P. Detecting population structure and recent demographic history in endangered livestock breeds: The case of the Italian autochthonous donkeys. *Anim. Genet.* **2013**, *44*, 69–78. [CrossRef]
6. Calero-Carretero, R.; Bravo-Delgado, J.A.; Bartolomé-García, P. Razas autóctonas extremeñas en peligro de extinción. In *Informe 2007 La Agricultura y la Ganadería Extremeñas*; Facultad de Ciencias Económicas y Empresariales, Escuelas de Ingenierías Agrarias, Universidad de Extremadura: Caja de Badajoz, Badajoz, España, 2007; pp. 203–226.
7. Jordana, J.; Folch, P. The endangered catalonian donkey breed: The main ancestor of the American ass or mammoth. *J. Equine Vet. Sci.* **1996**, *16*, 436–441. [CrossRef]
8. DAD-IS-FAO. Information System Domestic Animal Diversity. Available online: http://www.fao.org/dad-is/browse-by-country-and-species/en/ (accessed on 21 October 2020).
9. Scherf, B.D. *World Watch List for Domestic Animal Diversity*, 3rd ed.; FAO: Rome, Italy, 2000; p. 739.

10. Vita, D.; Passalacqua, G.; Di Pasquale, G.; Caminiti, L.; Crisafulli, G.; Rulli, I.; Pajno, G.B. Ass's milk in children with atopic dermatitis and cow's milk allergy: Crossover comparison with goat's milk. *Pediatr. Allergy Immunol.* **2007**, *18*, 594–598. [CrossRef]
11. Summers, P.M.; Shephard, A.M.; Hodges, J.K.; Kydd, J.; Boyle, M.S.; Allen, W.R. Successful transfer of the embryos of Przewalski's horses (*Equus przewalskii*) and Grant's zebra (*E. burchelli*) to domestic mares (*E. caballus*). *J. Reprod. Fertil.* **1987**, *80*, 13–20. [CrossRef]
12. Hinrichs, K. Assisted reproductive techniques in mares. *Reprod. Domest. Anim.* **2018**, *53*, 4–13. [CrossRef]
13. Squires, E.L.; Carnevale, E.M.; McCue, P.M.; Bruemmer, J.E. Embryo technologies in the horse. *Theriogenology* **2003**, *59*, 151–170. [CrossRef]
14. Squires, E.L.; McCue, P.M. Cryopreservation of Equine Embryos. *J. Equine Vet. Sci.* **2016**, *41*, 7–12. [CrossRef]
15. Pérez-Marín, C.C.; Vizuete, G.; Vazquez-Martinez, R.; Galisteo, J.J. Comparison of different cryopreservation methods for horse and donkey embryos. *Equine Vet. J.* **2018**, *50*, 398–404. [CrossRef]
16. Panzani, D.; Rota, A.; Crisci, A.; Kindahl, H.; Govoni, N.; Camillo, F. Embryo quality and transcervical technique are not the limiting factors in donkey embryo transfer outcome. *Theriogenology* **2012**, *77*, 563–569. [CrossRef]
17. Camillo, F.; Panzani, D.; Scollo, C.; Rota, A.; Crisci, A.; Vannozzi, I.; Balbo, S. Embryo recovery rate and recipients' pregnancy rate after nonsurgical embryo transfer in donkeys. *Theriogenology* **2010**, *73*, 959–965. [CrossRef]
18. Bottrel, M.; Mogas, T.; Pereira, B.; Ortiz, I.; Díaz-Jiménez, M.; Consuegra, C.; Hidalgo, M.; Morató, R.; Dorado, J. The cryoprotective effect of Ficoll 70 on the post-warming survival and quality of Cryotop-vitrified donkey embryos. *Theriogenology* **2020**, *148*, 180–185. [CrossRef]
19. Bottrel, M.; Ortiz, I.; Pereira, B.; Díaz-Jiménez, M.; Hidalgo, M.; Consuegra, C.; Morató, R.; Mogas, T.; Dorado, J. Cryopreservation of donkey embryos by the cryotop method: Effect of developmental stage, embryo quality, diameter and age of embryos. *Theriogenology* **2019**, *125*, 242–248. [CrossRef]
20. Pérez-Marín, C.C.; Vizuete, G.; Galisteo, J.J. Embryo recovery results in Hispano-Arabe horse and Spanish donkey breeds. *Livest. Sci.* **2017**, *206*, 76–81. [CrossRef]
21. Bottrel, M.; Fortes, T.; Ortiz, I.; Hidalgo, M.; Dorado, J. Establishment and maintenance of donkey-in-mule pregnancy after embryo transfer in a non-cycling mule treated with oestradiol benzoate and long-acting progesterone. *Span J. Agric. Res.* **2017**, *15*, e04SC01. [CrossRef]
22. Panzani, D.; Rota, A.; Marmorini, P.; Vannozzi, I.; Camillo, F. Retrospective study of factors affecting multiple ovulations, embryo recovery, quality, and diameter in a commercial equine embryo transfer program. *Theriogenology* **2014**, *82*, 807–814. [CrossRef]
23. Carnevale, E.M.; Ramirez, R.J.; Squires, E.L.; Alvarenga, M.A.; Vanderwall, D.K.; McCue, P.M. Factors affecting pregnancy rates and early embryonic death after equine embryo transfer. *Theriogenology* **2000**, *54*, 965–979. [CrossRef]
24. Panzani, D.; Vannozzi, I.; Marmorini, P.; Rota, A.; Camillo, F. Factors affecting recipients' pregnancy, pregnancy loss, and foaling rates in a commercial equine embryo transfer program. *J. Equine Vet. Sci.* **2016**, *37*, 17–23. [CrossRef]
25. Eldridge-Panuska, W.D.; Caracciolo Di Brienza, V.; Seidel, G.E., Jr.; Squires, E.L.; Carnevale, E.M. Establishment of pregnancies after serial dilution or direct transfer by vitrified equine embryos. *Theriogenology* **2005**, *63*, 1308–1319. [CrossRef]
26. Stout, T. Cryopreservation of Equine Embryos: Current state-of-the-art. *Reprod. Domest. Anim.* **2012**, *47*, 84–89. [CrossRef]
27. McKinnon, A.O.; Squires, E.L. Equine embryo transfer. *Vet. Clin. N. Am. Equine Pract.* **1988**, *4*, 305–333. [CrossRef]
28. Lagares, M.A.; Castanheira, P.N.; Amaral, D.C.G.; Vasconcelos, A.B.; Veado, J.C.C.; Arantes, R.M.E.; Stahlberg, R. Addition of ficoll and disaccharides to vitrification solutions improve in vitro viability of vitrified equine embryos. *Cryo-Letters* **2009**, *30*, 408–413.
29. Hinrichs, K. Assisted reproduction techniques in the horse. *Reprod. Fertil. Dev.* **2013**, *25*, 80–93. [CrossRef] [PubMed]
30. Perez-Marin, C.C.; Galisteo, I.; Perez-Rico, A.; Galisteo, J. Effects of breed, age, season, and multiple ovulations on cyclic, PGF2α-induced, and postpartum estrus characteristics in Spanish jennies. *Theriogenology* **2016**, *85*, 1045–1052. [CrossRef]

31. Davies Morel, M.C.; O'Sullivan, J.A. Ovulation rate and distribution in the thoroughbred mare, as determined by ultrasonic scanning: The effect of age. *Anim. Reprod. Sci.* **2001**, *66*, 59–70. [CrossRef]
32. Morel, M.C.; Newcombe, J.R.; Swindlehurst, J.C. The effect of age on multiple ovulation rates, multiple pregnancy rates and embryonic vesicle diameter in the mare. *Theriogenology* **2005**, *63*, 2482–2493. [CrossRef]
33. Fradinho, M.J.; Correia, M.J.; Grácio, V.; Bliebernicht, M.; Farrim, A.; Mateus, L.; Martin-Rosset, W.; Bessa, R.J.B.; Caldeira, R.M.; Ferreira-Dias, G. Effects of body condition and leptin on the reproductive performance of Lusitano mares on extensive systems. *Theriogenology* **2014**, *81*, 1214–1222. [CrossRef]
34. Taberner, E.; Medrano, A.; Peña, A.; Rigau, T.; Miró, J. Oestrus cycle characteristics and prediction of ovulation in Catalonian jennies. *Theriogenology* **2008**, *70*, 1489–1497. [CrossRef]
35. Henry, M.; Figueiredo, A.E.; Palhares, M.S.; Coryn, M. Clinical and endocrine aspects of the oestrous cycle in donkeys (*Equus asinus*). *J. Reprod. Fertil. Suppl.* **1987**, *35*, 297–303.
36. Blanchard, T.L.; Taylor, T.S.; Love, C.L. Estrous cycle characteristics and response to estrus synchronization in mammoth asses (*Equus asinus americanus*). *Theriogenology* **1999**, *52*, 827–834. [CrossRef]
37. Vandeplassche, G.M.; Wesson, J.A.; Ginther, O.J. Behavioral, follicular and gonadotropin changes during the estrous cycle in donkeys. *Theriogenology* **1981**, *16*, 239–249. [CrossRef]
38. Quaresma, M.; Payan-Carreira, R. Characterization of the estrous cycle of Asinina de Miranda jennies (*Equus asinus*). *Theriogenology* **2015**, *83*, 616–624. [CrossRef]
39. Ginther, O.J.; Scraba, S.T.; Bergfelt, D.R. Reproductive seasonality of the jenney. *Theriogenology* **1987**, *27*, 587–592. [CrossRef]
40. Allen, W.R.; Kydd, J.; Boyle, M.S.; Antczak, D.F. Between-species transfer of horse and donkey embryos: A valuable research tool. *Equine Vet. J.* **1985**, *17*, 53–62. [CrossRef]
41. Vendramini, O.M.; Bruyas, J.F.; Fieni, F.; Battut, I.; Tainturier, D. Embryo transfer in Poitou donkeys, preliminary results. *Theriogenology* **1997**, *47*, 409. [CrossRef]
42. Peña-Alfaro, C.E.; Barros, L.O.; Carneiro, G.F.; Gastal, M.O.; Gastal, E.L. Embryo transfer in Pega donkeys (*Equus asinus*) in Brazil. *J. Equine Vet. Sci.* **2014**, *34*, 185. [CrossRef]
43. Squires, E.; Barbacini, S.; Matthews, P.; Byers, W.; Schwenzer, K.; Steiner, J.; Loomis, P. Retrospective study of factors affecting fertility of fresh, cooled and frozen semen. *Equine Vet. Educ.* **2006**, *18*, 96–99. [CrossRef]
44. Camillo, F.; Vannozzi, I.; Rota, A.; Romagnoli, S.; Aria, G. Comparison of embryo recovery rates from two years old and mature mares. In *Havemeyer Foundation Monograph Series 3, Proceedings of the 5th International Symposium on Equine Embryo Transfer, Saari, Finland, 6–9 July 2000*; Katila, T., Wade, J.F., Eds.; R & W Publications (Newmarket) Limited: Suffolk, UK, 2001; p. 86.
45. Squires, E.L. Embryo transfer challenges and perspectives. *Rev. Bras. Reprod. Anim.* **2013**, *37*, 105–107.
46. Carluccio, A.; Panzani, S.; Tosi, U.; Faustini, M.; De Amicis, I.; Veronesi, M.C. Efficacy of hCG and GnRH for inducing ovulation in the jenny. *Theriogenology* **2007**, *68*, 914–919. [CrossRef]
47. Fleury, J.J.; Alvarenga, M.A. Effects of collection day on embryo recovery and pregnancy rates in a nonsurgical equine embryo transfer program. *Theriogenology* **1999**, *51*, 261. [CrossRef]
48. Stout, T.A. Equine embryo transfer: Review of developing potential. *Equine Vet J.* **2006**, *38*, 467–478. [CrossRef] [PubMed]
49. Pessoa, M.A.; Cannizza, A.P.; Reghini, M.F.S.; Alvarenga, M.A. Embryo Transfer Efficiency of Quarter Horse Athletic Mares. *J. Equine Vet. Sci.* **2011**, *31*, 703–705. [CrossRef]
50. Carnevale, E.M.; Beisner, A.E.; McCue, P.M.; Bass, L.D.; Squires, E.L. Uterine changes associated with repeated inseminations and embryo collections in mares. In Proceedings of the 51st Annual Convention of American Association of Equine Practitioner, Seattle, WA, USA, 3–5 December 2005; pp. 202–203.
51. Losinno, L.; Aguilar, J.; Lisa, H. Impact of multiple ovulations in a commercial equine embryo transfer programme. In *Havemeyer Foundation Monograph Series 3, Proceedings of the 5th International Symposium on Equine Embryo Transfer, Saari, Finland, 6–9 July 2000*; Katila, T., Wade, J.F., Eds.; R & W Publications (Newmarket) Limited: Suffolk, UK, 2001; pp. 81–83.

52. Riera, F.L.; Roldán, J.E.; Hinrichs, K. Ipsilateral double ovulation is associated with reduced embryo recovery in mares. In *Havemeyer Foundation Monograph Series 14, Proceedings of the 6th International Symposium on Equine Embryo Transfer, Rio de Janeiro, Brazil, 4–6 August 2004*; Alvarenga, M., Wade, J.F., Eds.; R & W Comunications: Suffolk, UK, 2005; pp. 108–109.
53. Carmo, M.T.; Alvarenga, M.A.; Losinno, L.; Aquilar, J.J. Oocyte transport to the oviduct of superovulated mares. *Anim. Reprod. Sci.* **2006**, *94*, 337–339.
54. Ono, T.; Isobe, T.; Morita, Y.; Do, L.T.K.; Tanihara, F.; Taniguchi, M.; Takagi, M.; Otoi, T. Effects of parity and season on pregnancy rates after the transfer of embryos to repeat-breeder Japanese Black beef cattle. *Arch. Anim. Breed* **2016**, *59*, 45–49. [CrossRef]

Publisher's Note: MDPI stays neutral with regard to jurisdictional claims in published maps and institutional affiliations.

animals

MDPI

Article

High Exposure to *Toxoplasma gondii* and *Neospora* Spp. in Donkeys in Israel: Serological Survey and Case Reports

Sharon Tirosh-Levy [1,2], Amir Steinman [1], Avital Minderigiu [1], Ori Arieli [1], Igor Savitski [2], Ludmila Fleiderovitz [2], Nir Edery [3], Gili Schvartz [1,4] and Monica Leszkowicz Mazuz [2,*]

1 Koret School of Veterinary Medicine, The Robert H. Smith Faculty of Agriculture, Food and Environment, The Hebrew University of Jerusalem, Rehovot 7610001, Israel; sharontirosh@gmail.com (S.T.-L.); amirst@savion.huji.ac.il (A.S.); avitalmind@gmail.com (A.M.); oriarieli6@gmail.com (O.A.); giliun@gmail.com (G.S.)
2 Division of Parasitology, Kimron Veterinary Institute, Bet Dagan 50250, Israel; igors@moag.gov.il (I.S.); ludaf@moag.gov.il (L.F.)
3 Division of Pathology, Kimron Veterinary Institute, Bet Dagan 50250, Israel; nire@moag.gov.il
4 Division of Virology, Kimron Veterinary Institute, Bet Dagan 50250, Israel
* Correspondence: monical@moag.gov.il; Tel.: +972-3-968-1690

Received: 16 September 2020; Accepted: 14 October 2020; Published: 19 October 2020

check for updates

Simple Summary: *Toxoplasma gondii* and *Neospora* spp. are major pathogenic parasites of animals worldwide, with the first also affecting humans. These parasites have two-host life cycles, with the cat and the dog being the definitive hosts of *T. gondii* and *N. caninum*, respectively. Both parasites can infect various animal species, as intermediate hosts, in which they form tissue cysts and may cause abortions and neurological disease. Both parasites have been reported in wild and domestic animals in Israel. This study aimed to evaluate the serologic exposure of donkeys to these parasites. A total of 98 donkeys were examined. Half of them ($n = 49$) were from animal shelters in Israel, and the rest ($n = 49$) were working donkeys from the Palestinian Authority. Anti-*T. gondii* antibodies and anti-*Neospora* spp. antibodies were found in 94% and in 70% of the donkeys, respectively. In addition, two cases of donkeys presenting *N. caninum* tissue cysts, which were detected during post-mortem examination, were described. This is the first report of the exposure of donkeys to *Toxoplasma gondii* and *Neospora* spp. in the area. The exposure of donkeys to both parasites was considerably higher than the exposure of other species in the area and may be the result of poor husbandry conditions and higher exposure to infection. These results indicate that donkeys may have an important role in the maintenance and transmission of these parasites.

Abstract: *Toxoplasma gondii* and *Neospora* spp. are closely related cyst-forming coccidian parasites, which infect various animal species and have considerable zoonotic and economic implications, respectively. Both parasites are endemic in Israel and have been reported to infect wild and domestic animals. This study was conceived to evaluate the serologic exposure of donkeys to these parasites. Serum samples were collected from 98 donkeys. Half of them ($n = 49$) were from animal shelters in Israel, and the rest ($n = 49$) were working donkeys from the Palestinian Authority. The donkeys were screened for the presence of anti-*Toxoplasma* and anti-*Neospora* antibodies by immunofluorescence antibody tests (IFATs). The seroprevalence of *T. gondii* and *Neospora* spp. was 94% and 70%, respectively, and 69% of the donkeys were exposed to both parasites. In addition, *N. caninum* tissue cysts were documented in two donkeys during post-mortem examination. This is the first report of the exposure of donkeys to *T. gondii* and *Neospora* spp. in the area. The high prevalence found in this study suggests that donkeys may have a role in the maintenance of these parasites in the area, thus serving as a source of infection for the definitive hosts.

Keywords: *Toxoplasma gondii*; *Neospora*; donkey; IFAT; serology

1. Introduction

Cyst-forming coccidian parasites, mainly *Toxoplasma gondii* and *Neospora* spp., are major pathogens of animals with worldwide distribution. These closely related apicomplexan intracellular parasites have heteroxenous life cycles, with felids being the definitive hosts of *T. gondii* and canids of *N. caninum*. Both parasites infect various mammalian species as intermediate hosts in which they form tissue cysts [1].

Toxoplasmosis may cause abortions, fetal damage or neurologic disease in a wide range of animals and humans. It is a prominent cause of abortion in sheep and a common zoonosis [1–4]. The clinical significance of *T. gondii* infection in horses is unclear. Nevertheless, horses may be exposed to or infected by the *T. gondii* parasite with no apparent clinical signs, and may pose a zoonotic risk through the consumption of infective horse or donkey meat [5–7]. In Israel, the seroprevalence of *T. gondii* was evaluated as 20–60% in humans [8], 25% in sheep [9], 36% in dogs [10], 17% in cats [11], 43% in crows, 40% in Griffon vultures [12] and 2.5% in horses [13].

Neosporosis causes abortion and neurologic disease in various animals. *Neospora caninum* is a major cause of abortion and economic loss in the cattle industry, and has been reported in horses [1,14]. *Neospora hughesi* is known to affect only equids and was isolated from cases of neurological disease in horses [14,15]. These two species are indistinguishable when using standard serological methods [16]. Horses may be infected with both parasites, and since most surveys were based on serology, the reported prevalence in horses and donkeys was of *Neospora* spp. In Israel, the seroprevalence of *Neospora* spp. was evaluated as 51% in cattle [17], 1.3–67% in wild animals [18] and 12% in horses [19]. The seroprevalence in aborting mares and cows was significantly higher than in the general population [17,19].

Domestic donkeys (*Equus africanus acinus*) in Israel are used as draught animals, riding animals, as pets and in petting zoos, and therefore are in close contact with humans. In other countries, donkeys are also used for their meat as food, and certain donkey-derived items are also important for traditional Chinese medicinal purposes [20]. Since donkeys are closely related to horses, they may be affected by similar pathogens. Exposure to both parasites has been reported in donkeys worldwide [5,6,21–27]. However, little is known about the role of donkeys in the epidemiology of these parasites. In Israel, data concerning the exposure of donkeys to infectious diseases are scarce. The aim of this study was to assess whether donkeys in Israel are exposed to *T. gondii* and *Neospora* spp. protozoan parasites, and to report two neosporosis cases of affected donkeys.

2. Materials and Methods

2.1. Sample Collection for Serological Survey

The sample size was calculated according to the prevalence of *Neospora* spp. in horses in Israel (12%) [19] using WinPepi 11.43®. A sample size of 77 donkeys met the criteria for an expected prevalence under 30%, with a relative error of ≤20% and 95% confidence level.

Since no data are available for estimating the distribution of donkeys in Israel, it was unfeasible to design a sample that would reliably represent the donkey population in the area. Therefore, donkeys were sampled at two donkey shelters in Israel that receive donkeys from different locations (49 donkeys), and at three locations in the Palestinian Authority (PA) to which working animals were brought to receive veterinary care given through a humanitarian association (49 donkeys).

Blood collections were performed with owners' consent, and the study was approved by the Internal Research Review Committee of the Koret School of Veterinary Medicine—Veterinary Teaching Hospital (KSVM-VTH/23_2014). Blood was collected from the jugular vein of each animal into a sterile vacuum tube without anticoagulant. Sera were obtained from the clotted blood samples by centrifugation (4000× *g* for 10 min) and stored at −20 °C until processing. During sample collection, data for each donkey were recorded including the farm's location, sex, age, and origin of the donkey, when available. At the time of sampling, all donkeys were apparently healthy according to both the owners and the veterinarians who collected the samples.

2.2. Serological Screening Using Immunofluorescence Antibody Test (IFAT)

Serological screening for the presence of anti-*T. gondii* antibodies was conducted on 1:2 serial dilutions of the sera, starting at 1:64 as a cut-off value for screening [28,29], up to a final dilution of 1:16,384.

Serological screening for the presence of anti-*Neospora* spp. antibodies was conducted on 1:2 serial dilutions of the sera, starting at 1:50 as a cut-off value for screening [19,25,29], up to a final dilution of 1:3200.

In-house antigens were prepared as previously described [30]. Briefly, free tachyzoites were obtained from an infected Vero cell culture, separated by centrifugation, diluted in phosphate buffered saline (PBS), dropped onto slides and stored at −80 °C until use. The slides were thawed at 37 °C for 30 min and air dried. The sera were diluted in PBS with 1% bovine serum albumin (BSA). A volume of 35 µL of serum was added to each antigen drop well and incubated in a humid chamber at 37 °C for 30 min. The slides were washed for 10 min in carbonate buffer (pH 9, diluted 1:4 in distilled water) before the application of 35 µL of anti-horse fluorescein isothiocyanate (FITC) secondary antibody (Sigma-Aldrich, St. Louis, MO, USA) diluted 1:80 with PBS–BSA, and incubation at 37 °C for 30 min in a humid chamber. The slides were later washed for 10 min in carbonate buffer, dried, mounted with glycerol/carbonate buffer (50%) and examined under a fluorescence microscope. Positive and negative control serum samples from positive horses were added to each run.

2.3. Statistical Analysis of Serology Results

Risk factors associated with exposure to either parasite or both parasites were assessed separately and included the farm, the geographical area (Israel versus the Palestinian Authority), the donkey's sex and age. Association with nominal independent variables was assessed by using the χ^2 test or Fisher's exact test, as appropriate, and odds ratios were calculated. Association with quantitative parameters was assessed using *t*-tests. Association between potential risk factors and antibody titer was assessed using ANOVA. Statistical significance was set at $p < 0.05$. The analysis was performed using the SPSS 22.0® and Win Pepi 11.43® statistical software.

2.4. Sample Collection, Histopathology and Polymerase Chain Reaction (PCR) of Clinical Samples

Skeletal muscle tissue from the triceps brachi muscle was collected from two donkeys (Donkeys 1 and 2) and was sent to the Kimron Veterinary Institute pathology department for post-mortem examination. Brain tissue was collected from one of these donkeys (Donkey 1), which was reported to have neurological signs prior to euthanasia, and stored at −20 °C until processing. Muscle tissue samples were transferred into 4% formaldehyde solution before the preparation of histological slides with hematoxylin and eosin (H&E) staining.

DNA was extracted from the paraffin-embedded muscle tissue dissolved in xylene and washed with PBS, using the DNeasy blood and tissue kit (QIAGEN, Hilden, Germany) according to the manufacturer's instructions. The presence of coccidian parasites was confirmed using PCR targeting the small-subunit rRNA (COC1 F: AAGTATAAGCTTTTATACGGCT, COC2 R: CACTGCCACGGTAGTCCAATAC), and the classification of the species was achieved by sequence- and species-specific PCRs targeting *Besnoitia besnoiti*, *Hammonidia*, *N. caninum*, *Sarcocystis* spp., *T. gondii* and *Trypanosoma* spp. (the primers and targets are specified in Table 1).

RNA was extracted from brain tissue, and the presence of West Nile virus (WNV) RNA was tested for and confirmed by real-time reverse transcription PCR (qRT-PCR) and further amplified and sequenced, as previously described [31].

Table 1. The PCR primers used to identify the coccidian species of the tissue cysts found in donkeys' muscle tissue.

Organism	Primer	Sequence	Amplicon Size (bp)	Target Gene	Reference
Besnoitia spp.	ITS1-F	TGACATTTAATAACAATCAACCCTT	250	ITS	[32]
	ITS1-R	GGTTTGTATTAACCAATCCGTGA			
	Bes-F	ATTGGGACCGTTTTGTGG		ITS	
	Bes-R	CCTCTCGAGGCTACAAGTCG			
	Bes-F2	CCTCCTCACTCTGCTATCACG	750	(nested)	
	Bes-R2	TTCCACTGGTAACGCCTCT			
Sarcocyst spp.	71-F	CGGATCGCATTATGACCTTT		18S rRNA	
	894-R	GGTGCAGGAGAAGTCAAGGA			
	317-F	ATTGGAATGATGGGAATCCA	300	(nested)	
	548-R	TGCCACCAACACAATGAAGT			
Toxoplasma gondii	Tox4	CGCTGCAGGGAGGAAGACGAAAG	500	Non-coding	[33]
	Tox5	CGCTGCAGACACAGTGCATCTGG			
Hammondia spp.	JS4	CGAAATGGGAAGTTTTGTGAAAC	270	ITS	[34]
	JS5	CAGCAGCTACATACGTAGA			
Neospora spp.	476-F	CTGCTGACGTGTCGTTGT		NC5	[35]
	1014-R	CATCTACCAGGCCGCTCTTC			
	631-F	GCGTCAGGGTGAGGACAGTG	279	(nested)	
	910-R	CTCTCCGTTCGCCAGCAGTG			
	ITS1D-F	TACCGATTGAGTGTTCCGGTG		ITS	[36]
	ITS1D-R	GCAATTCACATTGCGTTTCGC			
	ITS1Di-F	CGTAACAAGGTTTCCGTAGG	480	(nested)	
	ITS1Di-R	TTCATCGTTGCGCGAGCCAAG			
Trypanosoma spp.	ITS1	GATTACGTCCCTGCCATTTG		ITS	
	ITS2	TTGTTCGCTATCGGTCTTCC			
	ITS3	GGAAGCAAAAGTCGTAACAAGG	1200	(nested)	
	ITS4	TGTTTTCTTTTCCTCCGCTG			

All positive PCR products were sent for sequencing by HyLabs (Rehovot, Israel). Sequences were evaluated using the Chromas software (Technelysium Pty Ltd., Tewantin, QLD, Australia, version 2.6) and assembled using the MEGA7 software (http://www.megasoftware.net, version 7.0.18). Consensus sequences were created for each gene from both donkeys and were submitted to GenBank.

3. Results

3.1. Serologic Exposure to T. gondii and Neospora Spp. in Donkeys

The study population comprised donkeys from Israel ($n = 49$) and the Palestinian Authority ($n = 49$). Sixty of the donkeys were males (61%), and 38 were females (39%). Age was available for 70 of the donkeys and ranged between four months and 25 years (mean = 7.6 years, median = 7.0 years, standard deviation = 5.1 years). All donkeys were apparently healthy during blood collection.

Anti-*T. gondii* antibodies were detected in 92 of 98 donkeys (94%). The antibody titers ranged between 1:64 and 1:16,384 (Figure 1a), while high titers (≥1:256) were found in 66% of the animals. No significant risk factors for exposure were identified.

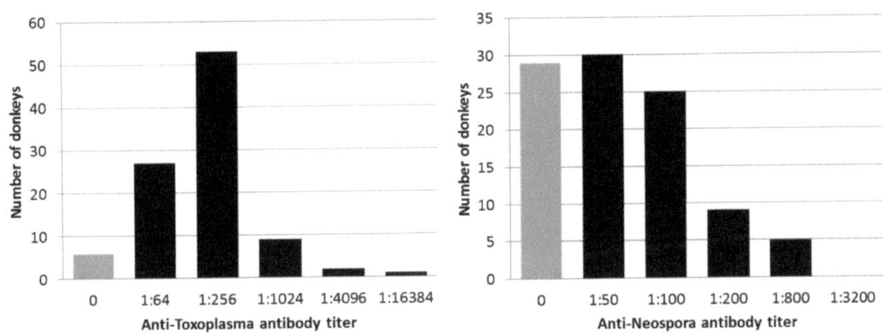

Figure 1. The distribution of anti-*T. gondii* (**a**) and anti-*Neospora* spp. (**b**) antibody titers in 98 donkeys, as detected by immunofluorescence antibody tests (IFATs).

Anti-*Neospora* spp. antibodies were detected in 69 of 98 donkeys (70%). The antibody titers ranged between 1:50 and 1:800 (Figure 1b), while high titers (≥1:200) were found in 14% of the animals. No significant risk factors for exposure were identified. High antibody titers (≥1:200) were associated with one farm in Israel. In this farm, 10 out of 25 donkeys had high antibody titers, representing 71.4% (10 of 14) of the donkeys with high antibody titers ($p > 0.001$, odds ratio (OR) = 11.5, 95% confidence interval (CI) = 2.76–55.29).

The majority of donkeys (68/98, 69.4%) were exposed to both parasites. An additional 24 donkeys (24.5%) were exposed only to *Toxoplasma gondii*, one donkey (1%) was exposed only to *Neospora* spp., and five donkeys (5.1%) were not exposed to any of these parasites. No significant risk factors for co-exposure were identified.

3.2. Clinical Cases of Neosporosis in Donkeys

Parasitic tissue cysts were identified in the skeletal muscles of two donkeys sent for post-mortem examination (Figure 2). Both donkeys were sent from the same animal shelter. The first donkey (Donkey 1) was over 30 years old and presented with neurological signs and severe weakness before euthanasia. The second donkey (Donkey 2) was over 20 years old, and his left thoracic limb had been amputated several years prior to his death from unrelated causes. After his demise, the right thoracic limb was sent for evaluation for any potential effect of the amputation on the contralateral limb. The donkey that had presented with neurological signs (Donkey 1) was diagnosed as being infected with WNV after viral RNA was isolated and sequenced from its brain tissue (MT828577). In both donkeys, multifocal tissue cysts containing parasites were observed (Figure 2). *Neospora caninum* DNA was identified in the skeletal muscle tissue of both donkeys and was confirmed by the sequencing of two different species-specific target genes (NC5 (MT831977) and ITS1 (MT826198), over 97% homology with species-specific sequences in GenBank). Specific PCRs targeting *Besnoitia* spp., *Hammonidia*, *T. gondii*, *Neospora* spp., *Sarcocystis* spp., *T. gondii* and *Trypanosoma* spp. were negative, and the presence of these related cyst-forming parasites was ruled out.

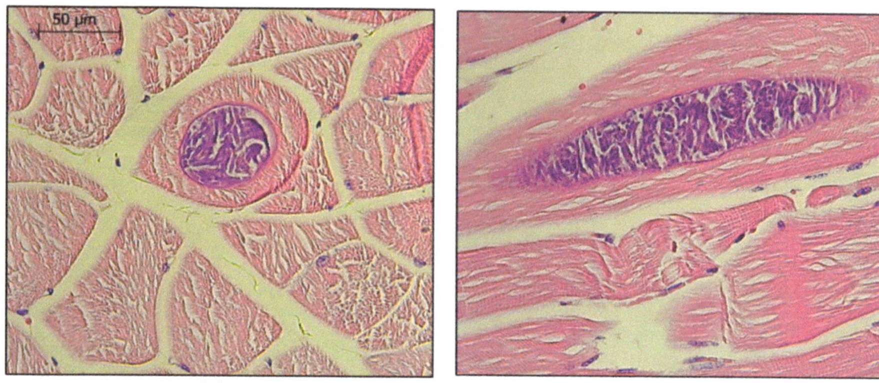

(a) (b)

Figure 2. *Neospora caninum* tissue cysts in the skeletal muscles of the front limb of Donkey 1: (**a**) transverse section; (**b**) longitudinal section. Giemsa, ×1000.

4. Discussion

The seroprevalence of both *T. gondii* and *N. caninum* in donkeys in Israel is high, and higher than the recorded prevalence in any other mammalian species in the area, including horses [13,18]. Donkeys in Israel are sometimes kept as burden animals, being more prevalent in Arab and Bedouin villages and often receiving little veterinary care, in comparison to horses. Half of the donkeys in this survey were sampled in animal shelters that receive neglected donkeys from various locations, while the other half were sampled in Arab villages in the Palestinian authority, by a veterinarian giving free veterinary care through a humanitarian organization. The high exposure to both parasites may be the result of the poor sanitation associated with low-income populations, which may increase the chance of exposure to oocysts in water sources or the environment. The association between low income and higher exposure to *T. gondii* has been previously described in humans and in horses [7,37,38]. In addition, in poor sanitary conditions, stray dogs and cats may have access to and feed on donkey carcasses, thus enhancing the maintenance and transmission of Neosporosis and Toxoplasmosis in these areas.

The seroprevalence of *T. gondii* in horses varies between countries and ranges between 1.2% in Sweden [39] and 71.2% in Iran [40]. The differences in prevalence may be associated with housing, stable hygiene and feeding practices [7,41]. The prevalence of *T. gondii* was higher in donkeys (72.7%) than in horses (27.7%) in a recent study from Brazil [42], as well as in other studies from Spain [23] and Pakistan [43]. However, since there are fewer studies evaluating both horses and donkeys in similar cohorts, it is difficult to determine whether these differences reflect a higher susceptibility of donkeys, or merely reflect differences in sanitation or management practices between these species. Since horses are considered to be naturally resistant to *T. gondii* infection [42], it is possible that they develop lower antibody titers that are not detected by serological tests.

In Israel, *T. gondii* seroprevalence was significantly higher in the Arab population (60.4%) than in Bedouins (27.5%) or Jews (19.9%) [8], and similar to the reported prevalence in Lebanon (62.2%) [44]. The seroprevalence was age-dependent and reached 96% in Arabs over 60 years of age [8].

The differences within ethnic groups in Israel may be attributable to a combination of economic status, exposure to animals and climate [8]. Since donkeys in Israel are more abundant in Arab settlements, the high seroprevalence may reflect similar conditions for exposure.

Although the consumption of donkey or horse meat is not common in this area, infection from contaminated meat is still a potential source for zoonotic transmission. Viable *T. gondii* parasites have been isolated from horse and donkey meat intended for human consumption worldwide [5,6,45–47], with a possible link to human disease [46].

In a recent study [48], an association between *T. gondii* seropositivity and the prevalence of impaired cognitive function was demonstrated in humans. Researchers theorize that behavior manipulation increases the parasite's likelihood of transmission by manipulating the host to engage in risky behaviors so that the host is likely to be preyed upon, particularly by a feline [48]. This was demonstrated in rats [49]. In this study, behavioral changes were not observed in any of the animals. To the best of our knowledge, behavioral changes have never been reported in equids in regard to *T. gondii* infection.

Neosporosis is not considered zoonotic, and its main impact is economic, mostly due to its effect on the reproduction of cattle and small ruminants, which are considered as the main intermediate hosts [14,50]. Therefore, fewer data are available regarding its prevalence in equines. In Israel, the prevalence in horses (12%) [19] was considerably lower than our findings in donkeys (70%). In neighboring Jordan, the reported seroprevalence in horses was 32% [51], and that in small ruminants was 63% [52]. Both dogs and donkeys often accompany small ruminant flocks in the Middle East, which may increase the chance of infection of both the primary and secondary hosts. In 1998, a new species of *Neospora*, *N. hughesi*, was identified in a horse from California [15]. Since then, *N. hughesi* has been reported only in horses and mostly from North America; however, since the two *Neospora* species are indistinguishable serologically, the global distribution of *N. hughesi* is unclear [14]. In the two cases of tissue cysts described here, parasites were classified as *N. caninum* based on two loci. In addition, *N. caninum* was identified in several cases of *Neospora*-induced abortion in mares in Israel (Mazuz et al., unpublished data), while *N. hughesi* has never been described in the area.

In horses, both toxoplasmosis and neosporosis had been associated with neurological disease, while neosporosis had also been described as a cause of abortions and neonatal disease [53,54]. However, reports of equine clinical cases are rare, and the majority of seropositive horses are asymptomatic. The clinical significance of high exposure to these parasites in donkeys is unclear. All donkeys in this study were apparently healthy, and to the best of our knowledge, clinical toxoplasmosis or neosporosis in donkeys has never been reported in Israel or elsewhere. Although one of the clinical cases in this report (Donkey 1) exhibited marked neurological signs, the cause of the neurological disease was determined as WNV, which is the most common cause of neurological disease in horses in the area [55]. Nevertheless, the identification of *N. caninum* tissue cysts in muscle specimens from two donkeys suggests that donkeys may be susceptible to clinical infection.

5. Conclusions

This is the first epidemiological survey investigating the exposure of donkeys to *T. gondii* and *N. caninum* in Israel. The high seroprevalence of both parasites in donkeys may reflect a high susceptibility of this species or high exposure due to poor husbandry conditions. The pathological findings of *N. caninum* tissue cysts in two donkeys suggest that donkeys may also be clinically infected and highlight the potential of donkeys to be a source of infection to other animal species. The higher exposure of donkeys in relation to other animal species in these areas suggests that donkeys could be used as sentinels to monitor exposure to these important, and potentially zoonotic, parasites.

Author Contributions: Conceptualization, A.S., M.L.M. and S.T.-L.; formal analysis, S.T.-L.; investigation, A.M., G.S., L.F., N.E. and I.S.; resources, S.T.-L., M.L.M., A.S. and O.A.; writing—original draft preparation, S.T.-L.; writing—review and editing, A.S. and M.L.M.; supervision, M.L.M. All authors have read and agreed to the published version of the manuscript.

Funding: This research received no external funding.

Acknowledgments: The authors wish to thank the workers of the donkey shelters and the workers in the veterinary clinics in the Palestinian Authority for their assistance and collaboration.

Conflicts of Interest: The authors declare no conflict of interest.

References

1. Dubey, J.P.; Lindsay, D.S. Neosporosis, toxoplasmosis, and sarcocystosis in ruminants. *Vet. Clin. North. Am. Food Anim. Pract.* **2006**, *22*, 645–671. [CrossRef] [PubMed]
2. Benavides, J.; Fernandez, M.; Castano, P.; Ferreras, M.C.; Ortega-Mora, L.; Perez, V. Ovine toxoplasmosis: A new look at its pathogenesis. *J. Comp. Pathol.* **2017**, *157*, 34–38. [CrossRef] [PubMed]
3. Dubey, J.P. Toxoplasmosis in sheep-The last 20 years. *Vet. Parasitol.* **2009**, *163*, 1–14. [CrossRef]
4. Dubey, J.P. *Toxoplasmosis of Animals and Humans*; CRC Press: Boca Raton, FL, USA, 2016.
5. Alvarado-Esquivel, C.; Alvarado-Esquivel, D.; Dubey, J.P. Prevalence of *Toxoplasma gondii* antibodies in domestic donkeys (*Equus asinus*) in Durango, Mexico slaughtered for human consumption. *BMC Vet. Res.* **2015**, *11*, 6. [CrossRef]
6. Cong, W.; Chen, L.; Shan, X.F.; Qian, A.D.; Meng, Q.F. First genetic characterization of *Toxoplasma gondii* infection in donkey meat slaughtered for human consumption in Shandong province, eastern China. *Infect. Genet. Evol.* **2018**, *61*, 1–3. [CrossRef] [PubMed]
7. Li, X.; Ni, H.B.; Ren, W.X.; Jiang, J.; Gong, Q.L.; Zhang, X.X. Seroprevalence of *Toxoplasma gondii* in horses: A global systematic review and meta-analysis. *Acta Trop.* **2020**, *201*, 105222. [CrossRef] [PubMed]
8. Markovich, M.P.; Shohat, T.; Riklis, I.; Avni, R.; Yujelevski-Rozenblit, D.; Bassal, R.; Cohen, D.; Rorman, E. Seroepidemiology of *Toxoplasma gondii* infection in the Israeli population. *Epidemiol. Infect.* **2014**, *142*, 149–155. [CrossRef] [PubMed]
9. Shkap, V.; Pipano, E.; Marcus, S.; Rapoport, E. The prevalence of *Toxoplasma gondii* antibodies in sheep and cattle in Israel. *Isr. J. Vet. Med.* **1992**, *47*, 100.
10. Baneth, G.; Shkap, V.; Savitsky, I.; Pipano, E. The prevalence of antibodies to *Toxoplasma gondii* in dogs in Israel. *Isr. J. Vet. Med.* **1996**, *51*, 31–34.
11. Salant, H.; Spira, D.T. A cross-sectional survey of anti-*Toxoplasma gondii* antibodies in Jerusalem cats. *Vet. Parasitol.* **2004**, *124*, 167–177. [CrossRef]
12. Salant, H.; Hamburger, J.; King, R.; Baneth, G. *Toxoplasma gondii* prevalence in Israeli crows and Griffon vultures. *Vet. Parasitol.* **2013**, *191*, 23–28. [CrossRef]
13. Aharonson-Raz, K.; Baneth, G.; Lopes, A.P.; Brancal, H.; Schallig, H.; Cardoso, L.; Steinman, A. Low Seroprevalence of *Leishmania infantum* and *Toxoplasma gondii* in the Horse Population in Israel. *Vector Borne Zoonotic Dis.* **2015**, *15*, 726–731. [CrossRef]
14. Dubey, J.; Hemphill, A.; Calero-Bernal, R.; Schares, G. *Neosporosis in Animals*; CRC Press: Boca Raton, FL, USA, 2017.
15. Marsh, A.E.; Barr, B.C.; Packham, A.E.; Conrad, P.A. Description of a new *Neospora* species (Protozoa: Apicomplexa: Sarcocystidae). *J. Parasitol.* **1998**, *84*, 983–991. [CrossRef] [PubMed]
16. Marsh, A.E.; Howe, D.K.; Wang, G.; Barr, B.C.; Cannon, N.; Conrad, P.A. Differentiation of *Neospora hughesi* from *Neospora caninum* based on their immunodominant surface antigen, SAG1 and SRS2. *Int. J. Parasitol.* **1999**, *29*, 1575–1582. [CrossRef]
17. Mazuz, L.M.; Fish, L.; Molad, T.; Savitsky, I.; Wolkomirsky, R.; Leibovitz, B.; Shkap, V. *Neospora Caninum* as Causative-Pathogen of Abortion in Cattle. *Isr. J. Vet. Med.* **2011**, *66*, 14–18.
18. Mazuz, M.L.; Alvarez-Garcia, G.; King, R.; Savisky, I.; Shkap, V.; Ortega-Mora, L.M.; Gutierrez-Exposito, D. Exposure to *Neospora* spp. and *Besnoitia* spp. in wildlife from Israel. *Int. J. Parasitol. Parasites Wildl.* **2018**, *7*, 317–321. [CrossRef] [PubMed]
19. Kligler, E.B.; Shkap, V.; Baneth, G.; Mildenberg, Z.; Steinman, A. Seroprevalence of *Neospora* spp. among asymptomatic horses, aborted mares and horses demonstrating neurological signs in Israel. *Vet. Parasitol.* **2007**, *148*, 109–113. [CrossRef] [PubMed]
20. Bennett, R.; Pfuderer, S. The Potential for New Donkey Farming Systems to Supply the Growing Demand for Hides. *Animals* **2020**, *10*, 718. [CrossRef]
21. Cong, W.; Nie, L.B.; Qin, S.Y.; Wang, W.L.; Qian, A.D.; Meng, Q.F. Prevalence of *Neospora* spp. in donkeys in China. *Parasite* **2018**, *25*. [CrossRef]

22. Galvao, C.M.M.D.; Rezende-Gondim, M.M.; Chaves, A.C.R.; Schares, G.; Ribas, J.R.L.; Gondim, L.F.P. Brazilian donkeys (*Equus asinus*) have a low exposure to *Neospora* spp. *Rev. Bras. Parasitol.* **2015**, *24*, 340–344. [CrossRef] [PubMed]
23. Garcia-Bocanegra, I.; Cabezon, O.; Arenas-Montes, A.; Carbonero, A.; Dubey, J.P.; Perea, A.; Almeria, S. Seroprevalence of *Toxoplasma gondii* in equids from Southern Spain. *Parasitol. Int.* **2012**, *61*, 421–424. [CrossRef]
24. Gennari, S.M.; Esmerini, P.D.; Lopes, M.G.; Soares, H.S.; Vitaliano, S.N.; Cabral, A.D.; Pena, H.F.J.; Horta, M.C.; Cavalcante, P.H.; Fortes, K.P.; et al. Occurrence of antibodies against *Toxoplasma gondii* and its isolation and genotyping in donkeys, mules, and horses in Brazil. *Vet. Parasitol.* **2015**, *209*, 129–132. [CrossRef] [PubMed]
25. Gennari, S.M.; Pena, H.F.D.; Lindsay, D.S.; Lopes, M.G.; Soares, H.S.; Cabral, A.D.; Vitaliano, S.N.; Amaku, M. Prevalence of antibodies against *Neospora* spp. and *Sarcocystis neurona* in donkeys from northeastern Brazil. *Rev. Bras. Parasitol.* **2016**, *25*, 109–111. [CrossRef] [PubMed]
26. Machacova, T.; Bartova, E.; Di Loria, A.; Sedlak, K.; Guccione, J.; Fulgione, D.; Veneziano, V. Seroprevalence and risk factors of *Neospora* spp. in donkeys from Southern Italy. *Vet. Parasitol.* **2013**, *198*, 201–204. [CrossRef] [PubMed]
27. Nazir, M.M.; Ayaz, M.M.; Ahmed, A.N.; Rasheed, I.; Faraz, A.; Akram, Q.; Althtar, S.; Maqbool, A.; Tabassum, S.; Zheng, Y.D.; et al. Prevalence and risk factors for IgG antibodies to *Neospora* spp. in three types of equids from Southern Punjab, Pakistan. *Acta Trop.* **2018**, *188*, 240–243. [CrossRef] [PubMed]
28. Dubey, J.P.; Murata, F.H.A.; Cerqueira-Cezar, C.K.; Kwok, O.C.H. *Toxoplasma gondii* infections in horses, donkeys, and other equids: The last decade. *Res. Vet. Sci.* **2020**, *132*, 492–499. [CrossRef]
29. Moreira, T.R.; Sarturi, C.; Stelmachtchuk, F.N.; Andersson, E.; Norlander, E.; de Oliveira, F.L.C.; Machado Portela, J.; Marcili, A.; Emanuelson, U.; Gennari, S.M.; et al. Prevalence of antibodies against *Toxoplasma gondii* and *Neospora* spp. in equids of Western Para, Brazil. *Acta Trop.* **2019**, *189*, 39–45. [CrossRef]
30. Shkap, V.; Reske, A.; Pipano, E.; Fish, L.; Baszler, T. Immunological relationship *between Neospora caninum* and *Besnoitia besnoiti*. *Vet. Parasitol.* **2002**, *106*, 35–43. [CrossRef]
31. Schvartz, G.; Farnoushi, Y.; Berkowitz, A.; Edery, N.; Hann, S.; Steinman, A.; Lublin, A.; Erster, O. Molecular characterization of the re-emerging *West Nile Virus* in avian species and equids in Israel, and pathological description of the disease. *Parasit. Vectors* **2018**, in press.
32. Cortes, H.C.; Reis, Y.; Gottstein, B.; Hemphill, A.; Leitao, A.; Muller, N. Application of conventional and real-time fluorescent ITS1 rDNA PCR for detection of *Besnoitia besnoiti* infections in bovine skin biopsies. *Vet. Parasitol.* **2007**, *146*, 352–356. [CrossRef]
33. Homan, W.L.; Vercammen, M.; De Braekeleer, J.; Verschueren, H. Identification of a 200- to 300-fold repetitive 529 bp DNA fragment in *Toxoplasma gondii*, and its use for diagnostic and quantitative PCR. *Int. J. Parasitol.* **2000**, *30*, 69–75. [CrossRef]
34. Slapeta, J.R.; Koudela, B.; Votypka, J.; Modry, D.; Horejs, R.; Lukes, J. Coprodiagnosis of *Hammondia heydorni* in dogs by PCR based amplification of ITS 1 rRNA: Differentiation from morphologically indistinguishable oocysts of *Neospora caninum*. *Vet. J.* **2002**, *163*, 147–154. [CrossRef] [PubMed]
35. Fish, L.; Mazuz, M.; Molad, T.; Savitsky, I.; Shkap, V. Isolation of *Neospora caninum* from dairy zero grazing cattle in Israel. *Vet. Parasitol.* **2007**, *149*, 167–171. [CrossRef] [PubMed]
36. Wobeser, B.K.; Godson, D.L.; Rejmanek, D.; Dowling, P. Equine protozoal myeloencephalitis caused by *Neospora hughesi* in an adult horse in Saskatchewan. *Can. Vet. J.* **2009**, *50*, 851–853. [PubMed]
37. Rostami, A.; Riahi, S.M.; Contopoulos-Ioannidis, D.G.; Gamble, H.R.; Fakhri, Y.; Shiadeh, M.N.; Foroutan, M.; Behniafar, H.; Taghipour, A.; Maldonado, Y.A.; et al. Acute *Toxoplasma* infection in pregnant women worldwide: A systematic review and meta-analysis. *PLoS Negl. Trop. Dis.* **2019**, *13*, e0007807. [CrossRef] [PubMed]
38. Rostami, A.; Riahi, S.M.; Gamble, H.R.; Fakhri, Y.; Nourollahpour Shiadeh, M.; Danesh, M.; Behniafar, H.; Paktinat, S.; Foroutan, M.; Mokdad, A.H.; et al. Global prevalence of latent toxoplasmosis in pregnant women: A systematic review and meta-analysis. *Clin. Microbiol. Infect.* **2020**, *26*, 673–683. [CrossRef] [PubMed]
39. Jakubek, E.B.; Lunden, A.; Uggla, A. Seroprevalences of *Toxoplasma gondii* and *Neospora* sp. infections in Swedish horses. *Vet. Parasitol.* **2006**, *138*, 194–199. [CrossRef] [PubMed]
40. Hajialilo, E.; Ziaali, N.; Harandi, M.F.; Saraei, M.; Hajialilo, M. Prevalence of anti-*Toxoplasma gondii* antibodies in sport horses from Qazvin, Iran. *Trop. Anim. Health Prod.* **2010**, *42*, 1321–1322. [CrossRef]

41. Tenter, A.M.; Heckeroth, A.R.; Weiss, L.M. *Toxoplasma gondii*: From animals to humans. *Int. J. Parasitol.* **2000**, *30*, 1217–1258. [CrossRef]
42. Munhoz, A.D.; Souza, M.A.; Costa, S.C.L.; Freitas, J.S.; Silva, A.N.D.; Lacerda, L.C.; Cruz, R.D.S.; Albuquerque, G.R.; Pereira, M.J.S. Factors associated with the distribution of natural *Toxoplasma gondii* infection among equids in Northeastern Brazil. *Rev. Bras. Parasitol. Vet.* **2019**, *28*, 283–290. [CrossRef]
43. Saqib, M.; Hussain, M.H.; Sajid, M.S.; Mansoor, M.K.; Asi, M.N.; Fadya, A.A.; Zohaib, A.; Sial, A.U.; Muhammad, G.; Ullah, I. Sero-epidemiology of equine toxoplasmosis using a latex agglutination test in the three metropolises of Punjab, Pakistan. *Trop. Biomed.* **2015**, *32*, 276–285. [PubMed]
44. Bouhamdan, S.F.; Bitar, L.K.; Saghir, H.J.; Bayan, A.; Araj, G.F. Seroprevalence of *Toxoplasma* antibodies among individuals tested at hospitals and private laboratories in Beirut. *J. Med. Liban* **2010**, *58*, 8–11.
45. Shaapan, R.M.; Ghazy, A.A. Isolation of *Toxoplasma gondii* from horse meat in Egypt. *Pak. J. Biol. Sci.* **2007**, *10*, 174–177. [CrossRef]
46. Pomares, C.; Ajzenberg, D.; Bornard, L.; Bernardin, G.; Hasseine, L.; Darde, M.L.; Marty, P. Toxoplasmosis and horse meat, France. *Emerg. Infect. Dis.* **2011**, *17*, 1327–1328. [CrossRef] [PubMed]
47. Pastiu, A.I.; Gyorke, A.; Kalmar, Z.; Bolfa, P.; Rosenthal, B.M.; Oltean, M.; Villena, I.; Spinu, M.; Cozma, V. *Toxoplasma gondii* in horse meat intended for human consumption in Romania. *Vet. Parasitol.* **2015**, *212*, 393–395. [CrossRef]
48. Wiener, R.C.; Waters, C.; Bhandari, R. The association of *Toxoplasma gondii* IgG and cognitive function scores: NHANES 2013-2014. *Parasitol. Int.* **2020**, *78*, 102123. [CrossRef] [PubMed]
49. da Silva, R.C.; Langoni, H. *Toxoplasma gondii*: Host-parasite interaction and behavior manipulation. *Parasitol. Res.* **2009**, *105*, 893–898. [CrossRef]
50. Reichel, M.P.; Ayanegui-Alcérreca, M.A.; Gondim, L.F.; Ellis, J.T. What is the global economic impact of *Neospora caninum* in cattle—the billion dollar question. *Int. J. Parasitol.* **2013**, *43*, 133–142. [CrossRef] [PubMed]
51. Talafha, A.Q.; Abutarbush, S.M.; Rutley, D.L. Seroprevalence and Potential Risk Factors Associated with *Neospora* spp. Infection among Asymptomatic Horses in Jordan. *Korean J. Parasitol.* **2015**, *53*, 163–167. [CrossRef]
52. Abo-Shehada, M.N.; Abu-Halaweh, M.M. Flock-level seroprevalence of, and risk factors for, *Neospora caninum* among sheep and goats in northern Jordan. *Prev. Vet. Med.* **2010**, *93*, 25–32. [CrossRef]
53. James, K.E.; Smith, W.A.; Packham, A.E.; Conrad, P.A.; Pusterla, N. *Toxoplasma gondii* seroprevalence and association with equine protozoal myeloencephalitis: A case-control study of Californian horses. *Vet. J.* **2017**, *224*, 38–43. [CrossRef] [PubMed]
54. Lindsay, D.S. Neosporosis: An emerging protozoal disease of horses. *Equine Vet. J.* **2001**, *33*, 116–118. [CrossRef] [PubMed]
55. Aharonson-Raz, K.; Lichter-Peled, A.; Tal, S.; Gelman, B.; Cohen, D.; Klement, E.; Steinman, A. Spatial and temporal distribution of *West Nile virus* in horses in Israel (1997–2013)—from endemic to epidemics. *PLoS ONE* **2014**, *9*, e113149. [CrossRef] [PubMed]

animals

MDPI

Article

A Pilot Serosurvey for Selected Pathogens in Feral Donkeys (*Equus asinus*)

Erin L. Goodrich [1,*], Amy McLean [2] and Cassandra Guarino [1]

1 Cornell University College of Veterinary Medicine, Animal Health Diagnostic Center, Ithaca, NY 14853, USA; cg82@cornell.edu
2 Department of Animal Science, University of California, Davis, CA 95616, USA; acmclean@ucdavis.edu
* Correspondence: elg25@cornell.edu; Tel.: +607-253-3972

Received: 31 August 2020; Accepted: 30 September 2020; Published: 2 October 2020

check for updates

Simple Summary: This study aimed to assess the pathogen exposure status of recently captured feral donkeys from Death Valley National Park, California. Assays to detect the presence of antibodies to equine herpesvirus 1, equine influenza virus, West Nile virus, and *Borrelia burgdorferi* (the causative agent of Lyme disease) were performed on serum samples from these feral donkeys. The results indicate that this population is mostly naïve and likely susceptible to these common equid pathogens upon removal from the wild.

Abstract: Recent removal and relocation of feral donkeys from vast public lands to more concentrated holding pens, training facilities, and offsite adoption locations raises several health and welfare concerns. Very little is known regarding the common equid pathogens that are circulating within the feral donkey population in and around Death Valley National Park, California, USA. The aim of this study was to utilize serologic assays to assess previous exposure of these donkeys to equine herpesvirus 1 (EHV-1), equine influenza (EIV), West Nile virus (WNV), and *Borrelia burgdorferi* (the causative agent of Lyme disease). The results of this study indicate that this feral equid population is mostly naïve and likely susceptible to these common equid pathogens upon removal from the wild.

Keywords: *Borrelia burgdorferi*; burro; donkey; equid; equine herpesvirus; equine influenza; West Nile virus

1. Introduction

The feral donkey (*Equus asinus*), or burro, was introduced to North America as a result of Spanish colonization in the sixteenth century. Being well adapted to thriving in desert environments, and useful as a mode of transportation for people and equipment, these donkeys later became utilized by miners in geographical regions that would later become United States of America's (US) National Parks, such as Death Valley National Park [1]. As is the case with many introduced species, feral donkeys are the subject of much debate regarding the perceived associated positive and negative ecological impacts. Some studies allege these feral burros are problematic due to competition with native animals for limited resources, changes to vegetation, damage to soils, and negative impacts on springs [2,3]. Other researchers have studied the feral burros of the Sonoran Desert and found beneficial ecological impacts where they are noted to dig deep groundwater wells that are subsequently also utilized by several other species of mammals and birds [3,4]. One strategy employed to maintain appropriate management levels by both the US Department of the Interior's Bureau of Land Management (BLM) and the United States Department of Agriculture's Forest Service is removal of excess donkeys from public rangelands. Many of these donkeys subsequently become available for adoption into private care, while others go on to live in BLM-managed off-range facilities. Free-roaming donkeys in resource-limited environments, such as Death Valley, tend to spend their lives in small groups or pairs, or sometimes even in solitude, unlike the large herd formations that are typical of other equines in the wild [5]. Capturing these donkeys from the wild and placing them in groups with other unfamiliar animals likely produces some stress. This combination of stress and co-mingling creates an increased potential for viral reactivation (as in the case of equine herpesvirus 1 (EHV-1)) and pathogen transmission for communicable diseases such as EHV-1 and equine influenza (EIV) [6]. Moreover, there is a scarcity of scientific literature available regarding the common pathogen exposures and subsequent humoral immune responses of these feral donkeys in the wild, which has a direct impact on their risk of developing clinical disease upon capture, co-mingling, and relocation. Nasal samples were taken from this same population of donkeys to detect viral and bacterial pathogen levels using quantitative real-time PCR (qPCR). The study found little evidence of EHV1, 4, EIV, or *Streptococcus equi* subspecies *equi*, but the samples did contain DNA from asinine herpes virus 2, 3, and 5 and *Streptococcus equi* subspecies *zooepidemicus* [7].

Due to the vast land area, the desert climate, and the tendency for donkeys to roam in small groups or even alone, we hypothesized that the feral donkeys recently captured from Death Valley National Park would be naïve to many common pathogens that circulate in equid populations throughout the United States, including equine herpesvirus 1 (EHV-1), equine influenza (EIV), West Nile virus (WNV), and *Borrelia burgdorferi* (the causative agent of Lyme disease). The viral pathogens were selected as they are all commonly circulating among our domestic equids in the US and elsewhere, with the potential to cause significant clinical disease, and all have readily available serological assays to indicate previous exposure [8–11]. *B. burgdorferi* was selected, as it is a common infection in horses living in endemic areas, and the geographic range of the vector continues to expand across the United States [12].

EHV-1 and EIV are both contagious pathogens capable of being transmitted among equids, especially those in close proximity. Likewise, equid populations commonly demonstrate exposure to WNV and *B. burgdorferi*; however, these two pathogens require competent vectors for transmission. Commercial equine vaccines are available for EHV-1, EIV, and WNV, although they have not been evaluated for use in donkeys. Likewise, commercial *B. burgdorferi* vaccines exist for use in dogs; however, some research suggests they may offer protection against infection in horses as well, although antibody response may be short-lived, and more frequent boosters may be required [13,14]. Our objective was to assess antibody levels to these common equid pathogens in recently captured feral burros from Death Valley as a means of understanding their risk for disease development and transmission. The results from this study can be used to inform protocol design surrounding the handling of feral donkeys upon removal from the wild, especially with regard to preventive medicine, vaccination practices, and movement to other geographical regions.

2. Materials and Methods

Blood was collected from 98 feral donkeys removed from the Death Valley National Park range (Shoshone, CA, USA) on three separate occasions. The first set of samples was collected in early November 2018 from 51 donkeys that were captured within 10 days. A second (n = 35) cohort was sampled in late November, and a third (n = 12) was sampled in December before leaving short-term holding. Each of the animals in these later two groups was sampled within 5 days of capture. Donkeys were co-mingled with adult females and foals held together and separated from the adult male group. A 10 mL blood sample was collected by venipuncture of the jugular vein. All animals were observed by veterinarians prior to sampling. Body condition score (BCS) on a 5-point scale [15], sex (n = 49 males, n = 49 females), and approximate age were recorded based on that veterinary evaluation. The protocol was approved by the University of California Davis Institutional Animal Care and Use Committee #20611.

The equine herpesvirus type 1 risk evaluation assay (Animal Health Diagnostic Center (AHDC), Cornell University, Ithaca, NY, USA) was performed as previously described [16]. Briefly, a fluorescent bead bound to an EHV-1 recombinant protein, glycoprotein C (gC), was incubated for 30 min with serum diluted 1:400 in phosphate buffered saline ((PBN) with 1% (w/v) bovine serum albumin (BSA) and 0.05% (w/v) sodium azide). Multiscreen®HTS plates (Millipore, Danvers, MA, USA) were used for incubations, wash steps were performed with phosphate buffered saline containing 0.05% (v/v) Tween 20 (PBS-T) using an ELx50 plate washer (Biotek Instruments Inc, Winooski, VT, USA), and all incubations were performed at room temperature. Serum antibodies bound to beads were detected with either biotinylated goat anti-horse IgG (H+L) antibody (Jackson Immunoresearch Laboratories, West Grove, PA, USA) or biotinylated anti-equine IgG4/7 (RRID: AB_ 2820277), incubated for 30 min. Streptavidin-PE (Invitrogen™, Carlsbad, CA, USA), incubated for 30 min, was used to label the detection antibody. Finally, beads were suspended in PBN and analyzed in a Luminex IS 100 instrument (Luminex Corp, Austin, TX, USA). The data were reported as median fluorescent intensities (MFI). The two values obtained from the assay were used to categorize animals as "high risk", "moderate risk", "low risk", and "very low risk" of contracting EHV-1 if exposed to the virus [16].

The equine influenza virus (EIV) hemagglutination inhibition assay (HAI) assay (AHDC, Cornell University, Ithaca, NY, USA) was performed as previously described [17], using an equine influenza strain isolated in New York in 2018, A/Eq/NY/192142/18 (H3N8). Briefly, serum samples were heat-inactivated at 56 °C for 30 min and treated with 0.016M potassium periodate solution, then subsequently neutralized with glycerol to prevent nonspecific reactions. Nonspecific hemagglutinins were removed by pre-incubation with turkey red blood cells for 30 min. Two-fold serial dilutions of each prepared serum, beginning at 1:4, were incubated with constant concentration of virus for 30 min at room temperature, and turkey red blood cells were then added. Following 30–45 min incubation at room temperature, wells were examined for the presence or absence of viral induced hemagglutination. The antibody titer of the serum was reported as the reciprocal of the last dilution of serum, which completely inhibited hemagglutination. Samples were run in duplicate with half titers reported when the end point was between two dilutions. Positive, negative, and serum sample controls were run concurrently. Historic data from diagnostic submissions requesting EIV HAI serologic testing (AHDC, Cornell University), totaling 916 equine samples submitted from 2013 to 2017 and 18 Equidae samples submitted from 2012 to 2020 were used to compare to expected results in domestic populations.

The West Nile virus (WNV) antigen capture IgG and IgM ELISAs (AHDC, Cornell University, Ithaca, NY, USA) were performed as previously described [18], with anti-equine IgM (RRID:AB_2737323) or anti-IgG1/3 (RRID:AB_2737325) as capture antibody, and biotinylated anti-WNV E-protein (RRID:AB_2744504) as the detection antibody. Briefly, Nunc MaxiSorp™ plates (Invitrogen™, Carlsbad, CA, USA) were coated overnight at 4 °C with 4 μg/mL of the capture antibody in Carbonate buffer, pH 9.65. Plates were washed between each incubation step with phosphate buffered saline containing 0.05% (v/v) Tween 20 (PBS-T). Each test and control sample was diluted 1:100 in tris-buffered saline containing 0.05% (v/v) Tween 20 (TBS-T) with 5% (w/v) nonfat dry milk and incubated for 1 h at 37 °C. Cell-culture-derived WNV antigen and control antigen were added to the positive and negative wells of each sample, respectively, and incubated overnight at 4 °C; inactivated WNV antigen was prepared from Verocells infected with a chimeric WNV [19], and control antigen was prepared from non-infected cells. Detection antibody was incubated for 1 h at 37 °C, followed by streptavidin-peroxidase (Jackson ImmunoResearch, West Grove, PA, USA) for 1 h at 37 °C. Plates were developed with TMB (KPL, Gaithersburg, MD, USA). Results were reported as a positive/negative (P/N) ratio—the optical density (OD) of the WNV antigen wells, divided by the OD of the negative antigen wells.

The Lyme Multiplex assay (AHDC, Cornell University, Ithaca, NY, USA) was performed as previously described [20]. Briefly, three sets of fluorescent beads bound to recombinant OspA, OspC, and OspF proteins, respectively, were incubated for 30 min with serum diluted 1:400 in PBN. Multiscreen HTS plates (Millipore, Danvers, MA, USA) were used for incubations, wash steps were performed with PBS-T using an ELx50 plate washer (Biotek Instruments Inc, Winooski, VT, USA), and all incubations were performed at room temperature. Serum antibodies bound to beads were detected with biotinylated goat anti-horse IgG (H + L) antibody (Jackson Immunoresearch Laboratories, West Grove, PA, USA), incubated for 30 min. Streptavidin-PE (Invitrogen™, Carlsbad, CA, USA), incubated for 30 min, was used to label the detection antibody. Finally, beads were suspended in PBN and analyzed in a Luminex IS 100 instrument (Luminex Corp, Austin, TX, USA). The data were reported as median fluorescent intensities (MFI), and cut-off values for "negative", "equivocal", and "positive" were determined based on values from horse serum [20]. Historic data from diagnostic submissions requesting *B. burgdorferi* serologic testing on serum (AHDC, Cornell University) totaling 5468 equine samples submitted the summer of 2018 and 1271 Equidae samples submitted 2011 to 2020 were used to compare expected results in domestic populations.

3. Results

Approximate ages of the 98 donkeys removed from the Death Valley National Park ranged from 4 months to 20 years, with a median of 6 years (48 females and 48 males). Of the 70 animals for which BCS data were approximated, two were <2, and the remaining 68 had a mean BCS of 3.08, with a median of 3.0 on a scale of 1–5 [21]. Results of the serologic assays are summarized in Table 1.

Table 1. Summary of Serologic Assay Results. The number of animals with serum antibodies that were reactive, moderately reactive, or non-reactive against each of the pathogens investigated.

Item	EHV-1	EIV	WNV	Lyme
Reactive	3	0	0	6
Moderately Reactive	29	9	0	1
Non-Reactive	60	61	92	85

A total of 92 serum samples were tested in the EHV-1 risk evaluation assay. In total, 3% (3/92) of the donkeys that were tested had antibody values that would categorize them as "low risk" for infection with EHV-1, 32% (29/92) demonstrated evidence of exposure to an equine herpes virus and would be categorized as "moderate risk" for infection with EHV-1, and the remaining 65% (60/92) of the burros were categorized as "high risk" for EHV-1 infection at the time of sampling.

Sufficient serum was available for testing 70 of the donkeys for evidence of exposure to EIV. Equine influenza H3N8 antibody titers were not detected at a minimum dilution of 1:4 in 13% (9/70) of donkeys, the highest titer of 12 was found in 13% (9/70) of donkeys, and the remaining 74% (52/70) of donkeys had a titer of 6 (Figure 1). Of the 18 historic domestic Equidae submission, only two samples had a titer of <8.

All of the donkeys tested negative for both IgM and IgG antibodies to West Nile virus. In total, 7% (6/92) of the donkeys tested positive for at least one of the three antibodies to *B. burgdorferi*. Five of those animals were positive for antibodies against the chronic infection marker, OspF, and one donkey was positive for only the early infection marker, OspC. One additional donkey had an "equivocal" value for OspF. In the domestic equidae submissions, 48% (616/1271) of animals tested positive for the infection markers OspC and/or OspF, compared to only 27% (1480/5468) of domestic equines.

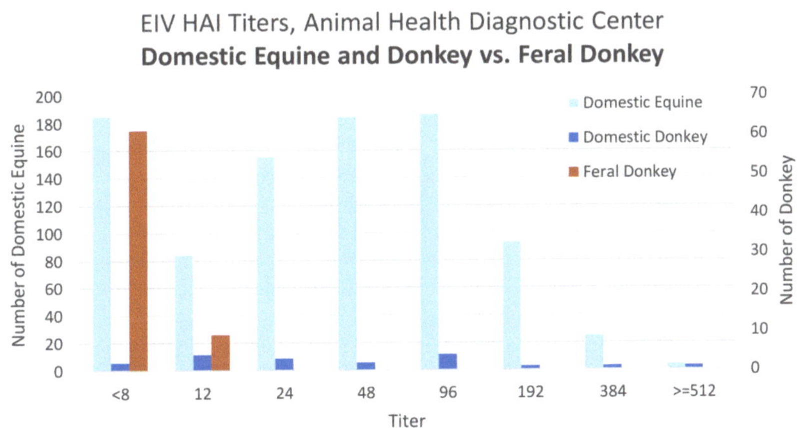

Figure 1. Equine Influenza Virus Hemagglutination Inhibition Titers (EIV HAI). Results for 70 feral donkeys from Death Valley in November and December, 2018, are compared to those obtained for over 900 domestic equines and Equidae (vaccinated and unvaccinated) tested at the AHDC, Cornell University, Ithaca, NY from 2013–2018.

4. Discussion

Our hypothesis was that equids removed from Death Valley National Park would have had limited contact with common infectious disease agents that are frequently found in equid populations in the USA and would be serologically naïve. The serologic survey conducted in this study supports this hypothesis.

A significant portion of this population of donkeys was found to be at high risk for infection with EHV-1, a common pathogen ubiquitous in most equine populations. EHV-1 can affect all members of the *Equidae* family, including donkeys. Infection with EHV-1 can manifest as a range of clinical presentations, from mild respiratory illness, to neurologic sequelae that can be fatal, and EHV-1 infection of pregnant mares during gestation can lead to late-term abortion [9,22–24]. EHV-1 typically results in a latent infection with subsequent re-activation and associated clinical disease and viral shedding during periods of stress, such as traveling and co-mingling [9,23]. A limitation of this testing is the specificity for EHV-1; while infection with asinine herpesvirus 3 (EHV-8) or EHV-9 might generate cross-reactive antibodies based on homology to the target antigen in the EHV-1 assay, evidence of infection with other herpesvirus strains, including EHV-4, might not be detected. However, asinine herpesvirus 2, 3, and 5 were detected in nasal swabs using qPCR in this population of donkeys [7]. Numerous vaccines are commercially available to protect horses and other equids from EHV-1 infection, including killed virus vaccines, a live virus vaccine, and formulations approved for use during pregnancy to prevent abortion, although there are no current vaccine claims to protect against the neurologic form of disease [25].

EIV infection in horses typically causes acute febrile respiratory disease, which can frequently be complicated by secondary bacterial infection, especially in unvaccinated animals [26]. EIV infection in donkeys can cause pyrexia, cough, nasal discharge, and lethargy, similar to the signs seen in infected horses [24]. However, donkeys appear to be more susceptible than horses to infection with EIV and to developing more severe clinical disease, including secondary bronchopneumonia [27,28]. In 1963, an H3N8 strain of influenza was identified as the cause of a major epidemic in equines in the state of Florida, and in the following years, it spread throughout the USA and Europe [29]. The virus used in the EIV HAI assay was a Florida clade 1 virus isolated in NY that matches isolates recently found in the USA [30]. EIV HAI result titers below 8 are typically considered negative, indicating high risk for infection if exposed to EIV, and titers of 12 are low positive. All donkeys in this study, including the few donkeys with low positive EIV HAI titers, would likely be completely susceptible to infection with equine influenza if exposed. One limitation of this assay is the specificity for the H3N8 strain—if a novel strain were circulating in these donkeys, this assay may not detect those antibodies; however, these animals would likely still be susceptible to the commonly circulating strains in the USA. Numerous EIV vaccines are commercially available in the USA, including a modified live vaccine, a canary pox vector vaccine, and inactivated virus vaccines [31].

WNV has been endemic in North America since 1999 and can be transmitted to susceptible hosts through the bite of an infected *Culex* mosquito [32]. Birds typically serve as the predominant reservoir for WNV, and humans and equids are considered dead-end hosts [33]. Although only about 10% of unvaccinated, exposed equids usually develop associated neurologic disease, the clinical manifestation can be quite severe, including fever, ataxia, weakness, recumbency, muscle fasciculations, and death, with a case fatality rate of 30–40% [34–37]. WNV infection is diagnosed by identifying WNV-specific IgM in horse serum [38]. Following infection or vaccination, horses will develop WNV-specific IgG [18]. None of the donkeys in this population had evidence of previous exposure to WNV and could be completely susceptible to infection if transported to a region where WNV infected mosquitos are present. Numerous vaccines are currently available to protect horses and other equids from WNV infection [39].

In North America, *B. burgdorferi* is the most common cause of Lyme disease, with Ixodid ticks serving as the vector for transmission [12]. Clinical signs of Lyme disease in equids are not well described owing to the fact that cause and effect are very difficult to document; however, stiffness, lameness, hyperesthesia, and behavior changes are often described [40]. The associated clinical syndromes that have been well described in equids, although rare, include neuroborreliosis, uveitis, and cutaneous pseudolymphoma [12]. Seroprevalence studies suggest that *B. burgdorferi* exposure among equids is common throughout many regions of the United States and that the range is likely increasing [12]. In addition, these studies indicate that it is common for clinically normal horses living in endemic regions to have antibodies against *B. burgdorferi* [12]. Given the sensitivity of *Ixodes* ticks to desiccation [41] and the rugged, arid environment of Death Valley, it was surprising to find that a small number of donkeys had evidence of infection with *B. burgdorferi*. However, *B. burgdorferi* has been identified in ticks in areas of the Mojave desert surrounding Death Valley [42]. There is no significant information available regarding the susceptibility of donkeys to Lyme disease; however, the high seroprevalence in the diagnostic submissions from domestic Equidae suggests that these wild donkeys will be at risk for infection with *B. burgdorferi* if they are moved to a Lyme endemic area. There is not currently an approved vaccine to protect against infection with *B. burgdorferi* in horses; however, in endemic areas, some clinicians will use a canine Lyme vaccine off-label [14].

Throughout this study, comparisons were made utilizing historic diagnostic data, which have limitations. These limitations in some instances include unknown clinical history provided to the AHDC, and testing for specific indications which may include clinical disease or to measure vaccine response. The comparison serologic data in this study were not sorted in any way to account for these limitations. In addition, this study is not a comprehensive serologic survey of all possible equid pathogens; however, the results of this study and one other study do indicate that these wild equid populations may be extremely naïve and susceptible to many common pathogens upon removal from the wild, and there appears to be an increase in pathogen detection upon introduction to long-term holding facilities [7].

5. Conclusions

Given the fact that donkeys in the wild tend to remain in pairs or very small groups and that they often live in vast arid landscapes, the findings in this study may be applicable to feral donkeys in similar locations. The apparently naïve nature of these donkeys to common pathogens supports the need for exceptional care and biosecurity measures upon removing wild equids from their native habitat. Transportation vehicles and holding pens should be appropriately cleaned prior to movement in order to minimize risk of exposure to common equine pathogens. Further work is needed to understand the incidence of clinical disease from these pathogens in these populations upon removal from the wild, and co-mingling, in order to determine the optimal vaccination recommendations.

Author Contributions: Conceptualization, E.L.G, A.M., and C.G.; Data curation, C.G.; Formal analysis, E.L.G. and C.G.; Funding acquisition, E.L.G., A.M., and C.G.; Investigation, E.L.G. and A.M.; Methodology, E.L.G., A.M., and C.G.; Project administration, A.M.; Resources, E.L.G. and C.G.; Software, C.G.; writing—original draft, E.L.G.; writing—review and editing, E.L.G., A.M., and C.G. All authors have read and agreed to the published version of the manuscript.

Funding: This research received no external funding.

Acknowledgments: We would like to acknowledge B. Wagner and E. Dubovi for access to reagents and materials for the serologic assays. We would also like to acknowledge A. Rollins for her technical expertise.

Conflicts of Interest: The authors declare no conflict of interest.

References

1. Tiller, B. *Feral Burro Populations: Distribution and Damage Assessment*; Pacific Northwest National Laboratory: Richland, WA, USA, 1997. [CrossRef]
2. Sanchez, P.G. Impact of Feral Burros on the Death Valley Ecosystem. In Proceedings of the 6th Vertebrate Pest Conference, Anaheim, CA, USA, 5–7 March 1974.
3. Marshal, J.P.; Bleich, V.C.; Andrew, N.G. Evidence for interspecific competition between feral ass *Equus asinus* and mountain sheep *Ovis canadensis* in a desert environment. *Wildl. Biol.* **2008**, *14*, 228–236. [CrossRef]
4. Lundgren, E.J.; Ramp, D.; Ripple, W.J.; Wallach, A.D. Introduced megafauna are rewilding the Anthropocene. *Ecography* **2018**, *41*, 857–866. [CrossRef]
5. Evans, L.; Crane, M. *The Clinical Companion of the Donkey*, 1st ed.; Matador: Sidmouth, Devon, UK, 2018.
6. Seeber, P.A.; Quintard, B.; Sicks, F.; Dehnhard, M.; Greenwood, A.D.; Franz, M. Environmental stressors may cause equine herpesvirus reactivation in captive Grévy's zebras (*Equus grevyi*). *PeerJ.* **2018**, *6*, e5422. [CrossRef] [PubMed]
7. Jerele, S.; Davis, E.; Mapes, S.; Pusterla, N.; Navas González, F.J.; Iglesias Pastrana, C.; Abdelfattah, E.M.; McLean, A. Survey of Serum Amyloid A and Bacterial and Viral Frequency Using qPCR Levels in Recently Captured Feral Donkeys from Death Valley National Park (California). *Animals* **2020**, *10*, 1086. [CrossRef]
8. Yang, H.; Xiao, Y.; Meng, F.; Sun, F.; Chen, M.; Cheng, Z.; Chen, Y.; Liu, S.; Chen, H. Emergence of H3N8 equine influenza virus in donkeys in China in 2017. *Vet. Microbiol.* **2018**, *214*, 1–6. [CrossRef]
9. Thiemann, A.K. Respiratory disease in the donkey. *Equine Vet. Educ.* **2012**, *24*, 469–478. [CrossRef]
10. Long, M.T. West Nile Virus and Equine Encephalitis Viruses: New Perspectives. *Vet. Clin.* **2014**, *30*, 523–542. [CrossRef]
11. Silva, A.S.G.; Matos, A.C.D.; da Cunha, M.A.C.R.; Rehfeld, I.S.; Galinari, G.C.F.; Marcelino, S.A.C.; Saraiva, L.H.G.; Martins, N.R.d.S.; Maranhão, R.d.P.A.; Lobato, Z.I.P.; et al. West Nile virus associated with equid encephalitis in Brazil, 2018. *Transbound. Emerg. Dis.* **2019**, *66*, 445–453. [CrossRef]
12. Divers, T.J.; Gardner, R.B.; Madigan, J.E.; Witonsky, S.G.; Bertone, J.J.; Swinebroad, E.L.; Schutzer, S.E.; Johnson, A.L. *Borrelia burgdorferi* Infection and Lyme Disease in North American Horses: A Consensus Statement. *J. Vet. Intern. Med.* **2018**, *32*, 617–632. [CrossRef]
13. Chang, Y.-F.; Novosol, V.; McDonough, S.P.; Chang, C.-F.; Jacobson, R.H.; Divers, T.; Quimby, F.W.; Shin, S.; Lein, D.H. Vaccination against Lyme Disease with recombinant *Borrelia burgdorferi* outer-surface protein A (rOspA) in horses. *Vaccine* **1999**, *18*, 540–548. [CrossRef]
14. Guarino, C.; Asbie, S.; Rohde, J.; Glaser, A.; Wagner, B. Vaccination of horses with Lyme vaccines for dogs induces short-lasting antibody responses. *Vaccine* **2017**, *35*, 4140–4147. [CrossRef]
15. Research Dept. The Donkey Sanctuary, Condition Scoring and Weight Estimation of the Donkey. Available online: https://www.thedonkeysanctuary.org.uk/sites/uk/files/2018-12/condition-scoring-and-weight-estimation-of-the-donkey-20181211.pdf (accessed on 11 March 2020).

16. Perkins, G.; Babasyan, S.; Stout, A.E.; Freer, H.; Rollins, A.; Wimer, C.L.; Wagner, B. Intranasal IgG4/7 antibody responses protect horses against equid herpesvirus-1 (EHV-1) infection including nasal virus shedding and cell-associated viremia. *Virology* **2019**, *531*, 219–232. [CrossRef]
17. Cullinane, A. Section 3.5.7. Equine Influenza, 2.1. Haemagglutination Inhibition Test. Manual of Diagnostic Tests and Vaccines for Terrestrial Animals 2019. World Organization for Animal Health (OIE). Available online: https://www.oie.int/fileadmin/Home/eng/Health_standards/tahm/3.05.07_EQ_INF.pdf (accessed on 21 February 2020).
18. Khatibzadeh, S.M.; Gold, C.B.; Keggan, A.E.; Perkins, G.A.; Glaser, A.L.; Dubovi, E.J.; Wagner, B. West Nile virus-specific immunoglobulin isotype responses in vaccinated and infected horses. *Am. J. Vet. Res.* **2014**, *76*, 92–100. [CrossRef] [PubMed]
19. Komar, N.; Langevin, S.; Monath, T.P. Use of a Surrogate Chimeric Virus To Detect West Nile Virus-Neutralizing Antibodies in Avian and Equine Sera. *Clin. Vaccine Immunol.* **2009**, *16*, 134–135. [CrossRef] [PubMed]
20. Wagner, B.; Freer, H.; Rollins, A.; Erb, H.N.; Lu, Z.; Gröhn, Y. Development of a multiplex assay for the detection of antibodies to *Borrelia burgdorferi* in horses and its validation using Bayesian and conventional statistical methods. *Vet. Immunol. Immunopathol.* **2011**, *144*, 374–381. [CrossRef] [PubMed]
21. Burden, F. Practical feeding and condition scoring for donkeys and mules. *Equine Vet. Educ.* **2012**, *24*, 589–596. [CrossRef]
22. Lunn, D.P.; Davis-Poynter, N.; Flaminio, M.J.B.F.; Horohov, D.W.; Osterrieder, K.; Pusterla, N.; Townsend, H.G.G. Equine Herpesvirus-1 Consensus Statement. *J. Vet. Intern. Med.* **2009**, *23*, 450–461. [CrossRef]
23. Negussie, H.; Gizaw, D.; Tessema, T.S.; Nauwynck, H.J. Equine Herpesvirus-1 Myeloencephalopathy, an Emerging Threat of Working Equids in Ethiopia. *Transbound. Emerg. Dis.* **2017**, *64*, 389–397. [CrossRef]
24. Barrandeguy, M.E.; Carossino, M. Infectious Diseases in Donkeys and Mules: An Overview and Update. *J. Equine Vet. Sci.* **2018**, *65*, 98–105. [CrossRef]
25. AAEP Equine Herpesvirus (Rhinopneumonitis) 2019. Available online: https://aaep.org/guidelines/vaccination-guidelines/risk-based-vaccination-guidelines/equine-herpesvirus-rhinopneumonitis (accessed on 1 October 2020).
26. Timoney, P.J. Equine influenza. *Comp. Immunol. Microb.* **1996**, *19*, 205–211. [CrossRef]
27. Mendoza, F.J.; Toribio, R.E.; Perez-Ecija, A. Donkey Internal Medicine—Part II: Cardiovascular, Respiratory, Neurologic, Urinary, Ophthalmic, Dermatology, and Musculoskeletal Disorders. *J. Equine Vet. Sci.* **2018**, *65*, 86–97. [CrossRef]
28. Rickards, K.J.; Thiemann, A.K. Respiratory Disorders of the Donkey. *Vet. Clin. North Am. Equine Pract.* **2019**, *35*, 561–573. [CrossRef] [PubMed]
29. Van Maanen, C.; Cullinane, A. Equine influenza virus infections: An update. *Vet. Q.* **2002**, *24*, 79–94. [CrossRef] [PubMed]
30. *Cornell University College of Veterinary Medicine EIVHAI H3N8 Test Data*; Cornell University: Ithaca, NY, USA, 2019.
31. AAEP Equine Influenza 2019. Available online: https://aaep.org/guidelines/vaccination-guidelines/risk-based-vaccination-guidelines/equine-influenza (accessed on 1 October 2020).
32. Hayes, E.B.; Komar, N.; Nasci, R.S.; Montgomery, S.P.; O'Leary, D.R.; Campbell, G.L. Epidemiology and Transmission Dynamics of West Nile Virus Disease. *Emerg. Infect. Dis.* **2005**, *11*, 1167–1173. [CrossRef] [PubMed]
33. Colpitts, T.M.; Conway, M.J.; Montgomery, R.R.; Fikrig, E. West Nile Virus: Biology, Transmission, and Human Infection. *Clin. Microbiol. Rev.* **2012**, *25*, 635–648. [CrossRef]
34. Leblond, A.; Hendrikx, P.; Sabatier, P. West Nile Virus Outbreak Detection Using Syndromic Monitoring in Horses. *Vector-Borne Zoonot.* **2007**, *7*, 403–410. [CrossRef]
35. García-Bocanegra, I.; Arenas-Montes, A.; Jaén-Téllez, J.A.; Napp, S.; Fernández-Morente, M.; Arenas, A. Use of sentinel serosurveillance of mules and donkeys in the monitoring of West Nile virus infection. *Vet. J.* **2012**, *194*, 262–264. [CrossRef]
36. Ostlund, E.N.; Andresen, J.E.; Andresen, M. West Nile Encephalitis. *Vet. Clin.* **2000**, *16*, 427–441. [CrossRef]
37. Bunning, M.L.; Bowen, R.A.; Cropp, B.C.; Sullivan, K.G.; Davis, B.S.; Komar, N.; Godsey, M.; Baker, D.; Hettler, D.L.; Holmes, D.A.; et al. Experimental Infection of Horses with West Nile virus. *Emerg. Infect. Dis.* **2002**, *8*, 380–386. [CrossRef]

38. Wagner, B.; Glaser, A.; Hillegas, J.M.; Erb, H.; Gold, C.; Freer, H. Monoclonal antibodies to equine IgM improve the sensitivity of West Nile virus-specific IgM detection in horses. *Vet. Immunol. Immunopathol.* **2008**, *122*, 46–56. [CrossRef]
39. AAEP West Nile Virus 2019. Available online: https://aaep.org/guidelines/vaccination-guidelines/core-vaccination-guidelines/west-nile-virus (accessed on 1 October 2020).
40. Divers, T.J. Equine Lyme Disease. *J. Equine Vet. Sci.* **2013**, *33*, 488–492. [CrossRef]
41. Sonenshine, D.E. Range Expansion of Tick Disease Vectors in North America: Implications for Spread of Tick-Borne Disease. *Int. J. Environ. Res. Public Health.* **2018**, *15*, 478. [CrossRef] [PubMed]
42. Foley, J.; Ott-Conn, C.; Worth, J.; Poulsen, A.; Clifford, D. An *Ixodes minor* and *Borrelia carolinensis* enzootic cycle involving a critically endangered Mojave Desert rodent. *Ecol. Evol.* **2014**, *4*, 576–581. [CrossRef] [PubMed]

animals

MDPI

Article

Welfare Assessment and Identification of the Associated Risk Factors Compromising the Welfare of Working Donkeys (*Equus asinus*) in Egyptian Brick Kilns

Shaaban F. Farhat [1,*], Amy K. McLean [2,3,*] and Hamdy F. F. Mahmoud [4,5]

1 Egyptian Society for Protection and Welfare of Working Animal (ESPWWA), Cairo 11865, Egypt
2 Department of Animal Science, University of California Davis, Davis, CA 95617, USA
3 World Donkey Breed Project, University of Cordoba, 14014 Cordoba, Spain
4 Department of Statistics, Virginia Polytechnic Institute and State University, Blacksburg, VA 24061, USA; ehamdy@vt.edu
5 Department of Statistics, Mathematics, and Insurance, Faculty of Commerce, Assiut University, Assiut 71515, Egypt
* Correspondence: shaban_fayez@yahoo.com (S.F.F.); acmclean@ucdavis.edu (A.K.M.); Tel.: +1-706-296-8743 (A.K.M.)

Received: 13 July 2020; Accepted: 31 August 2020; Published: 9 September 2020

check for updates

Simple Summary: Working donkeys suffer from many welfare challenges associated with, for example, physical health, poor living conditions, and unfair treatment. The aim of this study is to assess the welfare of working donkeys in the El-Saf brick kilns, identifying the health risk factors, establishing welfare regulations, enacting legislation, and implementing welfare strategies aimed at improving the quality of life of donkeys and owners within communities. The study found that working donkeys in Egypt suffer from many types of wounds associated with parts of the harness, such as the saddle, breeching, and neck collar, and with excessive force/beating, the shaft of the cart, and improper tethering. They often live in unhealthy housing situations, and a high percentage suffer from aggressive behavior. The study found an association between these health risks, behavioral parameters, and body condition in Egyptian working donkeys. Body condition was affected by multiple factors, including the number of hours worked/day, the number of donkeys/kilns, the distance from loading to unloading bricks in an oven, and the amount of concentrated food/donkey.

Abstract: Donkeys are a cornerstone in human existence, having played an important role throughout history in different economic activities, such as working in brick kilns in Egypt. This study was conducted from January 2017 to the end of April 2017 in the El-Saf brick kilns, which are located to the south of the Giza Governorate and 57 Km away from Cairo. Physical clinical health and behavior data were collected from 179 donkeys spanning over a random sample of 20 brick kilns selected from the El-Saf brick kilns. Behavioral, physical health, harness, and environmental parameters were assessed and recorded. The study found that 80 ± 3% (n = 179) of kiln donkeys have some type of wound, and the most serious wound is a beating wound (49 ± 3.7%), which is caused by drivers hitting the donkeys. The drivers are mostly children, who have insufficient knowledge, skills, and attitudes to effectively communicate with their donkeys and no motivation to enhance the welfare of these equids. Other wounds are related to the harness, such as the breeching (10 ± 2.2%), saddle (43 ± 3.7%), neck collar (40 ± 3.6%), and shaft of the cart (12 ± 2.4%). A poor body condition was seen in 56 ± 3.7% of kiln donkeys. A correlation in terms of the prevalence of wounds was found between the body condition (p-value < 0.01) and/or cleanliness of the harness. There was a negative association between the body condition and wound prevalence in brick kilns (Pearson coefficient of correlation −0.71). The physical enviromental factors that affect the body condition of working donkeys are the working hours of donkeys/day, the number of donkeys in a kiln, the distance from loading to the oven, and the concentrated food/donkey (p-value < 0.01). These three variables can explain 78.85% of the variability in body conditions based on a 1–5 scale. In addition to health parameters, behavior parameters, such as the donkeys' general attitude, reaction to observers, and chin contact are associated with the body condition (p-value < 0.01). As a consequence, it is important for the owners of working donkeys to pay attention to their body condition in order to avoid compromising their body condition and welfare.

Keywords: Donkeys; welfare; behavior; working equids; brick kilns

1. Introduction

Donkeys are a cornerstone in human existence and have played an important role throughout much history in different economic activities. The world donkey population is estimated to be about 43 million, with approximately 95% of them used for work in developing countries carrying out domestic activities, as well as tasks relating to agriculture, transport, and different industries (e.g., construction) [1–3]. It is not uncommon to find working donkeys suffering from many welfare-related problems, including wounds, poor body condition, respiratory diseases, parasites, a poor dental condition, and lameness [3–6]. Welfare problems do not stop with physical ailments; many working donkeys experience compromised mental states, such as a fear of humans and even depression [4]. Poor welfare is commonly associated with working donkeys in Africa and the Middle East. Their poor welfare has been linked to both harsh working conditions [3,5] and to handlers with insufficient knowledge of general husbandry and properly caring for working donkeys, such as wound management, harness care and fitting, watering, nutritional requirements, appropriate shelter arrangements, and veterinary services [6–9].

Approximately 3.7% of the world donkey population is found in Egypt, with close to 1.6 million animals [10]. Donkeys have a long history in Egypt, having been first domesticated there around 3000–4000BC [11]. Recently, the increasing human population has caused an increase in construction activities, including the building of new cities, and this has increased the demand for bricks. Consequently, there has been an increase in the demand for working donkeys in Egypt. Despite the increase of mechanization in brick kilns, donkeys are still well deserving of the name the "beasts of burden", and are heavily relied upon for production. In the El-Saf brick kilns, where the Egyptian Society for the Protection and Welfare of Working Animals (ESPWWA) is located, the donkey is the preferred working equid. Donkeys are responsible for transporting green dried bricks from the loading area to the ovens inside by pulling overloaded carts, which are driven by children. Donkeys working in these brick kilns face multiple challenges. They often receive little nutrition and veterinary care, and they are often over-worked, overloaded, and wear ill-fitted, poorly designed harnesses [4]. These deficiencies in care lead to many welfare problems, including harness-induced injuries, dehydration, wounds from beating, lameness, and other health problems [12]. There is no training for donkeys in performing such hard work. The average load of bricks weighs 6 to 12 times the body weight of a donkey found working in the kilns. Apart from pulling heavy loads, the donkeys work in a very harsh environment for long hours, pulling loads on difficult and uneven terrain and being exposed to extreme temperatures from the heat radiating from the ovens. Furthermore, the handlers are young children or adolescents, who have little knowledge of how to properly work, care for, and communicate with donkeys. Thus, the donkeys are subjected to many conditions that compromise their welfare. Despite their invaluable contribution to brick kilns and to sustaining over 250 workers/brick kiln, donkeys are the most neglected part of the industry, with brick kiln owners considering donkeys as machines with a low economic value. The welfare of working donkeys is very important not only for the health and survival of those animals, but also for the livelihood of the people who depend on them [9]. A welfare assessment of working equids is crucial for establishing welfare regulations, enacting legislation, and implementing welfare strategies aimed at improving the quality of life of owners and equids within communities [13]. The use of a combination of physical, behavioral, and mental parameters has been commonly used in the welfare assessment of working equids [5,14–17]. The ESPWWA faces many challenges in assessing the welfare of donkeys in brick kilns, including the fact that the work in brick kilns is temporary, and there is a dynamic movement of handlers from kilns. This variation of the welfare of donkeys in brick kilns may help us to understand the risk factors, including management practices that can lead to the compromise or improvement of donkey welfare. Studies focused on identifying the conditions that have a severe and negative impact on welfare, along with research that focuses on how to successfully implement changes that improve donkey welfare, are important. However, the current research in both of these areas is limited [18]. The aim of this study is to assess the welfare of working donkeys in the El-Saf brick kilns and identify the associated risk factors that may compromise the welfare of working donkeys in many Egyptian brick kilns.

2. Materials and Methods

2.1. Study Area

The study was conducted from January 2017 to the end of April 2017 in the El-Saf brick kilns, south of Giza Governorate, 57 Km from Cairo, Egypt. There are 250 brick kilns in El-Saf, with 120 operating kilns. Twenty kilns (16.6% of all kilns) were randomly selected from the operating kilns for the study.

The operating kilns are home to 1350 male donkeys and 55 mules, and 179 donkeys from this population were surveyed for this study. No female donkeys work in the brick kilns. The 20 selected kilns had no mules, so the random sample in this study consisted of 179 male donkeys.

2.2. Behavior Parameter Examination

Physical, clinical, and behavioral data were collected from the 179 donkeys in the random sample of 20 brick kilns selected from the El-Saf brick kilns. The behavior parameters were measured first, as soon as the donkey had completed its work [19]. The following measurements were assessed and recorded: general behavior (alert, apathetic, or depressed), response to an observer (unfamiliar person), and acceptance of chin contact. The sequence and procedure for taking the behavior measurements was as follows: to assess the alertness of the donkey (alert, apathetic, or depressed), and to observe the donkey from a distance of at least 3 m and for up to 10 s. Then, the observer walked slowly at an angle of 30° (based on visual observation) toward the head of the donkey and stopped approximately 30 cm from the head of the donkey to assess the response of the donkey at the moment that the observer stopped. Next, the donkey's reactions that fall within the following categories were recorded: discovering, not moving, avoiding, running away, and showing an aggressive reaction. Lastly, the observer put his hand very gently under the donkey's chin, enough to take some weight but not so much as to lift the head. If the donkey took his head away from the hand, the observer would not pursue it and recorded the acceptance of chin contact (acceptance or non-acceptance). After initiating the physical contact, the observer started the physical and health examination.

2.3. Physical Health Parameter Examination

The physical examination started from the head and progressed toward the rear of the working donkey. Any abnormalities or skin lesions were recorded. In addition to the number of wounds, the area of the wounds (multiplying the width by the length) were estimated as well. The wounds were classified into two categories, harness-induced wounds, and excessive force/beating wounds. The harness wounds were recorded and related to the anatomical part of the donkey's body and the part of the harness where the lesions were found, for example, lesions on the back caused by the saddle; wounds on the shoulders or top of the neck from the neck collar; wounds on the chest or sides from the shaft of the cart; and lesions on the hind legs or hips from the breeching. Lesions and wounds inflicted by the driver from excessive force were described as beating wounds. Figure 1 shows how the harness should fit (Figure 1a), and the other images (Figure 1b–d) display different types of wounds related to ill-fitting harnesses. Figure 2a–c shows wounds caused by excessive force and beating, limb wounds from improper tethering, and shaft wounds caused by the cart. Other physical parameters were measured, such as the age, estimated from the mandibular incisor occlusal table appearance [20]. The body condition was recorded using a scale from 1–5, where 1 stands for poor, and 5 stands for obese (1–5) [21]. The body weight was measured by measuring the body height (from the ground to the top of the withers) and by measuring the circumference of the heart girth. Then, the results were plotted on a nomogram [22]. Figure 3 shows two donkeys, one with a good body condition (Figure 3a) and a working donkey with a poor body condition (Figure 3b).

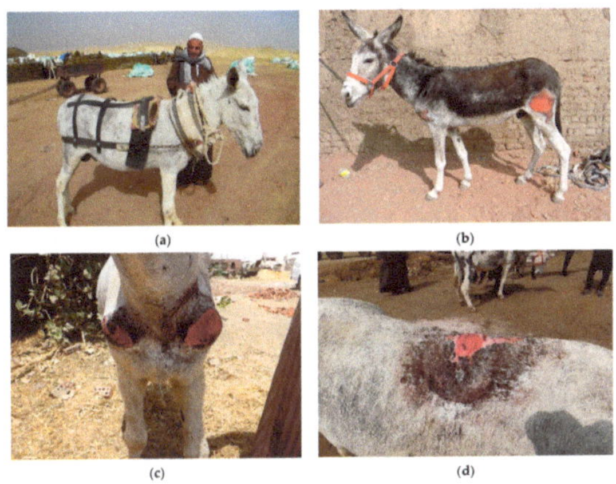

Figure 1. This figure shows a donkey with a full harness and different types of wounds caused by the harness: (**a**) a full harness setting, (**b**) a breaching wound, (**c**) a neck collar wound, and (**d**) a saddle wound.

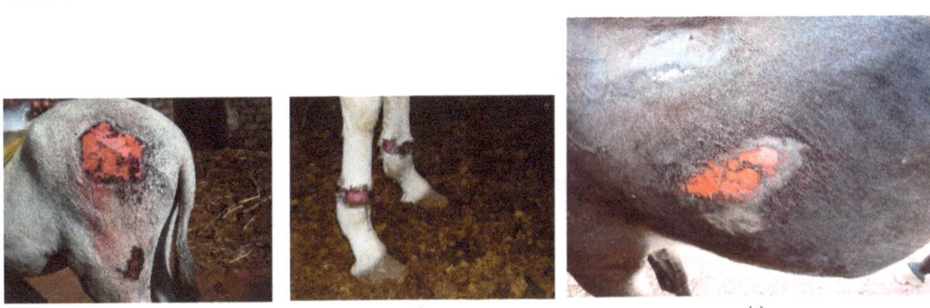

Figure 2. (**a**) A wound from excessive force/beating, (**b**) a limb wound from improper tethering, and (**c**) a shaft wound from the cart.

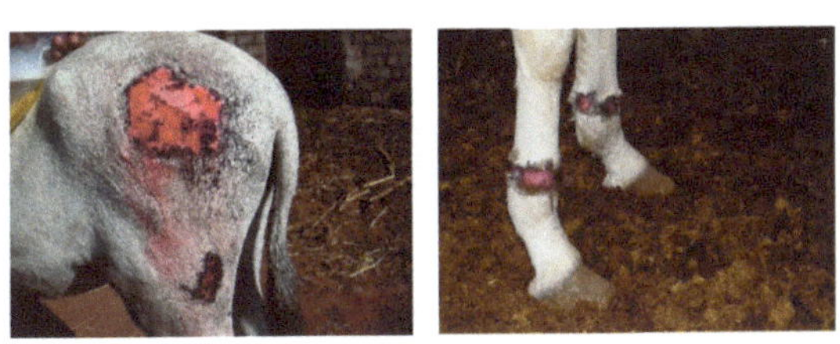

Figure 3. (**a**) Body condition was assessed using a scale of 1 (poor) to 5 (obese). The donkey in (**a**) would be considered to have a good body condition (3) and no wounds, and the donkey in (**b**) would be considered to have a poor body condition (2) and multiple lesions from the harness and excessive force.

2.4. Physical Environmental Parameters Assessment

In addition to physical and behavioral parameters, environmental parameters were assessed and recorded, such as the harness, cart, housing, watering, feeding, roads, and the load of bricks. During the period when the donkey was working, the following parameters were assessed: time working, harness condition, harness cleanliness, and the fitness of the harness. The condition of the harness was scored from 1 (very bad) to 5 (very good), as described in Table 1. The cleanliness of the harness was assessed from 1 (bad) to 3 (good) (Table 2). Figure 4a shows an example of an unacceptable or bad neck collar. The fitting of the hitching point, neck collar, and breeching were assessed and noted as either fitting or not fitting the animal. The cart condition was assessed by giving an overall rating to the cart's wheel bearings, tires, and shafts. The cart condition was scored from 1 (bad) to 3 (good), as described in Table 3. The road from the loading area to the oven was evaluated according to the amount of debris from bricks. A bad rating was given if the path had a lot of broken bricks that hindered the movement of the carts; a fair rating was given if a small amount of broken bricks were found in the path; and the path was considered to be good if the path was free of brick debris. Roads were assessed based on their slope. A bad rating was applied if there was a negative slope (traveling to or from the kiln); it was fair if a slope was found on part of the path; and it was good if a slope toward the oven was present. The water was assessed based on several parameters, such as clean or unclean, taste (salty or sweet). If the water was found to be salty or unclean, it was considered to be bad. Water found to be clean and sweet was scored as a good water source. Figure 4b shows unclean water. Donkey housing was assessed in terms of the floor condition (bad if it is muddy, dirty with urine and fecal matter; good if it is clean and dry) and ventilation (bad if there are not enough windows; good if there are enough windows). Brick loads were measured three times, first at the beginning of the workday, two hours later, and before the end of the workday. The average brick load was recorded. Figure 5 shows a cart loaded with bricks. Other parameters were assessed, such as the hours worked/day/kiln, the number of ovens in a kiln, the distance between the loading point and the oven, and the average amount of concentrate food provided for each donkey.

Table 1. The harness condition scores.

Grade	Description
1 = very bad	Poor contact surfaces due to excessive wear, broken parts, or bad construction (despite being new), and the use of non-equine-friendly materials, such as wire, plastics, and nylon, in areas that are in physical contact with the donkey
2 = bad	Well used, but developing problems with contact surfaces
3 = fair	Well used, but in a good condition and with good contact surfaces for the donkey
4 = good	A harness level that falls between 3 and 5
5 = very good	Well made, not well used, with good contact surfaces for the donkey

Table 2. Harness cleanliness scores.

Grade	Description
1 = bad	Encrusted and dried on bodies (blood, mud, feces, etc.) and likely to cause injury
2 = fair	Dusty, but not abrasive
3 = good	Totally dirt free

Table 3. Cart condition scores.

Grade	Description
1 = bad	Very poor, with no bearing left, and the shaft is narrow, causing injuries to donkeys' sides. It may contain sharp objects, and the tires are deflated
2 = fair	The wheel bearings have space to move from one side to the other, the shaft is tight, and the tires are not completely deflated, but the pressue inside is low
3 = good	The wheel bearings have no space to move from side to side, the wheels easily rotate, the distance between the two shafts is wide enough to prevent injuries to the donkeys' sides, and the tires are not deflated

(a) (b)

Figure 4. Examples of parameters compromising working donkey welfare including (**a**) a dirty neck collar, and (**b**) unclean water.

Figure 5. A cart loaded with brick cargo; the driver hits the donkey on its rump.

2.5. Statistical Analysis

To assess the welfare and identify the risk factors compromising the welfare of donkeys, as well as explore the association between these factors, different types of analyses were performed. Descriptive analyses were performed with numerical variables, such as the percentage, means, standard errors, median, maximum, and minimum. A chi-square test was performed to explore the association between variables and factors. T-tests were performed to compare donkeys with wounds to donkeys without wounds and considered to be healthy. ANOVA and Tuckey pairwise multiple comparisons were performed to compare the impact of body condition related to different types of wounds and/or behavioral parameters, and to detect groups that are statistically different in terms of body condition. Multiple linear regression was used to find the factors that impacted body condition. Two types of statistical software were used to analyze and graph the data. Minitab version 19 (Minitab, Inc., State College, PA, USA) and R version 3.6.3 (Free Software Foundation, Inc., Boston, MA, USA).

3. Results

3.1. Physical Health Parameters

The average of the numerical parameters was calculated and is displayed as the mean ± standard error. The average age of the donkeys in this study was 13.23 ± 0.32 years, and their body condition was 2.41 ± 0.06. The incidence of different types of wounds was calculated and is displayed as the percentage ± standard error. The study found that 80 ± 3% of kiln donkeys had some type of wounds, and 49 ± 3.7% had lesions or wounds caused by excessive force or beating. For the harness-related wounds, 43 ± 3.7% of kiln donkeys had wounds caused by the saddle, and 40 ± 3.6% had wounds in the neck region from the collar. These last two wounds were the result of unfit, dirty harnesses and the poor body condition of the working donkeys, as we found in this study. Wounds were found on 16 ± 2.7% of the donkeys' limbs, and of these wounds, 10 ± 2.2% were caused by the breeching, and 12 ± 2.4%

were caused by the shaft of the cart. Other wounds represented 13.5 ± 0.82%, which were found on the donkeys working in the kilns.

For the wound area, Table 4 shows a numerical summary of the area of the different wound types in cm^2 as the mean ± standard error. The serious wounds included both the wounds/lesions caused by the shaft (89.60 ± 31.2 cm^2) and breeching wounds (77.80 ± 40.3 cm^2). However, few donkeys had wounds that covered large areas; therefore, the median measure was also reported. The highest median was found for beating wounds (30 cm^2), followed by breeching wounds (19 cm^2) and shaft wounds (16 cm^2).

Table 4. Numerical summary of the area of different wound types (cm^2).

Wound Type	Mean ± SE [1]	Median	Minimum	Maximum
Saddle	25.86 ± 3.60	13.00	0.13	150.0
Beating	47.95 ± 6.26	30.00	1.75	365.5
Limb	35.5 ± 17.80	8.50	0.20	385.0
Neck	23.32 ± 4.75	7.50	0.25	197.5
Breeching	77.80 ± 40.3	19.10	0.80	625.0
Shaft	89.60 ± 31.2	16.80	0.30	577.0

[1] SE = standard error of the mean.

The chi-square test was used to test the association between the different types of wounds and the cleanliness, condition, and fit of the harness. We found that neck collar, saddle, and breeching wounds were associated with the level of cleanliness ($p < 0.01$). The neck collar wounds were associated with the condition and fit ($p < 0.01$). For the association between the most serious wounds, i.e., beating wounds, and other types of wounds, the chi-square test found a correlation between beating wounds and saddle and breeching wounds ($p < 0.01$). In addition, anssociation was found between wounds caused by beating and the overall condition of the harness including fit, condition, and cleanliness ($p < 0.01$). A relationship was found between the cleanliness of the saddle and breeching ($p< 0.01$) and the fit of the neck collar.

The results found that 56 ± 3.7% of the donkeys had body condition scores of 2 or less. A t-test was used to study the association between body condition and skin wounds. A significant difference ($p < 0.01$) was found in body condition, where injured donkeys had body condition scores of 2.3 ± 0.06, and healthy donkeys had a higher BCS (Body Condition Score) (2.9 ± 0.17). This suggests that there was an association between body condition and injuries or wounds, which means that donkeys received better treatment, healthier food, and improved care, and they had an overall better body condition and fewer injuries. For each type of wound, the body condition of the donkeys was found to be as follows: donkeys with saddle wounds (2.22 ± 0.08) and without saddle wounds (2.56 ± 0.9); donkeys with limb wounds (1.98 ± 0.11) and without limb wounds (2.49 ± 0.07); and donkeys with neck collar wounds (2.17 ± 0.08) and without collar wounds (2.57 ± 0.08). The scatter plot in Figure 6 shows that there was a negative association between the two parameters, wound prevalence and BCS. Additionally, the Pearson coefficient of correlation was −0.71 between body condition and wound prevalence in brick kilns. The regression equation was BSC = 144.2 − 25.78 (wound prevalence). The p-value of the intercept and slope are 1.8e-08 and 0.0003, respectively, and the coefficient of determination $R^2 = 0.52$.

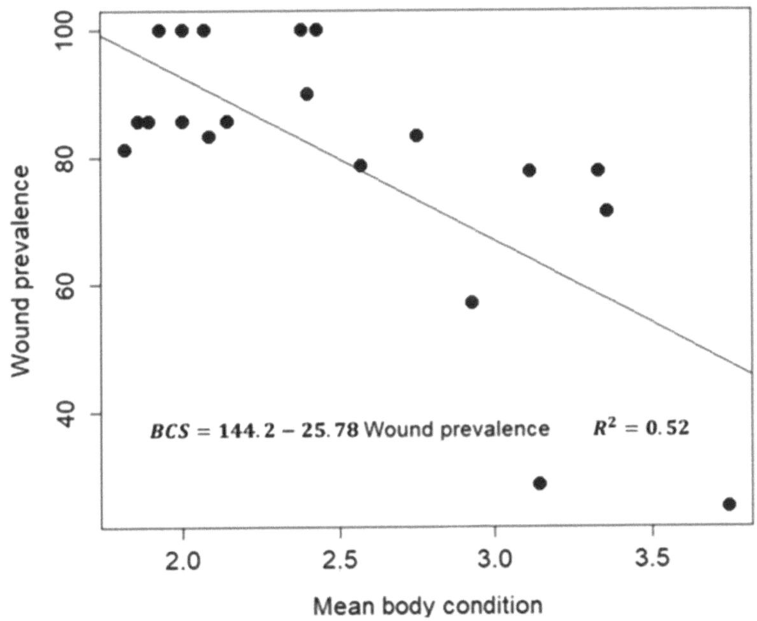

Figure 6. A scatter plot of the mean body condition and wound prevalence (%) in the 20 brick kilns selected, along with a regression line.

3.2. *Behavioral Parameter Assesment*

Table 5 shows that 82% (63 + 19%) of the kiln donkeys were apathetic or depressed, 68% (24 + 7 + 37%) were avoidant, ran away, or were aggressive, and 64% did not accept chin contact. To explore the association between body condition and behavior parameters, an ANOVA test was performed for the general attitude and reaction to an observer, and a t-test was performed for chin contact. A correlation was found between the general attitude and BCS ($p < 0.01$), reaction to observer and BCS ($p < 0.01$), and chin contact and BCS. Tuckey's pairwise comparison test was used to compare the difference between behavior and reactions based on BCS (refer to Table 5). When measuring behavioral responses in donkeys there was a correlation between response and BCS. Donkeys that expressed an alert behavior had a BCS of 2.95 ± 0.14. A correlation between BCS and response to observers was found by donkeys being more reactive to observers had a higher BCS of 3.42 ± 0.14. A correlation was found between chin contact acceptance and BCS. The donkeys that allowed chin contact had a higher BCS (2.68 ± 0.11), compared to donkeys that did not accept chin contact (2.27 ± 0.07). Figure 7 shows the body condition means of the statuses of the behavior parameters along with the standard error and shows the groups that are significantly different based on Tuckey's multiple comparisons as well.

The results suggest that the higher the BCS, the better the behavior of the donkey. We found an association between donkeys being injured and BCS, suggesting that body condition has an impact on the physical and behavioral conditions of working donkeys. Last, a correlation between donkeys with wounds from excessive force/beating and behavior was found (p-value < 0.05).

Table 5. Numerical summary of body condition for different statuses of general attitude, reaction to an observer, and chin contact, along with the statistical test values and *p*-values of the t-test and ANOVA test.

Parameter	Donkey Status	N (%)	Mean ± SE [1]	Test Value	*p*-Value
General attitude	Alert	32 (18%)	2.95 ± 0.14 [a]	F-value = 10.38	<0.001 **
	Apathetic	113 (63%)	2.34 ± 0.07 [b]		
	Depressed	34 (19%)	2.14 ± 0.14 [b]		
Reaction to observer	Discovering	13 (7%)	3.42 ± 0.14 [a]	F-value = 6.65	<0.001 **
	Not moving	44 (25%)	2.46 ± 0.13 [b]		
	Avoiding	43 (24%)	2.35 ± 0.12 [b]		
	Running away	13 (7%)	2.35 ± 0.17 [b]		
	Being aggressive	66 (37%)	2.24 ± 0.09 [b]		
Chin contact	Accepting	64 (36%)	2.68 ± 0.11 [a]	T-value = 3.24	0.002 **
	Not accepting	115 (64%)	2.27 ± 0.07 [b]		

[1] SE = standard error of the mean, ** means significant at the 1% significance level, shared letters are not significantly different groups, based on the Tuckey's multiple comparison method. [a] refers to significant different status compared to the others and [b] refers to nonsignificant values in each group.

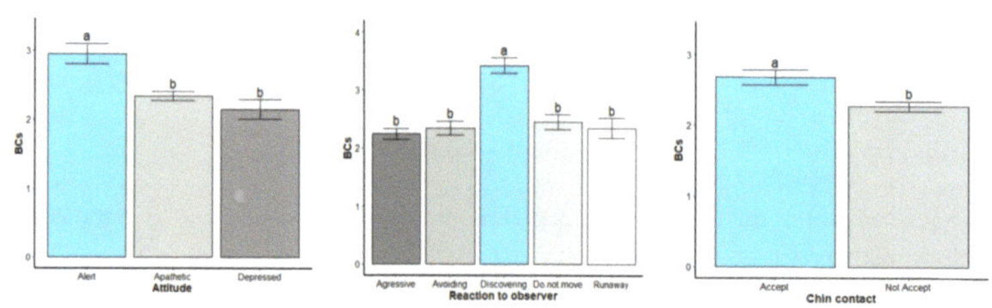

(a) (b) (c)

Figure 7. Error bars for the statuses of the behavior parameters: (**a**) the mean body condition of alert, apathetic, and depressed kiln donkeys, along with the mean ± standard error of the mean; (**b**) the mean body condition of donkeys that were aggressive and avoidant and that reacted to the observer by discovering, not moving, and running away, along with the mean ± standard error of the mean; (**c**) the mean body condition of donkeys who accepted and did not accept chin contact, along with the mean ± standard error of the mean. The bars labeled with different letters indicate a significant difference based on the Tuckey's test, $p < 0.05$, for attitude and response to observer, and the *t*-test for chin contact.

3.3. Physical Environmental Parameters

To explore the significant environmental parameters that have impacted the welfare of donkeys in brick kilns, including the BC, a multiple linear regression model was performed, with body condition as a response variable and all other environmental parameters included as explanatory variables. Table 6 shows the regression results and a numerical summary of the significant parameters. It was found that BCS was affected by three significant parameters (*p*-value < 0.05): hours worked/day, the number of donkeys working/brick kiln, and the cleanliness of the road. These three variables were able to explain 78.85% (R^2 = 78.85%) of the variability seen in the body conditions. However, no significant difference was seen in the following parameters with BCS: production of bricks, which relates to how the drivers (children) are paid (this could have led to an increase in the pressure or beatings imposed on the donkeys), and/or the number of ovens/kilns.

Table 6. Numerical summary of the environmental parameters (median and mean ± standard error), and the coefficient values, t-values, and *p*-values of the multiple linear regression.

Parameter	Mean ± SE [1]	Median	Coef [2]	T-Value	*p*-Value
Working hours/day	6.23 ± 0.19	6.0	−0.32	−2.75	0.016 *
Number of donkeys	8.95 ± 0.60	7.5	0.22	4.20	0.001 **
Distance	54.70 ± 3.53	53.5	−0.02	−2.40	0.031 *
Concrete amount/donkey	2.52 ± 0.20	2.15	0.48	4.85	0.000 **

[1] SE = standard error of the mean, * means significant at the 5% significance level, and ** means significant at the 1% significance level, [2] Coef = coefficient value

As for the housing condition, we found that 65 ± 0.11% of the working donkeys drink salty or/and unclean water. The results suggest that the water condition has a significant impact on the body condition (*p*-value < 0.01). Another housing factor found that 85 ± 0.08% of working donkeys live in areas where there is mud, urine, and fecal matter on the floor of the donkey corrals/paddocks/stalls. Most of the donkeys' housing (70 ± 0.10%) had poor ventilation and few windows. The floor conditions and ventilation have a significant impact on the body condition (the *p*-values are 0.011 and 0.000, respectively).

4. Discussion

Based on an ESPWWA estimation, there are more than 1000 brick kilns in Egypt, most of which are in the Giza and Qalyubia governorates. These brick kilns are not similar; they are different in terms of the mechanization, types of working equids, responsibilities of the stakeholders, and ownership of the donkeys. In the El-Saf kilns, the ownership of the donkeys belongs to the brick kiln owners, and the donkey drivers are children or adolescents, who are usually illiterate and do not have enough knowledge to properly communicate with the working equids. They live in villages near the brick kilns, and most of them are relatives of the owners or foreman. Working in these brick kilns is a temporary job for them. There is a dynamic movement of the handlers from kilns to other locations, and all stakeholders in the brick kilns are under pressure to work in order to meet the daily brick production target, even in harsh working conditions and bad weather.

Wounds or lesions have been considered to be one of the most prevalent and severe welfare problems facing working donkeys [3–5,23,24]. Wounds and lesions may have many origins, but little research has focused on the cause of wounds and their relationship to specific welfare parameters. This is true for donkeys working in brick kilns in various locations in Egypt. Some of the expected causes of wounds are improperly fitted and designed harnesses (saddles, collars), poor materials used to make the harness (natural and synthetic), poor communication between handlers (beating wounds) and their donkeys, aggressive behavior from other donkeys (e.g., bites and kicks), and improper husbandry/management practices [4,5].

This study revealed that the overall wound prevalence in working donkeys in brick kilns was high (80%), and it is by far the highest prevalence recorded, compared with other studies. For example, 77.5% was reported by Curran et al. [8], 79.4% was reported by Biffa and Woldemeskel [23], 48.9% was reported by Moltumo et al. [24], 42.2% was reported by Birhan et al. [25], 54.9% was reported by Fikru et al. [26], 59% was reported by Burn et al. [27] in Jordan, and 54% was reported by Sells et al. [28] in Morocco. This difference in the prevalence is attributed to the difference in environmental conditions, type of work, and the harness system for the donkeys in brick kilns. Brick kiln donkeys have the highest prevalence of the most severe lesions [5], which lead to severe pain and chronic suffering [29].

They often receive inadequate nutrition and veterinary care; they are overworked and overloaded; and they wear poorly fitted, insufficiently padded harnesses and are hitched to poorly designed harness [4], which may lead to many welfare problems, including harness-induced injuries, dehydration, and beating wounds [25]. In the El-Saf brick kilns, there is no training for donkeys to do such hard work, where sometimes they pull brick loads weighing 6 to 12 times their body weight. The work is often prolonged and strenuous under high environmental temperatures, with heat radiating from the kilns themselves. The terrain is harsh, there is limited access to clean fresh water, and the donkeys receive relatively little husbandry or veterinary care. There is a large proportion of working donkeys that are suffering from injuries due to poor handling by their handlers [29,30].

This study revealed that 49% of donkeys have wounds from excessive force or from being beaten, which is considered to be the most serious wound for donkeys. Pritchard et al. [3] reported the prevalence of hindquarter wounds from mistreatment to be 12.1% in donkeys. Another study [25] reported the prevalence of lesions from mistreatment to be 25.2% in donkeys and 41.7% in mules. The difference reported by other researchers and what was found in this study is likely due to the overall health of the working equids, the species (donkey or mule), the type of work, and environmental conditions. Considering that most drivers in El-Saf kilns are children or adolescents with little knowledge or skills relating to how to communicate and care for donkeys, coupled with their poor attitude toward donkeys, it is no surprise that the donkeys are mistreated. The drivers have little to no motivation to enhance the welfare of these equids, and so we find donkeys to be the most neglected animal. Brick kiln owners consider donkeys as machines with a low economic value, and the stakeholders of brick kilns work under severe pressure to meet daily production targets. From personal observations, it was frequently observed that the foreman in the El-Saf kilns imposed additional pressure on the young handlers, which resulted in the children beating their donkeys to achieve production goals.

A decrease in injuries may have been recorded if properly designed, well-fitted, and comfortable harnesses had been used in the brick kilns, or if the drivers had access to training relating to how to construct such harnesses. We know from other studies that properly fitted harnesses allow working animals to pull the equipment to the best of their ability, without risk of injuries. Unfortunately, we found many harnesses that were poorly designed or ill-fitted, which relates to the inefficient transfer of power from the animal to the implement, fatigue, discomfort, and/or injury to the animal [29]. While harness wounds can be avoided and represent 70% of veterinary intervention for working equids in developing countries, similar conditions were observed in this study. We could improve the performance of working equids by improving the overall harness quality and fit, as well as the material used to make the harnesses. With such improvements, we would likely see a decrease in wounds in the El-Saf brick kilns, as reported in other studies [31].

There are no available data on the harnesses of working donkeys in Egyptian brick kilns. This study revealed the prevalence of wounds caused by harness in working donkeys in brick kilns was correlated with certain parts of the harness such as the pack saddle, neck collar, and breeching or the wound was caused by the shafts of the cart. It is difficult to compare the prevalence of harness wounds found in this study with that found in other studies due to the fact that the harness system used in Egyptian brick kilns is different from that used in other studied areas, such as Ethiopia and Morocco [32]. Donkeys working in the El-Saf brick kilns are found pulling overloaded carts/wagons with harnesses in poor condition and ill-fitting and the cart is poorly attached to the donkey (Figure 5).

The study showed that the frequency of the wounds found on the limbs were related to the harness quality (e.g., poor) and donkeys being overworked and overloaded. This study and a study by Pearson et al. [29] showed that the level of severity and location of the wounds was associated with parts of the harness, BCS, and/or mistreatment.

This study revealed a correlation between cleanliness and wounds in the following locations and parts of the harness: the neck collar, pack saddle, and breeching. In 2009, Sells et al. [28] found that the cleanliness of the pack saddles is a significant factor in developing a pack wound. The presence of dirt may contribute to the early stages of wound formation by increasing the abrasiveness of the material. A poorly designed or ill-fitted harness will result in fatigue, discomfort, or lesions on donkeys [33].

We found that there is an association between beating wounds and saddle and breeching wounds (p-value < 0.01). In addition, the association between beating wounds and the cleanliness, condition, and fit of different parts of the harness was explored, and it was found that it is associated with the cleanliness of the saddle and breeching and associated with the fitness of the neck collar as well. Moreover, because the poorly designed or ill-fitted and unclean harness leads to an inefficient transfer of power from the donkey to the cart [33], painful lesions, which reduce the work capacity of working donkeys [34], it will result in an increase in the frequency of being beaten by their handlers to meet the daily production target.

Regarding the BCS of working donkeys in brick kilns, it is found that BCS is an important animal-based indicator in assessing the health status and welfare of working equids [35,36], without determining whether the nutrient requirements are met or exceeded [37]. A healthy equid should be fit, neither not too fat nor too thin [38]. A poor body condition is a major welfare concern for working equids [5,39]. Thin donkeys have less natural padding (adipose tissue and muscle), which protects them from friction, pressure, and lesions caused by harnessing. The present study revealed that 56% of kiln donkeys have a BCS of 2 or less. This score is higher than what other studies have reported for donkeys in similar scenarios [24,25]. This difference may be attributed to a difference in the working and environmental conditions, management practice, working load, and type of work. In general, poor body condition scores of working equids may be attributed to malnutrition [40] and/or parasitic infestation, coupled with a heavy working load [41]. Working equids used for strenuous work in hotter weather and or hot and humid weather are likely to suffer from dehydration and a poor body condition [42].

A poor body condition is generally coupled with many underlying factors, such as exhaustion from being overworked [30] and chronic pain [43]. Donkeys in such conditions may suffer from negative mental states (e.g., depression) [44] and live a life that is generally characterized by poor welfare [45–47].

The study found a difference between injured donkeys and healthy donkeys in terms of BCS. This finding suggests that a higher BCS and likely an overall improved care may decrease the number and severity of wounds. Pritchard et al. [3] identified the highest correlation coefficient to be that between a low body condition score and wounds of the skin and deeper tissues (r = 0.37), because they may have less natural padding protecting them from pressure, friction and shear lesions caused by the harness. The same results were reported by Abdela et al. [40], who found that there was a significant association between body condition and occurrences of wounds (p-value = 0.000), and animals with a poor body condition are more likely to be wounded than animals found to be in a good body condition. This finding agrees with the report of Birhan et al. [25] and Tsega et al. [41], who reported a significant difference between a poor and good BCS.

Behavior can be modified by several variables, including the following: workload, working conditions, and housing environment, as reported in this study. There is a need to understand the behavior of working donkeys and how it is associated with different management and production demands in order to improve the conditions and welfare of donkeys. Behavioral tests, such as those used in this study, can begin to identify compromised behavior that is linked to poor welfare, from a low BCS to excessive injuries or animals that have become apathetic due to being overworked.

In the present study, the majority of the donkeys displayed behavioral signs associated with being apathetic or depressed, and some even showed signs of aggression. The results in our study, compared to other studies, such as that of Ali et al. [18], suggest the findings that apathetic, depression and aggressive behaviors are more prevalent. Pritchard et al. [35] suggested that behavior could be related to or attributed to the severity of work, environmental and working conditions, type of work, and prevalence of harness-induced wounds in donkeys. The results in this study showed that donkeys had multiple challenges to overcome in order to live a healthy life. Factors that influenced their welfare and were documented included poor nutrition, water sources, harnesses, mental states, and overall BCS, which led to a compromised welfare. Such parameters have been shown to have similar results in other working equid studies [4]. Brick kiln donkeys have the highest prevalence of the most severe lesions [42]. In the El-Saf brick kilns, as mentioned before, the prevalence of wounds was the highest, compared with previous reports, and the prevalence of beating wounds in the present study was higher than the prevalence of mistreatment lesions reported by Ali et al. [18]. Burn et al. [4] reported that working equids that suffer from severe physical injuries showed depression and unresponsive behavior. Unresponsiveness behavior is considered to be an indicator of several poor welfare problems, for example, fatigue due to a heavy workload [30,48] and chronic pain [30] in the El-Saf brick kiln. The donkeys and their handlers are generally found to work for more than 8 h without rest in extreme conditions that are hot and to be under a high production pressure. Donkeys are generally overloaded and pull carts averaging 2.25 tons, in addition to the weight of the handlers, while suffering from pain and open lesions. In many cases, we found donkeys displaying aggressive reactions toward observers due to the working conditions and the fact the donkeys were regularly beaten by their handlers. This resulted in continuously compromised human–donkey interactions, and the donkeys continued to develop behaviors, such as fearfulness and aggression toward humans, which is generally not commonly observed in donkeys [42]. Rousing et al. [43] reported that depressed and unresponsive donkeys experienced improper handling. Positive human–animal interactions are very important when it comes to working equids, as it facilitates daily activities, encourages positive human attitudes, and improves animal welfare [44]. Burn et al. [4] reported that a low BCS is correlated most strongly with unresponsive behavior. The causes of a low BCS are multifactorial and likely include malnutrition, overwork, parasitism, and disease, which can simultaneously cause behavioral sickness that leads to anorexia and the development of a poor BCS [38,44,45]. Both fatigue from overworking, being overloaded [16] coupled with chronic pain [30], and depression or learned helplessness [36] can lead to a compromised welfare [37–39]. In the present study, we found that donkeys with a higher BCS show more alert behaviors and are friendlier with humans.

5. Conclusions

There is a need for basic research to understand the management, environmental, and working conditions in brick kilns to be able to promote a sustainable improvement program for donkey welfare. The El-Saf brick kilns are similar in terms of the work and mechanization systems, but there are differences in terms of the management practices and resources available for the donkeys in each kiln. This study introduced methods to measure the welfare status of the working donkeys in the El-Saf brick kilns and how to identify associated risk factors that compromise their welfare. We have found that the variation in the level of welfare associated with the donkeys in each kiln, along with differences found in the management practices, can help us to better understand why some kilns have a better welfare than others. The behavior, along with the welfare, of many of the donkeys was found to be compromised. The highest incidence of wounds was caused by beating, improper handling, and/or poor communication, which all reflected the missing link in human–animal interactions. Future studies could possibly measure the attitudes of donkey handlers in relation to overall donkey welfare. Another possibility would be measuring the correlation of human welfare to donkey welfare in each kiln. We believe that increasing brick kiln managers' and workers' knowledge of proper donkey husbandry, care, management, harness design, and function, as well as providing them with information on working donkeys, could improve the overall welfare of working donkeys. A standard protocol would need to be developed and shared in training sessions with brick kiln managers and then handlers. A further investigation may be warranted to understand the attitudes and knowledge of donkey handlers concerning the donkeys in brick kilns. Then, a strategy showing that an improved welfare (e.g., a higher BCS) leads to fewer wounds or clean water and improves the production of the donkeys and handlers is essential. This would be key to gaining the assent of stakeholders. A step toward this goal is understanding the risk factors, which this study has accomplished. We can now a design an effective educational program for the El-Saf brick kilns, which can be applied to all Egyptian Brick kilns. This study suggests that improvements can easily be made through knowledge transfer, if human behavior change can be facilitated.

Author Contributions: Conceptualization, S.F.F. and H.F.F.M.; methodology, H.F.F.M.; software, S.F.F., H.F.F.M. and A.K.M.; validation, H.F.F.M..; formal analysis, H.F.F.M.; investigation, S.F.F.; resources, S.F.F. and H.F.F.M.;

data curation, S.F.F.; writing—original draft preparation, A.K.M. and H.F.F.M.; writing—review and editing, S.F.F.; visualization, supervision and project administration, S.F.F., H.F.F.M. and A.K.M. All authors have read and agreed to the published version of the manuscript.

Funding: The research received no additional funding. The data were collected as part of a service being offered to donkey owners and handlers working in the brick kiln region serviced by ESPWWA.

Acknowledgments: We are extremely grateful to all owners who permitted an examination of their donkeys and ESPWWA and Donkey Sanctuary UK for working hard to improve the welfare of these poor donkeys.

Conflicts of Interest: The authors declare no conflict of interest.

References

1. Blakeway, J.S. The Welfare of Donkeys. MSc Thesis, University of Edinburgh, Edinburgh, UK, 1994.
2. Leeb, C.; Henstridge, C.; Dewhurst, K.; Bazeley, K. Welfare assessment of working donkeys: Assessment of the impact of an animal health care project in West Kenya. *Anim. Welf.* **2003**, *12*, 689–694.
3. Pritchard, J.C.; Lindberg, A.C.; Main, D.C.J.; Whay, H.R. Assessment of the welfare of working horses, mules and donkeys, using health and behaviour parameters. *Prev. Vet. Med.* **2005**, *69*, 265–283. [CrossRef]
4. Burn, C.C.; Denison, T.L.; Whay, H.R. Relationships between behaviour and health in working horses, donkeys, and mules in developing countries. *Appl. Anim. Behav. Sci.* **2010**, *126*, 109–118. [CrossRef]
5. Burn, C.C.; Dennison, T.L.; Whay, H.R. Environmental and demographic risk factors for poor welfare in working horses, donkeys and mules in developing countries. *Vet. J.* **2010**, *186*, 385–392. [CrossRef]
6. Saul, C.; Siefert, L.; Opuda-Asibo, J. Disease and health problems of donkeys: A case study from eastern Uganda. In *Improving Donkey Utilization and Management, Proceedings of the Animal Traction Network for Eastern and Southern Africa Workshop, Debre Zeit, Ethiopia, 5–9 May 1997*; Starkey, P., Ed.; Animal Traction Development: Oxgate, UK, 1998; pp. 58–63.
7. de Aluja, A.S. The welfare of working equids in Mexico. *Appl. Anim. Behav. Sci.* **1998**, *59*, 19–29. [CrossRef]
8. Curran, M.; Feseha, G.; Smith, D. The impact of access to animal health services on donkey health and livelihoods in Ethiopia. *Trop. Anim. Health Prod.* **2005**, *37*, 47–65. [CrossRef]
9. Tesfaye, A.; Curran, M.M. A longitudinal survey of market donkeys in Ethiopia. *Trop. Anim. Health Prod.* **2005**, *37*, 87–100. [CrossRef] [PubMed]
10. Wells, D.; Krecek, R.C. Socioeconomic, health and management aspects of working donkeys in Moretele 1, North West Province, South Africa. *J. S. Afr. Vet. Ass.* **2001**, *72*, 37–43. [CrossRef] [PubMed]
11. Oussat, M. Welfare assessment of working donkeys in Nouakchott (Mauritania). In *Fifth International Colloquium on Working Equines the Future for Working Equines, Addis Ababa, Ethiopia, 30 October–2 November 2006*; Pearson, A., Muir, C., Farrow, M., Eds.; The Donkey Sanctuary: Sidmouth, UK, 2006; pp. 130–137.
12. Pearson, R.A.; Krecek, R.C. Delivery of health and husbandry improvements to working animals in Africa. *Trop. Anim. Health Prod.* **2006**, *38*, 93–101. [CrossRef] [PubMed]
13. Geiger, M.; Hovorka, A. Donkeys in development: Welfare assessments and knowledge mobilisation. *Dev. Pract.* **2015**, *25*, 1091–1104. [CrossRef]
14. Popescu, S.; Diugan, E.A. The relationship between behavioral and other welfare indicators of working horses. *J. Equine Vet. Sci.* **2013**, *33*, 1–12. [CrossRef]
15. Reix, C.E.; Burn, C.C.; Pritchard, J.C.; Barr, A.R.S.; Whay, H.R. The range and prevalence of clinical signs and conformation associated with lameness in working draught donkeys in Pakistan. *Equine Vet. J.* **2014**, *46*, 771–777. [CrossRef] [PubMed]
16. Tadich, T.; Escobar, A.; Pearson, R.A. Husbandry and welfare aspects of urban draught horses in the south of Chile. *Arch. Med. Vet.* **2008**, *40*, 267–273. [CrossRef]
17. Kay, G. On a mission: Caring for working equids abroad. *In Pract.* **2007**, *29*, 108–111. [CrossRef]
18. Ali, A.B.; El Sayed, M.A.; Matoock, M.Y.; Fouad, M.A.; Heleski, C.R. Are mules or donkeys better adapted for Egyptian brick kiln work? (Until we can change the kilns). *J. Vet. Behav.* **2015**, *10*, 158–165. [CrossRef]
19. Rossel, S.; Marshall, F.; Peters, J.; Pilgram, T.; Adams, M.D.; O'Connor, D. Domestication of the donkey: Timing, processes, and indicators. *Proc. Natl. Acad. Sci. USA* **2008**, *105*, 3715–3720. [CrossRef]
20. Muylle, S. Aging. In *Equine Dentistry*, 3rd ed.; Easley, J., Dixon, P.M., Schumacher, J., Eds.; Elsevier: Edinburgh, Scotland, 2010; 424p, Available online: https://www.elsevier.com/books/equine-dentistry/easley/978-0-7020-2980-6 (accessed on 1 June 2020).

21. Pearson, R.A.; Ouassat, M.A. *Guide to Live Weight Estimation and Body Condition Scoring of Donkeys*; University of Edinburgh, Centre for Tropical Veterinary Medicine: Edinburgh, Scotland, 2000.
22. Svendsen, E.D. *The Professional Handbook of the Donkey*, 3rd ed.; Whittet Books Limited: London, UK, 1997; 400p.
23. Biffa, D.; Woldemeskel, M. Causes and factors associated with occurrence of external injuries in working equids in Ethiopia. *Int. J. Appl. Res. Vet. Med.* **2006**, *4*, 1–7.
24. Moltumo, S.; Mathewos, M.; Fesseha, H.; Yirgalem, M. Assessment of welfare problems on working donkeys in Hosaena District, Hadiya Zone, Southern Ethiopia. *Vet. Med. Open J.* **2020**, *4*, 100–106. [CrossRef]
25. Birhan, G.; Chanie, M.; Tesfaye, T.; Kassa, A.; Mekonnen, B.; Wagaw, N. Incidence of wound and associated risk factors in working donkeys in Yilmana Densa District. *Glob. Vet.* **2014**, *13*, 133–140.
26. Fikru, A.; Tadese, A.; Gebreegziabher, Z. Prevalence of equine wound and associated risk factors in and around Kombolcha Town, North Ethiopia. *J. Vet. Sci. Technol.* **2015**, *6*, 1–4.
27. Burn, C.C.; Pritchard, J.C.; Farajat, M.; Twaissi, A.A.M.; Whay, H.R. Risk factors for strap related lesions in working donkeys at the World Heritage Site of Petra in Jordan. *Vet. J.* **2007**, *178*, 263–271. [CrossRef] [PubMed]
28. Sells, P.; Pinchbeck, G.; Mezzane, H.; Ibourki, J.; Crane, M. Pack wounds of donkeys and mules in the Northern High Atlas and lowlands of Morocco. *Equine Vet. J.* **2009**, *42*, 219–226. [CrossRef] [PubMed]
29. Pearson, R.; Simalenga, T.E.; Krecek, R.C. *Harnessing and Hitching Donkeys, Mules and Horses for Work*; University of Edinburgh, Centre for Tropical Veterinary Medicine: Edinburgh, Scotland, 2003; 34p.
30. Pearson, R.A.; Muir, C.J.; Farrow, M. *The Future of Working Equines*; University of Edinburgh, Centre for Tropical Veterinary Medicine: Edinburgh, Scotland, 2003.
31. Popescu, S.; Diugan, E.A.; Spinu, M. The interrelations of good welfare indicators assessed in working horses and their relationships with the type of work. *Vet. Sci.* **2014**, *96*, 406–414. [CrossRef] [PubMed]
32. Mekuria, S.; Mulachew, M.; Abebe, R. Management practices and welfare problems encountered on working equids in Hawassa town, Southern Ethiopia. *J. Vet. Med. Anim. Health* **2013**, *5*, 243–250.
33. Pritchard, J.C.; Burn, C.C.; Barr, A.R.S.; Whay, H.R. Validity of indicators of dehydration in working horses: A longitudinal study of changes in skin tent duration, mucous membrane dryness and drinking behaviour. *Equine Vet. J.* **2008**, *40*, 558–564. [CrossRef] [PubMed]
34. Pritchard, J.C.; Burn, C.C.; Whay, H.R. Haematological and serum biochemical reference values for apparently healthy working horses in Pakistan. *Res. Vet. Sci.* **2009**, *87*, 389–395. [CrossRef]
35. Pritchard, J.C.; Barr, A.R.S.; Whay, H.R. Validity of a behavioural measure of heat stress and a skin tent test for dehydration in working horses and donkeys. *Equine Vet. J.* **2006**, *38*, 433–438. [CrossRef]
36. Dunn, A.J.; Swiergiel, A.H.; de Beaurepaire, R. Cytokines as mediators of depression: What can we learn from animal studies? *Neurosci. Biobehav. Rev.* **2005**, *29*, 891–909. [CrossRef]
37. Aubert, A. Sickness and behaviour in animals: A motivational perspective. Neurosci. *Biobehav. Rev.* **1999**, *23*, 1029–1036. [CrossRef]
38. Millman, S.T. Sickness behaviour and its relevance to animal welfare assessment at the group level. *Anim. Welf.* **2007**, *16*, 123–125.
39. Weary, D.M.; Huzzey, J.M.; von Keyserlingk, M.A.G. Using behavior to predict and identify ill health in animals. *J. Anim. Sci.* **2009**, *87*, 770–777. [CrossRef] [PubMed]
40. Abdela, N.; Teshome, E.; Hassan, A.; Deressa, F.B. Prevalence and associated risk factors of equine wound in and around Asella Town, South Eastern Ethiopia. *J. Vet. Med. Anim. Health* **2017**, *9*, 63–71.
41. Tsega, A.M.; Worku, Y.; Tesfaye, T.; Nazir, S. Prevalence of wound and associated risk factors of Donkeys in Merawi District, North- Western Ethiopia. *J. Anim. Res.* **2016**, *6*, 765–771. [CrossRef]
42. Vollmayr, B.; Henn, F.A. Stress models of depression. *Clin. Neurosci. Res.* **2003**, *3*, 245–251. [CrossRef]
43. Rousing, T.; Bonde, M.; Sorensen, J.T. Aggregating welfare indicators into an operational welfare assessment system: A bottom-up approach. *Acta Agric. Scand. A Ann. Suppl.* **2001**, *30*, 53–57.
44. Kielland, C.; Skjerve, E.; Osteras, O.; Zanella, A.J. Dairy farmer attitudes and empathy towards animals are associated with animal welfare indicators. *J. Dairy Sci.* **2010**, *93*, 2998–3006. [CrossRef]
45. Hart, B.L. Biological basis of the behavior of sick animals. *Neurosci. Biobehav. Rev.* **1988**, *12*, 123–137. [CrossRef]
46. Dantzer, R.; Kelley, K.W. Twenty years of research on cytokineinduced sickness behavior. *Brain Behav. Immun.* **2017**, *21*, 153–160. [CrossRef]

47. El Dirdiri, N.I.; Damir, H.A.; Wahbi, A.A. Disease incidence in donkeys (*Equus asinus asinus*) with emphasis on strongyle infection. *Acta. Vet.* **1986**, *36*, 313–320.
48. Rodriguez-Maldonado, G. The principle problems in working donkeys in Mexico. In *Donkeys, Mules and Horses in Tropical Agricultural Development*; Fielding, D., Pearson, R.A., Eds.; University of Edinburgh, Centre for Tropical Veterinary Medicine: Edinburgh, Scotland, 1991; pp. 138–139.

animals

MDPI

Article

Vitrification of Donkey Sperm: Is It Better Using Permeable Cryoprotectants?

Manuel Hidalgo *, Maria Diaz-Jimenez, César Consuegra, Blasa Pereira and Jesús Dorado

Veterinary Reproduction Group, Department of Medicine and Animal Surgery, Faculty of Veterinary Medicine, University of Cordoba, 14071 Cordoba, Spain; mariadijm@gmail.com (M.D.-J.); mtc93vet@gmail.com (C.C.); blasypereiraaguilar@gmail.com (B.P.); jdorado@uco.es (J.D.)

* Correspondence: mhidalgo@uco.es; Tel.: +34-957-218-363

Received: 30 July 2020; Accepted: 17 August 2020; Published: 20 August 2020

Simple Summary: Conventional donkey sperm-freezing using permeable cryoprotectants has been successfully performed, and good sperm parameters have been obtained after thawing. Unfortunately, artificial insemination of jennies with cryopreserved semen has given unsatisfactory results. Vitrification by directly dropping the sperm into the liquid nitrogen following the spheres methodology has been developed in human beings as an alternative to conventional freezing. This technique has shown to be a species-specific methodology and the concentration of cryoprotectants should be optimized in donkeys. Additionally, in this study, a permeable cryoprotectant (glycerol) has been tested for the first time for donkey sperm vitrification. According to our findings, vitrification of donkey sperm was effectively carried out using an extender supplemented with sucrose or bovine serum albumin (BSA) as non-permeable agent. When glycerol, a permeable agent, was compared to sucrose 0.1 M and BSA 5%, sperm quality significantly decreased. Therefore, donkey sperm vitrification in the absence of permeable agents obtained better results and gives a new approach to create a pattern for future studies of fertility trials.

Abstract: Vitrification by direct exposure of sperm to liquid nitrogen is increasing in popularity as an alternative to conventional freezing. In this study, the effect of permeable cryoprotectant agents for donkey sperm vitrification was compared to an extender containing non-permeable cryoprotectants. First, three different concentrations of sucrose (0.1, 0.2, and 0.3 molar, M) and bovine serum albumin, BSA (1, 5, and 10%) were compared. Secondly, the concentration of non-permeable agents producing the most desirable results was compared to an extender containing glycerol as permeable agent. Vitrification was performed by dropping 30 µL of sperm suspension directly into LN2 and warming at 42 °C. Sperm motility (total, TM; and progressive, PM) and plasma membrane integrity, PMI (mean ± SEM) were statistically compared between treatments. Sucrose 0.1 M showed a significantly higher percentage of total sperm motility (21.67 ± 9.22%) than sucrose 0.2 M (14.16 ± 4.50%) and 0.3 M (8.58 ± 6.22%); and no differences were found in comparison to the control (19.71 ± 10.16%). Vitrification with sucrose 0.1 M or BSA 5% obtained similar results for TM (21.67 ± 9.22% vs. 19.93 ± 9.93%), PM (13.42 ± 6.85% vs. 12.54 ± 6.37%) and PMI (40.90 ± 13.51% vs. 37.09 ± 14.28); but both showed higher percentages than glycerol (TM = 9.71 ± 4.19%; PM = 5.47 ± 3.17%; PMI = 28.48 ± 15.55%). In conclusion, donkey sperm vitrification in spheres using non-permeable cryoprotectants exhibited better sperm motility and viability parameters after warming than sperm vitrification using extenders containing permeable cryoprotectants.

Keywords: donkey; sperm; vitrification; spheres; sucrose; BSA; glycerol

1. Introduction

According to the Food and Agricultural Organization, European donkey populations have diminished considerably in the last century [1]. According to the Spanish regulations, the Andalusian donkey breed is in danger of extinction, with a breeding population of 100 males and 436 females in the last update in 2020 [2]. In this sense, and considering the importance of environment and biodiversity preservation of domestic species resources, the creation of sperm banks contributes to preserve valuable genetic material from these endangered populations. Unfortunately, artificial insemination (AI) with cryopreserved donkey semen has resulted in poor fertility outcomes [3]. Different strategies have been developed to improve pregnancy rates, including the combination of different cryoprotectant agents (CPAs) [4], addition of seminal plasma to frozen-thawed donkey semen before AI [3], post-thaw centrifugation for cryoprotectant removal [5], study of the jennies endometrial response after AI [6,7], or the influence of different insemination protocols [8]. The low fertilizing capacity of cryopreserved donkey sperm has been attributed to the impact of permeable CPAs, but this hypothesis remains unclear [3,9,10].

In addition, the osmotic stress produced in the sperm cell during conventional freezing and thawing may induce structural and functional damage through the formation of ice crystals, and affects the fertilizing ability of cryopreserved sperm [11]. Vitrification is a cryopreservation method widely used for embryo, oocyte or tissue storage [12,13]. It involves the solidification of a solution, which turns into a glass-like state [14]. During vitrification, viscosity greatly increases, and intracellular or extracellular ice crystals are not formed because water does not precipitate [15]. A high concentration of permeable CPAs has been used to reach the high viscosity needed for oocyte and embryo vitrification [13]. However, this methodology has yet to be applied to the sperm cell due to its higher sensitivity to increasing concentrations of permeable CPAs [16,17]. Nevertheless, it has been demonstrated that the concentration of CPAs required for achieving vitrification is inversely related to the rate of cooling/warming. This means that if the sample is ultra-fast cooled (immersing small volume samples directly into LN_2), high concentration of permeable CPAs are not necessary to achieve a vitrified state, therefore avoiding their toxicity [13,18]. This methodology, combined with non-permeating substances such as proteins and/or carbohydrates, has been called 'kinetic sperm vitrification' or 'permeable cryoprotectant-free sperm vitrification' and has been successfully developed in human [15], dog [19], fish [20,21] wild ungulates [17,22], cats [23] and, recently, in stallions [24] and donkeys [25]. It is a simple, fast, and cost-effective method to cryopreserve sperm, even under field conditions, since a reasonably equipped laboratory is required, making it attractive for the conservation of wild or endangered species and genetically valuable animals distributed in different regions, as it usually happens in endangered donkey breeds. Taking into account that sperm vitrification has led to similar or an increase in sperm quality after warming in comparison to conventional freezing in stallion [24], it could be considered to be another alternative to improve sperm cryopreservation in donkeys.

The optimal concentration of CPAs seems to be species-specific and has been proposed as a key factor for sperm vitrification success, depending on the methodology used [24]. In a preliminary research, donkey sperm vitrified in straws showed significant higher sperm motility percentages when compared to vitrification in spheres [25]. However, only sucrose was tested as cryoprotectant agent, and a fixed concentration was employed for both methods derived from previous studies in other species. Consequently, the optimal concentration of non-permeable CPAs for donkey sperm vitrification using the spheres methodology has not been determined yet. Additionally, as previously stated, sperm vitrification has been developed using non-permeable CPAs [26] and a few attempts for sperm vitrification have been performed using a combination of permeable and non-permeable CPAs [27]. To the best of our knowledge, donkey sperm vitrification in spheres with the sole use of permeable CPAs has not been tested yet.

Therefore, the present study was designed to examine the effectiveness of different concentrations of sucrose and bovine serum albumin for donkey sperm vitrification in spheres, in comparison to a vitrification extender containing permeable CPAs by examining the post-thaw quality in vitro.

2. Materials and Methods

2.1. Animals, Semen Collection, and Processing

All procedures were approved by the Ethical Committee for Animal Experimentation of the University of Cordoba, in compliance with the Regional Government of Andalusia (project no. 31/08/2017/105) and the Spanish law for animal welfare and experimentation (RD 53/2013).

The jackasses were housed in individual paddocks at "Centro Rural Malpica" (Palma del Rio, Cordoba, Spain). They were fed with teff hay, oat grains and water "ad libitum". Semen was collected from four adult, healthy and fertile Andalusian donkeys using a Missouri artificial vagina (Minitüb, Tiefenbach, Germany) in the presence of a jenny in estrus. Semen was collected one or twice a week per donkey until a total of 16 ejaculates was completed (four ejaculates per donkey). Immediately after collection, volume (mL) and sperm concentration ($\times 10^6$ spermatozoa/mL) were measured in each gel-free semen sample using a graduated collector and a sperm photometer (Spermacue®, Minitüb, Tiefenbach, Germany) respectively. Sperm motility and viability parameters were evaluated as described below. Thereafter, sperm was extended 1:1 (*v/v*) with INRA96 (IMV Technologies, France) and centrifuged at 400× *g* for 7 min to remove seminal plasma. Sperm pellets were then re-extended in each vitrification media (see experimental design) to reach a final concentration of 200×10^6 spermatozoa/mL. Sperm suspensions were maintained at room temperature for 10 min and slowly cooled for 1 h at 5 °C into a sperm container (Equitainer, Hamilton Research, Inc. Ipswich, Massachusetts, USA) before the vitrification procedure was performed.

2.2. Vitrification and Warming

Sperm vitrification was carried out following the methodology previously described [24,25]. Briefly, a styrofoam box was loaded with LN_2 and five 30 µL drops from each treatment were plunged directly into the LN_2. A micropipette held at an angle of about 45° and a distance of 10 cm from the surface was used (Figure 1a,b). After contact with the LN_2 a sphere is immediately formed (Figure 1c). Spheres were then packaged into 1.8 mL cryotubes and stored LN_2 tanks. The warming procedure was performed by introducing the spheres one by one into two milliliters of extender (INRA-96) previously warmed to 42 °C. A gentle vortexing for a few seconds of each sample was done before centrifugation as described above. Sperm pellets were re-extended with INRA-96 after supernatant removal, to reach final concentration of 25×10^6 spermatozoa/mL for sperm evaluation. Sperm motility was objectively evaluated by using the Sperm Class Analyzer (SCA v.5.4.0, Microptic S.L., Barcelona, Spain) as previously described [4]. The following kinetic parameters were calculated by the system: total (TM, %) and progressive motility (PM, %); curvilinear (VCL, µm/s), straight line (VSL, µm/s) and average path velocities (VAP, µm/s), linearity (LIN, VSL/VCL × 100), straightness rate (STR, VSL/VAP × 100), wobble (WOB, VAP/VCL × 100), lateral head displacement amplitude (ALH, µm) and beat cross frequency (BCF, Hz). Sperm membrane integrity was assessed using Vital-Test commercial kit (Halotech DNA S.L., Madrid, Spain) for sperm staining following the manufacturer´s instructions. In brief, an aliquot of 10 µL of diluted semen was mixed with 1 µL propidium iodide stock solution and 1 µL of acridine orange stock solution and evaluated under epifluorescence microscopy. At least 200 spermatozoa were counted, and sperm with intact plasma membrane was recorded (PMI, %).

2.3. Experimental Design

2.3.1. Experiment 1. Vitrification of Donkey Sperm Using Different Concentrations of Sucrose

Sucrose (Sigma-Aldrich Corp., St. Louis, MI, USA) was added to a control base extender (Control) commonly used for horse sperm. This base extender contains egg yolk and antibiotics (Gent, Minitüb, Tiefenbach, Germany). Powder sucrose was weighted with a precision balance and mixed with the control extender by vortexing. Three concentrations of sucrose (mol/L, M) were compared: 0.1 M,

0.2 M and 0.3 M; each extender was then added to the sperm pellets and vitrified as described before. Post-warming sperm parameters were recorded and compared between treatments.

2.3.2. Experiment 2. Vitrification of Donkey Sperm Using Different Concentrations of Bovine Serum Albumin

Bovine serum albumin (BSA, Sigma-Aldrich, Sant Louis, MI, USA) was weighted and added to the same control base extender as described before at the following concentrations: 1%, 5% and 10%. Each sample was then vitrified, and post-warming analysis was carried out as previously described, comparing the results obtained among treatments.

2.3.3. Experiment 3. Comparison Between Permeable and Non-Permeable CPAs for Donkey Sperm Vitrification

Having identified the best concentration of non-permeable CPAs for donkey sperm vitrification in spheres, a commercial extender for stallion sperm-freezing containing permeable CPAs, in particular glycerol (Gent B, Minitüb, Tiefenbach, Germany), was compared for sperm vitrification. Post-warming sperm parameters were assessed for each treatment as described before.

2.4. Statistical Analysis

Results were analyzed using the Statistical Analysis Systems software (SAS version 9.0; SAS Institute Inc, Cary, NC, USA). All data was first tested for normality of the data distribution and homogeneity of variances using the Kolmogorov–Smirnov and Levene test, respectively. The sperm parameters assessed were compared between treatments using a repeated measured general lineal model (GLM). Animals and ejaculates were considered to be random factors. The Post-Hoc HSD Tukey test was used to compare differences between mean values. Results were expressed as mean ± standard deviation of the mean (SD). The level of significance was set at $p < 0.05$.

3. Results

Mean average parameters of ejaculates used in this study were as follows: gel-free volume of 112 ± 41 mL, sperm concentration of 191 ± 68 × 10^6 spermatozoa/mL, total motility 79.7 ± 12.8%, progressive motility 64.7 ± 15.4% and sperm with intact plasma membrane 58.7 ± 15.7%.

3.1. Experiment 1. Vitrification of Donkey Sperm Using Different Concentrations of Sucrose

Vitrification in spheres using a sucrose concentration of 0.1 M resulted in the greatest values ($p < 0.05$) for all the sperm motility variables assessed in comparison to the other sucrose concentrations (Table 1). There were no significant differences ($p > 0.05$) between control and sucrose treatments for the assessment of plasma membrane integrity. There were no differences between sucrose 0.1 M and the control for total (21.67 ± 9.22 vs. 19.71 ± 10.16) and progressive (13.42 ± 6.85 vs. 12.34 ± 8.13) sperm motility, respectively; however, sucrose 0.1 M showed a tendency to obtain higher values of sperm motility percentages.

Table 1. Vitrification of donkey semen samples (n = 16) using different concentrations of sucrose.

Sperm Parameters	Vitrification Extender				p-Values
	Control	Sucrose 0.1 M	Sucrose 0.2 M	Sucrose 0.3 M	
TM (%)	19.71 ± 10.16 ^{a,b}	21.67 ± 9.22 ^{a}	14.16 ± 4.50 ^{b}	8.58 ± 6.22 ^{c}	<0.001
PM (%)	12.34 ± 8.13 ^{a}	13.42 ± 6.85 ^{a}	7.69 ± 3.01 ^{b}	3.53 ± 4.42 ^{b}	<0.001
PMI (%)	33.66 ± 14.84	40.90 ± 13.50	39.67 ± 13.12	38.62 ± 10.32	>0.05
VCL (μm/s)	77.04 ± 19.70 ^{a}	82.16 ± 13.31 ^{a}	65.07 ± 13.04 ^{b}	44.69 ± 16.98 ^{c}	<0.001

Table 1. *Cont.*

Sperm Parameters	Vitrification Extender				p-Values
	Control	Sucrose 0.1 M	Sucrose 0.2 M	Sucrose 0.3 M	
VSL (µm/s)	62.52 ± 19.12 [a]	66.91 ± 12.48 [a]	50.42 ± 13.80 [b]	32.61 ± 15.43 [c]	<0.001
VAP (µm/s)	67.23 ± 19.51 [a,b]	71.96 ± 12.93 [a]	56.46 ± 11.74 [b]	36.75 ± 16.58 [c]	<0.001
ALH (µm)	2.34 ± 0.29 [a]	2.46 ± 0.43 [a]	2.04 ± 0.56 [b]	1.78 ± 0.67 [b]	<0.001
LIN (%)	80.02 ± 5.69 [a]	81.16 ± 6.19 [a]	76.52 ± 11.01 [a,b]	68.13 ± 17.85 [b]	<0.05
STR (%)	92.74 ± 2.80 [a]	92.84 ± 2.85 [a]	88.08 ± 11.32 [a,b]	84.44 ± 17.97 [b]	<0.05
WOB (%)	86.40 ± 4.32 [a]	87.34 ± 4.64 [a]	86.70 ± 3.79 [a]	78.35 ± 13.85 [b]	<0.05
BCF (Hz)	9.65 ± 0.82 [a]	9.64 ± 0.66 [a]	8.60 ± 1.79 [a,b]	7.68 ± 2.90 [b]	<0.001

Different letters ([a–c]) in the same row indicate significant differences. TM, total motility; PM, progressive motility; PMI, plasma membrane integrity; VCL, curvilinear velocity; VSL, straight line velocity; VAP, average path velocity; ALH, amplitude of lateral head displacement; LIN, linearity; STR, straightness; WOB, wobble; BCF, beat cross frequency; Control, control extender without sucrose. Values are expressed as mean ± standard deviation of the mean.

3.2. Experiment 2. Vitrification of Donkey Sperm Using Different Concentrations of Bovine Serum Albumin

The addition of different concentrations of BSA to the vitrification extender resulted in no significant differences ($p > 0.05$) in any of the sperm parameters assessed when compared to control samples (Table 2).

Table 2. Vitrification of donkey semen samples ($n = 16$) using different concentrations of bovine serum albumin (BSA).

Sperm Parameters	Vitrification Extender				p-Values
	Control	BSA-1%	BSA-5%	BSA-10%	
TM (%)	19.71 ± 10.16	19.51 ± 9.67	19.93 ± 8.93	15.52 ± 6.37	>0.05
PM (%)	12.34 ± 8.13	12.58 ± 7.65	12.54 ± 6.37	9.24 ± 4.39	>0.05
PMI (%)	33.66 ± 14.84	36.37 ± 11.36	37.09 ± 14.28	35.49 ± 13.28	>0.05
VCL (µm/s)	77.04 ± 19.70	83.74 ± 15.80	83.27 ± 16.31	79.69 ± 16.87	>0.05
VSL (µm/s)	62.52 ± 19.12	68.74 ± 14.78	68.19 ± 15.88	64.57 ± 17.07	>0.05
VAP (µm/s)	67.23 ± 19.51	74.54 ± 15.59	73.86 ± 16.43	69.91 ± 17.06	>0.05
ALH (µm)	2.34 ± 0.29	2.31 ± 0.36	2.46 ± 0.22	2.49 ± 0.31	>0.05
LIN (%)	80.02 ± 5.69	81.72 ± 4.40	81.09 ± 5.06	80.14 ± 5.74	>0.05
STR (%)	92.74 ± 2.77	92.12 ± 2.73	92.08 ± 2.48	91.96 ± 3.16	>0.05
WOB (%)	86.40 ± 4.31	88.67 ± 3.28	88.04 ± 3.95	87.06 ± 4.13	>0.05
BCF (Hz)	9.65 ± 0.82	9.67 ± 0.62	14.82 ± 21.65	15.16 ± 21.58	>0.05

TM, total motility; PM, progressive motility; PMI, plasma membrane integrity; VCL, curvilinear velocity; VSL, straight line velocity; VAP, average path velocity; ALH, amplitude of lateral head displacement; LIN, linearity; STR, straightness; WOB, wobble; BCF, beat cross frequency; Control, control extender without BSA; Values are expressed as mean ± standard deviation of the mean.

3.3. Experiment 3. Comparison Between Permeable and Non-Permeable CPAs for Donkey Sperm Vitrification

The sole use of glycerol for sperm vitrification in spheres decreased ($p < 0.05$) TM (9.71 ± 4.19%), PM (5.47 ± 3.17%) and PMI (28.48 ± 15.55%) compared to vitrification using sucrose 0.1 M and BSA 5% (Table 3). No significant differences ($p > 0.05$) were found in the remaining sperm parameters assessed.

Table 3. Comparison between permeable (glycerol) and non-permeable (sucrose and BSA) cryoprotectants for donkey sperm vitrification (n = 16).

Sperm Parameters	Vitrification Extender			p-Values
	Glycerol	Sucrose 0.1 M	BSA-5%	
TM (%)	9.71 ± 4.19 [b]	21.67 ± 9.22 [a]	19.93 ± 8.93 [a]	<0.001
PM (%)	5.47 ± 3.17 [b]	13.42 ± 6.85 [a]	12.54 ± 6.37 [a]	<0.001
PMI (%)	28.48 ± 15.55 [b]	40.90 ± 13.51 [a]	37.09 ± 14.28 [a]	<0.05
VCL (μm/s)	80.72 ± 17.14	82.16 ± 13.31	83.27 ± 16.31	>0.05
VSL (μm/s)	68.09 ± 18.06	66.91 ± 12.48	68.19 ± 15.88	>0.05
VAP (μm/s)	71.97 ± 17.55	71.96 ± 12.93	13.86 ± 16.43	>0.05
ALH (μm)	2.32 ± 0.54	2.46 ± 0.43	2.46 ± 0.22	>0.05
LIN (%)	83.04 ± 6.19	81.16 ± 6.19	81.09 ± 5.06	>0.05
STR (%)	93.68 ± 3.26	92.84 ± 2.85	92.08 ± 2.48	>0.05
WOB (%)	88.56 ± 4.26	87.34 ± 4.64	88.04 ± 3.95	>0.05
BCF (Hz)	9.47 ± 1.36	9.64 ± 0.66	14.82 ± 21.65	>0.05

Different letters ([a,b]) in the same row indicate significant differences. TM, total motility; PM, progressive motility; PMI, plasma membrane integrity; VCL, curvilinear velocity; VSL, straight line velocity; VAP, average path velocity; ALH, amplitude of lateral head displacement; LIN, linearity; STR, straightness; WOB, wobble; BCF, beat cross frequency; Values are expressed as mean ± standard deviation of the mean.

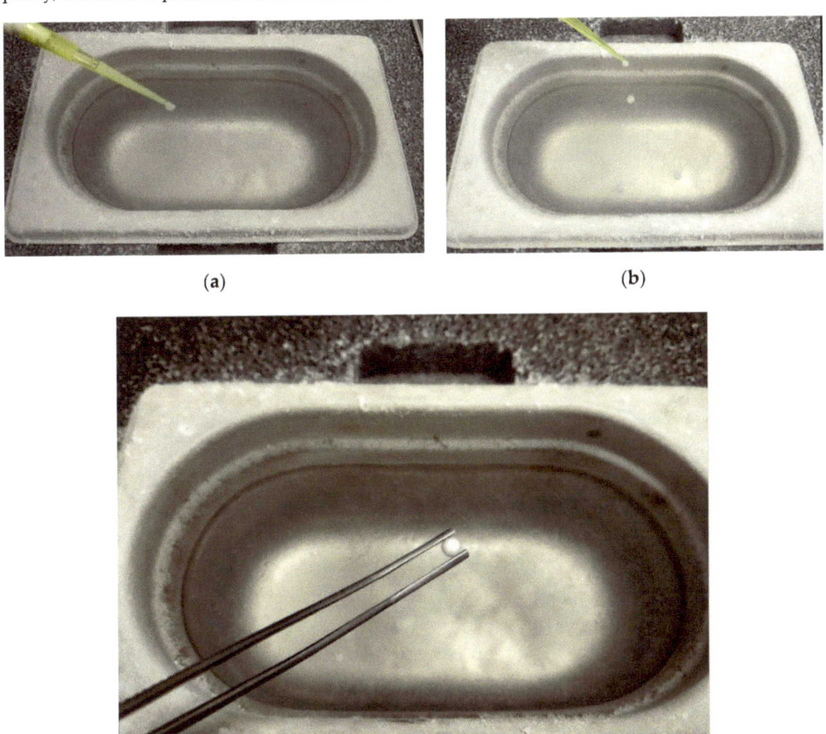

(a) (b) (c)

Figure 1. Images of the spermatozoa vitrification procedure, (**a**,**b**) 30 μL of spermatozoa suspension dropped into LN_2 using a micropipette, (**c**) sphere formed after sperm vitrification.

4. Discussion

Permeable CPAs have been conventionally employed for the cryopreservation of donkey sperm samples by slow freezing in donkey sperm, and by vitrification in spheres in other species, but they had not been employed for vitrification in donkeys yet. Non-permeable CPAs have been successfully used for sperm vitrification in different animal species [15,19,24,25,28]. It has been pointed out that the optimal concentration of non-permeable cryoprotectants varies between vitrification techniques. In this sense, studies comparing different concentrations of non-permeable cryoprotectants for sperm vitrification using the spheres technique in donkey sperm are so far lacking in the scientific literature.

In the first experiment of the present study, the addition of sucrose 0.1 M to the extender had a positive effect on sperm motility percentages in comparison to the control without sucrose. Sucrose concentrations were selected taking into consideration other reports of sperm vitrification in spheres in mammals [12,17,19,23–25,29], in which various amounts of sucrose between 0.02 M to 0.5 M were added. According to the results obtained in this study, the lowest sucrose concentration (0.1 M) resulted in the greatest values for sperm motility parameters in comparison to higher concentrations of sucrose (0.2 M and 0.3 M). These results are in agreement with previous reports in which different sucrose concentrations were compared, and higher sucrose concentrations (0.3 M and 0.5 M) resulted in significantly lower values for sperm motility in comparison to 0.1 M in wild goat [17] and sheep [22]. Similarly, sucrose 0.1 M has also been successfully used for sperm vitrification in spheres in 14 different wild endangered species [30] and in mouflon and fallow deer [31]. However, previous research of sperm vitrification using the spheres method have shown sucrose requirements to be slightly higher than 0.1 M in other species. In this regard, concentrations of 0.2 M [32] and 0.25 M [29,33] have been successfully employed in studies performed in human; 0.125 M in fish [20]; 0.2 M in cat [23], and 0.25 M in dog [19]. Interestingly, though, despite the phytogenic proximity between horse and donkey species [24], sucrose requirements for vitrification of stallion sperm showed to be much lower than that of in donkeys. It has been reported the upper limit of sucrose concentration to be 0.02 M in stallions [24], which is far lower than those used in the present study.

Our results once again reaffirm the previously described differences of cryoprotectant requirements between species [22,24]. A reasonable explanation for the diverse responses to vitrification extender is that species have different sperm cryosurvival, as described by other authors [12,27]; it seemed to be species-specific, which in turn may be a consequence of the cryostability and properties of the sperm cell: sperm size, water content, membrane fluidity, osmotic content and/or internal compaction [30,31,34]. In agreement with previous reports in other species [17,28], in the current study no differences were found between the different sucrose concentrations and the control extender regarding plasma membrane integrity.

Serum albumin has been traditionally added to the vitrification extender for sperm cryopreservation in several species. Thus, human serum albumin has been employed at 1% in human [29]; and BSA at 0.5% in rabbit [12]; 1% in equine [35], dog [19], fish [28], donkey [36] and wild goat [17]; and 2% in ram [37]. This molecule has shown to reduce oxidative stress [38] and to protect the sperm membrane against cryodamage, although the exact mechanism is still not clear [39]. Few studies have, however, tested different concentrations for sperm vitrification to determine the most adequate [24]. Moreover, to the best of the authors knowledge, the effect of BSA for sperm vitrification has always been examined in combination with sucrose and other cryoprotectants, but not by itself. Therefore, we aimed to test if the sole use of BSA could increase sperm quality after vitrification. Surprisingly, no differences among BSA concentrations nor with the control extender were found in any of the parameters assessed. However, the concentration of BSA 5% showed a tendency to obtain higher motility results.

In the last experiment, and considering that the use of 0.1 M of sucrose in the extender highly improved motility parameters, and BSA 5% also showed a tendency to improve motility results, both extenders were compared with an extender with a permeable CPA for donkey sperm vitrification. Glycerol was selected as the permeable agent because it has been widely and successfully employed for donkey sperm cryopreservation in previous studies [3,4,40–42], and could be considered to be a starting point to study the impact of permeable agents in donkey sperm vitrification. Glycerol has been previously included in vitrification extenders following the spheres or straws method, with positive results in sperm quality parameters after warming in several species such as ram [16] and sea bream [43]; dimethyl sulfoxide in salmon [21], and a combination of ethylene glycol and dimethyl sulfoxide in goat [44]. According to our results, the addition of glycerol to the vitrification extender significantly reduced sperm motility and plasma membrane integrity after warming; however, similar results were obtained when sperm vitrification was performed using either sucrose 0.1 M or BSA 5% as non-permeable agents. These results agree with previous reports in wild sheep [22], in which glycerol addition to the vitrification extender decreased sperm motility and plasma membrane integrity compared to the sole use of non-permeable CPAs. Nonetheless, other authors reported no motile or viable boar sperm after vitrification with only sucrose, neither in combination with permeable agents [27]. Sperm vitrification in spheres using only non-permeable agents have also been problematic in other species, obtaining few motile or viable sperm after warming in rabbit [12] and ram [37]. Interestingly, in a previous study performed in donkeys [42] in which authors compared between slow freezing using only non-permeable agents and the same freezing protocol but containing glycerol, no differences were found for sperm motility and DNA integrity. Therefore, it seems that glycerol may protect donkey sperm during slow freezing but no during vitrification. As mentioned before, these differences among studies can be explained in part by the cryopreservation method employed, an inadequate concentration and/or type of non-permeable agent [12], sperm cryosurvival regarding the species [27], as well as the lack of equilibration temperature, which have shown to be essential for sperm vitrification in spheres [24].

5. Conclusions

The present study showed that donkey sperm could not be vitrified in small volumes (spheres) using only glycerol as permeable CPAs. Vitrification using non-permeable CPAs (sucrose 0.1 M and BSA 5%) enhanced sperm motility and viability after warming. Further studies will concentrate on evaluating combinations of permeable and non-permeable CPAs for donkey sperm vitrification, assessing a wider range of sperm parameters after warming.

Author Contributions: Conceptualization, M.H., M.D.-J. and J.D.; methodology, M.D.-J.; software, M.D.-J.; validation, M.H., M.D.-J. and J.D.; formal analysis, M.D.-J., C.C. and B.P.; investigation, M.D.-J., C.C. and B.P.; resources, M.H. and J.D.; data curation, M.D.-J., M.H. and J.D.; writing—original draft preparation, M.D.-J., M.H. and J.D.; writing—review and editing, M.H. and J.D; visualization J.D. and M.H.; supervision, J.D. and M.H.; project administration, J.D. and M.H.; funding acquisition, J.D. and M.H. All authors have read and agreed to the published version of the manuscript.

Funding: This research was funded by MINECO ("Ministerio de Economía Industria y Competitividad", Spain), grant number AGL-2013-42726-R. M. Diaz-Jimenez is supported by a FPU ("Formación de Profesorado Universitario") fellowship from Spanish MECD ("Ministerio de Educación, Cultura y Deporte", Spain).

Acknowledgments: The authors are extremely grateful to "Centro Rural Malpica" for providing animals and facilities.

Conflicts of Interest: The authors declare no conflict of interest. The funders had no role in the design of the study; in the collection, analyses, or interpretation of data; in the writing of the manuscript, or in the decision to publish the results.

References

1. Camillo, F.; Rota, A.; Biagini, L.; Tesi, M.; Fanelli, D.; Panzani, D. The Current Situation and Trend of Donkey Industry in Europe. *J. Equine Vet. Sci.* **2018**, *65*, 44–49. [CrossRef]
2. FAO. DAD-IS, Domestic Animal Diversity Information System Breeds by Species and Country. 2018. Available online: http://www.fao.org/dad-is/browse-by-country-and-species/en/ (accessed on 7 March 2020).
3. Rota, A.; Panzani, D.; Sabatini, C.; Camillo, F. Donkey jack (Equus asinus) semen cryopreservation: Studies of seminal parameters, post breeding inflammatory response, and fertility in donkey jennies. *Theriogenology* **2012**, *78*, 1846–1854. [CrossRef] [PubMed]

4. Acha, D.; Hidalgo, M.; Ortiz, I.; Gálvez, M.J.; Carrasco, J.J.; Gómez-Arrones, V.; Dorado, J. Freezability of Andalusian donkey (Equus asinus) spermatozoa: Effect of extenders and permeating cryoprotectants. *Reprod. Fertil. Dev.* **2015**, *28*, 1990–1998. [CrossRef] [PubMed]
5. Ortiz, I.; Dorado, J.; Morrell, J.M.; Crespo, F.; Gosálvez, J.; Gálvez, M.J.; Acha, D.; Hidalgo, M. Effect of single-layer centrifugation or washing on frozen–thawed donkey semen quality: Do they have the same effect regardless of the quality of the sample? *Theriogenology* **2015**, *84*, 294–300. [CrossRef]
6. Vilés, K.; Rabanal, R.; Rodríguez-Prado, M.; Miró, J. Effect of ketoprofen treatment on the uterine inflammatory response after AI of jennies with frozen semen. *Theriogenology* **2013**. [CrossRef]
7. Vilés, K.; Rabanal, R.; Rodríguez-Prado, M.; Miró, J. Influence of seminal plasma on leucocyte migration and amount of COX-2 protein in the jenny endometrium after insemination with frozen–thawed semen. *Anim. Reprod. Sci.* **2013**, *143*, 57–63. [CrossRef]
8. Oliveira, J.V.D.; Oliveira, P.V.D.L.F.; Melo e Oña, C.M.; Guasti, P.N.; Monteiro, G.A.; Sancler da Silva, Y.F.R.; Papa, P.D.M.; Alvarenga, M.A.; Dell'Aqua Junior, J.A.; Papa, F.O. Strategies to improve the fertility of fresh and frozen donkey semen. *Theriogenology* **2016**, *85*, 1267–1273. [CrossRef]
9. Vidament, M.; Vincent, P.; Martin, F.X.; Magistrini, M.; Blesbois, E. Differences in ability of jennies and mares to conceive with cooled and frozen semen containing glycerol or not. *Anim. Reprod. Sci.* **2009**, *112*, 22–35. [CrossRef]
10. Trimeche, A.; Renard, P.; Tainturier, D. A procedure for poitou jackass sperm cryopreservation. *Theriogenology* **1998**, *50*, 793–806. [CrossRef]
11. Peña, F.J.; Macías García, B.; Samper, J.C.; Aparicio, I.M.; Tapia, J.A.; Ortega Ferrusola, C. Dissecting the molecular damage to stallion spermatozoa: The way to improve current cryopreservation protocols? *Theriogenology* **2011**, *76*, 1177–1186. [CrossRef]
12. Rosato, M.P.; Iaffaldano, N. Cryopreservation of rabbit semen: Comparing the effects of different cryoprotectants, cryoprotectant-free vitrification, and the use of albumin plus osmoprotectants on sperm survival and fertility after standard vapor freezing and vitrification. *Theriogenology* **2013**, *79*, 508–516. [CrossRef] [PubMed]
13. Isachenko, V.; Isachenko, E.; Katkov, I.I.; Montag, M.; Dessole, S.; Nawroth, F.; Van Der Ven, H. Cryoprotectant-free cryopreservation of human spermatozoa by vitrification and freezing in vapor: Effect on motility, DNA integrity, and fertilization ability. *Biol. Reprod.* **2004**, *71*, 1167–1173. [CrossRef] [PubMed]
14. Fahy, G.M.; MacFarlane, D.R.; Angell, C.A.; Meryman, H.T. Vitrification as an approach to cryopreservation. *Cryobiology* **1984**, *21*, 407–426. [CrossRef]
15. Isachenko, E.; Isachenko, V.; Sanchez, R.; Katkov, I.I.; Kreienberg, R. Cryopreservation of spermatozoa: Old routine and new perspectives. In *Principles and Practice of Fertility Preservation*; Cambridge University Press: Cambridge, UK, 2011; pp. 176–198. [CrossRef]
16. Jiménez-Rabadán, P.; García-Álvarez, O.; Vidal, A.; Maroto-Morales, A.; Iniesta-Cuerda, M.; Ramón, M.; del Olmo, E.; Fernández-Santos, R.; Garde, J.J.; Soler, A.J. Effects of vitrification on ram spermatozoa using free-egg yolk extenders. *Cryobiology* **2015**, *71*, 85–90. [CrossRef] [PubMed]
17. Pradiee, J.; Esteso, M.C.; Lopez-Sebastián, A.; Toledano-Díaz, A.; Castaño, C.; Carrizosa, J.A.; Urrutia, B.; Santiago-Moreno, J. Successful ultrarapid cryopreservation of wild Iberian ibex (Capra pyrenaica) spermatozoa. *Theriogenology* **2015**, *84*, 1513–1522. [CrossRef]
18. Isachenko, E.; Isachenko, V.; Katkov, I.I.; Dessole, S.; Nawroth, F. Vitrification of mammalian spermatozoa in the absence of cryoprotectants: From past practical difficulties to present success. *Reprod Biomed Online* **2003**, *6*, 191–200. [CrossRef]
19. Sánchez, R.; Risopatrón, J.; Schulz, M.; Villegas, J.; Isachenko, V.; Kreinberg, R.; Isachenko, E. Canine sperm vitrification with sucrose: Effect on sperm function. *Andrologia* **2011**, *43*, 233–241. [CrossRef]
20. Merino, O.; Risopatrón, J.; Sánchez, R.; Isachenko, E.; Figueroa, E.; Valdebenito, I.; Isachenko, V. Fish (Oncorhynchus mykiss) spermatozoa cryoprotectant-free vitrification: Stability of mitochondrion as criterion of effectiveness. *Anim. Reprod. Sci.* **2011**, *124*, 125–131. [CrossRef]
21. Figueroa, E.; Merino, O.; Risopatron, J.; Isachenko, V.; Sanchez, R.; Effer, B.; Isachenko, E.; Farias, J.G.; Valdebenito, I. Effect of seminal plasma on Atlantic salmon (Salmo salar) sperm vitrification. *Theriogenology* **2015**, *83*, 238–245. [CrossRef]

22. Pradiee, J.; Esteso, M.C.; Castaño, C.; Toledano-Díaz, A.; Lopez-Sebastián, A.; Guerra, R.; Santiago-Moreno, J. Conventional slow freezing cryopreserves mouflon spermatozoa better than vitrification. *Andrologia* **2017**, *49*, 4. [CrossRef]
23. Swanson, W.F.; Bateman, H.L.; Vansandt, L.M. Urethral catheterization and sperm vitrification for simplified semen banking in felids. *Reprod. Domest. Anim.* **2017**, *52*, 255–260. [CrossRef] [PubMed]
24. Hidalgo, M.; Consuegra, C.; Dorado, J.; Diaz-Jimenez, M.; Ortiz, I.; Pereira, B.; Sanchez, R.; Crespo, F. Concentrations of non-permeable cryoprotectants and equilibration temperatures are key factors for stallion sperm vitrification success. *Anim. Reprod. Sci.* **2018**, *196*, 91–98. [CrossRef] [PubMed]
25. Diaz-Jimenez, M.; Dorado, J.; Pereira, B.; Ortiz, I.; Consuegra, C.; Bottrel, M.; Ortiz, I.; Hidalgo, M. Vitrification in straws conserves motility features better than spheres in donkey sperm. *Reprod. Domest. Anim.* **2018**, *53*, 1–3. [CrossRef] [PubMed]
26. Diaz-Jimenez, M.; Dorado, J.; Consuegra, C.; Ortiz, I.; Pereira, B.; Carrasco, J.J.; Gomez-Arrones, V.; Domingo, A.; Hidalgo, M. Optimization of donkey sperm vitrification: Effect of sucrose, sperm concentration, volume and package (0.25 and 0.5 mL straws). *Anim. Reprod. Sci.* **2019**, *204*, 31–38. [CrossRef]
27. Arraztoa, C.C.; Miragaya, M.H.; Chaves, M.G.; Trasorras, V.L.; Gambarotta, M.C.; Neild, D.M. Porcine sperm vitrification II: Spheres method. *Andrologia* **2017**, *49*. [CrossRef]
28. Merino, O.; Sanchez, R.; Risopatron, J.; Isachenko, E.; Katkov, I.; Figueroa, E.; Valdebenito, I.; Mallmann, P.; Isachenko, V. Cryoprotectant-free vitrification of fish (Oncorhynchus mykiss) spermatozoa: First report. *Andrologia* **2012**, *44*, 390–395. [CrossRef]
29. Isachenko, E.; Isachenko, V.; Weiss, J.M.; Kreienberg, R.; Katkov, I.I.; Schulz, M.; Lulat, A.G.M.I.; Risopatrón, M.J.; Sánchez, R. Acrosomal status and mitochondrial activity of human spermatozoa vitrified with sucrose. *Reproduction* **2008**, *136*, 167–173. [CrossRef]
30. O'Brien, E.; Esteso, M.C.; Castaño, C.; Toledano-Díaz, A.; Bóveda, P.; Martínez-Fresneda, L.; López-Sebastián, A.; Martínez-Nevado, E.; Guerra, R.; López Fernández, M.; et al. Effectiveness of ultra-rapid cryopreservation of sperm from endangered species, examined by morphometric means. *Theriogenology* **2019**, *129*, 160–167. [CrossRef]
31. Bóveda, P.; Esteso, M.C.; Castaño, C.; Toledano-Díaz, A.; López-Sebastián, A.; Muñiz, A.; Prieto, P.; Mejía, O.; Ungerfeld, R.; Santiago-Moreno, J. Slow and ultra-rapid freezing protocols for cryopreserving mouflon (Ovis musimon) and fallow deer (Dama dama) epididymal sperm. *Anim. Reprod. Sci.* **2018**, *192*, 193–199. [CrossRef]
32. Zhu, J.; Jin, R.T.; Wu, L.M.; Johansson, L.; Guo, T.H.; Liu, Y.S.; Tong, X.H. Cryoprotectant-free ultra-rapid freezing of human spermatozoa in cryogenic vials. *Andrologia* **2014**, *46*, 642–649. [CrossRef]
33. Riva, N.S.; Ruhlmann, C.; Iaizzo, R.S.; Marcial López, C.A.; Martínez, A.G. Comparative analysis between slow freezing and ultra-rapid freezing for human sperm cryopreservation. *JBRA Assisted Reprod.* **2018**, *22*, 331–337. [CrossRef] [PubMed]
34. Katkov, I.; Bolyukh, V.; Chernetsov, O.; Dudin, P.; Grigoriev, A.; Isachenko, V.; Isachenko, E.; Lulat, A.; Moskovtsev, S.; Petrushko, M.; et al. Kinetic Vitrification of Spermatozoa of Vertebrates: What Can We Learn from Nature. In *Current Frontiers in Cryobiology*, 1st ed.; Katkov, I., Ed.; InTechOpen: Rijeka, Croatia, 2012; pp. 3–40. [CrossRef]
35. Consuegra, C.; Crespo, F.; Bottrel, M.; Ortiz, I.; Dorado, J.; Diaz-Jimenez, M.; Pereira, B.; Hidalgo, M. Stallion sperm freezing with sucrose extenders: A strategy to avoid permeable cryoprotectants. *Anim. Reprod. Sci.* **2018**, *191*, 85–91. [CrossRef] [PubMed]
36. Diaz-Jimenez, M.; Pereira, B.; Ortiz, I.; Consuegra, C.; Partyka, A.; Dorado, J.; Hidalgo, M. Effect of different extenders for donkey sperm vitrification in straws. *Reprod. Domest. Anim.* **2017**, *4*, 55–57. [CrossRef] [PubMed]
37. Arando, A.; Gonzalez, A.; Delgado, J.V.; Arrebola, F.A.; Perez-Marin, C.C. Storage temperature and sucrose concentrations affect ram sperm quality after vitrification. *Anim. Reprod. Sci.* **2017**, *181*, 175–185. [CrossRef] [PubMed]
38. Uysal, O.; Korkmaz, T.; Tosun, H. Effect of bovine serum albumine on freezing of canine semen. *Indian Vet. J.* **2005**, *82*, 97–98.
39. Cabrita, E.; Sarasquete, C.; Martínez-Páramo, S.; Robles, V.; Beirão, J.; Pérez-Cerezales, S.; Herráez, M.P. Cryopreservation of fish sperm: Applications and perspectives. *Appl. Ichthyol.* **2010**, *26*, 623–635. [CrossRef]
40. Ortiz, I.; Dorado, J.; Acha, D.; Galvez, M.J.; Urbano, M.; Hidalgo, M. Colloid single-layer centrifugation improves post-thaw donkey (Equus asinus) sperm quality and is related to ejaculate freezability. *Reprod. Fertil. Dev.* **2015**, *27*, 332–340. [CrossRef]

41. Flores, E.; Taberner, E.; Rivera, M.M.; Peña, A.; Rigau, T.; Miró, J.; Rodríguez-Gil, J.E. Effects of freezing/thawing on motile sperm subpopulations of boar and donkey ejaculates. *Theriogenology* **2008**, *70*, 936–945. [CrossRef]
42. Diaz-Jimenez, M.; Dorado, J.; Ortiz, I.; Consuegra, C.; Pereira, B.; Gonzalez-De Cara, C.A.; Aguilera, R.; Mari, G.; Mislei, B.; Love, C.C.; et al. Cryopreservation of donkey sperm using non-permeable cryoprotectants. *Anim. Reprod. Sci.* **2018**, *189*, 103–109. [CrossRef]
43. Zilli, L.; Bianchi, A.; Sabbagh, M.; Pecoraro, L.; Schiavone, R.; Vilella, S. Development of sea bream (Sparus aurata) semen vitrification protocols. *Theriogenology* **2018**, *110*, 103–109. [CrossRef]
44. Daramola, J.O.; Adekunle, E.O.; Iyasere, O.S.; Oke, O.E.; Sorongbe, T.A.; Iyanda, O.A.; Kehinde, A.R.; Aluko, S.P.; Olaoye, I.O.; Gbadebo, O.E.; et al. Effects of coconut milk alone or supplementation with pyridoxine in tris-extenders on viability of buck spermatozoa during vitrification. *Small Rumin. RES* **2016**, *136*, 208–213. [CrossRef]

animals

MDPI

Article

Development of a Donkey Grimace Scale to Recognize Pain in Donkeys (*Equus asinus*) Post Castration

Emma K. Orth [1], Francisco J. Navas González [2,3,*], Carlos Iglesias Pastrana [2,3], Jeannine M. Berger [4], Sarah S. le Jeune [5], Eric W. Davis [6] and Amy K. McLean [3,7,*]

1. Department of Animal Biology, University of California Davis, Davis, CA 95617, USA; emma.kira.orth@gmail.com
2. Department of Genetics, Faculty of Veterinary Sciences, University of Córdoba, 14071 Córdoba, Spain; ciglesiaspastrana@gmail.com
3. The Worldwide Donkey Breeds Project, University of Córdoba, 14071 Córdoba, Spain
4. San Francisco SPCA, Society for the Prevention of Cruelty to Animals, San Francisco, CA 94115, USA; drb@sacvetbehavior.com
5. Department of Surgical and Radiological Sciences, School of Veterinary Medicine, University of California Davis, Davis, CA 95617, USA; sslejeune@ucdavis.edu
6. International Animal Welfare Training Institute, School of Veterinary Medicine, University of California Davis, Davis, CA 95617, USA; ewdavis@ucdavis.edu
7. Department of Animal Science, University of California Davis, Davis, CA 95617, USA

* Correspondence: fjng87@hotmail.com (F.J.N.G.); acmclean@ucdavis.edu (A.K.M.)

Received: 11 July 2020; Accepted: 10 August 2020; Published: 13 August 2020

Simple Summary: Donkeys originally evolved as a desert animal, and unlike the horse (which flees or runs away from danger), the donkey fights to avert danger. Hence, donkeys are more stoic and tend to express fear, pain, and discomfort in more subtle ways than horses. For owners and practitioners, it can prove to be challenging to identify donkeys in a state of pain or discomfort until the animal has reached an advanced degree of disease, at which point veterinary intervention may be too late. This study aims to identify signs of pain from both facial and body parameters in donkeys undergoing a surgical procedure. Scores were based on noted signs of discomfort/pain from the observed body language of the face, such as the eyes, ears, nose, nostrils, and muzzle, along with their overall body appearance. The study demonstrated that developing a scoring system donkey grimace scale proved to be accurate in identifying discomfort related to pain. However, the accuracy of the scale can be influenced by the observers' gender, level of donkey knowledge, and experience.

Abstract: The objectives of this study were to establish a donkey ethogram, followed by a donkey grimace scale to be applied to donkeys pre- and post-castration and to test if there was a notable difference in scores based on observer knowledge, gender, and experience, which could reveal possible discomfort/pain. Nine healthy male adult donkeys were surgically castrated. Fifty-four photos were selected from frontal, lateral, and body views taken pre- and post-castration. Observers ranging from minimal to extensive knowledge and levels of experience based on education and hours/month spent with donkeys scored six photos/donkey on a scale of 0–2 (0 = not present, 1 = moderately present, 2 = obviously present). Scores were based on body language and facial parameters: Ears down, ears back, eye white showing, glazed look, orbital tightening, eyes round shape, nostril tension, eyes narrow shape, muzzle tension, and abnormal stance and overall perception of the animal being in pain. Level of experience and knowledge, as well as gender significantly ($p < 0.001$), affected observers' ability to accurately score images. The study suggests that the most significant indicators of pain in donkeys are overall appearance and abnormal body stance provided their sensitivity, specificity and accuracy values of 63.18%, 62.07%, and 62.60%, respectively.

Keywords: donkey; *Equus asinus*; facial expression; pain; grimace scale

1. Introduction

The misconception that donkeys do not feel pain to the degree that horses do, by suggesting donkeys have a higher level of pain tolerance, combined with a difficulty to identify indicators of pain in donkeys [1], has likely deprived many donkeys of receiving treatment when ill or injured [2,3]. Although there have been limited studies examining specific indicators of pain in donkeys, few behavioral signs of pain displayed in horses have not been described in donkeys [2–4]. Donkeys likely feel pain and yet display little signs of pain, but currently, there is no evidence that donkeys have a different pain tolerance to that of other equines [2,5,6].

Animal pain has been described as "an aversive sensory experience that elicits protective motor actions, results in avoidance and may modify species-specific traits of behavior, including social behavior" [5]. Since animals are unable to verbally communicate with humans, owners and veterinarians are left with the task of identifying when an animal is in pain. Signs of pain have been associated with specific behavioral indicators in the face and body in many animals from humans to livestock [7]. However, identifying pain in donkeys presents a specific challenge. Donkeys are often described as stoic animals, meaning they may endure pain or hardship without the display of feelings and complaints. Due to this behavior, a donkey may ultimately mask pain with behaviors, such as sham eating or by only showing slight indications of pain, such as a twitch to the tail or change in posture. These subtle behaviors may go unnoticed, thus concealing illness/injury [2,3,8,9]. These behavioral adaptations allow the donkey to appear normal, and likely decrease the risk of an ill/injured donkey being targeted by a predator. As useful as these behaviors and stoicism may be to a wild donkey, they make it difficult to identify pain when evaluating domesticated donkeys (e.g., working, companion or production donkeys) or feral donkeys now living in captivity.

As donkeys become increasingly popular as companion animals along with the use of donkeys for draft animals and growing use for production (milk, meat, and skin), there is a growing need to understand more about their behavior related to their overall well-being from both an owner and veterinary perspective. Previous studies have yielded facial expression ethograms [10] and composite ethograms (face and body) to assess pain in horses [5,11], as well as facial expression pain scales for donkeys and visual analogue score based on disease conditions observed [12,13]. The objectives of this study were to construct a donkey ethogram, followed by a donkey grimace scale to be applied to donkeys in order to determine whether there was a notable difference in scores pre- and post-castration, which could reveal possible discomfort/pain. The study also aimed to compare the reliability of observers with different levels of experience and knowledge with donkeys. It was hypothesized that a change would be seen in the pre- and post-castration score per donkey; more specifically, on average the post castration pictures would have higher scores, when compared to the pre-castration pictures, indicating a higher presence of pain behavior markers in donkeys post-castration. The hypothesis regarding the reliability of observers was that the more donkey experience and knowledge an observer had, the higher probability they would correctly identify pain markers in the images.

2. Materials and Methods

2.1. Animal Sample

Nine adult male standard donkeys (Standard is a breed of donkey recognized in the U.S. according to height and weight, 150 to 300 kg), ranging from 2 to 10 years old were photographed pre- and post-castration. All donkeys were clinically evaluated by the attending veterinarian (a board-certified surgeon with over thirty years of clinical experience working with and owning donkeys) and his team. Any donkey not deemed healthy based on physiological parameters and overall appearance did not undergo castration. The evaluation of pain based on the evidence reported by Dyson et al. [11], who suggested the recognition of certain behavioral features may act as potential indicators of musculoskeletal pain, which may enable the early recognition of certain distressing conditions in equids. The indicators proposed by Gleerup and Lindegaard [14] were considered as behavioral indicators of pain in the context of normal equine behavior. The donkeys were physiologically evaluated by veterinarians before their free-from-pain condition could be ensured, and these could be returned to their habitual housing. The review by Ashley et al. [2] was considered to adjust the information and protocols of identification of pain-related signs to the specific behavioral nature of donkeys. The overall appearance was evaluated following the premises in McLean et al. [15].

Donkeys were acclimated to the farm and arrived 10 days before surgery and housed in a group in paddocks with shelter, ad libitum hay, and free access to water. All donkeys were part of a private donkey rescue.

2.2. Photo Set and Selection Protocol

A total of 270 photographs were initially taken for the present study. All photos were taken with an Apple mini iPad (version ISO 11.4.1, Apple Computer, Cupertino, CA, USA). The photos were identified depending on whether they were taken pre- or post-castration. Five photos were taken of each donkey from the following positions: One frontal photograph, one lateral photograph from the right side, one lateral photograph from the left side, one body photograph from the right side and one body photograph from the left. Each donkey was photographed 48 h, 24 h, 0 h before surgery and 8 h, 24 h and 48 h post-surgery on the farm. Fifty-four photos were randomly selected (27 pre- and 27 post-castration) to build the materials used for the surveys as suggested by Dalla Costa et al. [16], seeking a balanced design to perform the statistical analysis.

2.3. Anesthesia and Surgery Protocol

Donkeys were anesthetized beginning in the morning by a standard xylazine/ketamine protocol [5]: xylazine hydrochloride 1 mg/kg and butorphanol tartrate 0.02 mg/kg IV for sedation, followed by ketamine 2 mg/kg and midazolam 0.03 mg/kg IV for induction of anesthesia. Donkeys requiring additional anesthesia to complete the procedure were given 0.5 mg/kg xylazine mixed with 1 mg/kg of ketamine. No catheter was used. All injections were administered IV in the jugular vein with a 3 cm, 18 ga needle. Flunixin meglumine (Merck Animal Health USA, De Soto, KS, USA) 1 mg/kg was administered IV intraoperatively as an analgesic drug to reduce inflammation and to relieve pain. Procaine Penicillin G (Pfizer, New York City, NY, USA) 18,000 IU/kg IM was administered intraoperatively for infection prophylaxis. All patients were placed in either right or left lateral recumbency, with the uppermost rear leg elevated and restrained with a soft rope. Their spermatic cords were identified by palpation, and 10 mL of 2% lidocaine was injected into each cord. Following a routine surgical prep using povidone-iodine surgical scrub (First Priority Inc., Elgin, IL, USA) and povidone-iodine solution rinse (First Priority Inc.), two incisions were made ventrally in the scrotum. The fascia covering the vaginal tunic was bluntly dissected, and each testis with an attached spermatic cord was exteriorized, while keeping the tunic closed ("closed technique"). The cremaster muscles were broken to decrease the diameter of the tissue to be ligated. A modification of the "Miller's" knot was used for transfixation and ligation of the spermatic cords. The ligation process was completed by using Coated Polyglactin 910 (Coated Vicryl Plus, Ethicon, Somerville, NJ, USA) and Triclosan suture (Coated Vicryl Plus, Ethicon). The spermatic cords were crushed and transected 1 cm distal to the ligature with Serra Emasculators, and spermatic cord stumps were examined for hemorrhage. Excess loose fascia that might extend from the incision site was removed, and incision stretched to maintain postoperative drainage. All surgeries were performed on the same day with the same experienced NGO team, including donkey surgeon, team, and anesthetics, and recovered on site. The surgeries were done in the field. No complications were found, nor any of the donkeys needed redosing during the application of the castration procedures. Castration took 5 min ± 5 min to perform. The whole procedure, including preparation, surgery, and recovery, took an average of 20 min ± 5 min per donkey.

2.4. Procedure for Volunteer Classification, Development of Donkey Ethogram, Survey and Sample Description

2.4.1. Observer Demographics

Twelve observers (41.7% males and 58.3% females) with ages ranging from 20 to 70 years volunteered to participate in the donkey ethogram training and castration image survey. Those who volunteered had various levels of donkey experience and knowledge and included veterinarians, veterinary students, researchers, and donkey owners or were a combination of the aforementioned descriptors. All levels of knowledge and experience were equally represented in their frequency in the sample. All observers were then placed into categories based on the amount of time they spent with donkeys (interaction); minimal (<4 h/month), intermediate (4–6 h/month), and extensive (>6 h/month), and knowledge: Minimal (up to 100 h of formal training, attended a symposium dedicated to donkeys or had a certified course in donkey related science), moderate (more than 100 h of formal training, attended two or more donkey symposiums or had more than one certified course in donkey related science), and extensive knowledge (more than 100 h of formal training, a degree or PhD in donkey-related science, and research/work with donkeys on an everyday basis). Observer sample distributed across knowledge and experience levels as follows—33.3% of the observers presented minimal interaction with donkeys and minimal knowledge, 25% of the observers presented intermediate interaction with donkeys and moderate knowledge, and 41.7% of the observers presented extensive interaction with donkeys and extensive knowledge. The observers in the sample were chosen seeking an almost-equal representation of sexes, interaction and knowledge levels, to prevent the bias derived from the potential overrepresentation of some categories over the rest.

2.4.2. Ethogram

An ethogram displaying key body language often used to evaluate pain in other animals was developed to train observers to evaluate this body language for signs of change (Refer to supplemental information for ethogram) [10,11,15]. The training module included photographs, written descriptions, and a voice-over to describe each descriptor and photograph. Changes in tension, posture, or position were displayed in the training material and described by written and voice descriptions. Signs of discomfort or pain, such as orbital tightening of eyes, muzzle tension, along with ear and body language were illustrated and described [2–4,17–20] (Figures 1–5). Each observer was required to complete the donkey ethogram training module before completing the study survey. From the ethogram training materials, nine facial and body language markers (ear frontal and side position, eye description (shape and tension), muzzle and nostril description (shape and tension), and stance) were then identified as possible pain indicators and were utilized for the development of the donkey grimace scale tested in the survey.

2.4.3. Survey

The survey included 54 photographs taken with an Apple mini iPad, six pictures per donkey: Lateral, frontal head, and lateral body view (one pre and post castration photograph/view). Each observer took the survey once per day on their own device (e.g., computer, laptop, tablet, or smartphone), for three consecutive days to ensure intraobserver reliability. Photographs were presented in a digital form and survey to the observers in the same order on each consecutive day. Observers were blinded to the treatment of each donkey photograph, meaning the observer did not know whether each picture was pre-castration or post-castration. The database comprised of 18,072 scores.

Figure 1. An ethogram describing ear position related to grimace scale (0—not present, 1—moderately present, and 2—obviously present). Ear positions that maybe associated with pain or discomfort include both ears back (**C**), down (**D**), one ear forward, one to the side I, One ear to the side and one back (**F**), one ear to the side and one down (**H**), and one forward and one down (**I**) along with other facial action units associated with pain. Ear positions that are less likely to be associated with pain would include both ears are erect to side (**A**), both ears are forward (**B**), one ear forward and one to the side (**E**), and possibly one ear forward and one ear back (**G**).

Figure 2. Ethogram showing nostril and muzzle tension presences. Tension is not present in picture (**A**) meaning the donkey is likely not in pain or discomfort. The tension associated with pain can be found in picture (**B**).

Figure 3. Ethogram displaying eye shape related to pain (0—not present, 1—moderately present, and 2—obviously present) with (**A**,**D**,**E**) suggesting discomfort or pain. Eyes that are round in shape (**B**) and almond shape (**C**) are likely not in pain or discomfort.

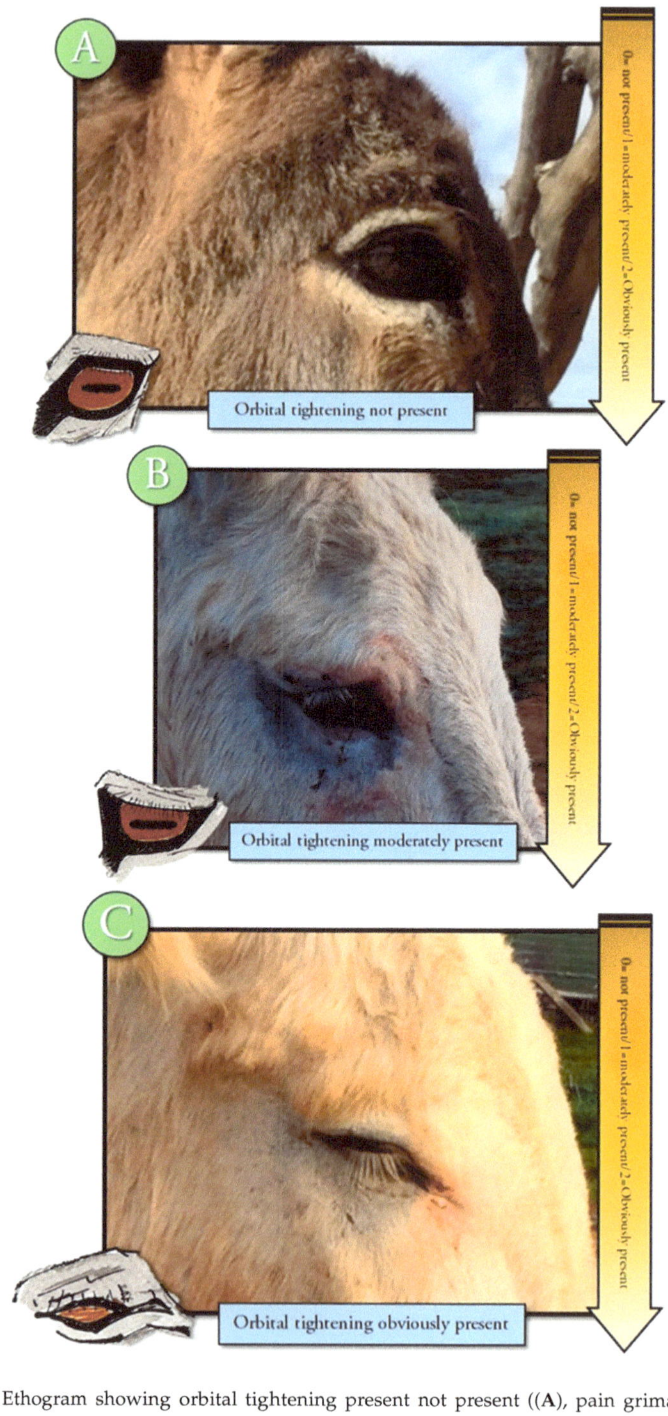

Figure 4. Ethogram showing orbital tightening present not present ((**A**), pain grimace score 0), moderately present ((**B**), pain score 1) and tightening present ((**C**), pain score 2).

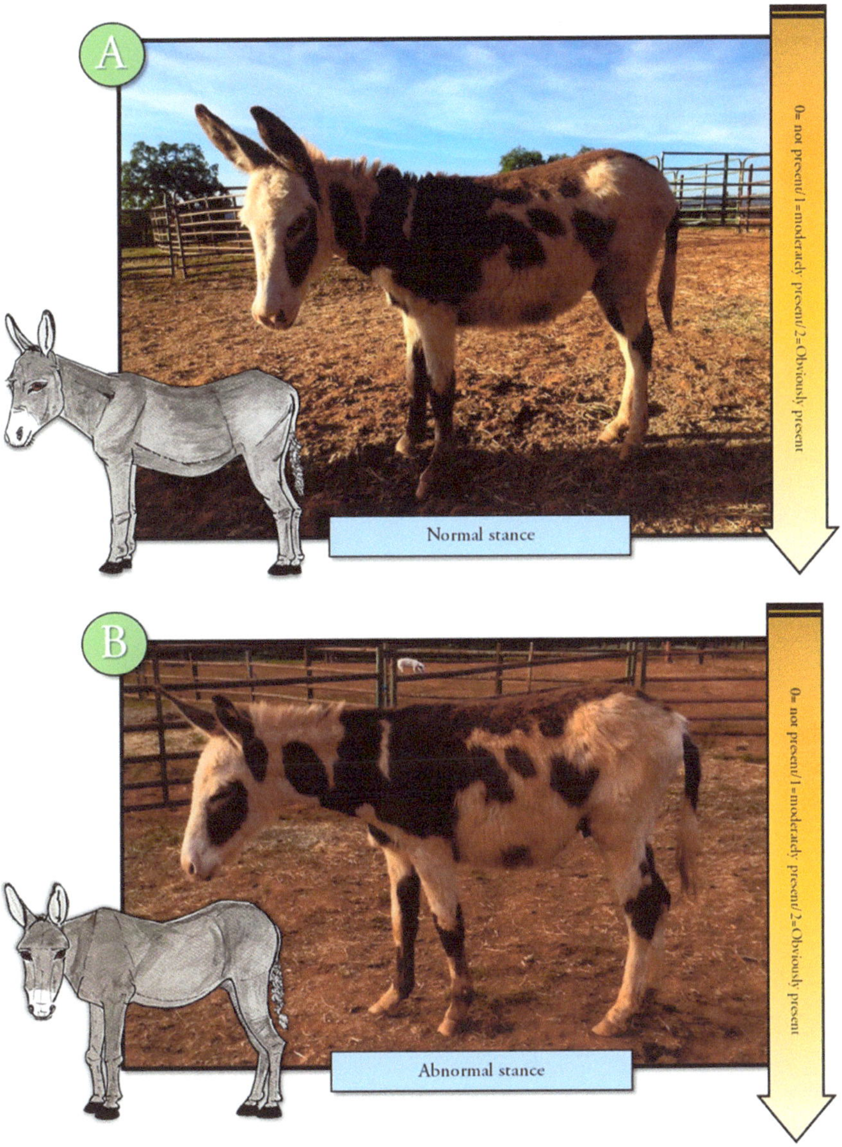

Figure 5. Ethogram displaying body stance, normal (**A**) and abnormal (**B**) which may be associated with discomfort or pain.

2.4.4. Sample Description

Observers were asked to score eleven parameters per photograph of each of the nine donkeys. The observer responded with a score ranging from 0 to 2 (0 being absent, 1 being moderately present, and 2 being obviously present) for each of the body language and facial parameters: Ears down, ears back, eye white showing, glazed look, orbital tightening, eyes round shape, nostril tension, eyes narrow shape, muzzle tension, and abnormal stance [18]. The observers were also given an "I don't know" option (in addition to the 0–2 scale) which they were instructed to select if they could not identify the donkey's response for that marker based on the image. Also, observers were asked to score their overall perception of the animal's pain status, while being blinded to the treatment (pre or post castration) present in the images (Figures 1–5).

2.5. Intra and Interobserver Reliability Testing

Interobserver reliability was assessed at two levels. First, the measurements of every observer were compared for days one to three, and second, measurements of scores across the total panel for the summation of all three days were compared. Intraobserver reliability was tested by comparing the responses of the same observer over three consecutive days scoring the survey. The intraclass correlation coefficient of the Reliability Analysis procedure of the Scale package in SPSS Statistics, Version 25.0, IBM Corp., Armonk, NY, USA, (2017) was used to assess inter (across observers) and intraobserver (for the observations of the same observer across the different days) reliability. Responses were then classified as successful or not, depending on whether the observers had identified signs of discomfort related to pain or not compared to pre-castration images [8,21]. Per clinical examination from attending veterinarian, all donkeys were considered to be in the absence of pain before castration, and thus, served as their control. Post castration pain status was determined after the diagnosis issued by the attending veterinarian. The pain status determined was used to evaluate the responses by each of the observers participating in the survey. The summaries of the results for inter and intraobserver reliability are shown in Tables S1 and S2.

2.6. Conditioning Factors Identification and Testing for the Probability of Success to Identify Post Procedure Pain-Related Signs

Before testing for the probability of success to identify signs of pain, Kolmogorov-Smirnov (Table S3) and Levene's test were performed to study data distribution and variance across groups to discard gross violations of normality and homoscedasticity assumptions. As both parametric assumptions had been grossly violated ($p > 0.01$), a nonparametric approach was considered. Frequency analysis and Pearson's Chi-squared were performed to test for differences in the probability of success to identify pain signs (ability to detect true positive signs of pain, true negative signs of pain, false-positive signs of pain and false-negative signs of pain) across factors: Castration status, observer gender, donkey knowledge, donkey interaction, body language. Frequency analysis and Pearson's Chi-squared were performed using the Crosstab task from the descriptive statistics procedure in SPSS Statistics, Version 25.0, IBM Corp. (2017). Pearson's chi-square test is used to determine whether there is a statistically significant difference between the expected frequencies and the observed frequencies in one or more categories of a contingency table [22], with each of the observation being exclusively categorized in one category (false or true positive or false or true negative [23]). As stated in Roscoe and Byars [24], when sampling from a uniformly distributed populations (as the one in our study), the Kolmogorov test is markedly and consistently conservative even under conditions most favorable to the test, and chi-square proved to be quite robust even with very small samples. Additionally, chi-square has proven to be robust in tests of goodness-of-fit to uniform with fractional expected frequencies, as it occurs in our study.

2.7. Overall and Specific Body Language Sensitivity, Specificity and Accuracy for Post Procedure Pain-Related Signs Identification

Overall sensitivity, specificity, and accuracy were computed to determine the efficacy of the technique to report a comprehensive, accurate assessment of pain-related body language signs across body areas and on the whole, as a way to identify potential body language which should be considered better predictors for pain-related conditions versus those which may report misleading information. Sensitivity is the percentage of donkeys' post castration being identified as in pain. Specificity is the percentage of donkeys pre castration being identified as not in pain. Accuracy is the likelihood of detecting either a true positive or true negative [25].

2.8. Declaration of Ethics

The study was reviewed and approved by the University of California Davis' Institutional Animal Care and Use Committee, but no protocol number was given, since only observations were being conducted by the research team. Castrations were scheduled and performed by a veterinary nongovernmental organization on-farm (in the field) independent of the study. No additional permission was needed following the recommendations of Royal Decree-Law 53/2013 and its credited entity, the Ethics Committee of Animal Experimentation from the University of Córdoba, given the application of the protocols present in this study followed the premises cited in the 5th section of its 2nd article, as the animals assessed were used for credited zootechnical use.

3. Results

3.1. Intra and Interobserver Reliability Testing

All observers showed a Cronbach's α parameter above 0.70 (average measures), suggesting that each observer reliably scored across all test days. Reliability coefficients (Cronbach's alpha) above 0.85 are generally regarded as high and those between 0.65 and 0.85 as moderate. The total panel across all three days reported a Cronbach's α parameter above 0.90 (average measures), suggesting total panel reliably scored across days. It was likely for the panel to score the same for the same photo on the three repeated sessions. Additionally, interobserver reliability for the whole panel was tested, and the value of 0.97 (average measures) suggests agreement between observers was excellent. Hence, any of the members is valid, reliable, and provides consistent enough measures to be considered independently.

3.2. Conditioning Factors Identification

There was a significant ($p < 0.001$) difference between female and male observers in their ability to correctly identify signs of pain after castration (Table S4). The relationship between knowledge and gender was studied using partial eta squared (ηp^2). Partial eta squared (ηp^2) was 0.08, which meant that the effect of gender accounted for 8% of the differences in knowledge plus associated error variance. Sample size and possible factor combinations were not sufficient to justify the performance of univariate follow-up statistical analyses to determine whether significant differences could be ascribed to the interaction between knowledge and gender—however, these factors separately, as some of the categories, could be misrepresented. The study of the interaction between knowledge and gender may imply the multiplication of the number of possibilities within the interaction factor. Hence, instead of considering two variables (knowledge and gender) with two (male and female) and three levels (minimal, intermediate and extensive knowledge), respectively, we would be considering one variable (knowledge level-gender) with six levels (male-minimal knowledge, female-minimal knowledge, male-intermediate knowledge, female-intermediate knowledge, male-extensive knowledge and female-extensive knowledge). Then, in the context of our sample and to maintain the balance in the representativity of each possibility, clustering both variables under an interaction term may reduce the frequency of cases across levels to two cases per level. This offers lower chances to compare than in a case in which the variables of knowledge level and gender are considered separately [26]. Males were slightly less likely to identify a true positive (identify pain in the post castration images) and to score a false-negative (fail to identify markers considered to be pain indicators) than female observers were (Figure 6). There was also a significant difference across levels of donkey knowledge in the observers' ability to correctly identify signs of pain in post castration images. The probability of scoring true positives was higher when observers had moderate or extensive knowledge of donkeys (Table S5). Conversely, observers with minimal knowledge of donkeys had higher a likelihood to score false-positives or false-negatives than those with moderate or extensive knowledge of donkeys. Observers with moderate knowledge of donkeys reported a slightly higher ability to accurately detect signs of pain after castration than those having extensive knowledge and considerably higher than those having minimal knowledge. However, observers having minimal or moderate knowledge of donkeys presented a higher likelihood to make a false-negative. On the contrary, observers presenting extensive knowledge of donkeys had a higher likelihood to make false-positives (Table S5). A significant difference across the levels of interaction with donkeys was seen in the results from the observers' ability to correctly identify signs of pain after castration. The pattern of scoring false-negatives, true negatives/positives and false-positives between observers who occasionally (<4 h/month), sometimes (4–6 h/month), and often (>6 h/month) spent time with donkeys followed the same trend as observers with minimal, intermediate, and extensive knowledge of donkeys (Table S6 and Figure 7).

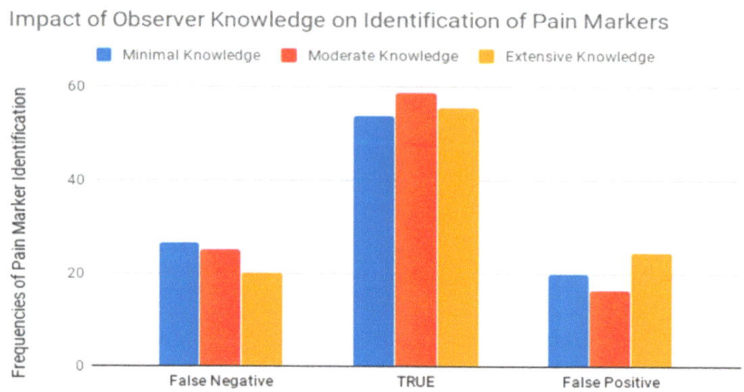

Figure 6. Frequencies of pain identification success possibilities across the different levels of donkey knowledge from minimal, moderate to extensive in both pre-and post-castration images. In medical diagnosis, test sensitivity represents the ability of a test to correctly identify those with the disease (true positive rate), whereas test specificity is the ability of the test to correctly identify those without the disease (true negative rate). Hence, the bar representing true values provides an overall direct measure of the quality (sensitivity/specificity) of the test in comparison to its potential drawbacks (false-positives or -negatives).

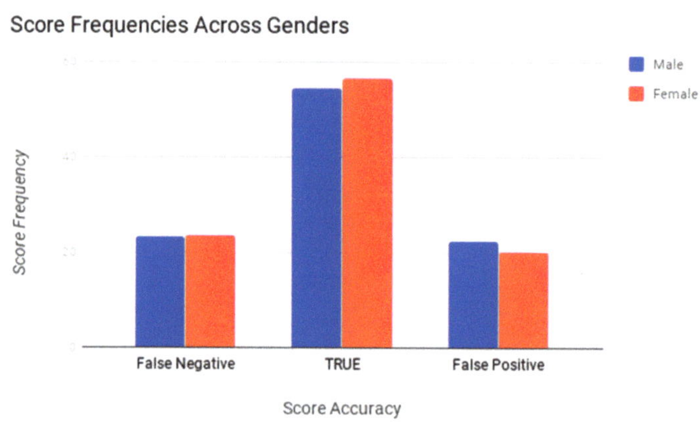

Figure 7. Frequencies of pain identification success possibilities across genders (male and female observers) identifying false-negative signs of pain, actual/true signs of pain and false-positive signs of pain (*p*- value 0.001). In medical diagnosis, test sensitivity is the ability of a test to correctly identify those with the disease (true positive rate), whereas test specificity is the ability of the test to correctly identify those without the disease (true negative rate). Hence, the bar representing true values provides an overall direct measure of the quality (sensitivity/specificity) of the text in comparison to its potential drawbacks (false-positives or -negatives).

3.3. *Testing for the Probability of Success to Identify Post Procedure Pain-Related Signs*

Significant differences ($p < 0.001$) were found for the distribution of the intensity of signs identified by the observers between pre castration and post castration images (Table 1). As reported in Table 1, 2.6% and 8.7% higher frequencies were found when moderately present signs were detected, and obviously present signs were detected for post castration images, respectively. Frequencies for the absence of signs were 14.3% higher for pre castration images. 'I don't know' response frequency was quite similar with only 0.30% higher frequency of 'I don't know' responses from post castration pictures (Figure 8).

Table 1. Summary for the results of frequency analyses and frequency differences between pre and post castration conditions.

Score Frequencies	I Don't Know	Not Present	Moderately Present	Obviously Present
Pre-castration	3.90%	54.00%	32.00%	10.20%
Post-castration	4.20%	42.30%	34.60%	18.90%
Frequency differences	Value	df	Asymptotic p-value (2-sided)	
Pearson Chi-Square	373.496	3	0.001	

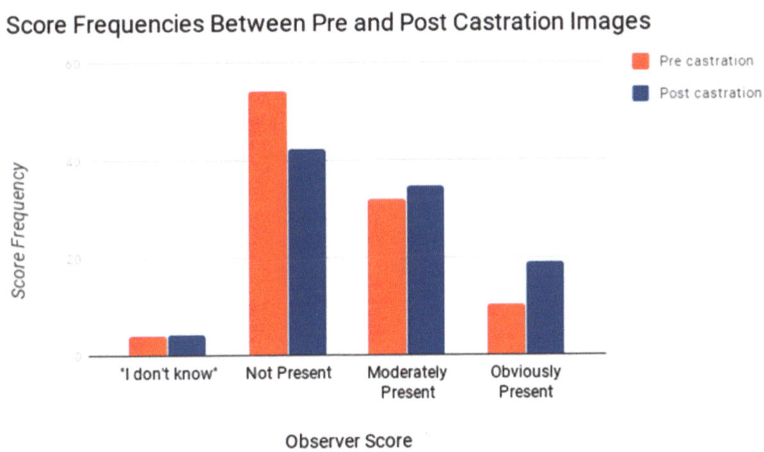

Figure 8. Comparing score responses related to signs of pain identification in pre- and post-castration images.

3.4. *Overall and Specific Body Language Sensitivity, Specificity and Accuracy for Post Procedure Pain-Related Signs Identification*

Significant differences were found for the probability of detecting signs of pain across the body language markers tested (Table S7). A change in stance and overall appearance reported the lowest rates of false-positives and false-negatives, improving the specificity and sensitivity of the technique. Eyes presenting a narrow shape, orbital tightening, and glazed were used to correctly classify 60% of cases, however, the possibility of making a false-positive increased in comparison to false-positive and negative frequency reported for abnormal stance or overall appearance. Nostril tension and eye whites showing described the opposite trend, although they were able to correctly classify around 54% of the cases. The nostril tension marker yielded 33% of false-positive cases, while the eye white showing marker yielded 40% of false-negatives. Contrastingly, nostril tension reported the lowest rate for false-negatives (12.40%) and eye whites showing reported the lowest rate for false-positives (6.90%). The sign reporting the least accurate results was eyes round shape; only 39% were classified correctly, with 35% being false-negatives, and 25% false-positives. Our results may be in line with those for sheep by Häger et al. [27], who reported an accuracy rate of 68.2%, and a false-positive rate of 22.7% (more common than false-negatives, 9.1%). These similar results may base on the suggestions by Larrondo et al. [28], who reported an increased difficulty in pain recognition in lambs which indeed conditions the use of analgesics. A slightly higher accuracy of 73.3% was reported for a Horse Grimace Scale, which may reflect a relative higher ease to identify signs [10], which has traditionally been supported by the literature [2]. Higher values have been reported by Reijgwart et al. [29] when using their Grimace scales methods for ferrets. These authors reported sensitivity and specificity values of 85% and 74%, respectively. The highest accuracy reported (80%) was achieved when whisker retraction was excluded, and only orbital tightening was considered. Lower rates of false-negatives together with higher rates of false-positives may be indicative of a rather cautious assessment of pain, due to the difficulty to identify pain-related behavioral signs in donkeys. Our results may complement those reported by Dai et al. [30], as it may be the ability of panel training methods to inform about the subtle signs reported by animals rather than the time needed to train the members of the panel, especially for signs, such as tension above the eye area, prominent strained chewing muscles, mouth strained and pronounced chin and strained nostrils. This could also be supported on the higher accuracy found in our DGS (Donkey Grimace Scale) for overall appearance and abnormal stance, which may indeed reflect a higher interobserver agreement, and indirectly indicative of a successful training period. Body language markers that have sensitivity or specificity below 50% should not be considered when aiming at assessing pain (for instance, eyes round shape) (Figure 9). The sensitivity of post castration images was 63.18% with detected pain. Significant differences were the distribution of the intensity of signs identified by observers between pre-castration and post-castration images. Higher frequencies of pain markers were found when moderately (2.6%) and obviously present signs (8.7%) were detected for post castration images. Table 2 reports a summary of the results for sensitivity, specificity and accuracy of each body language as a pain indicator and of the overall donkey grimace scale.

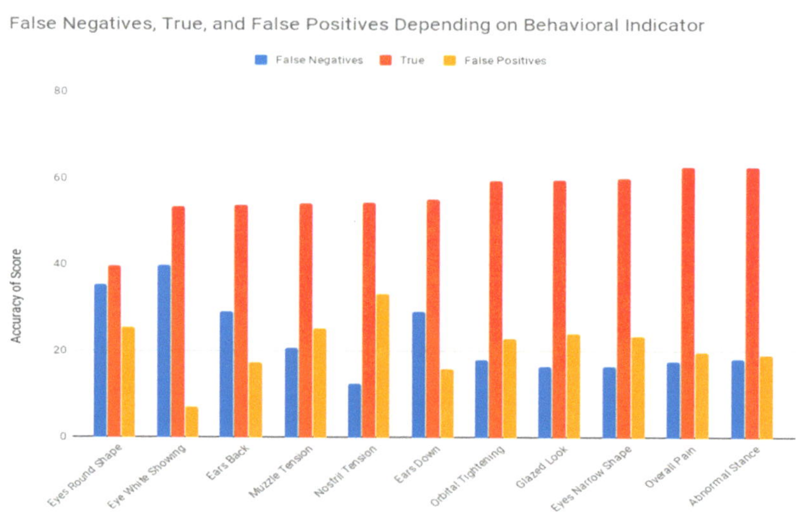

Figure 9. Accuracy comparison of frequency scores (false-negative, true, or false-positives) for different facial and body parameters related to signs of pain. Pre- and post-castration results were considered as the figure represent the quality of the donkey grimace scale used. In medical diagnosis, test sensitivity is the ability of a test to correctly identify those with the disease (true positive rate), whereas test specificity is the ability of the test to correctly identify those without the disease (true negative rate). Hence, the bar representing true values provides an overall direct measure of the quality (sensitivity/specificity) of the text in comparison to its potential drawbacks (false-positives or -negatives).

Table 2. Summary for the results of sensitivity, specificity and accuracy for each body language sign and overall donkey grimace scale.

	Sensitivity	Specificity	Accuracy
Ears Down	54.04%	56.91%	55.10%
Ears Back	51.51%	56.86%	53.65%
Eye White Showing	52.05%	59.88%	53.40%
Glazed Look	61.18%	58.26%	59.50%
Orbital Tightening	61.17%	57.70%	59.30%
Eyes Round Shape	41.24%	36.91%	39.50%
Nostril Tension	57.68%	53.18%	54.50%
Eyes Narrow Shape	61.77%	58.84%	60.10%
Muzzle Tension	54.51%	53.76%	54.10%
Abnormal Stance	60.91%	64.25%	62.70%
Overall Appearance/Donkey Grimace Scale	63.18%	62.07%	62.60%

4. Discussion

The original lack of scientific support in regard to the assessment of specific donkey welfare contrasts the advances that have recently appeared in the scene. This methodological revolution appeared as an effort to address the misattributions and misconceptions of the donkey species. Incorrect evaluation of donkey welfare often bases on an incorrect perception of the behavioural, physiological, pathological idiosyncrasies of the species, among others. In this context, complex validated welfare approaches [31] can benefit from the methodologies implemented by large scale applicable methods [32] to report replicable and still trustable results in contexts, such as medium- or low-income countries, for which donkey is a common and relatively frequent element. The present research can complement the aforementioned methods as, it focuses on the human ability to determine signs of pain, which according to literature has often been deemed difficult provided to the subtleness of donkeys to display signs of distress [2].

Castration is considered to be one of the most common surgical procedures practiced by veterinarians [10]. The present study measured indicators of pain in donkeys before and after the procedure by observing changes in facial units and body language with the objective that this tool could be easily applied and used to identify pain in donkeys. Previous studies have sought to identify indicators of pain in equids. One study measured similar facial action units in horses before and after castration and created a grimace scale to assist with recognizing signs of discomfort and pain, but this has not been achieved in donkeys until this study was conducted [10]. The horse study assessed pain in similar areas to this study, such as orbital tightening, muzzle and ear position and similar indicators were observed in donkeys in this study in the same facial areas.

Ashley et al. [2] reported the present paucity of scientific research into equine pain behavior in spite of the relevance that it is attributed to it in horse clinics. In their review involving papers published during a period of 20 years, the same authors would state, this lack of scientific attention to the horse's pain-related behavior is extensive and even less common in regards other equine species, such as the donkey or its hybrids. This becomes explicit in the review as among the nonspecific behavioral indicators of pain in equids, the review by Ashley et al. [2] offers an extensive description of horse pain-related behavior, such as considerable restlessness, agitation and anxiety, rigid stance and reluctance to move, lowered head carriage, fixed stare and dilated nostrils, clenched jaw, aggression towards own foal, handlers, other horses, objects and self. Oppositely, the same authors claim these signs cannot either be relied upon, are often misdiagnosed, are difficult to differentiate, are not reported or described, or appear in a more subtle display for the donkey, respectively. In regard to the signs of abdominal pain, the review reports deep groaning or vocalization, rolling, abdominal kicking, flank watching, stretching and dullness and depression, in horses. Oppositely, the first four aforementioned behavior was not reported or have not been described in the donkey, while, stretching, which has importantly been related to other signs with the purpose to evaluate analgesia, has been nonspecifically been ascribed to colic, and is often accompanied by difficulty in defecating or urinating. Dullness and depressive-like-behaviors have been commonly reported as the only observable pain-related behaviors in donkeys, which is often simultaneous to lethargy and reduced alertness, self-isolation or handlers' contact reluctance.

Among the behavioral indicators of limb and foot pain in equids collected by Ashley et al. [2] in their review mention weight-shifting between limbs, limb guarding, abnormal weight distribution, pointing, hanging and rotating limbs, abnormal movement and reluctance to move in horses. Later, the authors add these patterns involving the limbs and foot are normally unreported or not commonly seen in donkeys (we understand, if they are experiencing the same level of pain). Furthermore, these authors suggest literature may have reported reluctance to move to be weakly associated with limb pain in the donkey, while repeated episodes of lying down may offer a rather evident sign of the donkey being in distress. Last, Ashley et al. [2] report the set of most frequent behavioral indicators of head and dental pain in horses and donkeys may involve headshaking, abnormal bit behavior, altered eating; anorexia, quidding, food pocketing. Particularly, quidding partially chewed portions of feed may be indicative of severe chewing discomfort and can particularly lead to choking in older donkeys. Contextually, the information reported shows the noticeable lack of specific or updated information in regards how pain is behaviorally expressed by donkeys, which makes the development of a donkey grimace scale necessary in order to avoid the misconception of pain-related signs as a result of an incorrect comparison with other affine species, such as horses.

Another study measured pain thresholds in donkeys and horses with mechanical stimuli applied to their limb and recorded behavioral responses; the animals served as their own controls [3]. In the present study, a donkey grimace scale was developed and applied to pre- and post-castration images to measure a difference in facial action units and body language. A recent study included a facial grimace scale for donkeys experiencing acute and chronic pain from various anatomical locations (such as the head, limbs, and abdomen), and, compared their signs of pain with donkeys, assumed to not be in pain [1]. Granted, pain behavior may be dependent upon individual animal basis, and signs of discomfort may vary depending on the type of pain if it is acute versus chronic. Accordingly, donkeys in the present study served as their own control, following the model used in some previous pain studies [3,11,33]. When using the same donkey as a self-control, facial expression and body language could be observed and recorded before a painful procedure, castration, and after; thus, emphasizing changes in behaviors before and after the procedure. In donkeys, signs of discomfort and pain may be subtler; therefore, one must consider the whole donkey to be able to accurately identify the presence of pain [1,3,4]. Furthermore, being able to recognize these subtle differences may prove to be challenging to those who have limited experience working with donkeys. Regan et al. [3] observed that painful donkeys would swish their tails, hold their ears to the side (facing down) and exhibit changes to head posture. The ethogram created from the present study suggested that observing the overall body posture and reactions of donkeys may improve our chances of recognizing pain in donkeys, as has been indicated in a study by Grint et al. [4]. The results of the present study indicated there was a statistically significant difference in observer scores when using the donkey grimace scale for both the face and body between castration images. All observers had undergone an extensive training process to recognize signs of donkey discomfort from observations in facial action units and body stance and/or posture, thus increasing the likelihood that observers would be able to identify subtle pain indicators. Observer training is strongly recommended to assess and improve interobserver reliability if necessary and possible before trials start [26]. As suggested by Bell et al. [27], an improvement of interobserver reliability (ICC, κ or Cronbach's α) from 0.50 to 0.80 (below the levels found in the present study) within 3 to 5 training sessions (three sessions were used in our study) may be associated with a reduction of approximately 40% of the number of patients in each treatment group required to show a particular true score group difference. In a recently published study, researchers were able to correlate a change in donkeys' facial grimace, behavior, and physiology with head pain, lameness and post colic surgery [1,3,4]. A previous study observed donkeys during castration surgery as having the same or greater amount of pain when compared to ponies [5]. Other studies have indicated that many mammals, including sheep, pigs, cows, and horses, experience some degree of pain after castration [17–20]. Therefore, it was assumed that post-castration donkeys were not completely comfortable, despite receiving an analgesic drug, and that changes in scores from the donkey grimace scale seen after castration could be indicators of pain. One of the most accurate body language markers we found for detecting distress related to pain was the overall appearance of the donkey and body stance. Our study's results indicate the importance of observing more than just the face of the donkey to identify signs of pain. Facial indicators, such as ears, eyes, nostrils may be more easily influenced by external factors, such as other donkeys and the environment of the donkeys, thus it is important to look at the donkey as a whole for signs of behavior related to pain [21]. In previous studies, indicators of a donkey in pain or sick may include the following behavioral signs associated with pain, such as a slight twitch to the tail, but authors claimed they were subtler than those of horses [2,9,10,34–36]. Several studies have confirmed that indicators of a donkey in pain may include sham eating, chewing, generalized dullness, shifting weight to contralateral limb, unresponsive/decreased mobility of ears, twitching tail, and tail tucked [3–5]. The present study confirmed the importance of observing the overall appearance of the donkey to accurately identify pain. However, since the present study examined pictures and not the video of the donkeys, it was difficult to monitor ear mobility or sham eating.

Some studies have developed and tested a Horse Grimace Scale [10], an ethogram of facial expressions of ridden horses [36] and applied this ethogram to differentiate between sound and lame ridden horses [37]. Another study developed an ethogram for the body, face, and gait of the horse and aimed to differentiate between lame and non-lame ridden horses [11]. Similarly, to the aforementioned study, the current study also aimed to distinguish pain indicators based on the whole body in addition to facial indicators. Contrastingly, the present study examined donkeys with post-operative pain instead of horses with musculoskeletal pain. Although multiple studies have utilized facial action units to recognize signs of pain in horses, therefore, suggesting a similar scale, such as the scale used in this study could be effectively applied and used in donkeys. Recent studies have applied various pain scales, such as a Facial Expression Pain Scale, EQUUS-DONKEY-COMPASS and EQUUS-DONKEY-FAP, to groups of donkeys experiencing different kinds of pain, including acute orthopedic pain, colic, head-related pain, and post-operative pain and applied the same scales to a control group of healthy donkeys. Although these studies did not yield significant results in regards to post-operative donkeys, likely due to the variation in how long the donkey may have been experiencing pain or the stage of the disease, both studies yielded significant results for facial pain, colic, and lameness [6,12], and provided a basis for the implementation of a grimace scale to be used to identify pain indicators in donkeys.

Donkeys in the current study served as self-controls. The decision for using donkeys as self-controls was supported in previous studies for three main reasons. First, for ethical reasons, as well as to adhere to a high standard of welfare, a control group of donkeys undergoing castration without analgesic treatment was not used in this study [10]. This is based upon the claims made on Mair et al. [38], who reported that providing no analgesia for castration in equines is inhumane.

The second main reason this study was designed using donkeys as their own self-control was due to findings in previous donkey studies attempting to measure and define pain behavior and measure the clearance of analgesic agents. Donkeys were used as self-controls in a study that tested for the palliative effect of detomidine as a sedative and analgesic in cases where severe pain was reported during abdominal colic [39]. The severity of abdominal pain accompanying colic was determined prior to (time 0) and after drug administration (15, 30, 45 and 60 min) using a standardized scoring system to describe several clinical parameters common to equine colic. Among these clinical parameters, the presence or absence of body signs was considered using a scale from no evidence of signs of discomfort to severe evidence of signs of discomfort. This study, similar to our study, photographed the donkeys as various times [40]. For our study, additional times were set to take photos prior to castration, as a way to monitor the animals to make sure no evident changes occurred prior to castration at 48 h, 24 h, 0 h before surgery. Amin and Najim [40] also compared detomidine, ketamine and other anesthetics stated that full recovery was achieved at 40 min in donkeys. In regard to the metabolism of analgesics, donkeys may describe two different trends, one characterizing the period of time that takes for half-life to be reached, then afterwards the period of time needed for the analgesic to disappear from the body or at least its effects to not be present. Contextually, Grosenbaugh et al. [41], reported NSAIDs may be clinically effective longer in donkeys than in horses as indicated by the respective plasma t1/2 (half time in donkeys ranges from 0.75 to 4.5 h, with maximum limit only being half an hour longer than in horses). The study suggested a faster reduction of drug plasma levels after the drug had reached half its concentration (half-life) provided the notably increased clearance times of 1.78 (ml/kg bwt/h), which almost double those in horses [2]. Bearing this in mind and to ensure the effects of any of the drugs did not alter the donkeys' behavior, post castration photos were taken 8 h, 24 h and 48 h post-surgery. Another comparison to this study was the differences in scales—our study developed a scale that included a "don't know" possibility to be able to gather those cases in which observers were not able to identify the degree of expression of a certain body sign related to pain which was not included in other studies [41].

Finally, we considered a previous study [10], which indicated that horses receiving varying levels of flunixin meglumine as analgesic post castration still had a significantly higher composite pain scale score and horse grimace scale score than the respective scores of the control group that did not undergo a surgical procedure; furthermore, there was no significant difference in horse grimace scale score or composite pain scale score between the two treatment groups that received different amounts of flunixin as an analgesic at varying time intervals. Another option that may have been considered for a control would have been to have one group of donkeys undergo analgesic, but no surgery procedure and another group complete the surgery. However, based on the fact other studies had used a similar protocol to this study (animals serving as self-controls pre-/post-surgery) we did not see it necessary nor in the best interest of the donkey's welfare to place them under general anesthesia to simply photograph their face pre and post sedation.

Behavior and images for each donkey were recorded before the procedure and then after. Each donkey underwent the same medical procedure at the same time so the differences in scores and subsequently pain indicators should be due, in the majority, to the individual donkey's acute response, thus effects derived from factors other than the procedure itself were minimized. Often donkeys will not exhibit signs of pain or discomfort until the disease and/or condition is in advanced stages or is a chronic condition (e.g., laminitis), therefore the subtle signs of pain may go unnoticed. Even when the disease or condition becomes chronic, the signs may remain unnoticed, due to the donkey having become habituated to the discomfort. This study began to define signs of discomfort related to pain in an acute phase versus chronic, in hopes of providing a tool to assist practitioners and owners in noticing these subtle behavioral changes and ultimately lead to providing treatment for pain [2–4,9,10,34–36].

The ability to recognize signs of pain in this study was significantly dependent on gender (Table S4), experience or interaction with donkeys (Table S5), and knowledge level of observers (Table S6). These factors impacted the observer's ability to correctly score a post castration donkey as painful. Previous studies have indicated that female respondents, when surveyed, tend to provide a significantly higher pain score among cattle, dogs, cats, and horses across a variety of conditions and procedures, however, the studies in which this was observed did not include whether the scores of male versus female participants were more accurate [42–44]. The present study found a significant difference between the ability of male versus female observers to identify pain indicators after castration. More specifically, female observers were more likely than male observers to provide a higher score for post castration pictures. Furthermore, female observers were less likely to score a false-positive (identify pain markers in pre castration pictures—control group).

Results of the present study indicated that observers with more donkey experience and knowledge are less likely to miss indicators of pain (score false-negatives). This is likely because by spending more time observing donkeys, one may become more skilled at noticing behavioral deviations from "normal", and therefore, be more likely to pick up on subtle signs. These results suggest that practitioners and owners can become more successful at identifying when a donkey is painful by spending more time observing and becoming familiar with donkey behavior in general. Previous studies examining donkey behavior have confirmed that because donkeys demonstrate different clinical signs of pain/illness when compared to horses, previous knowledge and an understanding of the donkey can be crucial to correct examination, diagnosis and medical treatment of the donkey [9,10,21,35,36]. Another study indicated that painful conditions in donkeys, such as laminitis, are often overlooked by the owner and veterinarian, due to the donkey's stoicism [44]. One survey of donkey owners and veterinarian surgeons investigated how pain is identified and then treated in donkeys [33]. The study found that a majority of the responses from veterinarians administered horse rate and doses of analgesic drugs to control pain in donkeys which likely is not effective. Therefore, in addition to considering the variations in the obviousness of pain behavior between horses and donkeys, once a veterinarian can identify clinical signs of a painful donkey, the treatment of whatever ails the donkey must likely be approached slightly differently to the way it would be treated for a horse [9,10,21,35]. When pain is identified in donkeys, it is possible that therapeutic levels of analgesic drugs are not met, due to species differences; previous studies have shown that donkeys have the ability to metabolize analgesics at a faster rate than a horse of similar size [45,46].

Practitioners and owners should be aware of differences between donkey and horse medication dosages, and appropriate donkey dosages should be administered [9,10,21,35,36]. The present study provided all donkeys with an injectable nonsteroidal anti-inflammatory and analgesic drug, flunixin meglumine, during the surgery. This may have influenced pain indicators seen post-surgery, though donkeys clear the drug much more rapidly than horses or mules. The sedative, xylazine, the sedative analgesic drug, butorphanol, and the dissociative anesthetic, ketamine, also have short term analgesic effects [47]. While these drugs might affect the donkey's pain perception, they would be unlikely to change responses 24 h after surgery [47]. For this reason, even with the administration of this commonly used analgesic drug to the donkeys in this study, signs of discomfort and pain were still noted as early as eight hours post-castration. It is important to reiterate that the donkey's stoicism does not diminish the donkey's pain or distress, hence it is important for owners and veterinarians to observe what normal donkey behavior looks like. The present study has shown that there are behavioral indicators of pain in faces and bodies of donkeys (abnormal stance and overall appearance of the donkey proved to be particularly accurate); identification of these markers may require observation by a more experienced observer familiar with donkeys.

5. Conclusions

The ability to recognize subtle body language signals may prove to be challenging to those who have limited experience working with donkeys. Observing the overall body posture and reactions of donkeys may improve our chances of recognizing pain in donkeys. This indicates the importance of observing more than just the face of the donkey to identify signs of pain. Facial indicators, such as ears, eyes, and nostrils, may be more easily influenced by external factors, such as other donkeys and the environment of the donkeys, thus it is important to look at the donkey as a whole for signs of behavior related to pain. Gender, experience or interaction with donkeys, and knowledge level of observers impacted the observer's ability to correctly score a post castration donkey as painful. Female observers were more likely than male observers to provide a higher score for post castration pictures and were less likely to score a false-positive. By spending more time observing donkeys, one may become more skilled at noticing behavioral deviations from "normal", and therefore, be more likely to pick up on subtle signs. Conclusively, identifying the appearance of a normal donkey may assist in decreasing the number of donkeys that go untreated, due to the subtleness of their behavioral pain indicators, as the biggest obstacle of treating pain in donkeys may be identifying the pain in the first place.

Supplementary Materials: The following are available online at http://www.mdpi.com/2076-2615/10/8/1411/s1, Table S1. Testing for Intra class correlation coefficient (Cronbach's α) for interobserver reliability, Table S2. Testing for Intra class correlation coefficient (Cronbach's α) for intraobserver reliability across days and for the total panel, Table S3. Testing for normality (Kolmogorov-Smirnov) for castration status, score and success to identify pain,

Table S4. Summary for the results of frequency analyses and frequency differences between female and male observers, Table S5. Summary for the results of frequency analyses and frequency differences across knowledge levels, Table S6. Summary for the results of frequency analyses and frequency differences across experience or interaction levels, Table S7. Summary for the results of frequency analyses and frequency differences across body language signs.

Author Contributions: Conceptualization, A.K.M.; Data curation, E.K.O., F.J.N.G., C.I.P. and A.K.M.; Formal analysis, F.J.N.G., C.I.P., J.M.B. and A.K.M.; Funding acquisition, A.K.M.; Investigation, E.K.O., Carlos Iglesias Pastrana, J.M.B., S.S.l.J., E.W.D. and A.K.M.; Methodology, F.J.N.G., S.S.l.J., E.W.D. and A.K.M.; Resources, E.W.D. and A.K.M.; Software, F.J.N.G.; Supervision, J.M.B. and A.K.M.; Validation, F.J.N.G.; Visualization, A.K.M.; Writing—original draft, E.K.O., F.J.N.G. and A.K.M.; Writing—review & editing, E.K.O., F.J.N.G., C.I.P., J.M.B., S.S.l.J., E.W.D. and A.K.M. All authors have read and agreed to the published version of the manuscript.

Funding: Funding to support the publication of this article was funded by the UC Davis Provost's Undergraduate Fellowship, grant number 39423.

Acknowledgments: The authors would like to thank Rural Veterinary Experience Training and Service (RVETs) for their time and services. We would like to thank The University of California Davis' Undergraduate Research Center and the Provost's Undergraduate Fellowship program.

Conflicts of Interest: The authors declare no conflict of interest.

References

1. Van Dierendonck, M.C.; Burden, F.A.; Rickards, K.; van Loon, J.P. Monitoring Acute Pain in Donkeys with the Equine Utrecht University Scale for Donkeys Composite Pain Assessment (EQUUS-DONKEY-COMPASS) and the Equine Utrecht University Scale for Donkey Facial Assessment of Pain (EQUUS-DONKEY-FAP). *Animals* **2020**, *10*, 354. [CrossRef] [PubMed]
2. Ashley, F.; Waterman-Pearson, A.; Whay, H. Behavioural assessment of pain in horses and donkeys: Application to clinical practice and future studies. *Equine Vet. J.* **2005**, *37*, 565–575. [CrossRef] [PubMed]
3. Regan, F.H.; Hockenhull, J.; Pritchard, J.C.; Waterman-Pearson, A.E.; Whay, H.R. Behavioural repertoire of working donkeys and consistency of behaviour over time, as a preliminary step towards identifying pain-related behaviours. *PLoS ONE* **2014**, *9*. [CrossRef] [PubMed]
4. Grint, N.J.; Beths, T.; Yvorchuk-St Jean, K.; Whay, H.R.; Murrell, J.C. Analysis of behaviors observed during mechanical nociceptive threshold testing in donkeys and horses. *J. Equine Vet. Sci.* **2017**, *50*, 102–109. [CrossRef]
5. Burden, F.; Thiemann, A. Donkeys are different. *J. Equine Vet. Sci.* **2015**, *35*, 376–382. [CrossRef]
6. Van Loon, J.P.; Van Dierendonck, M.C. Monitoring acute equine visceral pain with the Equine Utrecht University Scale for Composite Pain Assessment (EQUUS-COMPASS) and the Equine Utrecht University Scale for Facial Assessment of Pain (EQUUS-FAP): A scale-construction study. *Vet. J.* **2015**, *206*, 356–364. [CrossRef]
7. Prunier, A.; Mounier, L.; Le Neindre, P.; Leterrier, C.; Mormède, P.; Paulmier, V.; Prunet, P.; Terlouw, C.; Guatteo, R. Identifying and monitoring pain in farm animals: A review. *Animal* **2012**, *7*, 998–1010. [CrossRef]
8. Grint, N.; Johnson, C.; Clutton, R.; Whay, H.; Murrell, J. Spontaneous electroencephalographic changes in a castration model as an indicator of nociception: A comparison between donkeys and ponies. *Equine Vet. J.* **2015**, *47*, 36–42. [CrossRef]
9. Evans, L.; Crane, M. *The Clinical Companion of the Donkey*; Troubador Publishing: Leicester, UK, 2018.
10. Dalla Costa, E.; Minero, M.; Lebelt, D.; Stucke, D.; Canali, E.; Leach, M.C. Development of the Horse Grimace Scale (HGS) as a pain assessment tool in horses undergoing routine castration. *PLoS ONE* **2014**, *9*. [CrossRef]
11. Dyson, S.; Berger, J.; Ellis, A.D.; Mullard, J. Development of an ethogram for a pain scoring system in ridden horses and its application to determine the presence of musculoskeletal pain. *J. Vet. Behav.* **2018**, *23*, 47–57. [CrossRef]
12. Olmos, G.; Alvarado-Arellano, A.; Du Toit, N.; Burden, F.; Gregory, N. A Novel Approach of Pain Recognition and Assessment in Donkeys; Initial Results. In Proceedings of the 45th Congress of the International Society for Applied Ethology, Indianapolis, IN, USA, 31 July–4 August 2011.
13. Smalbroek, A. Facial Expression Pain Scale (FEPS) for Donkeys. Master's Thesis, Faculty of Veterinary Medicine, University of Utrecht, Utrecht, The Netherlands, 2018.
14. Gleerup, K.; Lindegaard, C. Recognition and quantification of pain in horses: A tutorial review. *Equine Vet. Educ.* **2016**, *28*, 47–57. [CrossRef]

15. McLean, A.K.; Heleski, C.R.; Yokoyama, M.T.; Wang, W.; Doumbia, A.; Dembele, B. Improving working donkey (Equus asinus) welfare and management in Mali, West Africa. *J. Vet. Behav.* **2012**, *7*, 123–134. [CrossRef]
16. Dalla Costa, E.; Stucke, D.; Dai, F.; Minero, M.; Leach, M.C.; Lebelt, D. Using the horse grimace scale (HGS) to assess pain associated with acute laminitis in horses (Equus caballus). *Animals* **2016**, *6*, 47. [CrossRef] [PubMed]
17. Molony, V.; Kent, J.; Robertson, I. Assessment of acute and chronic pain after different methods of castration of calves. *Appl. Anim. Behav. Sci.* **1995**, *46*, 33–48. [CrossRef]
18. Molony, V.; Kent, J.; Hosie, B.; Graham, M. Reduction in pain suffered by lambs at castration. *Vet. J.* **1997**, *153*, 205–213. [CrossRef]
19. Hay, M.; Vulin, A.; Génin, S.; Sales, P.; Prunier, A. Assessment of pain induced by castration in piglets: Behavioral and physiological responses over the subsequent 5 days. *Appl. Anim. Behav. Sci.* **2003**, *82*, 201–218. [CrossRef]
20. Heleski, C.; Cinq-Mars, D.; Merkies, K.; Stämpfli, H.; Cottee, S.Y.; de Wit, J. *Code of Practice for the Care and Handling of Equines: Review of Scientific Research on Priority Issues*; Equine Code of Practice Scientists' Committee: Ottawa, ON, Canada, 2012.
21. Van Loon, J.; Van Dierendonck, M. Objective pain assessment in horses (2014–2018). *Vet. J.* **2018**, *242*, 1–7. [CrossRef]
22. Landis, J.R.; Koch, G.G. A review of statistical methods in the analysis of data arising from observer reliability studies (Part II). *Stat. Neerl.* **1975**, *29*, 151–161. [CrossRef]
23. Greenwood, P.E.; Nikulin, M.S. *A Guide to Chi-Squared Testing*; John Wiley & Sons: Hoboken, NJ, USA, 1996.
24. Roscoe, J.T.; Byars, J.A. An Investigation of the Restraints with Respect to Sample Size Commonly Imposed on the Use of the Chi-Square Statistic. *J. Am. Stat. Assoc.* **1971**, *66*, 755–759. [CrossRef]
25. Zhu, W.; Zeng, N.; Wang, N. Sensitivity, specificity, accuracy, associated confidence interval and ROC analysis with practical SAS implementations. *NESUG Proc. Health Care Life Sci.* **2010**, *19*, 67.
26. Navas González, F.J.; León Jurado, J.M.; Delgado Bermejo, J.V. Interpretation of Interactions between Categorical Variables in Categorical Regression Models for Meta-Analysis of Fixed Effects of Animal Models. In Proceedings of the XIX Simposio Iberoamericano Sobre Conservación y Utilización de Recursos Zoogenéticos, Riobamba, Ecuador, 22–26 October 2017.
27. Häger, C.; Biernot, S.; Buettner, M.; Glage, S.; Keubler, L.; Held, N.; Bleich, E.; Otto, K.; Müller, C.; Decker, S. The Sheep Grimace Scale as an indicator of post-operative distress and pain in laboratory sheep. *PLoS ONE* **2017**, *12*, e0175839. [CrossRef] [PubMed]
28. Larrondo, C.; Bustamante, H.; Gallo, C. Sheep farmers' perception of welfare and pain associated with routine husbandry practices in Chile. *Animals* **2018**, *8*, 225. [CrossRef] [PubMed]
29. Reijgwart, M.L.; Schoemaker, N.J.; Pascuzzo, R.; Leach, M.C.; Stodel, M.; de Nies, L.; Hendriksen, C.F.M.; van der Meer, M.; Vinke, C.M.; van Zeeland, Y.R.A. The composition and initial evaluation of a grimace scale in ferrets after surgical implantation of a telemetry probe. *PLoS ONE* **2017**, *12*, e0187986. [CrossRef] [PubMed]
30. Dai, F.; Leach, M.; MacRae, A.M.; Minero, M.; Dalla Costa, E. Does Thirty-Minute Standardised Training Improve the Inter-Observer Reliability of the Horse Grimace Scale (HGS)? A Case Study. *Animals* **2020**, *10*, 781. [CrossRef]
31. Raw, Z.; Rodrigues, J.B.; Rickards, K.; Ryding, J.; Norris, S.L.; Judge, A.; Kubasiewicz, L.M. Equid Assessment, Research and Scoping (EARS): The Development and Implementation of a New Equid Welfare Assessment and Monitoring Tool. *Animals* **2020**, *10*, 297. [CrossRef]
32. Sommerville, R.; Brown, A.F.; Upjohn, M. A standardised equine-based welfare assessment tool used for six years in low and middle income countries. *PLoS ONE* **2018**, *13*, e0192354. [CrossRef]
33. Grint, N.; Murrell, J.; Whay, H. Investigating the opinions of donkey owners and veterinary surgeons towards pain and analgesia in donkeys. *Equine Vet. Educ.* **2015**, *27*, 365–371. [CrossRef]
34. Matthews, N.; Taylor, T.; Hartsfield, S. Anaesthesia of donkeys and mules. *Equine Vet. Educ.* **2005**, *15*, 102–107. [CrossRef]
35. Mullard, J.; Berger, J.M.; Ellis, A.D.; Dyson, S. Development of an ethogram to describe facial expressions in ridden horses (FEReq). *J. Vet. Behav.* **2017**, *18*, 7–12. [CrossRef]
36. McLean, A.K.; Navas González, F.J.; Canisso, I.F. Donkey and Mule Behavior. *Vet. Clin. N. Am. Equine* **2019**, *35*, 575–588. [CrossRef]

37. Dyson, S.; Berger, J.M.; Ellis, A.D.; Mullard, J. Can the presence of musculoskeletal pain be determined from the facial expressions of ridden horses (FEReq)? *J. Vet. Behav.* **2017**, *19*, 78–89. [CrossRef]
38. Mair, T.; Love, S.; Schumacher, J.; Smith, R.K.; Frazer, G. *Equine Medicine, Surgery and Reproduction*; Elsevier Health Sciences: Amsterdam, The Netherlands, 2013.
39. Jöchle, W. Field trial evaluation of detomidine as a sedative and analgesic in horses with colic. *Equine Vet. J.* **1989**, *21*, 117–120. [CrossRef]
40. Amin, A.; Najim, I. Use of Detomidine, Butorphanol and Ketamine for Induction of General Anesthesia in Donkeys. *Al-Anbar J. Vet. Sci.* **2011**, *4*, 1–8.
41. Grosenbaugh, D.; Reinemeyer, C.; Figueiredo, M. Pharmacology and therapeutics in donkeys. *Equine Vet. Educ.* **2011**, *23*, 523–530. [CrossRef]
42. Williams, V.; Lascelles, B.; Robson, M. Current attitudes to, and use of, peri-operative analgesia in dogs and cats by veterinarians in New Zealand. *N. Z. Vet. J.* **2005**, *53*, 193–202. [CrossRef]
43. Laven, R.; Huxley, J.; Whay, H.; Stafford, K. Results of a survey of attitudes of dairy veterinarians in New Zealand regarding painful procedures and conditions in cattle. *N. Z. Vet. J.* **2009**, *57*, 215–220. [CrossRef]
44. Waran, N.; Williams, V.; Clarke, N.; Bridge, I. Recognition of pain and use of analgesia in horses by veterinarians in New Zealand. *N. Z. Vet. J.* **2010**, *58*, 274–280. [CrossRef]
45. Matthews, N.; van Loon, J.P. Anesthesia, sedation, and pain management of donkeys and mules. *Vet. Clin. N. Am. Equine* **2019**, *35*, 515–527. [CrossRef]
46. Lizarraga, I.; Sumano, H.; Brumbaugh, G. Pharmacological and pharmacokinetic differences between donkeys and horses. *Equine Vet. Educ.* **2004**, *16*, 102–112. [CrossRef]
47. Coakley, M.; Peck, K.E.; Taylor, T.S.; Matthews, N.S.; Mealey, K.L. Pharmacokinetics of flunixin meglumine in donkeys, mules, and horses. *Am. J. Vet. Res.* **1999**, *60*, 1441–1444.

animals

MDPI

Article

Preference by Donkeys and Goats among Five Mediterranean Forest Species: Implications for Reducing Fire Hazard

Jordi Bartolomé [1,*], Jordi Miró [2,*], Xavier Panadès [3], Maria José Broncano [1], Josefina Plaixats [1], Teresa Rigau [2], Maria José Milán [1] and Elena Baraza [4]

1 Departament de Ciència Animal i dels Aliments, Grup de Recerca en Remugants, Universitat Autònoma de Barcelona, 08193 Bellaterra, Spain; mj.broncano@creaf.uab.cat (M.J.B.); josefina.plaixats@uab.cat (J.P.); mariajose.milan@uab.cat (M.J.M.)
2 Departament de Medicina i Cirurgia Animal, Universitat Autònoma de Barcelona, 08193 Bellaterra, Spain; teresa.rigau@uab.cat
3 Museu de la Conca Dellà, C/ del Museu 4, 25650 Isona, Spain; xpan100@gmail.com
4 Departament de Biologia, Universitat de les Illes Balears, Cra. de Valldemossa km 7.5, 07122 Palma, Spain; elena.baraza@uib.es
* Correspondence: jordi.bartolome@uab.cat (J.B.); jordi.miro@uab.cat (J.M.); Tel.: +34-3-5811496 (J.B.); +34-3-5811899 (J.M.)

check for updates

Received: 29 June 2020; Accepted: 28 July 2020; Published: 30 July 2020

Simple Summary: Donkeys and goats are animals adapted to graze in understories, and thus contribute to the prevention of forest fires. In this work, the preferences of donkeys and goats have been determined for five key plant species of the Mediterranean forest, where large forest fires have increased as a result of global change. Using a multiple selection test, it has been observed that both species can complement each other, since donkeys consume more fine fuel, such as *Brachypodium retusum*, and goats, highly flammable woody species, such as *Pinus halepensis*. In this way, browsing becomes an ecosystem service, which, in the case of donkeys, can even help prevent their extinction.

Abstract: During the second half of the 20th century, European countries experienced an increase in their forest area due to the global change. Consequently, there has been an increase in large forest fires, mainly in the Mediterranean basin, and this has forced the development of several types of prevention programs. One of them is the control of the understory by livestock. In this sense, browsing with a combination of donkeys and goats could be a good option, as both animals usually feed on forest species. However, little is known about their preferences for the key species of the Mediterranean forest. Using a cafeteria test, the preferences and consumption of both animals have been determined for five typical species of the Mediterranean forest, such as *Quercus ilex*, *Pinus halepensis*, *Phillyrea latifolia*, *Rubus ulmifolius*, and *Brachypodium retusum*. Results showed that donkeys and goats could act complementarily in the reduction of the fuel biomass of forests. Donkeys appear to act more on fine fuel, such as *B. retusum*, and goats on the more pyrophyte species, in this case *P. halepensis*. In addition, given that donkeys are at severe risk of extinction in Europe, this role of providing ecosystem services could contribute to their conservation. Despite this study only showing that goats and donkeys would consume all five presented plant species and that there are some differences in consumption during a short-term test, it constitutes a useful first step for conservation and fire prevention in the Mediterranean forests.

Keywords: cafeteria test; *Quercus ilex*; *Pinus halepensis*; *Phillyrea latifolia*; *Rubus ulmifolius*; *Brachypodium retusum*

1. Introduction

There has been massive depopulation in rural areas due to the intense socio-economical changes since the industrial revolution, as well as the reduction of subsistence and traditional agriculture, forestry, and animal husbandry [1]. One of the consequences of this process has been that Europe, North America, and part of Asia experienced an increase in their forest area during the second half of the 20th century [2]. The forests have accumulated into a thick carpet of biomass with highly combustible undergrowth, which ultimately has contributed to the recurrent and virulent forests fires in the Mediterranean basin [3]. In recent decades, fires in Mediterranean Europe have become larger and more frequent [4]. In the past, livestock used to graze the undergrowth, even if it was dedicated to logging, looking for the scarce food resources that the forest produces, and contributing to diminishing the fire hazard.

Grazing programs in forests from the private and public sectors have currently been implemented in some Mediterranean areas, with the explicit intention of reducing fire risk [5–7]. These have shown that grazing management of forest vegetation not only prevents fires, but also preserves or increases ecosystem biodiversity, activates the rural economies (e.g., ecotourism, controlled hunting), and enhances scenic qualities [8,9]. In addition, control of the undergrowth by livestock is considered the most cost-effective treatment, even though it requires certain investments (fences, water supplies, forage supplementation), and occasionally needs to be combined with another method [10,11].

Different livestock species have been used as a control tool for the Mediterranean understory. Small ruminants, such as sheep and goats, have often been used due to their hardiness and ability to include a wide variety of woody species in their diet [12]. Due to their more selective nature, and their ability to tolerate many secondary compounds [13], goats are surely the most widely used animal for this purpose (e.g., Mancilla-Leytón et al. [5]; Lovreglio at et al. [6]). For instance, several organizations have recently appeared in California, dedicated to renting their goats to clear public and private land in order to prevent fire risk (e.g., https://www.citygrazing.org/). However, limitations in the cut diameter of bites, the difficulty of penetrating very dense understories, and the avoidance of grasses by goats [6] have led, in many cases, to larger livestock to be preferred for forest management. Bovines and equines can browse in thicker branches than small ruminants, and add the greatest effect of trampling on vegetation. However, the cattle effect is limited by their preference for herbaceous plants [14] and by the quality of the food offered by the Mediterranean understory [15]. The equines are well suited to reduce grass encroachment as they can graze in low-quality fodder [16,17]. It has been observed that horses and goats grazing together in a Mediterranean scrub reduced plant phytovolume significantly [18]. Lamoot et al. [19] showed that donkeys preferred a grassy habitat for foraging in all seasons. Donkeys have also been used successfully to control flammable grasses in the Mediterranean scrub [20]. Along these lines, the combination of donkeys and goats seems to be a good option to control the biomass of the Mediterranean understory. However, donkeys were the livestock most affected by the industrial revolution, and with the upheaval in agriculture, most European donkey breeds are disappearing [21]. In this sense, providing the species with an ecosystem service role could strengthen its survival. Finally, the success of this type of management will depend on the preferences of both animals for the most abundant and most flammable species in the forest. There is currently very little information on this aspect. For this reason, the objective of this work was to determine preferences, and compare the consumption of donkeys and goats for key plant species of the Western Mediterranean forests.

2. Materials and Methods

2.1. Plant Species

Five plants were offered to animals during the experiment: *Quercus ilex*, *Pinus halepensis*, *Phillyrea latifolia*, *Rubus ulmifolius*, and *Brachypodium retusum*. These are important components of the Western Mediterranean forests. They are all evergreen, so their twigs and leaves are available year-round. *Q. ilex* and *P. halepensis* are the most common tree species in these forests. Both are highly flammable [5,22]. In addition, *P. halepensis* is a pyrophyte species, with several traits that facilitate its dispersion after fire [23]. *Ph. latifolia* and *R. ulmifolius* are common shrubs in the understory. The first is a deep-rooted shrub distributed in the warmer and drier Mediterranean basin, and with good resilience after fires [24]. The second can become dominant after fire, especially if it is not affected by grazing [25]. Both facilitate the spread of fire from the understory to the canopy of trees. *B. retusum* is a grass usually abundant in the herbaceous layer, and constitutes fine fuel for the initiation of many fires [26].

2.2. Animals and Feeding Trials

The study was performed with seven Catalan donkeys (*Equus asinus*) and seven female Murciano–Granadina goats (*Capra hircus*) at the Experimental Farm of Autonomous University of Barcelona (Spain). Both are autochthonous breeds of the Iberian Peninsula. Animals were 3–8 years of age in the case of donkeys and 1–2 years in the case of goats, and their initial body weight was 375 ± 25 and 39 ± 6 kg for donkeys and goats, respectively. These animals were selected in a pre-experiment from two herds of 15 donkeys and 30 goats, obtaining the average values of preference on basic feeds, such as barley (*Hordeum vulgare*), alfalfa (*Medicago sativa*), and rye-grass (*Lolium multiflorum*), offered in the same proportion, of the whole herd. The seven animals that presented the values closest to the mean were chosen. All of them were in excellent condition and remained at a constant live weight throughout the experiment. No more animals were used in the experiment for logistical reasons. In order to avoid biases from acquired diet preferences [27], the selected animals had not previously grazed in a Mediterranean forest. In addition, they were given a 5-day adaptation period to adjust to the 3 × 4 m individual boxes environment. During this period, each morning the animals remained in the boxes for 15 min without any food supply.

Every day, animals were placed in the individual boxes, and were tested separately, and in turn. Each one could consume 500 g of fresh forage of each plant species for 15 min undisturbed. It was considered that 15 min was enough time for the animals to finish the offered food. The plant material consisted of bundles of branches less than 5mm in diameter, collected daily from the forest near the farm, and tied to a metal bar located within reach of the animals' mouths. Plant material in the box was randomly changed every day to avoid bias by side preferences of certain animals. Species availability allowed animals to make their own choices. The first choice was recorded as an estimation of preferred species [28,29]. An observer (the same person in all trials) standing in the neighboring enclosure recorded the choices made by each animal. Consumption was calculated by subtracting the remaining fresh weight from the initial weight. These cafeteria trials were conducted between 0800 and 1100 in the morning. The order in which animals were tested varied randomly every day to prevent conditioned patterns. The animals were within sight of companions during tests to minimize isolation and stress. The experiment was carried out for a period of fifteen consecutive days. After the feeding trial, animals were reincorporated to the corresponding stable, where the rest of the farm herd was. Once there, the herd received a basic diet of concentrated feed and barley straw in its feeder. All animals had free access to water and salt blocks. According to McArthur et al. [27], it is important for animals to receive a basic diet in order to test preferences between plant species under conditions in which nutritional or energetic requirements are satisfied.

2.3. Legal Requirements

The experiment complied with the Decree 214/97 of the Catalan Autonomous Government regarding the use of animals in scientific experiments. It was authorized by the Institutional Animal Care and Use Committee of the Autonomous University of Barcelona.

2.4. Statistical Analysis

Multivariate analysis of variance (MANOVA) with repeated measurements was used to determine whether there were any differences in the fresh weight consumed (dependent variable) among plant species and between animals throughout the 15 days of experimentation. Time (days) was the repeated factor, and the time × species interaction term was used to recognize differences among plant species in consumption patterns over time.

The analysis of the first plant species chosen every day by the animals was carried out using a logistic regression model for nominal responses. JMP7 software (Version 7.0: SAS Institute, Inc., Cary, NC, USA) was employed in this statistical analysis.

3. Results

There were no significant differences in the amounts of plant material consumed between both animal types (MANOVA, $p = 0.1915$), despite a clear difference in size between both species. This can probably be explained by the difference in the consumption rate between both species. Thus, Hoffmann et al. [30] and Lamoot et al. [19] got ranges of 8 to 15 and 9 to 12 bites per minute respectively for donkeys, and Negui et al. [31] and Rodrigues et al. [32] got ranges of 15 to 31 and 19 to 24 bites per minute, respectively, for goats. Figures 1 and 2 show the evolution of the consumption of each species over time by goats and donkeys, respectively.

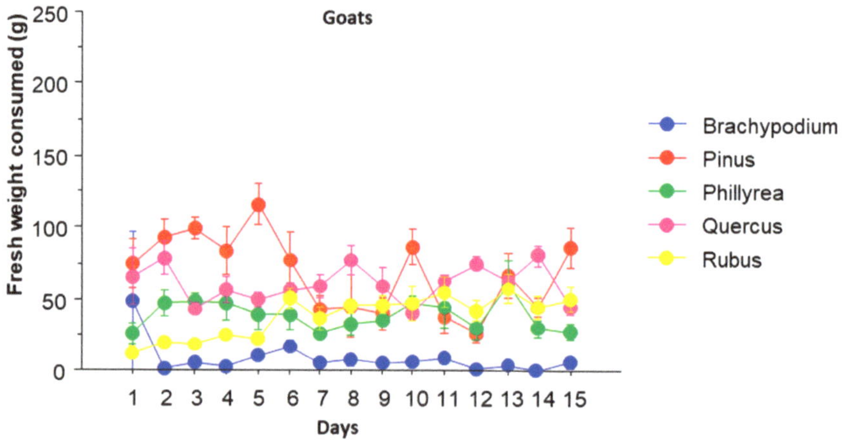

Figure 1. Daily average of fresh weight consumed by goats in the choice tests ($n = 7$). Lines crossing the circles are standard errors.

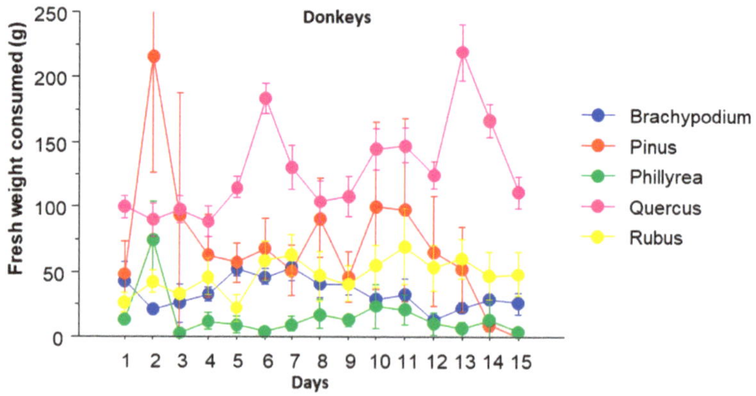

Figure 2. Daily average of fresh weight consumed by donkeys in the choice tests ($n = 7$). Lines crossing the circles are standard errors.

Significant differences were obtained in the consumption of the different species, with *Q. ilex* being the most consumed, and *B. retusum* the least (MANOVA, $p < 0.0001$). Significant differences were also obtained for the interactions between animal type and plant species (MANOVA, $p < 0.0001$), animal type and time (MANOVA, $p = 0.0004$), time and plant species (MANOVA, $p < 0.0001$), and for the interactions of the three factors (MANOVA, $p < 0.0001$).

The species that was consumed the most by donkeys were *Q. ilex*, and by goats, *P. halepensis* and *Q. ilex*. The species least consumed by donkeys was *Ph. latifolia*, and by goats it was *B. retusum*. Donkeys consumed more *B. retusum* and *Q. ilex* than goats, and goats more *Ph. latifolia* than donkeys. The amount of plant consumed throughout the time is more constant in goats than in donkeys. The only plant species in which the amount consumed did not vary significantly throughout the 15 days was *Ph. latifolia*. The consumption of *Q. ilex* and *R. ulmifolius* increased during the firsts five days and then remained high, with a peak towards the last days. The amount of *B. retusum* oscillated, with a maximum consumed the first day, and a minimum consumed the second, then increased again until the sixth day and from that decreased again. Finally, the consumption of *P. halepensis* was decreasing with small oscillations throughout the period.

Goats consumed more *Ph. latifolia* than donkeys for almost the whole period, and more *P. halepensis* mainly during the first half of the period. On the other hand, donkeys ate much more *B. retusum* than goats, with the only exception of the first day. They also ate more *Q. ilex* during the whole period.

There was no significant difference observed in the first choice of the animals. Clearly, *Q. ilex* was the first chosen in most times, and *B. retusum* was the last (Figure 3). Despite this, the first choice of goats was more constant between animals and between days than in donkeys. A couple of donkeys never chose *P. halepensis* as a first choice, another never chose *B. retusum*, and another one never chose *Ph. latifolia*. On the other hand, there was a donkey that chose this last species for most of the days. Donkeys chose *Ph. latifolia* as the first option, mainly during the first days, *Q. ilex* mainly during the second half of the period, and *B. retusum* mainly during the intermediate days.

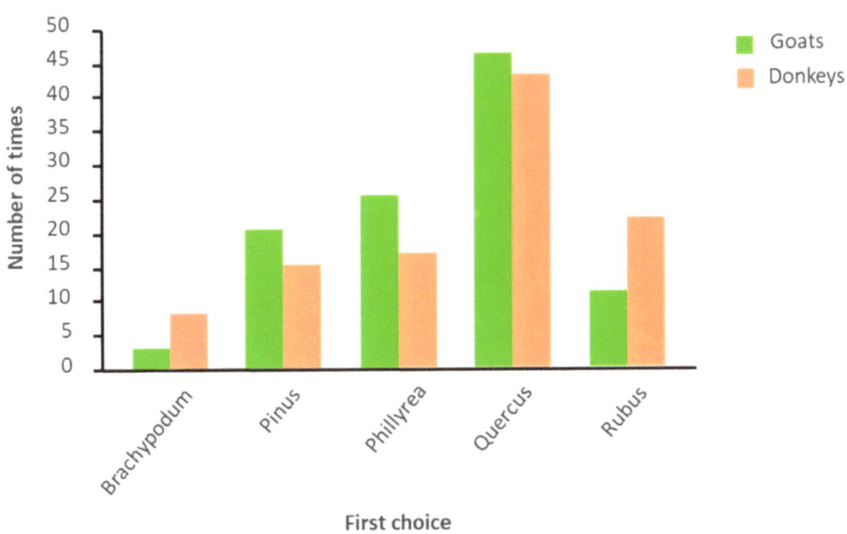

Figure 3. Total number of times each plant species was chosen as the first option.

4. Discussion

This study is an encouraging first step in a much longer process to establish if goats and donkeys could really reduce fire hazards in Mediterranean forests. From the results obtained, it can be concluded that donkeys and goats could act complimentarily in the reduction of the fuel biomass of forests. Both species have a long history of browsing on Mediterranean shrub ranges, and both animals consumed all the plant species considered, although in different ways. The higher consumption and higher preference for *Q. ilex*, as well as the lower consumption and lower preference for *B. retusum* could be related to their nutritional value. The dominant woody species in the Mediterranean basin are generally of low nutritional quality and contain secondary metabolites, such as tannins, terpenes, and volatile oils [33,34]. Nonetheless, these woody species are often selected by grazing animals, because their leaves have more protein and less fiber than leaves and stems of grasses [35,36]. *Q. ilex* is known to have a higher crude protein content and lower fiber content than *B. retusum* [34]. *R. ulmifolius* also has an acceptable crude protein content [37], and this could possibly explain the consumption increase over the days in both animals, as happened with *Q. ilex*. The higher consumption of *Q. ilex* by donkeys could perhaps be explained by the differences in the bite size of both animals. However, both species have physical defenses, small spines on the leaves of *Q. ilex* and spikes on the stems of *R. ulmifolius*. Several studies have reported on the influence of spinescence in shrubs on the feeding behavior of mammalian browsers [38,39]. Apparently, in this case, these defenses do not represent an impediment for either animal.

Goats consumed *P. halepensis* at the same level as *Q. ilex*, despite the content of secondary compounds that this species contains. *P. halepensis* is a terpene-storing plant [40]. Something similar occurs with *Ph. latifolia*, with a high number of biologically active substances, such as glycosides and terpenes [41], and it was consumed more by goats than by donkeys. This is surely due to the salivary secretions of goats containing proline-rich proteins that bind some of these secondary compounds, thus alleviating their aversive effects [13].

On the other hand, the fact that donkeys consumed more *B. retusum* could be related to their ability to eat foods with high fiber content and low nutritional quality. Equids can compensate for low-quality fodder through increased fodder uptake [42]. The ability of donkeys to reduce biomass of easily flammable grasses has been highlighted by Gulias et al. [20], in the case of the tall grass, *Ampelodesmos mauritanica*. Control of dominant tall grasses by other non-ruminants, such as horses, was also shown in sub-Mediterranean mountain pastures in Italy [43]. For its part, the preference in goats seems to be more affected by the fiber content than by secondary compounds [44].

Another interesting result is the fact that goats, compared to donkeys, had more constant consumption and preferences over time, and with less variability between individuals. This is surprising because goats tended to select a mixed diet when different food species were simultaneously available, and this has been suggested as a foraging strategy for reducing the risk of toxicity [45].

Finally, some limitations of this study, related to the fact that the study was carried out under controlled conditions and not in the field, should be explicitly mentioned. This includes the fact that there may have been novelty effects of using animals who had never consumed those plant species. The relative attractiveness may vary when animals have more options, such offered by an actual forest situation. In addition, animals were tested individually, looking for possible differences between individuals, but it should be recognized that social effects could alter feed consumption in highly gregarious animals like goats and donkeys in field conditions.

5. Conclusions

In summary, considering the consumption and preferences of both animals, the use of mixed herds of donkeys and goats as a tool for controlling biomass fuel allowed us to predict an effect on key species in fires in the Mediterranean forest. Donkeys acted on fine fuel, such as *B. retusum*, while goats acted on the more pyrophyte species, such as *P. halepensis*. However, caution is required in extrapolating these results to free-ranging situations. Obviously, much larger studies (more animals, exposed to the feed for much longer each day and considering Mediterranean seasonality) are essential to provide a more realistic idea of how actual implementation would work.

Finally, it is worth noting that since the donkey can be a good manager of the Mediterranean forest, this is another possible strategy for the conservation of endangered breeds.

Author Contributions: Conceptualization: J.B., J.M., J.P., and T.R.; methodology: J.M., T.R., and J.B.; validation: J.M.; investigation: J.B., J.M., T.R., J.P., X.P., and M.J.M.; resources: J.B. and J.M.; data curation: M.J.B. and E.B.; writing—original draft preparation: X.P. and J.B.; writing—review and editing: J.B. and J.M.; supervision: J.B., J.M., T.R., J.P., X.P., M.J.M, E.B., and M.J.B.; project administration: J.B. All authors have read and agreed to the published version of the manuscript.

Funding: This research was funded by Ministry of Economy and Competitiveness (MINECO) of Spain through the project CGL2015-70449-R.

Acknowledgments: The authors wish to thank Ramón Casals and Ramón Costa for their technical assistance in handling the experimental animals, and in the overall conducting of the experiment.

Conflicts of Interest: The authors declare no conflict of interest.

References

1. Collantes, F.; Pinilla, V. *Peaceful Surrender: The Depopulation of Rural Spain in the Twentieth Century*; Cambridge Scholars Publishing: Newcastle upon Tyne, UK, 2011; pp. 10–15.
2. Keenan, R.J.; Reams, G.A.; Achard, F.; de Freitas, J.V.; Grainger, A.; Lindquist, E. Dynamics of global forest area: Results from the FAO Global Forest Resources Assessment. *For. Ecol. Manag.* **2015**, *352*, 9–20. [CrossRef]
3. Lasanta-Martínez, T.; Vicente-Serrano, S.M.; Cuadrat-Prats, J.M. Mountain Mediterranean landscape evolution caused by the abandonment of traditional primary activities: A study of the Spanish Central Pyrenees. *Appl. Geogr.* **2005**, *25*, 47–65. [CrossRef]
4. Pausas, J.G.; Fernández-Muñoz, S. Fire regime changes in the Western Mediterranean Basin: From fuel-limited to drought-driven fire regime. *Clim. Change* **2012**, *110*, 215–226. [CrossRef]
5. Mancilla-Leytón, J.M.; Pino Mejías, R.; Martín Vicente, A. Do goats preserve the forest? Evaluating the effects of grazing goats on combustible Mediterranean scrub. *Appl. Veg. Sci.* **2013**, *16*, 63–73. [CrossRef]
6. Lovreglio, R.; Meddour-Sahar, O.; Leone, V. Goat grazing as a wildfire prevention tool: A basic review. *IForest* **2014**, *7*, 260–268. [CrossRef]
7. Mena, Y.; Ruiz-Mirazo, J.; Ruiz, F.A.; Castel, J.M. Characterization and typification of small ruminant farms providing fuelbreak grazing services for wildfire prevention in Andalusia (Spain). *Sci. Total Environ.* **2016**, *544*, 211–219. [CrossRef] [PubMed]
8. Gutierrez-Peña, R.; Mena, Y.; Ruiz, F.A.; Delgado-Pertíñez, M. Strengths and weaknesses of traditional feeding management of dairy goat farms in mountain areas. *Agroecol. Sustain. Food Syst.* **2016**, *40*, 736–756. [CrossRef]
9. Speed, J.D.M.; Austrheim, G. The importance of herbivore density and management as determinants of the distribution of rare plant species. *Biol. Conserv.* **2017**, *205*, 77–84. [CrossRef]
10. Ruiz-Mirazo, J. La prevención de incendios forestales mediante pastoreo controlado en Andalucía. *Ganaderia* **2008**, *55*, 46–51.
11. Casals, P.; Baiges, T.; Bota, G.; Chocarro, C.; de Bello, F.; Fanlo, R.; Sebastià, M.T.; Taull, M. Silvopastoral Systems in the Northeastern Iberian Peninsula: A Multifunctional Perspective. In *Agroforestry in Europe*; Rigueiro-Rodróguez, A., McAdam, J., Mosquera-Losada, M.R., Eds.; Springer: Dordrecht, The Netherlands, 2009; Volume 6, pp. 161–181.
12. Bartolomé, J.; Franch, J.; Plaixats, J.; Seligman, N.G. Diet selection by sheep and goats on Mediterranean heath-woodland range. *J. Range Manag.* **1998**, *51*, 383–391. [CrossRef]
13. Hofmann, R.R. Evolutionary steps of ecophysiological adaptation and diversification of ruminants: A comparative view of their digestive system. *Oecologia* **1989**, *78*, 443–457. [CrossRef] [PubMed]
14. Calleja, J.A.; Escolà, M.; Carvalho, J.; Forcadell, J.M.; Serrano, E.; Bartolomé, J. Cattle Grazing Fails to Control Shrub Encroachment in Mediterranean Landscapes. *Rangel. Ecol. Manag.* **2019**, *72*, 803–811. [CrossRef]
15. Teruel-Coll, M.; Pareja, J.; Bartolomé, J.; Serrano, E.; Mentaberre, G.; Cuenca, R.; Espunyes, J.; Pauné, F.; Calleja, J.A. Effects of boom and bust grazing management on vegetation and health of beef cattle used for wildfire prevention in a Mediterranean forest. *Sci. Total Environ.* **2019**, *665*, 18–22. [CrossRef] [PubMed]
16. Süss, K.; Schwabe, A. Sheep versus donkey grazing or mixed treatment: Results from a 4-year field experiment in Armerio-Festucetum trachyphyllae sand vegetation. *Phytocoenologia* **2007**, 135–160. [CrossRef]

17. Schwabe, A. What are the long-term effects of livestock grazing in steppic sandy grassland with high conservation value? Results from a 12-year field study. *Tuexenia* **2013**, *33*, 189–212.
18. Bartolomé-Filella, J.; Josep Broncano-Atienza, M.; Plaixats-Boixadera, J. Efecte de la pastura amb equí i cabrum en el fitovolum d'un matollar mediterrani. *Quaderns Agraris* **2016**, *40*, 55–72.
19. Lamoot, I.; Vandenberghe, C.; Bauwens, D.; Hoffmann, M. Grazing behaviour of free-ranging donkeys and Shetland ponies in different reproductive states. *J. Ethol.* **2005**, *23*, 19–27. [CrossRef]
20. Gulías, J.; Mairata, A.; Frontera, J.; Janer, I.; Jaume, J.; Cifre, J. Firebreak maintenance with equines in Serra de Tramuntana mountains (Mallorca, UNESCO world heritage). In *Mountain Pastures and Livestock Farming Facing Uncertainty: Environmental, Technical and Socio-Economic Challenges*; Casasús, I., Lombardi, G., Eds.; Options Méditerranéennes: Zaragoza, Spain, 2016; pp. 333–336.
21. Kugler, W.; Grunenfelder, H.P.; Broxham, E. *Donkey Breeds in Europe: Inventory, Description, Need for Action, Conservation*; Report 2007/2008; Monitoring Institute for Rare Breeds and Seeds in Europe: St. Gallen, Switzerland, 2008; pp. 13–14. Available online: https://www.doc-developpement-durable.org/file/Elevages/Anes/donkey%20(1).pdf (accessed on 18 March 2020).
22. El-Amine Henaoui, S.; Bouazza, M.; Hassiba, S. Rural livelihoods diversity—the coping strategies in Wa West District in Northern Ghana. *J. Biol. Agric. Healthc.* **2013**, *3*, 101–113.
23. Ne'eman, G.; Goubitz, S.; Nathan, R. Reproductive traits of *Pinus halepensis* in the light of fire—A critical review. *Plant EcoL.* **2004**, *171*, 69–79. [CrossRef]
24. Goudelis, G.; Ganatsas, P.P.; Spanos, I.; Karpi, A. Effect of repeated fire on plant community recovery in Penteli, central Greece. In *Eco-and Ground Bio-Engineering: The Use of Vegetation to Improve Slope Stability*; Springer: New York, NY, USA, 2007; Volume 103, pp. 337–343.
25. Mazzoleni, S.; Rego, F.; Giannino, F.; Vincenot, C.E.; Pezzatti, G.B.; Legg, C. Ecosystem modelling: Vegetation and disturbance. In *Environmental Modelling: Finding Simplicity in Complexity*; Wainwright, J., Mulligan, M., Eds.; John Wiley & Sons Ltd.: Chichester, UK, 2004; pp. 171–186.
26. Vilà-Cabrera, A.; Saura-Mas, S.; Lloret, F. Effects of fire frequency on species composition in a Mediterranean shrubland. *Écoscience* **2008**, *15*, 519–528. [CrossRef]
27. McArthur, C.; Goodwin, A.; Turner, S. Preferences, selection and damage to seedlings under changing availability by two marsupial herbivores. *For. Ecol. Manag.* **2000**, *139*, 157–173. [CrossRef]
28. Litviatis, J. Investigating food habits of terrestrial vertebrates. In *Research Techniques in Animal Ecology: Controversies and Consequences*, 2nd ed.; Boitani, L., Fuller, T.K., Eds.; Columbia University Press: New York, NY, USA, 2000; pp. 165–190.
29. Vallentine, J.F. *Grazing Management*, 2nd ed.; Academic Press: Cambridge, MA, USA, 2001; pp. 261–302.
30. Hoffmann, M.; Cosyns, E.; Deconinck, M.; Lamoot, I.; Zwaenepoel, A. Donkey Diet in a Flemish Coastal Dune Area in the First Year of Grazing. Available online: https://www.environment.nsw.gov.au/resources/coasts/coastal-dune-mngt-manual.pdf (accessed on 18 March 2020).
31. Negi, G.C.S.; Rikhari, H.C.; Ram, J.; Singh, S.P. Foraging Niche Characteristics of Horses, Sheep and Goats in an Alpine Meadow of the Indian Central Himalaya. *J. Appl. Ecol.* **1993**, *30*, 383–394. [CrossRef]
32. Rodrigues, M.M.; de Oliveira, M.E.; de Moura, R.L.; de Rufino, M.O.A.; Silva, W.K.A.; do Nascimento, M.P.S.C.B. Forage intake and behavior of goats on Tanzania-grass pasture at two regrowth ages. *Acta Sci. Anim. Sci.* **2013**, *35*, 37–41. [CrossRef]
33. Tisserand, J.L.; Alibes, X. Feeds of the Mediterranean área. In *Ruminant Nutrition. Recommended Allowance and Feed Tables*; Jarrige, R., Ed.; INRA and John Libbey Eurotext: Paris, France, 1989; pp. 305–323.
34. Bartolomé, J.; Baraza, E.; Rita, J.; Serrano, E.; Pareja, J.; Ibáñez, M.; Capó, M.; Alomar, D.; Domenech, O. Cabra I forest: La qualitat dels recursos dels bosc per a la cabra mallorquina. In *Recerca I Gestió Dins L'àmbit Cinegètic*; Pons, G.X., Barceló, A., Muñoz, M., del Valle, L., Seguí, B., Eds.; Mallorca-Societat d'Història Natural de les Balears: Palma de Mallorca, Spain, 2019; Volume 28, pp. 95–100.
35. Holechek, J.L.; Pieper, R.D.; Herbel, C.H. *Range Management: Principles and Practices*; Prentice Hall: Englewood Cliffs, NJ, USA, 1989; pp. 23–25.
36. Papachristou, T.G.; Nastis, A.S. Nutritive value of diet selected by goats grazing on kermes oak shrublands with different shrub and herbage cover in Northern Greece. *Small Rumin. Res.* **1993**, *12*, 35–44. [CrossRef]

37. Decandia, M.; Molle, G.; Sitzia, M.; Cabiddu, A.; Ruiu, A.; Pampiro, F.; Pintus, A. Effect of polyethylene glycol on browsing behaviour and performance of late lactating goats. In *Sheep and Goat Nutrition: Intake, Digestion, Quality of Products and Rangelands*; Ledin, I., Morand-Fehr, P., Eds.; Cahiers Options Méditerranéennes, CIHEAM: Zaragoza, Spain, 2000; Volume 52, pp. 147–150.
38. Cooper, S.M.; Owen-Smith, N. Effects of plant spinescence on large mammalian herbivores. *Oecologia* **1986**, *68*, 446–455. [CrossRef]
39. Milewski, A.V.; Young, T.P.; Madden, D. Thorns as induced defenses: Experimental evidence. *Oecologia* **1991**, *86*, 70–75. [CrossRef]
40. Blanch, J.S.; Peñuelas, J.; Sardans, J.; Llusià, J. Drought, warming and soil fertilization effects on leaf volatile terpene concentrations in *Pinus halepensis* and *Quercus ilex*. *Acta Physiol. Plant.* **2009**, *31*, 207–218. [CrossRef]
41. Recuero Carretero, C.; Díaz Lanza, A.M.; Fernández Matellano, L.; Rumbero Sánchez, A.; Villaescusa Castillo, L. Phytochemical Analysis of *Phillyrea latifolia*, L., a New Source of Oleuropeoside. *Z. Naturforsch. C J. Biosci.* **2018**, *56*, 353–356. [CrossRef]
42. Duncan, P.; Foose, T.J.; Gordon, I.J.; Gakahu, C.G.; Lloyd, M. Comparative nutrient extraction from forages by grazing bovids and equids: A test of the nutritional model of equid/bovid competition and coexistence. *Oecologia* **1990**, *84*, 411–418. [CrossRef]
43. Catorci, A.; Gatti, R.; Cesaretti, S. Effect of sheep and horse grazing on species and functional composition of sub-Mediterranean grasslands. *Appl. Veg. Sci.* **2012**, *15*, 459–469. [CrossRef]
44. Alonso-Díaz, M.A.; Torres-Acosta, J.F.J.; Sandoval-Castro, C.A.; Hoste, H.; Aguilar-Caballero, A.J.; Capetillo-Leal, C.M. Is goats' preference of forage trees affected by their tannin or fiber content when offered in cafeteria experiments? *Anim. Feed Sci. Technol.* **2008**, *141*, 36–48. [CrossRef]
45. Duncan, A.J.; Young, S.A. Can goats learn about foods through conditioned food aversions and preferences when multiple food options are simultaneously available? *J. Anim. Sci.* **2002**, *80*, 2091–2098. [PubMed]

animals

MDPI

Article

Survey of Serum Amyloid A and Bacterial and Viral Frequency Using qPCR Levels in Recently Captured Feral Donkeys from Death Valley National Park (California)

Sara Jerele [1], Eric Davis [1], Samantha Mapes [2], Nicola Pusterla [2], Francisco Javier Navas González [3,4,*], Carlos Iglesias Pastrana [3,4], Essam Mahmoud Abdelfattah [5] and Amy McLean [4,6,*]

1 International Animal Welfare Training Institute, School of Veterinary Medicine, University of California Davis, Davis, CA 95617, USA; jerele.sara@gmail.com (S.J.); ewdavis@ucdavis.edu (E.D.)
2 Equine Viral Disease Laboratory, Department of Pathology, Microbiology, and Immunology, School of Veterinary Medicine, University of California Davis, Davis, CA 95617, USA; smmapes@ucdavis.edu (S.M.); npusterla@ucdavis.edu (N.P.)
3 Department of Genetics, Faculty of Veterinary Sciences, University of Córdoba, 14071 Córdoba, Spain; ciglesiaspastrana@gmail.com
4 The Worldwide Donkey Breeds Project, Faculty of Veterinary Sciences, University of Córdoba, 14071 Córdoba, Spain
5 Department of Animal Hygiene, and Veterinary Management, Faculty of Veterinary Medicine, Benha University, Qalubiya Governorate 13511, Egypt; essam.abdelfattah@fvtm.bu.edu.eg
6 Department of Animal Science, University of California Davis, Davis, CA 95616, USA
* Correspondence: fjng87@hotmail.com (F.J.N.G.); acmclean@ucdavis.edu (A.M.); Tel.: +34-651-679-262 (F.J.N.G.); +1706-296-8743 (A.M.)

Received: 5 May 2020; Accepted: 22 June 2020; Published: 23 June 2020

Simple Summary: This study aimed to measure the inflammatory marker (Serum Amyloid A; SAA) derived from such pathogenic processes specifically bacterial and virus found in recently captured feral donkeys. Quantitative Real-Time PCR (qPCR) was performed to detect and quantify RNA and DNA viruses and bacterial DNA. Behavior, body condition score, nasal discharge, and coughing were found to be related in cases of Asinine Herpesvirus 2 (AHV-2) and *Streptococcus equi* subspecies *zooepidemicus* DNA. SAA concentrations descended with age, with foals presenting higher concentrations. Positive tests differed for AHV-2 and *Streptococcus equi* spp. *zooepidemicus* between sampling moments. In conclusion, donkeys caught in the wild may not be a source of disease for domestic horses. However, the transmission of some pathogens, such as *Streptococcus equi* subspecies *zooepidemicus*, and/or AHV-2, may occur.

Abstract: Feral donkey removal from state land has raised concerns in terms of disease transmission between equine species. Disease outbreaks may occur as a result of the relocation of animals to new environments. Virus and bacteria DNA load and serum amyloid A derived from the pathogenic processes that they involve were measured in recently captured donkeys. Blood and nasal swabs were collected from 85 donkeys (Death Valley National Park, Shoshone, California); 24 were retested after 30/60 days in the Scenic (Arizona) long-term holding facility co-mingled with feral donkeys from Arizona and Utah. Quantitative Real-Time PCR (qPCR) was performed to detect viral and bacterial genomic material (equine influenza A [EIV], equine rhinitis A and B viruses, AHV-2, AHV-3, AHV-5 and EHV-1, EHV-4, *Streptococcus equi* subspecies *equi* and *zooepidemicus*,). Significant relations between behavior, body condition score, nasal discharge, and coughing were found in donkeys for which AHV-2 and *Streptococcus zooepidemicus* DNA was detected. Higher SAA concentrations were found in foals. AHV-2 and *Streptococcus zooepidemicus* DNA concentrations significantly differed between sampling moments ($p < 0.05$). In conclusion, donkeys do not appear to be a substantial risk for disease transmission to horses but could be if they carried strangles or other processes in which AHV-2 and *Streptococcus zooepidemicus* were involved.

Keywords: donkeys; burro; serum amyloid A; *Streptococcus equi*; Asinine Herpesvirus; PCR

---

1. Introduction

The feral burro or donkey population of Death Valley presumably originates from the descendants of domesticated donkeys that were brought to the southwestern deserts by miners in the USA over 100 years ago [1]. These hardy animals found an ideal environment in Death Valley, with enough feed and water and no natural predators. In the past century, the burro has become part of the megafauna of this inhospitable area, though some believe their effect on native flora and biota is deleterious [2]. Like the Bureau of Land Management (BLM) on other state lands, the National Parks Service (NPS) has been tasked with the management of all wild animals living in the Death Valley National Park. As part of their mission, the NPS assesses burro (donkey) populations and has made the managerial decision to reduce herd sizes with a nonprofit capture and relocate the burros [3].

As little is known about the overall health of feral donkeys and the prevalence of pathogens, specifically in the category of infectious equine viruses and bacteria in their populations [4]; it is critical to conduct field health assessments of donkeys to predict early potential diseases before relocation. In this regard, Serum Amyloid A (SAA), is an acute-phase apolipoprotein, synthesized by the liver as a response to inflammation derived from infection or injury [5–7]. A new SAA test kit provides rapid on-site SAA blood concentrations (StableLab, Sligo, Ireland), which may be of help when assessing the health condition of feral donkeys or in field conditions. Normal concentrations of plasma SAA are virtually undetectable [7]; rapid increasing concentrations of SAA, from ten to 1000 times that of normal levels, indicate some form of inflammation or infection [8]. Countries such as the USA have run surveillance programs for equine infectious respiratory pathogens [9]. These programs focus on the prevalence and epidemiology of important viral and bacterial agents, such as equine influenza virus (EIV), equine herpesvirus-1 and -4 (EHV-1 and EHV-4), and *Streptococcus equi* subspecies *equi* in horses (*Equus caballus*). However, such studies have not been performed on donkeys.

Contextually, *Streptococcus equi* subspecies *zooepidemicus* (*S. zooepidemicus*) and subspecies *equi* (*S. equi*) infect equids worldwide but are especially prevalent in donkeys [10]. *S. zooepidemicus*, a commensal organism in the equine nasopharynx, can occasionally invade the respiratory mucosa, resulting in purulent rhinitis, bronchitis, lymphadenopathy, and pneumonia in horses and donkeys of all ages [10,11]. Furthermore, single strains with *S. zooepidemicus* phenotypes similar to those found in the nasopharynx of healthy horses have been reported to cause pneumonia in donkeys [12]. Donkeys have been reported to act as asymptomatic chronic carriers of *S. zooepidemicus*, with conditions being characterized by guttural pouch infection or intermittent nasal discharge. Indeed, clinical observations [13] have suggested an increased likelihood of guttural pouch disease in asses.

The *herpesviridae* family plays an important role in the prevalence and epidemiology of equid respiratory disease. This viral family sub-divides into the *alpha*- (EHV-1, -3, and -4) [14], *beta*-, and *gamma*- (EHV-2 and -5) *herpesviridae* which can infect all equids [15]. Donkeys and hybrids (mules and hinnies) may not be a source of herpesviruses transmission to other equids [16], but host a range of specific Asinine Herpesviruses [17]. Asses are susceptible to EHV-1 and -4, with respiratory and abortion being described in the literature. Three of the herpesviruses found in donkeys have been isolated and identified as similar to EHV-1, -7, and -8 [11–16,18] and have been described and designated as both equid (EHV) and Asinine Herpesviruses (AHV) [18]. Asinine Herpesvirus 2 (AHV-2), also denoted as EHV-7 [18], closely resembles gamma-herpesviruses EHV-2 and -5 [16], can cause clinical signs of respiratory infection, and can also be isolated from clinically healthy donkeys [18].

Limited information is available regarding AHV-3 (designated as EHV-8) [17,18] AHV-4, AHV-5 [19], and AHV-6 [20]. The GG nucleotide sequences-assisted phylogenetic assessment addressed a shorter evolutionary distance between AHV-3 and EHV-1 than between either virus or EHV-4. These outcomes contribute to the hypothesis that AHV-3, or other affine or related viruses, are the precursors of EHV-1 and only adapted to horses relatively recently [17].

Thiemann [21] described AHV-5 isolates from lungs displaying interstitial pneumonia and remarkable syncytial cell formation which did not affect lung tissue in donkeys, which presented evidence of pneumonia with a bacterial or verminous underlying etiology [19]. Furthermore, this gamma herpesvirus was similar to that of a donkey identified with neurological disease [22]. AHV-4 and -5 isolates have been obtained from lungs belonging to donkeys coursing pulmonary fibrosis. Donkey's pulmonary fibrosis is similar to horses' multinodular pulmonary fibrosis, which has been ascribed to the action of EHV-5. Positive cases of alpha herpesviruses EHV-1 and -4 from animals presenting respiratory and neurological conditions have been clinically described as well [21].

Stress from relocation and social comingling present an increased likelihood for outbreaks of latent infections [23]. Typically described clinical signs, such as a harsh cough, pyrexia, serous nasal secretion, and lethargy have been reported, as also described for horses. Donkey herpesviruses may differ from that of horses, although the guidelines for diagnosis, treatment, and control remain the same [21]. However, horse EHV-1 and EHV-4 vaccination viability have not been reported in donkeys. Coinfection of some of *S. zooepidemicus* group C and Asinine Herpesvirus 5 (AHV-5) with *Dictyocaulus arnfieldi* has been hypothesized to contribute to the clinical disease, but without empirical or experimental evidence [24].

This study aimed to measure the frequency of potential pathogens in recently captured feral donkeys from Death Valley National Park, testing for the potentially existing correlations of pathogen loads with respiratory, ocular, behavioral, ambulatory clinical signs and inflammatory markers (serum amyloid A) to assess whether changes in pathogen presence after donkeys are co-mingled with domestic equids is warranted.

2. Materials and Methods

2.1. Animal Sample and Sampling

Whole blood and nasal discharge were sampled from 85 feral donkeys (37 jennies and 48 jacks) removed from Death Valley National Park range in Shoshone, California (GW4C+56 Skidoo, California, United States/ 36°29′58.7″ N 117°03′07.7″ W). A 12 mL blood sample was collected by venipuncture of the jugular vein by syringe for further analysis. Nasal secretions were collected on a synthetic swab with a plastic applicator and stored in a small zip-lock plastic bag at −20 °C until extraction. All jennies and foals were moved to a long-term holding facility in Scenic, Arizona (176 miles away) and co-mingled with other wild donkey populations from Arizona and Utah (sampling moment one, 4–5 November 2018 (group one) and 4–5 December 2018 (group two)). Twenty-four donkeys were retested for SAA and nasal secretions (NS) (sampling moment two, around 30/60 days later, 7 January 2019). The second sampling (moment two) was set at 30/60 days to ensure a minimum period of 21 days had been left as a way to cover the incubation period of all potential pathogens present. Clinical examination was performed on all animals before sampling by two operators. Invasive procedures were limited to blood sampling and nasal swabs. Temperature Pulse Respiration (TPR) was not performed as abnormally elevated values could be expected. In this context, the correct assessment of health status could have been distorted, as a result of the lack of handling and human interaction of the feral donkeys participating in the study until the moment of the sampling. Visual examination comprised records on body condition score (BCS), gender (37 jennies and 48 jacks), age (81 adults of ≥2 years old, and 4 foals, 1 jenny foal and 3 male foals, which were still nursing), the presence of skin/hair conditions (absent, alopecia, photosensitivity/photodermatitis, minor wound, deep wound), the presence of lameness (1 lame, 2 not lame when visually observed walking from holding pen to chute), the presence of either ocular or nasal discharge (1 present and 2 not present), the presence of coughing (1 present if animal was coughing and 2 not present if animal was not coughing), the presence of abnormal breathing (1 if abnormal breathing was observed during sample collection and 2 if it was not present), and behavior signs (alert, lethargic, aggressive, fearful). BCS was determined following a 1 to 5 scoring system, 1 poor to 5 obese, described in Valle et al. [25] and Polidori and Vincenzetti [26] and represented in Figure 1. The presence of lameness was assessed through visual examination of the holding pen where they were kept to the area were these were sampled in the chutes. Since the donkeys were feral/wild animals, lameness was evaluated by only visual observation: 1 if donkey was lame and displayed signs of lameness like bobbing head, irregular gait, non-weight bearing limb were visually observed; or 2 if no irregular gait or obvious visual signs were observed by the clinicians. Behavior signs were evaluated following the premises in Navas González et al. [27] by visually observing the face (eye, nostril and ear position), head, neck and body posture and language. Cronbach's alpha parameter to test for interobserver reliability reported agreement based on internal consistency was over 0.8 for the observations of the two operators performing a clinical examination, hence their information could be considered objective enough to be included in the study.

Figure 1. Body condition score in donkeys as described in Valle et al. [25] and Polidori and Vincenzetti [26].

2.2. SAA Levels

Whole blood was analyzed for serum amyloid A (SAA) by using a commercial stable-side kit (StableLab, Sligo, Ireland). Whole blood was taken directly from the syringe and no additive was added. SAA concentrations were obtained from the kit after 10 min post-sampling on-field, and the values for SAA concentrations expressed in mg/L were recorded. Further information on the procedures used to test for SAA can be found in Kay et al. [7].

2.3. Load Quantification for Viruses and Bacteria

Nasal swabs were collected from the left nostril as the position in which the animal was restrained made it more accessible. Immediately after collection, blood samples were collected and transferred to the lab in portable refrigerators. Nasal swabs were collected in virus transport media (VTM) and stored at −80 °C until testing. The median time-lapse from collection to processing was 2 days (Interquartile range 2–3 days). Additional information about sample preparation, submission, and process can be found in the q-PCR diagnostic submission packet (University of California Davis' Veterinary Medicine PCR Laboratory [28]).

Once at the lab, swabs were placed in a 15 mL conical with 500 μL phosphate-buffered saline, vortexed and spun at 4000 rpm to obtain a cell pellet. Total nucleic acid was extracted from the swabs using a QIAcube HT automated nucleic acid (ANA) workstation (Qiagen, Valencia, CA, USA) according to the manufacturer's instructions for the QIAamp 96 DNA QIAcube HT Kit (Qiagen).

Real-time PCR (quantitative PCR, qPCR) was performed using nasal secretions for the detection and quantification of different microbial agents: *Streptococcus equi* subspecies *equi, Streptococcus equi* subspecies *zooepidemicus*, equine influenza type A (H3N8), equine Rhinitis A, equine Rhinitis B, AHV-2, AHV-3, AHV-5, Equine Herpesvirus 1, Equine Herpesvirus 1 (EHV-1, neuropathogenic), Equine Herpesvirus- 1 EHV-1(non-neuropathogenic), and Equine Herpesvirus-4 (EHV-4).

2.4. Quantitative PCR Systems

Primer Express software (Thermo Fisher Scientific, Carlsbad, CA, USA) was used to design two primers and an internal hydrolysis fluorescent-labeled probe (5′ end, reporter dye fluorescein amidite (FAM) (6-carboxyfluorescein), 3′ end, quencher NFQMGB (Non-Fluorescent Quencher Minor Grove Binding) for each target gene (Table 1). Unique species detection was confirmed by a Basic Local Alignment Search Tool (BLAST) of each amplicon. Ten-fold dilutions of DNA testing positive for the target genes were used to validate qPCR (real-time or quantitative polymerase chain reaction) systems. The dilutions were analyzed in triplicate and a standard curve plotted against the dilutions. Amplification efficiencies were computed through the formula $E = 10^{1/-s}-1$ (slope of the standard curve). A minimum of 90% efficiencies was required to pass validation.

Table 1. Target genes used for qPCR-assay pathogen loads testing.

Assay Name	Gene, NCBI [a] Accession #	Assay Location (bp)	Amplicon Length (bp)
Asinine Herpesvirus 2	Polymerase, EU165547	100	81
Asinine Herpesvirus 3	Glycoprotein B, U24184	740	145
Asinine Herpesvirus 5	Polymerase, AY054993	600	64
Equine Herpesvirus 1	Glycoprotein B, NC_001491	400	89
Equine Herpesvirus 1, neuropathogenic	ORF 30, KF644574	200	92
Equine Herpesvirus 1, non-neuropathogenic	ORF 30, KX101095	200	92
Equine Herpesvirus 4	Glycoprotein B, AF030027	440	77
Streptococcus equi subspecies *equi*	M Protein, AF012927	150	185
Streptococcus equi subspecies *zooepidemicus*	ITS, EU860336	80	88
Influenza A (H3N8)	Hemagglutinin Precursor, EF541443	350	200
Equine rhinitis A virus	RNA polymerase, X96870	150	111
Equine rhinitis B virus	RNA polymerase, X96871	350	87
Glyceraldehyde-3-phosphate dehydrogenase	GAPDH, AF097179	60	105

[a] NCBI: National Center for Biotechnology Information.

2.5. RT-Reaction and Quantitative PCR

The cDNA synthesis was performed following the modified protocol of manufacture prescriptions of the Quantitect Reverse transcription kit (Qiagen). Ten microliters of RNA were digested with 1 μL of gDNA wipeout buffer by incubation at 42 °C for 5 min and then briefly centrifuged. A total of 1 μL of digested RNA and the real-time PCR housekeeping gene was used to test for genomic DNA contamination. Then, 0.5 μL of Quantitect Reverse Transcriptase, 2 μL Quantitect RT (Real-Time) buffer, 0.5 μL RT primer mix, 0.5 μL 20 pmol random primers (Invitrogen) were added and brought up to a final volume of 20 μL and incubated at 42 °C for 40 min. The samples were inactivated at 95 °C for 3 min, chilled, and 80 μL of water was added.

Each PCR reaction contained 20× primer and probes for the respective qPCR system, with a final concentration of 400 nM for each primer and 80 nM for the probe and commercially available PCR master mix (TaqMan Universal PCR Mastermix, Applied Biosystems) containing 10 mM Tris-HCl (pH 8.3), 50 mM KCl, 5 mM $MgCl_2$, 2.5 mM deoxynucleotide triphosphates, 0.625 U AmpliTaq Gold DNA polymerase per reaction, 0.25 U AmpErase UNG per reaction and 5 μL of the diluted gDNA and cDNA sample in a final volume of 12 μL. The samples were placed in 384 well plates and amplified in an automated fluorometer (ABI PRISM 7900 HTA FAST, ABI). ABI's standard amplification conditions were used: 2 min at 50 °C, 10 min at 95 °C, 40 cycles of 15 s at 95 °C and 60 s at 60 °C. Fluorescent signals were collected during the annealing temperature and CT (cycle threshold) values extracted with a threshold of 0.1 and baseline values of 3–12.

2.6. Statistical Analyses

Field observations were collected at two moments (sampling moments one and two). The population membership changed over time from sampling moment one to sampling moment two but retained certain common participants, which are known as partially overlapping samples. As suggested by Derrick et al. [29], the assumptions of partially overlapping samples t-test with Welch's degrees of freedom are the same as those for Welch's test (normality and heteroscedasticity [30]). When a non-gross violation of the normality assumption has occurred, Welch's test power could be comparable to t-test power if homoscedasticity is still met, as suggested Rasch et al. [31]. Still, in cases of unequal variances and skewness values < 3, Welch's test has been reported to keep 20% robustness. Parametric assumptions were tested to determine the best set of tests to statistically process data. All parametric assumption testing tests, except for the normality tests, were performed using SPSS Statistics for Windows, Version 25.0, IBM Corp. [32]. Normality assumptions were tested using StataCorp Stata version 14.2 one-way ANOVA routine of the compare samples task of SPSS Statistics for Windows, Version 25.0, IBM Corp. [32], and partially overlapping samples t-test with Welch's degrees of freedom using the "Partially overlapping" package by Derrick [33] of RStudio 1.1.463 by the R Studio Team [34] were used to detect differences in the mean of viral and bacterial DNA quantities.

Pearson correlations were used to determine the relationship between SAA maximal concentration expressed in mg/L and microbial DNA quantity using the Bivariate procedure from the correlate task from SPSS Statistics for Windows, Version 25.0, IBM Corp. [32]. Afterward, categorial regression (CATREG) was performed to determine best-fitting linear predictive models for SAA maximal concentrations and AHV-2 and *S. zooepidemicus* DNA quantities using combinations of the five significant independent factors (sampling moment, BCS, behavior, nasal discharge presence, and coughing). Categorical regression was performed using the Optimal Scaling procedure from the Regression task from SPSS Statistics for Windows, Version 25.0, IBM Corp. [32].

2.7. Ethical Approval

This study was approved by the University of California Davis' Animal Care and Use Committee IACUC #20611.

3. Results

Parametric assumptions were tested to decide whether a parametric or nonparametric approach should be implemented. Homoscedasticity (Levene's test for equality of error variance, $p < 0.05$) assumption for DNA quantity across the different categories for all the factors studied (respiratory, ocular, behavioral, ambulatory clinical signs, and SAA maximal approximate concentration) was violated. Normality was assumed (Shapiro–Wilk's Francia W, $p > 0.001$) as our sample was obtained out of a presumably normal population, hence a partially overlapping samples t-test with Welch's degrees of freedom was carried out to determine the differences in the means across factor levels for each of the pathogens. Summaries of the descriptive statistics for both sampling moments one and two are reported in Tables 2 and 3, respectively.

The summary of the results for the partially overlapping samples t-test to determine the differences in the means in levels for clinical signs, the SAA maximal concentration expressed in mg/L, qPCR microbial load detection, and controls between sampling moment one and two are shown in Table S1. Significant differences across the levels for BCS, behavior signs, nasal discharge, and coughing presence were found ($p < 0.05$) between sampling moments one and two (Table S1). In these regards, for pathogens quantification and controls between sampling moments one and two, significant differences were found for AHV-2 aka EHV7, and *Streptococcus equi* subspecies *zooepidemicus*.

Table 2. Summary of descriptive statistics for clinical examination signs, Serum Amyloid A (SAA) maximal concentration expressed in mg/L, qPCR-assay microbe loads, and controls at sampling moment one.

	Parameters	Mean	SEM	SD	Skewness	Kurtosis
Clinical Examination Signs	Body condition score (BCS)	3.12	0.09	0.83	−0.15	−0.96
	Behavior signs	2.11	0.04	0.38	1.14	3.13
	Skin/hair condition	1.29	0.10	0.94	3.21	9.14
	Lameness presence	1.96	0.02	0.19	−5.13	24.88
	Ocular discharge presence	1.82	0.04	0.38	−1.73	1.01
	Nasal discharge presence	1.66	0.05	0.48	−0.68	−1.57
	Abnormal breathing presence	1.07	0.03	0.26	3.41	9.88
	Coughing presence	1.93	0.03	0.26	−3.41	9.88
SAA	SAA maximal concentration (mg/L)	10.80	3.72	34.27	5.42	33.75
Pathogen Load qPCR-Assay	Asinine Herpesvirus 2 (AHV-2) aka EHV7	39.69	0.14	1.30	−4.33	18.30
	Asinine Herpesvirus 3 (AHV-3) aka EHV8	39.74	0.14	1.30	−5.92	36.83
	Asinine Herpesvirus 5 (AHV-5)	34.39	0.50	4.59	−0.51	−0.38
	Equine Herpesvirus 1 (EHV-1)	40.00	0.00	0.00	N/A	N/A
	Equine Herpesvirus 1 (EHV-1) neuropathogenic	40.00	0.00	0.00	N/A	N/A
	Equine Herpesvirus 1 (EHV-1) non-neuropathogenic	40.00	0.00	0.00	N/A	N/A
	Equine herpesvirus 4 (EHV-4)	40.00	0.00	0.00	N/A	N/A
	Streptococcus equi subspecies equi	39.94	0.04	0.39	−6.77	46.45
	Streptococcus equi subspecies *zooepidemicus*	35.04	0.44	4.08	-0.75	0.07
	Influenza AH3N8	40.00	0.00	0.00	N/A	N/A
	Equine rhinitis A virus	40.00	0.00	0.00	N/A	N/A
	Equine rhinitis B virus	40.00	0.00	0.00	N/A	N/A
Controls	Glyceraldehyde 3 phosphate dehydrogenase (First control)	29.24	0.38	3.49	0.83	1.91
	Glyceraldehyde 3 phosphate dehydrogenase (Second control)	32.13	0.41	3.78	0.27	−0.87
	Fibrinogen	0.44	0.09	0.50	4.56	23.39

SEM: Standard error of the mean; N/A: Not applicable as all animals reported the same values. Skewness Standard Error was 0.26 and Kurtosis Standard Error was 0.52 for all variables except for Fibrinogen, whose values were 0.41 and 0.81, respectively. BCS (1–5), Behavior signs (1 alert, 2 apathetic, 3 lethargic, 4 aggressive, 5 fearful, 6 withdrawn), Lesions (1 present, 2 not present), Lameness (1 present, 2 not present), Ocular discharge (1 present, 2 not present), Nasal discharge (1 present, 2 not present), Breathing (1 normal, 2 abnormal), coughing (1 present, 2 not present), SAA maximal concentration expressed in mg/L, qPCR-assay pathogen loads (40 = zero viral or bacterial agents present; lower numbers indicate higher number of DNA found), and controls at sampling moment one in 53 donkeys including jacks, jennies and foals.

The BCS reported a 0.69 higher score in sampling moment one compared to sampling moment two. Behavior signs reported a 0.15 higher score, which translated into a higher trend to display alert or fearful responses; nasal discharge reported a significant increased higher frequency of 0.24 in sampling moment one, while this increased higher trend was reported for coughing, which was more frequent in sampling moment one than two (0.80 value for estimate difference in the means, as shown in Table S1). When the qPCR results were compared from the initial sampling to the final sampling, there was a significant increase in the mean of 1.66 and 2.00, for AHV-2 aka EHV7 and Streptococcus *equi* subspecies *zooepidemicus*, respectively.

Afterward, a CATREG (Categorical Regression)was performed using five independent significant factors in Table S1 (sampling moment, body condition score, behavior, nasal discharge presence, and coughing) as predictors and AHV-2 and *Streptococcus zooepidemicus* qPCR findings as the dependent variables. Table S2 reports a model summary of stepwise linear regression for AHV-2 and *Streptococcus equi* subspecies *zooepidemicus* qPCR-assay. The standardized coefficients (β) for predictive factors are listed in Table S3 and were used to build predictive equations.

Table 3. Summary of descriptive statistics for clinical examination signs, SAA maximal concentration expressed in mg/L, qPCR-assay for organisms and controls at sampling moment two.

	Parameters	Mean	SEM	SD	Skewness	Kurtosis
Clinical Examination Signs	Body condition score	2.43	0.08	0.37	0.97	1.81
	Behavior signs	1.96	0.07	0.36	−0.65	6.34
	Skin/Hair condition	1.00	0.00	0.00	N/A	N/A
	Lameness presence	2.00	0.00	0.00	N/A	N/A
	Ocular discharge presence	1.87	0.07	0.34	−2.42	4.21
	Nasal discharge presence	1.42	0.10	0.50	0.36	−2.05
	Abnormal breathing presence	1.13	0.07	0.34	2.42	4.21
	Coughing presence	1.13	0.07	0.34	2.42	4.21
SAA	SAA maximal concentration (mg/L)	17.42	10.73	52.58	4.43	20.53
Pathogen Load qPCR-Assay	Asinine Herpesvirus 2 (AHV-2) aka EHV7	38.03	0.90	4.40	−2.57	6.18
	Asinine Herpesvirus 3 (AHV-3) aka EHV8	39.94	0.06	0.31	−4.90	24.00
	Asinine Herpesvirus 5 (AHV-5)	32.65	0.89	4.36	−1.37	2.84
	Equine Herpesvirus 1 (EHV-1)	40.00	0.00	0.00	N/A	N/A
	Equine Herpesvirus 1 (EHV-1) neuropathogenic	40.00	0.00	0.00	N/A	N/A
	Equine Herpesvirus 1 (EHV-1) non-neuropathogenic	40.00	0.00	0.00	N/A	N/A
	Equine Herpesvirus 4 (EHV-4)	40.00	0.00	0.00	N/A	N/A
	Streptococcus equi subspecies equi	40.00	0.00	0.00	N/A	N/A
	Streptococcus equi subspecies *zooepidemicus*	33.04	0.69	3.38	-0.65	0.77
	Influenza AH3N8	40.00	0.00	0.00	N/A	N/A
	Equine rhinitis A virus	40.00	0.00	0.00	N/A	N/A
	Equine rhinitis B virus	40.00	0.00	0.00	N/A	N/A
Controls	Glyceraldehyde 3 phosphate dehydrogenase (First control)	23.61	0.97	4.76	2.04	5.02
	Glyceraldehyde 3 phosphate dehydrogenase (Second control)	29.37	0.85	4.17	1.45	1.75
	Fibrinogen	0.63	0.54	2.67	4.73	22.74

N/A: Not applicable as all animals reported the same values. Skewness Standard Error was 0.47 and Kurtosis Standard Error was 0.92 for all variables. BCS (1-5), Behavior signs Behavior signs (1 alert, 2 apathetic, 3 lethargic, 4 aggressive, 5 fearful, 6 withdrawn), Lesions (1 present, 2 not present), Lameness (1 present, 2 not present), Ocular discharge (1 present, 2 not present), Nasal discharge (1 present, 2 not present), Breathing (1 normal, 2 abnormal), coughing (1 present, 2 not present), SAA maximal concentration expressed in mg/L, qPCR-assay microbe loads (40 = zero viral or bacterial present; lower numbers indicate higher number of DNA found), and controls at sampling moment two in 30 donkeys (28 jennies and 4 foals) in long-term holding facilities.

The CATREG reported that none of the factors was linearly related with AHV-2 load. However, *Streptococcus equi* subspecies *zooepidemicus* significantly depended on the independent variables of behavior, coughing presence, and sampling moment. Concretely, the CATREG reported the model predicting for AHV-2, explained 20.5% of the variability in qPCR quantifications for such pathogen (Table S3). Considering the value of the bootstrap (1000) estimate of standard error, we could identify a nonlinear monotonic relationship between AHV-2 qPCR pathogen load and nasal discharge presence.

The model predicting for *Streptococcus equi* subspecies *zooepidemicus* was able to capture 15.2% of the variability in qPCR results (Table S3). In this case, and considering the value of the bootstrap (1000) estimate of standard error, we could identify a significant linear monotonic relationship between *Streptococcus equi* subspecies *zooepidemicus* PCR quantities and the behavior, coughing and sampling moment. The resulting predictive regression equation for *Streptococcus equi* subspecies *zooepidemicus* qPCR detection is as follows:

$$Zy'_{\text{Streptococcus equi subspecies zooepidemicus}} = \beta_{\text{behavior}} * Z_{\text{behavior}} + \beta_{\text{coughing}} * Z_{\text{coughing}} + \beta_{\text{samplingmoment}} * Z_{\text{samplingmoment}} \quad (1)$$

where $Zy'_{\text{Streptococcus equi subspecies zooepidemicus}}$ is the phenotypic record of *Streptococcus equi* subspecies *zooepidemicus* PCR quantities, β is the standardized coefficient for each independent factor as marked by a sub-index for the whole population, and Z is the specific value for that same factor for each individual. β is the standardized coefficients and can be found in Table S3. Only significant coefficients were considered for the regression equation as they identified the variable which holds a linear relationship with the amount of DNA tested by qPCR. The analysis of Pearson correlations reported a significant linear correlation between age and the SAA maximal approximate concentration of 0.561 ($p < 0.001$), while the rest of the factors seemed not to be correlated to SAA maximal approximate concentration expressed in mg/L.

4. Discussion

Donkeys in general are stoic animals that often do not respond in the same manner to disease as horses. When working with feral donkey and attempting to diagnose the health status without diagnostic tools or the ability to fully perform clinical exams, the methodology used in this study may provide an adequate approach to identifying healthy and/or sick animals. Donkeys gathered and relocated to short-term and or long-term facilities will likely go through increased stress. During the relocation process to new environments, incoming donkeys co-mingling with many animals for multiple locations may further expose donkeys to pathogens. This exposure in a stressed state may cause disease outbreaks [35]. Jennies relocated to a long-term holding facility and co-mingled with donkeys from multiple locations showed an increase in AHV-2 and *Streptococcus zooepidemicus*. Equine herpesviruses (EHVs) can affect all members from the *Equidae* family by establishing acute [36] and latent infections [37]; hence, transmission does not face any totally effective preventive interspecific barrier. Over an animal's lifetime, latent herpes infections can reoccur or new acute infections can occur [36].

Generally, few clinical signs of infection or suggested guidelines for prevention and control in donkeys have been made [36] for Asinine Herpesviruses. However, one study suggested that donkeys subjected to stress (e.g., relocation or transport) are at higher risk of the recrudescence of latent infections [38]. The use of such stable side tests like SAA may assist in detecting such disease at an early stage. In this study, we found an increase in SAA in young donkeys less than a year of age along with an increase in *Streptococcus zooepidemicus* and AHV-2 after being moved to long-term holding.

Our findings also suggested that nasal discharge presence was the most relevant sign to be considered given its direct relationship with an increased AHV-2 pathogen DNA quantity. The present information provides further insight into the recognition of AHV-2 in previously undiagnosed donkeys. AHV-2 (EHV-7) and other barely characterized gamma-herpesviruses have been isolated from asymptomatic healthy equids (horses, donkeys and mules), and from donkeys presenting encephalitis or severe interstitial pneumonia conditions [39].

The detection of AHV-5 from respiratory fluids from nasal swabs of horses that did not present any clinical sign of respiratory disease has been described [40,41]. Similarly, the asymptomatic donkeys in this study could test positive for AHV-5. Presumably, they have similar infection and latency patterns as those reported for EHV-2 and EHV-5 [41]. Goehring [42] reported EHV-2 to be present in the respiratory fluids and PBMCs (peripheral blood mononuclear cells) of the majority of foals from birth to weaning. These findings were in conjunction with moderate clinical disease symptoms such as fever, nasal discharge, pharyngeal follicular hyperplasia, and mandibular lymphadenopathy. These symptoms may agree with our findings, which address nasal discharge as the most relevant sign to be found in cases of EHV-2 infection. Similarly, when both EHV-2 and EHV-5 were found, the disease presented a rather severe course, as also suggested by our positive results.

In this group of animals, *Streptococcus equi* ssp. *zooepidemicus* genome quantities were latent for a longer time, potentially making this commensal bacterium more prone to the active transmission when feral animals are introduced with other equids. Coughing has been reported to be the most relevant sign and to be linearly correlated to *Streptococcus equi* ssp. *zooepidemicus* DNA quantity. The increased levels of coughing as bacterial loads increase may be also the basis for greater discomfort, which translates into the animal adopting a more alert or fearful status as reported by our study, while appetite appeared unaffected. *Streptococcus equi* ssp. *zooepidemicus* has been considered an opportunistic commensal in horses, though it has also been identified in the guttural pouches of systemically ill donkeys with empyema and chondroids [43]. This may suggest the role of *Streptococcus zooepidemicus* as a primary pathogen in donkeys. Contextually, our results suggest that this bacterium may be more important when recently feral donkeys are introduced to domestic equines.

Some authors [44] have even provided evidence supporting the fact that some highly frequent beta-hemolytic streptococci other than *S. equi*, such as *S. equisimilis* and *S. zooepidemicus*, could be the causative agent of strangle-like disease also associating its presence with other epidemic upper respiratory diseases. This contradicts Timoney [10], who reported that the related species *S. equi* is more virulent and much more likely to cause lymphadenopathy and abscessation ("strangles") [10].

Donkeys and mules may be more susceptible to strangles with *S. zooepidemicus* acting as a primary pathogen, with a variable clinical presentation ranging from fever, purulent nasal discharge, and the abscessation of mandibular and retropharyngeal lymph nodes to mild persistent nasal discharge, as supported by our results. In this context, the first description of a strangles outbreak in donkey intensive farming systems was described in China in 2018 [45].

5. Conclusions

In conclusion, captured feral donkeys harbor some asinine viruses and *Streptococcus zooepidemicus* but not pathogens that might potentially infect other equine species. The comingling of wild-caught feral donkeys with domestic animals in a holding facility increases the levels of detectible *Streptococcus zooepidemicus* and AHV- 2 in the population, suggesting that captured wild donkeys may be at risk in holding facilities and during transport. Nasal discharge and coughing may be symptoms of the presence of potentially pathogenic organisms such as *Streptococcus zooepidemicus* and AHV- 2. The use of diagnostic testing tools such as stable side SAA kits and testing for pathogens through PCR nasal swab samples along with basic clinical exams may aide in the early detection of the disease and overall improve the health of donkeys. This study has contributed to our knowledge about SAA levels in healthy donkeys but additional research measuring SAA in clinically ill animals would be beneficial.

Supplementary Materials: The following are available online at http://www.mdpi.com/2076-2615/10/6/1086/s1. Table S1. Summary of the results for the partially overlapping samples t-test with Welch's degrees of freedom to determine differences in the means in categories within clinical examination signs, SAA maximal concentration expressed in mg/L, microbial loads qPCR-assay and controls between sampling moment one and two. Table S2. Model summary of stepwise linear regression for Asinine Herpesvirus (AHV-2) and *Streptococcus equi* subspecies *zooepidemicus* qPCR-assay. Table S3. Standardized coefficients for factors presenting a significant relationship with Asinine Herpesvirus (AHV-2) and Streptococcus *equi* subspecies *zooepidemicus* q-PCR microbial loads and significance within each predictive CATREG model.

Author Contributions: Conceptualization, E.D., N.P. and A.M.; Data curation, F.J.N.G. and C.I.P.; Formal analysis, F.J.N.G.; Funding acquisition, E.D. and A.M.; Investigation, S.J., S.B., E.M.A. and A.M.; Methodology, S.J., S.M., F.J.N.G., E.M.A., C.I.P. and A.M.; Project administration, E.D., N.P. and A.M.; Resources, E.D., N.P. and A.M.; Software, F.J.N.G. and C.I.P.; Supervision, E.D., N.P. and A.M.; Validation, F.J.N.G., E.D. and A.M.; Visualization, A.M.; Writing–original draft, F.J.N.G. and A.M.; Writing–review & editing, F.J.N.G., E.D., N.P. and A.M. All authors have read and agreed to the published version of the manuscript.

Funding: This research received no external funding.

Acknowledgments: Authors would like to thank Peaceful Valley Donkey Rescue, Mark Myers and staff members, Kyle and Amanda Nealey, Eagle Eye Donkey Rescue, Susie Reed, and Victor Urbala for donating their donations, time and services. Thanks to Henrich Anhold and Stable Lab for the StableLab kit donation. Special thanks to Lauren Brizgys for her contribution in the sampling.

Conflicts of Interest: The authors declare no conflict of interest.

References

1. Tiller, B. *Feral Burro Populations: Distribution and Damage Assessment*; Pacific Northwest National Laboratory: Richland, WA, USA, 1997.
2. Weaver, R.A. Feral burros and wildlife. In Proceedings of the 6th Vertebrate Pest Conference, Anaheim, CA, USA, 5–7 March 1974.
3. National Research Council. *Using Science to Improve the BLM Wild Horse and Burro Program: A Way Forward*; National Academies Press: Washington, DC, USA, 2013.

4. National Research Council/Commission on Natural Resources/National Research Council; Committee on Wild, Free-Roaming Horses Burros. *Wild and Free-Roaming Horses and Burros: Current Knowledge and Recommended Research*; National Academies Press: Washington, DC, USA, 1980.
5. Satué, K.; Calvo, A.; Gardón, J.C. Factors influencing serum amyloid type A (SAA) concentrations in horses. *Open. J. Vet. Med.* **2013**, *3*, 58. [CrossRef]
6. Nolen-Walston, R. How to interpret serum amyloid A concentrations. In Proceedings of the 65th Annual American Association of Equine Practitioners (AAEP), Denver, CO, USA, 7–11 December 2019; pp. 130–137.
7. Kay, G.; Tligui, N.; Semmate, N.; Azrib, R.; González, F.J.N.; Brizgys, L.; McLean, A. Determining factors and interspecific modeling for serum amyloid a concentrations in working horses, donkeys, and mules. *Res. Vet. Sci.* **2019**, *125*, 256–265. [CrossRef] [PubMed]
8. Jacobsen, S.; Andersen, P.H. The acute phase protein serum amyloid A (SAA) as a marker of inflammation in horses. *Equine Vet. Educ.* **2007**, *19*, 38–46. [CrossRef]
9. Pusterla, N.; Kass, P.H.; Mapes, S.; Johnson, C.; Barnett, D.; Vaala, W.; Gutierrez, C.; McDaniel, R.; Whitehead, B.; Manning, J. Surveillance programme for important equine infectious respiratory pathogens in the USA. *Vet. Rec.* **2011**, *169*, 12. [CrossRef] [PubMed]
10. Timoney, J.F. The pathogenic equine streptococci. *Vet. Res.* **2004**, *35*, 397–409. [CrossRef]
11. Wilson, W.D. Equine herpesvirus 1 myeloencephalopathy. *Vet. Clin. North Am. Equine Pract.* **1997**, *13*, 53–72. [CrossRef]
12. Van Maanen, C. Equine herpesvirus 1 and 4 infections: An update. *Vet. Q.* **2002**, *24*, 57–78. [CrossRef]
13. Allen, G.; Kydd, J.; Slater, J.; Smith, K. Recent advances in understanding the pathogenesis, epidemiology, and immunological control of equid herpesvirus-1 (EHV-1) abortion. Proceedings of the 8th International Conference on Equine Infectious Diseases. *J. Equine Vet. Sci.* **1999**, *8*, 129–146.
14. Reed, S.M.; Toribio, R.E. Equine herpesvirus 1 and 4. *Vet. Clin. North Am. Equine Pract.* **2004**, *20*, 631–642. [CrossRef]
15. Ostlund, E.N. The equine herpesviruses. *Vet. Clin. North Am. Equine Pract.* **1993**, *9*, 283–294. [CrossRef]
16. Fortier, G.; Van Erck, E.; Pronost, S.; Lekeux, P.; Thiry, E. Equine gammaherpesviruses: Pathogenesis, epidemiology and diagnosis. *Vet. J.* **2010**, *186*, 148–156. [CrossRef] [PubMed]
17. Ficorilli, N.; Studdert, M.; Crabb, B. The nucleotide sequence of Asinine Herpesvirus 3 glycoprotein G indicates that the donkey virus is closely related to equine herpesvirus 1. *Arch. Virol.* **1995**, *140*, 1653–1662. [CrossRef] [PubMed]
18. Bell, S.A.; Pusterla, N.; Balasuriya, U.B.; Mapes, S.M.; Nyberg, N.L.; Maclachlan, N.J. Isolation of a gammaherpesvirus similar to asinine herpesvirus-2 (AHV-2) from a mule and a survey of mules and donkeys for AHV-2 infection by real-time PCR. *Vet. Microbiol.* **2008**, *130*, 176–183. [CrossRef]
19. Kleiboeker, S.M.; Schommer, S.K.; Johnson, P.J.; Ehlers, B.; Turnquist, S.E.; Boucher, M.; Kreeger, J.M. Association of two newly recognized herpesviruses with interstitial pneumonia in donkeys *(Equus asinus)*. *J. Vet. Diagn. Investig.* **2002**, *14*, 273–280. [CrossRef] [PubMed]
20. Kleiboeker, S.M.; Turnquist, S.E.; Johnson, P.J.; Kreeger, J.M. Detection and nucleotide sequencing of a DNA-packaging protein gene of equine gammaherpesviruses. *J. Vet. Diagn. Investig.* **2004**, *16*, 67–74. [CrossRef] [PubMed]
21. Thiemann, A. Respiratory disease in the donkey. *Equine Vet. Educ.* **2012**, *24*, 469–478. [CrossRef]
22. Vengust, M.; Wen, X.; Bienzle, D. Herpesvirus-associated neurological disease in a donkey. *J. Vet. Diagn. Investig.* **2008**, *20*, 820–823. [CrossRef]
23. Browning, G.; Ficorilli, N.; Studdert, M. Asinine Herpesvirus genomes: Comparison with those of the equine herpesviruses. *Arch. Virol.* **1988**, *101*, 183–190. [CrossRef]
24. Goodrich, E.L.; Behling-Kelly, E. Clinical Pathology of Donkeys and Mules. *Vet. Clin. North Am. Equine Pract.* **2019**, *35*, 433–455. [CrossRef]
25. Valle, E.; Raspa, F.; Giribaldi, M.; Barbero, R.; Bergagna, S.; Antoniazzi, S.; Mc Lean, A.K.; Minero, M.; Cavallarin, L. A functional approach to the body condition assessment of lactating donkeys as a tool for welfare evaluation. *PeerJ* **2017**, *5*, e3001. [CrossRef]
26. Polidori, P.; Vincenzetti, S. Chapter 4. Farm management and feeding strategies for donkey milk production. In *Agricultural Research Updates*; Mandhatri, P.G.S., Ed.; Nova Science Publishers, Inc.: Hauppauge, NY, USA, 2017; p. 93.

27. Navas González, F.J.; Jordana Vidal, J.; León Jurado, J.M.; Arando Arbulu, A.; McLean, A.K.; Delgado Bermejo, J.V. Genetic parameter and breeding value estimation of donkeys' problem-focused coping styles. *Behav. Processes* **2018**, *153*, 66–76. [CrossRef] [PubMed]
28. Veterinary Medicine PCR Laboratory, UCDavis. q-PCR diagnostic submission packet. Available online: https://pcrlab.vetmed.ucdavis.edu/sites/g/files/dgvnsk6571/files/inline-files/DiagnosticPacket08.2019.pdf (accessed on 15 March 2020).
29. Derrick, B.; Toher, D.; White, P. How to compare the means of two samples that include paired observations and independent observations: A companion to Derrick, Russ, Toher and White (2017). *Tutor. Quant. Methods Psychol.* **2017**, *13*, 120–126. [CrossRef]
30. Moder, K. Alternatives to F-test in one way ANOVA in case of heterogeneity of variances (a simulation study). *Psychol. Test Assess. Model.* **2010**, *52*, 343–353.
31. Rasch, D.; Kubinger, K.D.; Moder, K. The two-sample t test: Pre-testing its assumptions does not pay off. *Stat. Pap.* **2011**, *52*, 219–231. [CrossRef]
32. IBM Corp. *IBM SPSS Statistics for Windows*; 25.0; IBM Corp: Armonk, NY, USA, 2017.
33. Derrick, B. The Partially Overlapping Samples t-test. Package 'Partiallyoverlapping' for RStudio. Available online: https://cran.r-project.org/web/packages/Partiallyoverlapping/index.html (accessed on 4 May 2020).
34. RStudio Team. *RStudio 1.1.463*; RStudio: Integrated Development for R. RStudio, Inc.: Boston, MA, USA, 2016.
35. Oliveira, F.G.; Cook, R.F.; Naves, J.H.F.; Oliveira, C.H.S.; Diniz, R.S.; Freitas, F.J.C.; Lima, J.M.; Sakamoto, S.M.; Leite, R.C.; Issel, C.J.; et al. Equine infectious anemia prevalence in feral donkeys from Northeast Brazil. *Prev. Vet. Med.* **2017**, *140*, 30–37. [CrossRef]
36. Hartley, C.A.; Dynon, K.J.; Mekuria, Z.H.; El-Hage, C.M.; Holloway, S.A.; Gilkerson, J.R. Equine gammaherpesviruses: Perfect parasites? *Vet. Microbiol.* **2013**, *167*, 86–92. [CrossRef]
37. Barrandeguy, M.E.; Carossino, M. Infectious diseases in donkeys and mules: An overview and update. *J. Equine Vet. Sci.* **2018**, *65*, 98–105. [CrossRef]
38. Edington, N.; Bridges, C.; Huckle, A. Experimental reactivation of equid herpesvirus 1 (EHV 1) following the administration of corticosteroids. *Equine Vet. J.* **1985**, *17*, 369–372. [CrossRef]
39. Osterrieder, K. Herpesvirales. In *Fenner's Veterinary Virology*, 5th ed.; MacLachlan, N.J., Dubovi, E.J., Eds.; Academic Press: Boston, MA, USA, 2015; pp. 189–216.
40. Rushton, J.O.; Kolodziejek, J.; Nell, B.; Nowotny, N. Prevalence of Asinine Herpesvirus type 5 (AsHV-5) infection in clinically normal Lipizzaner horses. *Vet. J.* **2014**, *200*, 200–203. [CrossRef]
41. Fortier, G.; Van Erck, E.; Fortier, C.; Richard, E.; Pottier, D.; Pronost, S.; Miszczak, F.; Thiry, E.; Lekeux, P. Herpesviruses in respiratory liquids of horses: Putative implication in airway inflammation and association with cytological features. *Vet. Microbiol.* **2009**, *139*, 34–41. [CrossRef]
42. Goehring, L.S. Donkeys. In *Robinson's Current Therapy in Equine Medicine*, 1st ed.; Sprayberry, K.A., Robinson, N.E., Eds.; Elsevier: Philadelphia, PA, USA, 2015; pp. 155–157.
43. Rickards, K.J.; Thiemann, A.K. Respiratory Disorders of the Donkey. *Vet. Clin. N. Am. Equine Pract.* **2019**, *35*, 561–573. [CrossRef] [PubMed]
44. Laus, F.; Preziuso, S.; Spaterna, A.; Beribe, F.; Tesei, B.; Cuteri, V. Clinical and epidemiological investigation of chronic upper respiratory diseases caused by beta-haemolytic Streptococci in horses. *Comp. Immunol. Microbiol. Infect. Dis.* **2007**, *30*, 247–260. [CrossRef] [PubMed]
45. Dong, J.; Gao, N.; Waller, A.; Cook, F.; Fan, S.; Yuan, D.; Du, Y.; Li, F.; Norimine, J.; Zhu, W. An outbreak of strangles associated with a novel genotype of Streptococcus equi subspecies equi in donkeys in China during 2018. *Equine Vet. J.* **2019**, *51*, 743–748. [CrossRef] [PubMed]

animals

MDPI

Article

Morphometric Characteristics of the Skull in Horses and Donkeys—A Pilot Study

Katrina Merkies [1,2,*], Georgios Paraschou [3] and Paul Damien McGreevy [4]

1 Department of Animal Biosciences, University of Guelph, Guelph, N1G 2W1 ON, Canada
2 Campbell Centre for the Study of Animal Welfare, University of Guelph, Guelph, N1G 2W1 ON, Canada
3 The Donkey Sanctuary, Sidmouth, Devon EX10 0NU, UK; georgios.paraschou@thedonkeysanctuary.org.uk
4 Sydney School of Veterinary Science, Faculty of Science, University of Sydney, Sydney, NSW 2006, Australia; paul.mcgreevy@sydney.edu.au
* Correspondence: kmerkies@uoguelph.ca

Received: 13 May 2020; Accepted: 5 June 2020; Published: 8 June 2020

check for updates

Simple Summary: Horses and donkeys are often lumped together but as separate species; they differ from each other in many ways. We examined the head characteristics of deceased horses and donkeys and measured various structures of both the skull and the brain. We calculated various ratios of skull measurements from horses and donkeys to be able to compare them to each other. We noted that donkey skulls have a larger forehead than horses. The olfactory bulb was smaller in donkeys and was rotated more forward than in horses. Most strikingly, we noted that the hair whorl on the forehead of horses almost always corresponded with the location of the olfactory bulbs, but in donkeys, the hair whorl was located much further down the nose. While it is unclear why these differences exist, it may relate to some of the striking differences in behaviour and physiology that have also been noted between donkeys and horses.

Abstract: Horses and donkeys belong to the genus Equus, but important differences exist between the species, many of which affect their management and welfare. This study compared skull morphology between horses and donkeys. Horse (n = 14) and donkey (n = 16) heads were obtained post-mortem, sectioned sagittally close to the midline, and photographed for subsequent measurement of various skull structures. Skull, cranial, nasal, and profile indices were calculated for topographical comparisons between the species. The olfactory bulb area (OBA), OB pitch (the angle between the hard palate and the OB axis), and whorl location (WL) were also measured. A General Linear Model determined the main effect of species with Sidak's multiple comparisons of species' differences among the various measurements. There was no species difference in cranial or nasal indices ($p > 0.13$), but donkeys had a larger cranial profile than horses ($p < 0.04$). Donkeys had a smaller OBA ($p < 0.05$) and a steeper OB pitch ($p < 0.02$) than horses. The WL corresponded to the level of the OB in horses but was extremely rostral in donkeys ($p < 0.0001$). These results show clear differentiation in skull morphology between horses and donkeys. This may be useful in validating other physiological and behavioural differences between horses and donkeys.

Keywords: olfactory bulb; skull morphology; breed differences; hair whorl

1. Introduction

Since equids were domesticated, over 5000 years ago, humans have given them diverse roles ranging from work and transportation to sport and leisure. The selection of breeding animals has focused on specific traits such as colour, conformation, behaviour, and personality. Over time, such selection pressures have given rise to a wide variety of breeds with highly diverse physical characteristics from bulky draft horses to fine-boned Arabians, to miniature horses, to donkeys and mules.

Equine intraspecific diversity is evident in size, shape, colouration, conformation and, behavioural suitability for a particular purpose. Manifestations of selection pressures in various breeds have been documented in reports that focus on colour [1], genetics [2], anatomy [3], physiology [4], reproduction [5], trainability [6], personality [7], and behaviour [8]. Even within a breed, conformational differences are evident depending on the geographical region of origin [9,10].

Breed types can be broadly categorised according to their typical conformational characteristics based on genetic heritage. Cold-blooded horses refer to those from draft lineage, generally exhibiting a so-called coarse head, short neck, broad chest, and muscular body [9,11] compared to hot-blooded horses who descend mainly from Thoroughbred and Arabian lines. Thoroughbreds demonstrate longer withers, and Arabians have more sloping shoulders than draft horses [12]. Hot-blooded horses (e.g., Arabians) have a more concave head profile, while warm- to cold-blooded horses (e.g., Standardbreds) have relatively more convex profiles and longer skulls [13]. Warmbloods, originating from agricultural stock, changed their breeding goals over time to a lighter sport horse type, presenting a conformational mix between hot- and cold-bloods [14].

Breed types also show specific variations in behaviour and temperament. The Standardbred, as an example of a cold-blooded-type horse, is less reactive and quicker to habituate to aversive stimuli than more refined breeds, such as Thoroughbreds [7]. Cold-blooded horses are easier to handle [15] while Thoroughbreds are said to be more nervous [16], and take longer to learn a training task than other breeds [17]. Breeders of the Austrian Noriker draft horse have a firm belief in the effect of coat colour on temperament [9], for example, black Norikers are considered more lively.

Horses and donkeys belong to the same family of Equidae. However, within the genus Equus, there are clear variances in anatomical structure and behaviour of the various species. While donkeys and Standardbred racehorses both typically exhibit convex nasal profiles, the physiology and behaviour of these two equids is markedly different. Donkeys, evolving from arid climates, are more affected by intemperate weather than horses, but are less reactive to biting insects [18]. Donkeys can thrive on lower quality forage than horses, but this also puts them at risk for metabolic diseases when donkey owners feed them like horses [19–21]. While horses show more obvious behavioural responses to pain than donkeys [22,23], both donkeys and mules show more metabolic changes indicative of stress than horses when subjected to transportation and mixing at auction houses [24].

It has been suggested that behavioural differences among dog breeds may be predicted by skull size and shape [25–27]. Thus, phenotypic and behavioural dissimilarities among different horse breeds may also be reflected through skull morphology. The objective of this preliminary research is to determine differences in skull morphology between the Standardbred (as a cold-blooded representative) and donkeys, as a starting point for further investigation into the brain organisation that may underlie behavioural differences.

2. Materials and Methods

2.1. Subjects and Head Preparation

No approval for the use of animals in research was obtained as the research dealt only with material from deceased subjects. All horses and donkeys used in this research were euthanized for reasons unrelated to the research project. Heads of mature Standardbred-type horses (n = 14) and mature donkeys (n = 16) were obtained post-mortem. Donkeys ranged in size from standard to so-called mammoth. Ages of the horses were estimated from an assessment of their dentition while ages of the donkeys were known. The estimated age of the horses ranged from 3–16 years (mean 7.5 ± 4.5 years). The age of the donkeys ranged from 3–38 years (mean 21 ± 10 years). The skulls were separated from the body by cutting through the proximal end of the first cervical vertebrae using a band-saw. Each skull was sectioned sagittally as close to the midline as possible, and each section was weighed.

No other preparation was performed on the skulls. Coloured pins were inserted to mark topographical landmarks (occipital crest, olfactory bulb, orbital notch, incisive bone), and a ruler was placed along the hard palate to reference a horizontal dimension and provide a known length for measurement calibration. Photographs of the sectioned skulls were taken (Canon Powershot SD940 IS 12.1MP; Ottawa, ON, Canada) for retrospective measurement of various structures using the open-source software Image J (v1.46r; Madison, WI, USA).

2.2. Skull Measurements

Skull measurements included: cranial length (measured from the occipital crest to the frontonasal suture line), nasal length (measured from the rostral point of the incisive bone to the frontonasal suture line), skull length (sum of cranial length + nasal length), zygomatic width (sum of left and right sections measured from the zygomatic bone to the midline), cranial width (sum of left and right sections measured from the orbital notch to the midline along the frontonasal suture line) (Figure 1A), mandibular depth (measured from the orbital notch to the widest part of the mandible; Figure 1B), and whorl location (described by the facial distance between the rostral tip of the olfactory bulb (OB) and the centre of the forehead whorl). The frontonasal suture line was estimated by a line perpendicular to the orbital notch. Brain measurements included the OB pitch (the angle between the hard palate and the OB's longitudinal axis), the brain pitch (the angle between the hard palate and the longitudinal axis of the cerebral hemispheres) (Figure 2), and the OB area estimated from the right olfactory bulb in situ. For skull length, nasal length, cranial length, mandibular depth, OB pitch, and brain pitch, averages were calculated from the measurements from the left and right skull sections. Raw data are available in Table S1.

From these measurements, the following indices were calculated as described by Evans and McGreevy [13], using the relevant averages from the above measurements:

$$\text{skull index (SI)} = \text{zygomatic width} \times 100/\text{skull length}$$

$$\text{cranial index (CI)} = \text{cranial width} \times 100/\text{cranial length}$$

$$\text{nasal index (NI)} = \text{zygomatic width} \times 100/\text{nasal length}$$

$$\text{mandibular index (MI)} = \text{mandibular depth} \times 100/\text{skull length}$$

nasal profile area (NPA) = rectangular area defined by 80 mm vertical line from the orbital notch to the rostral point of incisive bone (Figure 1B)

$$\text{nasal profile index (NPI)} = \text{NPA}/\text{nasal length}$$

cranial profile area = rectangular area defined by 80 mm vertical line from the orbital notch to the occipital crest (Figure 1B)

$$\text{cranial profile index (CPI)} = \text{CPA}/\text{cranial length}$$

For NPA and CPA, averages were calculated from the measurements from the left and right skull sections.

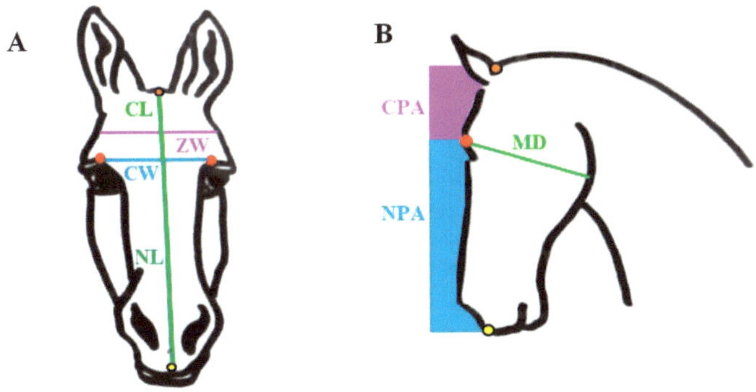

Figure 1. Diagram of the various measurements taken from horse (n = 14) and donkey (n = 16) skulls. The orbital notch is marked by the red dot. The occipital crest is marked by the orange dot, and the incisive bone is marked by the yellow dot. (**A**): The frontonasal suture line is estimated by a horizontal line joining the left and right orbital notches to determine cranial width = CW blue line. Cranial length = CL light green line; nasal length = NL dark green line; zygomatic width = ZW purple line. (**B**): mandibular depth = MD green line, nasal profile area = NPA blue area, cranial profile area = CPA purple area. See text for details of each measure.

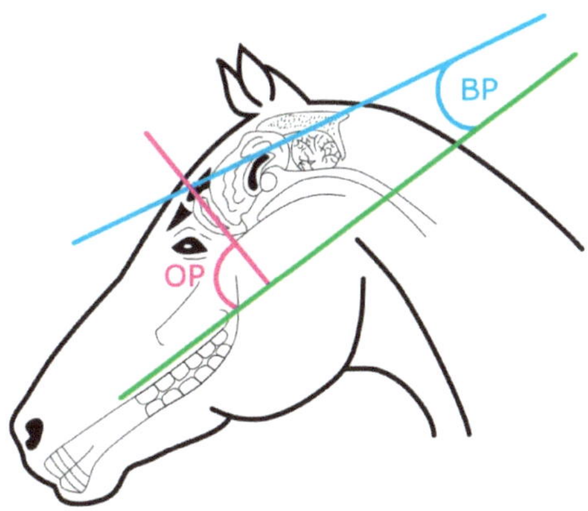

Figure 2. Diagram of brain measurements on horse (n = 14) and donkey (n = 16) skulls. The olfactory pitch (OP) was measured as the angle between the hard palate (green line) and the longitudinal axis of the olfactory bulb (purple line). The brain pitch (BP) was measured as the angle between the hard palate and the longitudinal axis of the brain (blue line).

2.3. Statistical Analyses

All brain measures and calculations were analysed using SAS (v9.4, SAS Institute, Cary, NC, USA). Pearson correlations determined any relationships between skull morphology and brain measurements. A General Linear Model determined the main effect of species with Sidak's multiple comparisons of species differences among the various measurements.

3. Results

3.1. Horses

The results of the morphometric measurements of the horse skulls are presented in Table 1. There was large variability in the measurements among individual horses. For example, NI had a mean of 76.1 ± 14.4, with measurements ranging from 63.6 to 108.3. The brain pitch and olfactory pitch also exhibited a wide range (2.3° to 176.5° and 51.4°to 75.9°, respectively). The olfactory bulb pitch showed a trend of being correlated to cranial width (X^2 (1, n = 12) = −0.55, p = 0.063). The location of the whorl aligned to the location of the olfactory bulbs within 1.5 cm. The whorl location tended to be correlated weakly to nasal length (X^2 (1, n = 14) = −0.49, p = 0.078) and brain pitch (X^2 (1, n = 14) = 0.48, p = 0.08).

Table 1. Morphometric measurements (mean ± SD) of horse (n = 12) and donkey (n = 14) skulls. *P* values were calculated using a General Linear Model, with a value <0.05 considered significantly different.

Variable	Horse	Donkey	*p*-Value
	Mean	Mean	
Head weight (kg)	22.5 ± 4.9	13.6 ± 3.7	<0.0001 *
Skull length (cm)	53.6 ± 2.6	46.6 ± 5.0	0.0080 *
Cranial length (cm)	23.6 ± 1.4	20.4 ± 2.7	0.0002 *
Nasal length (cm)	30.4 ± 3.9	27.5 ± 3.1	0.1919
Zygomatic width (cm)	22.6 ± 1.8	21.9 ± 2.2	0.0776
Cranial width (cm)	19.0 ± 0.8	16.3 ± 2.4	<0.0001 *
Mandibular depth (cm)	27.2 ± 1.1	24.5 ± 2.4	0.0019 *
Skull index	42.2 ± 3.2	46.2 ± 3.4	0.0025 *
Cranial index	80.9 ± 6.2	77.2 ± 9.1	0.1399
Nasal index	76.0 ± 14.5	79.6 ± 7.5	0.6600
Mandibular index	50.8 ± 2.6	52.2 ± 5.2	0.5791
Cranial profile index	8.8 ± 1.0	7.6 ± 1.2	0.0335 *
Nasal profile index	9.1 ± 2.8	7.3 ± 0.8	0.8359
Olfactory bulb area (cm^2)	2.3 ± 1.3	1.98 ± 1.3	0.0478 *
Olfactory bulb pitch (°)	66.0 ± 11.8	56.6 ± 1318	0.0163 *
Brain pitch (°)	18.3 ± 45.7	16.7 ± 37.8	0.4338
Whorl location (cm)	0.1 ± 1.5	8.0 ± 6.5	0.0001 *

* denotes the statistical difference between horses and donkeys at the level of $p < 0.05$.

3.2. Donkeys

The results of the morphometric measurements of the donkey skulls are presented in Table 1. A greater range in measurements was seen in the donkeys as the animals in the current sample ranged in size from standard to so-called mammoth. In particular, skull length and cranial length varied widely (40.3–54.9 cm and 16.3–24.9 cm, respectively). The location of the whorl was almost exclusively located quite rostrally. The whorl location was correlated to NPI (X^2 (1, n = 12) = 0.66, $p < 0.019$), skull length (X^2 (1, n = 14) = 0.75, $p < 0.002$) and nasal length (X^2 (1, n = 14) = 0.74, p = 0.0023).

3.3. Comparison of Measurements between Donkeys and Horses

Table 1 shows the statistical outcomes of the skull morphometrics between species. Donkeys had smaller heads than horses, as reflected in lower skull weights (F (1.18) = 26.38, $p < 0.0001$), a smaller cranial width (F (1.15) = 42.67, $p < 0.0001$), and mandibular depth (F (1.21) = 13.05, p = 0.0019). Donkeys had shorter heads than horses, mostly due to a shorter cranial length (F (1.23) = 51.49, p = 0.0002), given that there was no statistical difference in their nasal lengths (F (1.23) = 2.09, $p > 0.19$). The calculated indices provide a better comparison between the species because donkeys ranged in size much more than the horses did. There was no species difference in cranial index, nasal index, or mandibular index ($p > 0.13$), but donkeys had a larger skull index than horses (F (1.15) = 14.01, p = 0.0025). Donkeys had a smaller CPI than horses (F (1.21) = 7.54, $p < 0.034$), but there was no difference in NPI (F (1.21) = 0.05, $p > 0.83$). Donkeys also had a smaller OB area (F (1.13) = 4.96, $p < 0.05$) and a smaller OB pitch (F (1.15) = 7.60, p = 0.0163) than horses, although there was no difference in the brain pitch (F (1.23) = 0.69, $p > 0.43$). The greatest difference was seen in the location of the whorl, which corresponded to the level of the OB bulb in horses, but was located extremely rostrally in donkeys (F (1.21) = 56.28, $p < 0.0001$). Age ($p > 0.08$) and sex ($p > 0.09$) were not statistically significant for any of the measures.

4. Discussion

These preliminary results show that donkeys and Standardbred horses have similarly shaped heads, although donkeys have smaller heads and a more distinct forehead than horses. Donkeys also have smaller olfactory bulb areas than horses. While the orientation of the brain does not significantly differ between horses and donkeys, the olfactory bulbs in donkeys are rotated more rostrally. Facial whorls in donkeys are located lower down the face while in the current series of horses, they are in close proximity to the olfactory bulb. Anatomical differences between horses and donkeys have been noted in the jaws and teeth [19], but this is the first report of morphometric differences in skull morphology.

Values obtained for horse skull measurements in the current study corresponded quite closely to those previously obtained for Arabians [28], and Standardbreds [13]. However, Evans and McGreevy [13] demonstrated that Standardbreds had longer and wider skulls than Arabians or Thoroughbreds. This difference might be attributed to differing measurements since Cervantes et al. [28] did not indicate landmarks from which they conducted their head length and width measurements. The relatively large head of the Standardbred may have accentuated the differences in skull length and width between the Standardbred horses and donkeys measured in the current study. The use of indices helped to account for these variations in absolute size, and given that there was no difference between Standardbred horses and donkeys in the nasal, cranial, or mandibular indices, it appears that donkeys and Standardbred horses have similar ratio aspects. However, the skull index did differ, showing donkeys to have somewhat shorter and narrower faces than Standardbred horses. In contrast, the nasal profile index did not differ between Standardbred horses and donkeys, indicating they both exhibit a more convex profile. That said, donkeys did exhibit more convexity than Standardbred horses in the cranial profile, as is often manifested by their distinct forehead (Figure 3). Interestingly, while others have indicated a greater variability in measurements of horse skulls due to the nasal component [9,13,29,30], here we show greater variability in the cranial component between Standardbred horses and donkeys.

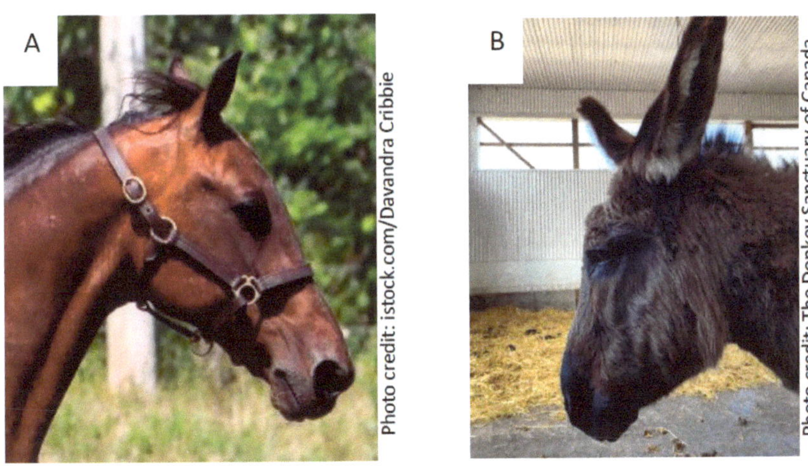

Figure 3. Comparison between the Standardbred (**A**) and donkey (**B**) head profiles. Donkeys have shorter and narrower faces than Standardbreds, but they did not differ in the nasal, cranial or, mandibular indices. While the nasal profile index also did not differ, donkeys did exhibit a smaller cranial profile index, as portrayed by the distinct forehead in the donkey.

Selective breeding in dogs has similarly resulted in great diversity in conformation across different breeds, notably seen in the range of facial features. Skull length can range from 7–28 cm [31,32]. Brachycephaly has resulted not only in a frontal rotation of the brain of domestic dogs [32] but in shifts of typical behaviour. For dog owners, this implies higher levels of fear and increased chasing behaviour in long-nosed dogs suitable for hunting, while short-nosed dogs appear to have the capacity for increased visual focus on human cues suitable for companion animals [26,27]. These behavioural differences may be the result of accompanying differences in both brain organization and retinal structure. In dogs, the concentration of retinal ganglion cells is positively correlated to the skull shape and the eye size, with a more concentrated distribution in the equivalent of the area centralis in short-nosed dogs, possibly indicating increased visual acuity in the centre of the visual field [31]. In contrast, long-skulled dogs have a visual streak with ganglion cells distributed across the retina in a band that suggests good peripheral vision. Correspondingly, horse breeds with longer heads display increased retinal ganglion cell density along with their equivalent of a visual streak [33].

Standardbred horses are predominantly bay, which may even be a contributing factor to their skull morphometrics. While the Austrian Noriker draft horse has larger absolute skull measurements than Standardbred horses due to its larger overall size, researchers have demonstrated differences in head size in that breed that are dependent on colour; bay or black horses have longer noses than tobiano Noriker horses, and roans have smaller heads [9].

The current study provides the first reported finding of an external landmark for the position of the olfactory bulbs, which are always found in close proximity to the facial whorl in horses, while in donkeys, the whorl is located extremely rostral to the olfactory bulbs. Despite having a relatively larger cranial space, donkeys; nevertheless, have smaller olfactory bulbs than Standardbred horses. Olfaction plays a key role in social interactions, such that in rodents, the removal of olfactory cues through cage cleaning causes an increase in anxiety and aggression [34]. Olfaction is the second-most represented behaviour between jennies and their new-born foals (only behind visual observation) [35], resulting in changes in the neural structure of the olfactory bulb affecting memory that drives maternal recognition and bonding critical to survival [34]. In honey bees, the olfactory bulb is responsible for identifying odours that stimulate memory to help relocate food sources [36].

Olfactory stimuli are processed in the amygdala, which is also a centre for processing emotional stimuli [37]. The amygdala, among other roles, regulates feeding behaviour through odour intensity, enhances memory performance through emotions elicited from odours, particularly in response to aversive stimuli [38], and is important in associative learning [39,40]. The hippocampus also responds to olfactory stimuli and plays a role in long-term memory, emotional experience, and response to stress [38]. For example, mice show higher activity in the dorsal hippocampus when detecting familiar odours in a novel environment, but this ability is impaired with olfactory lesions [41]. The significance of the olfactory bulb size and angle to species differences is unclear. However, horses have significantly larger olfactory bulbs than humans and are able to detect a wider range of odours [42]. Further studies in equids on the impact of olfactory memory on behavioural responses in relation to the location of their olfactory bulbs may prove interesting. Given that population laterality in nostril use has been reported in horses [43], it may pay to explore the evidence for a similar attribute in donkeys.

It appears that in Standardbred horses, at least, the location of the olfactory bulb can be inferred by the forehead whorls. This may have relevance to documented temperament differences based on whorl location. In cattle, whorls located above the eye line (that can be drawn between the medial canthi) are related to individuals who are more reactive when placed in a squeeze chute [44]. In horses, forehead hair whorls have been used to predict the temperament of a horse since ancient times [45]. Anecdotally, horses with whorls located below the eye level indicate so-called intelligence, while whorls located in the centre of the forehead present a so-called uncomplicated personality [45]. While whorls can vary greatly in size and shape, the location of the forehead whorl is a highly heritable trait ($h = 0.83$ [46]), and horses with whorls located above the eye line are harder to manage [47]. Associations between the development of neural tissue and hair whorls have been suggested as the processes occur simultaneously during embryonic development [48]. Furthermore, the direction of hair in the whorl pattern correlates with laterality, with clockwise whorls predicting left side bias and counter clockwise whorls predicting right-side bias in horses presented with a novel object [48]. To the authors' knowledge, there are no published studies of any relationship between whorl location and temperament in donkeys, but the notion that lower whorls are related to calmer temperaments may align with anecdotal observations that donkeys, in general, are less reactive than horses.

Temperament in horses is a very important trait related to a horse's usefulness for its selected purpose [49]. Researchers have demonstrated specific temperament traits related to dog-specific breeds [50,51], so it follows that specific horse breeds would also tend to display certain behavioural traits. Temperament has been investigated in a number of horse breeds, and researchers have demonstrated a more reactive temperament in Thoroughbreds and Arabians than in cold-blooded horses [52,53]. Thoroughbreds take longer to perform a stepping backward test than other breeds [17], indicating some relationship between breed and cognition. The hot-blooded ancestry of Thoroughbreds and Arabians contributes to higher levels of anxiousness, excitability, sociability, and inquisitiveness compared to Quarter Horses, Appaloosas, and Irish Draft Horses [7,16]. These breeds also exhibit particular conformation as reinforced by stud books and breed standards, thus conformation and temperament are quite likely genetically linked [54].

Temperament in donkeys differs from horses in that they are generally less overtly reactive, leading them to be defined as stoic [23]. Donkeys generally live alone or in very small groups, thus do not exhibit typical "herd" behaviour like horses [19]. They may be territorial and display aggression toward other species sharing their space [19]. While generally regarded as a flight animal, donkeys' resort to fight behaviour much more readily than horses. Donkeys may also assume a "freeze" posture when presented with frightening stimuli, which may be defined as "stubborn" by some [19]. However, although donkeys may take more time, they are more adept at solving puzzles than horses [55]. Similarly, Baragli and Regolin [56] showed donkeys to be proficient at finding a hidden object while Standardbred horses in the same study failed to do so. Anatomically, donkeys differ from horses with their large ears, short neck, and small feet [19], and coupled with differences in skull morphometrics shown in the current study; this may explain some of the behavioural differences between the two species. However, it is important to bear in mind that environment and individual experience have a significant impact on how an animal responds to any particular stimuli.

5. Conclusions

This pilot study compared the skull morphology of horses and donkeys, revealing that donkeys have a more distinct forehead, and yet have smaller olfactory bulbs rotated more rostrally compared to horses. Further, the facial whorl in donkeys is not associated with the location of the olfactory bulbs as it is in horses. These results may be linked to other documented differences between horses and donkeys, particularly behavioural differences. Future research exploring possible links between skull morphology and behaviour may result in better prediction of behaviours based on more than merely breed or species. Thus, horses and donkeys of a given morphotype could be better matched to a job or discipline suitable to their temperament, improving their welfare, increasing the safety of humans working with them, and reducing behavioural wastage due to miscommunication with humans.

Supplementary Materials: The following are available online at http://www.mdpi.com/2076-2615/10/6/1002/s1, Table S1: Raw morphometric data from horse and donkey skulls.

Author Contributions: Conceptualization, K.M., P.D.M.; Methodology, K.M., P.D.M.; Validation, K.M.; Formal analysis, K.M.; Investigation, K.M., G.P.; Resources, G.P., P.D.M.; Data curation, K.M.; Writing—original draft preparation, K.M.; writing—review and editing, G.P.; K.M.; P.D.M; Visualization, K.M., P.D.M.; Supervision, P.D.M.; Project administration, K.M., P.D.M. All authors have read and agreed to the published version of the manuscript.

Funding: This research received no external funding.

Acknowledgments: The authors extend gratitude to the Anatomy staff at the Sydney School of Veterinary Science for assistance in procuring the horse skulls. Anna Harrison, a former welfare veterinarian at The Donkey Sanctuary, is warmly thanked for facilitating our work with donkeys.

Conflicts of Interest: The authors declare no conflict of interest.

References

1. Grilz-Seger, G.; Reiter, S.; Neuditschko, M.; Wallner, B.; Rieder, S.; Leeb, T.; Jagannathan, V.; Mesarič, M.; Cotman, M.; Pausch, H.; et al. A genome-wide association analysis in noriker horses identifies a snp associated with roan coat color. *J. Equine Vet. Sci.* **2020**, *88*, 102950. [CrossRef] [PubMed]
2. Aberle, K.S.; Hamann, H.; Drögemüller, C.; Distl, O. Genetic diversity in german draught horse breeds compared with a group of primitive, riding and wild horses by means of microsatellite dna markers. *Anim. Genet.* **2004**, *35*, 270–277. [CrossRef] [PubMed]
3. Robert, C.; Valette, J.-P.; Denoix, J.-M. Longitudinal development of equine forelimb conformation from birth to weaning in three different horse breeds. *Vet. J.* **2013**, *198*, e75–e80. [CrossRef] [PubMed]
4. Malinowski, K.; Christensen, R.A.; Hafs, H.D.; Scanes, C.G. Age and breed differences in thyroid hormones, insulin-like growth factor (igf)-i and igf binding proteins in female horses. *J. Anim. Sci.* **1996**, *74*, 1936–1942. [CrossRef] [PubMed]
5. Gastal, E.L.; Gastal, M.O.; Beg, M.A.; Neves, A.P.; Petrucci, B.P.L.; Mattos, R.C.; Ginther, O.J. Miniature ponies: Similarities and differences from larger breeds in follicles and hormones during the estrous cycle. *J. Equine Vet. Sci.* **2008**, *28*, 508–517. [CrossRef]
6. Janczarek, I.; Stachurska, A.; Kedzierski, W.; Wilk, I. Responses of horses of various breeds to a sympathetic training method. *J. Equine Vet. Sci.* **2013**, *33*, 794–801. [CrossRef]
7. Lloyd, A.S.; Martin, J.E.; Bornett-Gauci, H.L.I.; Wilkinson, R.G. Horse personality: Variation between breeds. *Appl. Anim. Behav. Sci.* **2008**, *112*, 369–383. [CrossRef]
8. McGreevy, P.D.; Thomson, P.C. Differences in motor laterality between breeds of performance horse. *Appl. Anim. Behav. Sci.* **2006**, *99*, 183–190. [CrossRef]
9. Druml, T.; Baumung, R.; Sölkner, J. Morphological analysis and effect of selection for conformation in the noriker draught horse population. *Livest. Sci.* **2008**, *115*, 118–128. [CrossRef]
10. Goodwin, D.; Levine, M.; McGreevy, P.D. Preliminary investigation of morphological differences between ten breeds of horses suggests selection for paedomorphosis. *J. Appl. Anim. Welf. Sci.* **2008**, *11*, 204–212. [CrossRef]
11. Folla, F.; Sartori, C.; Guzzo, N.; Pigozzi, G.; Mantovani, R. Genetics of linear type traits scored on young foals belonging to the italian heavy draught horse breed. *Livest. Sci.* **2019**, *219*, 91–96. [CrossRef]
12. Merkies, K.; Alebrand, J.; Harwood, B.; LaBarge, K.; Scott, L. Investigation into thoracic asymmetry in ridden horses. *Comp. Exerc. Physiol.* **2020**, *16*, 55–62. [CrossRef]
13. Evans, K.E.; McGreevy, P.D. Conformation of the equine skull: A morphometric study. *J. Vet. Med. Ser. C Anat. Histol. Embryol.* **2006**, *35*, 221–227. [CrossRef] [PubMed]
14. Nolte, W.; Thaller, G.; Kuehn, C. Selection signatures in four german warmblood horse breeds: Tracing breeding history in the modern sport horse. *PLoS ONE* **2019**, *14*, e0215913. [CrossRef] [PubMed]
15. Górecka-Bruzda, A.; Jastrzebska, E.; Sosnowska, Z.; Jaworski, Z.; Jezierski, T.; Chruszczewski, M.H. Reactivity to humans and fearfulness tests: Field validation in polish cold blood horses. *Appl. Anim. Behav. Sci.* **2011**, *133*, 207–215. [CrossRef]
16. Sackman, J.E.; Houpt, K.A. Equine personality: Association with breed, use, and husbandry factors. *J. Equine Vet. Sci.* **2019**, *72*, 47–55. [CrossRef]
17. Fenner, K.; Freire, R.; McLean, A.; McGreevy, P. Behavioral, Demographic, and Management influences on equine responses to negative reinforcement. *J. Vet. Behav.* **2019**, *29*, 11–17. [CrossRef]
18. Proops, L.; Osthaus, B.; Bell, N.; Long, S.; Hayday, K.; Burden, F. Shelter-seeking behavior of donkeys and horses in a temperate climate. *J. Vet. Behav.* **2019**, *32*, 16–23. [CrossRef]
19. Burden, F.; Thiemann, A. Donkeys are different. *J. Eq. Vet. Sci.* **2015**, *35*, 376–382. [CrossRef]
20. Davis, E. Donkey and mule welfare. *Vet. Clin. N. Am.-Eq. Pract.* **2019**, *35*, 481–491. [CrossRef]
21. McLean, A.K.; Navas González, F.J.; Canisso, I.F. Donkey and mule behavior. *Vet. Clin. N. Am.-Eq. Pract.* **2019**, *35*, 575–588. [CrossRef] [PubMed]
22. Grint, N.J.; Beths, T.; Yvorchuk-St Jean, K.; Whay, H.R.; Murrell, J.C. Analysis of behaviors observed during mechanical nociceptive threshold testing in donkeys and horses. *J. Equine Vet. Sci.* **2017**, *50*, 102–109. [CrossRef]
23. van Dierendonck, M.C.; van Loon, J.P.A.M.; Burden, F.A.; Rickards, K. Monitoring acute pain in donkeys with the equine utrecht university scale for donkeys composite pain assessment (equus-donkey-compass)

and the equine utrecht university scale for donkey facial assessment of pain (Equus-Donkey-Fap). *Animals* **2020**, *10*, 354. [CrossRef] [PubMed]
24. Corrales-Hernández, A.; Mota-Rojas, D.; Guerrero-Legarreta, I.; Roldan-Santiago, P.; Rodríguez-Salinas, S.; Yáñez-Pizaña, A.; de la Cruz, L.; González-Lozano, M.; Mora-Medina, P. Physiological responses in horses, donkeys and mules sold at livestock markets. *Int. J. Vet. Sci. Med.* **2018**, *6*, 97–102. [CrossRef] [PubMed]
25. Georgevsky, D.; Carrasco, J.J.; Valenzuela, M.; McGreevy, P.D. Domestic dog skull diversity across breeds, breed groupings, and genetic clusters. *J. Vet. Behav.* **2013**, *9*, 228–234. [CrossRef]
26. McGreevy, P.D.; Georgevsky, D.; Carrasco, J.; Valenzuela, M.; Duffy, D.L.; Serpell, J.A. Dog behavior co-varies with height, bodyweight and skull shape. *PLoS ONE* **2013**, *8*. [CrossRef]
27. Stone, H.R.; McGreevy, P.D.; Starling, M.J.; Forkman, B. Associations between domestic-dog morphology and behaviour scores in the dog mentality assessment. *PLoS ONE* **2016**, *11*, e0149403. [CrossRef]
28. Cervantes, I.; Baumung, R.; Molina, A.; Druml, T.; Gutiérrez, J.P.; Sölkner, J.; Valera, M. Size and shape analysis of morphofunctional traits in the spanish arab horse. *Livest. Sci.* **2009**, *125*, 43–49. [CrossRef]
29. Radinsky, L. Ontogeny and phylogeny in horse skull. *Evolution* **1984**, *38*, 1–15. [CrossRef]
30. Komosa, M.; Molinski, K.; Godynicki, S. The variability of cranial morphology in modern horses. *Zoolog. Sci.* **2006**, *23*, 289–298. [CrossRef]
31. McGreevy, P.; Grassi, T.D.; Harman, A.M. A strong correlation exists between the distribution of retinal ganglion cells and nose length in the dog. *Brain. Behav. Evol.* **2004**, *63*, 13–22. [CrossRef] [PubMed]
32. Roberts, T.; McGreevy, P.; Valenzuela, M. Human induced rotation and reorganization of the brain of domestic dogs. *PLoS ONE* **2010**, *5*, e11946. [CrossRef] [PubMed]
33. Evans, K.E.; McGreevy, P.D. The distribution of ganglion cells in the equine retina and its relationship to skull morphology. *J. Vet. Med. Ser. C Anat. Histol. Embryol.* **2007**, *36*, 151–156. [CrossRef] [PubMed]
34. Nielsen, B.L. Making Sense of It All: The importance of taking into account the sensory abilities of animals in their housing and management. *Appl. Anim. Behav. Sci.* **2018**, *205*, 175–180. [CrossRef]
35. Mazzatenta, A.; Veronesi, M.C.; Vignola, G.; Ponzio, P.; Carluccio, A.; De Amicis, I. Behavior of martina franca donkey breed jenny-and-foal dyad in the neonatal period. *J. Vet. Behav.* **2019**, *33*, 81–89. [CrossRef]
36. Simcock, N.K.; Gray, H.; Bouchebti, S.; Wright, G.A. Appetitive olfactory learning and memory in the honeybee depend on sugar reward identity. *J. Insect Physiol.* **2018**, *106*, 71–77. [CrossRef] [PubMed]
37. Patin, A.; Pause, B.M. Human amygdala activations during nasal chemoreception. *Neuropsychologia* **2015**, *78*, 171–194. [CrossRef]
38. Soudry, Y.; Lemogne, C.; Malinvaud, D.; Consoli, S.M.; Bonfils, P. Olfactory system and emotion: Common substrates. *Eur. Ann. Otorhinolaryngol. Head Neck Dis.* **2011**, *128*, 18–23. [CrossRef]
39. Cardinal, R.N.; Parkinson, J.A.; Hall, J.; Everitt, B.J. Emotion and motivation: The role of the amygdala, ventral striatum, and prefrontal cortex. *Neurosci. Biobehav. Rev.* **2002**, *26*, 321–352. [CrossRef]
40. Corbit, L.H.; Leung, B.K.; Balleine, B.W. The role of the amygdala-striatal pathway in the acquisition and performance of goal-directed instrumental actions. *J. Neurosci.* **2013**, *33*, 17682–17690. [CrossRef]
41. Pena, R.R.; Medeiros, D. de C.; Guarnieri, L. de O.; Guerra, J.B.; Carvalho, V.R.; Mendes, E.M.A.M.; Pereira, G.S.; Moraes, M.F.D. Home-Cage odors spatial cues elicit theta phase/gamma amplitude coupling between olfactory bulb and dorsal hippocampus. *Neuroscience* **2017**, *363*, 97–106. [CrossRef] [PubMed]
42. Saslow, C.A. Understanding the perceptual world of horses. *Appl. Anim. Behav. Sci.* **2002**, *78*, 209–224. [CrossRef]
43. McGreevy, P.D.; Rogers, L. Motor and sensory laterality in thoroughbred horses. *Appl. Anim. Behav. Sci.* **2005**, *92*, 337–352. [CrossRef]
44. Grandin, T.; Deesing, M.J.; Struthers, J.J.; Swinker, A.M. Cattle with hair whorl patterns above the eyes are more behaviorally agitated during restraint. *Appl. Anim. Behav. Sci.* **1995**, *46*, 117–123. [CrossRef]
45. Meyer, J.F. What's in a whorl (hair swirls of horses). *Horse Rider* **2008**, *47*, 46–53.
46. Górecka, A.; Słoniewski, K.; Golonka, M.; Jaworski, Z.; Jezierski, T. Heritability of hair whorl position on the forehead in konik horses. *J. Anim. Breed. Genet.* **2006**, *123*, 396–398. [CrossRef]
47. Górecka, A.; Golonka, M.; Chruszczewski, M.; Jezierski, T. A note on behaviour and heart rate in horses differing in facial hair whorl. *Appl. Anim. Behav. Sci.* **2007**, *105*, 244–248. [CrossRef]
48. Shivley, C.; Grandin, T.; Deesing, M. Behavioral laterality and facial hair whorls in horses. *J. Equine Vet. Sci.* **2016**, *44*, 62–66. [CrossRef]

49. Graf, P.; König von Borstel, U.; Gauly, M. Importance of personality traits in horses to breeders and riders. *J. Vet. Behav.* **2013**, *8*, 316–325. [CrossRef]
50. Starling, M.J.; Branson, N.; Thomson, P.C.; McGreevy, P.D. "Boldness" in the domestic dog differs among breeds and breed groups. *Behav. Process.* **2013**, *97*, 53–62. [CrossRef]
51. Mehrkam, L.R.; Wynne, C.D.L. Behavioral Differences among breeds of domestic dogs (Canis Lupus Familiaris): Current status of the science. *Appl. Anim. Behav. Sci.* **2014**, *155*, 12–27. [CrossRef]
52. Janiszewska, J.; Ignor, J.; Cieśla, A. Einfluss eines 11-monatigen trainings auf die ergebnisse des "Ängstlichkeitstests" von jungen halbblut-hengsten. *Arch. Anim. Breed.* **2004**, *47*, 7–13. [CrossRef]
53. Borstel, U.; Pirsich, W.; Gauly, M.; Bruns, E. Repeatability and reliability of scores from ridden temperament tests conducted during performance tests. *Appl. Anim. Behav. Sci.* **2012**, *139*, 251–263. [CrossRef]
54. Kuhnke, S.; Bär, K.; Bosch, P.; Rensing, M.; Borstel, U.K.V. Evaluation of a novel system for linear conformation, gait, and personality trait scoring and automatic ranking of horses at breed shows: A pilot study in american quarter horses. *J. Equine Vet. Sci.* **2019**, *78*, 53–59. [CrossRef] [PubMed]
55. Osthaus, B.; Proops, L.; Hocking, I.; Burden, F. Spatial cognition and perseveration by horses, donkeys and mules in a simple a-not-b detour task. *Anim. Cogn.* **2013**, *16*, 301–305. [CrossRef]
56. Baragli, P.; Regolin, L. Cognition tests in equids (Equus Caballus and Equus Asinus). In Proceedings of the International Equine Science Meeting, Regensburg, Germany, 3–5 October 2008.

animals

MDPI

Article

Anthelmintic Efficacy and Pharmacokinetics of Ivermectin Paste after Oral Administration in Mules Infected by Cyathostomins

Marilena Bazzano [1], Alessandra Di Salvo [2], Manuela Diaferia [2], Fabrizia Veronesi [2], Roberta Galarini [3], Fabiola Paoletti [3], Beniamino Tesei [1], Amy McLean [4], Vincenzo Veneziano [5] and Fulvio Laus [1,*]

1 School of Biosciences and Veterinary Medicine, University of Camerino, Via Circonvallazione, 62024 Matelica (MC), Italy; marilena.bazzano@unicam.it (M.B.); beniamino.tesei@unicam.it (B.T.)
2 Department of Veterinary Medicine, University of Perugia. Via S. Costanzo, 06126 Perugia, Italy; alessandra.disalvo@unipg.it (A.D.S.); manuela.diaferia@unipg.it (M.D.); fabrizia.veronesi@unipg.it (F.V.)
3 Istituto Zooprofilattico Sperimentale dell'Umbria e delle Marche "Togo Rosati", Via Salvemini, 06126 Perugia, Italy; r.galarini@izsum.it (R.G.); f.paoletti@izsum.it (F.P.)
4 Department of Animal Science, University of California Davis, Davis, CA 95616, USA; acmclean@ucdavis.edu
5 Department of Veterinary Medicine and Animal Productions, University of Naples Federico II, 80137 Naples, Italy; vinvene@unina.it
* Correspondence: fulvio.laus@unicam.it; Tel.: +39-07-3740-3403

Received: 8 April 2020; Accepted: 23 May 2020; Published: 28 May 2020

check for updates

Simple Summary: Mules and donkeys are often treated as horses from a therapeutic point of view. This approach could be dangerous due to species differences in drug pharmacokinetics which could reflect on the drug effectiveness. Ivermectin is a commonly used anthelmintic compound due to the broad spectrum of activity. The improper use of ivermectin (i.e., dosage, route of administration) could cause a lack of parasite control and contribute to development of drug resistance. Studies on the pharmacokinetics and efficacy of antiparasitic molecules in mules are limited, although these drugs are crucial for the welfare of these equines. The aim of the present study was to evaluate the efficacy and pharmacokinetics of ivermectin administered to mules at the same dosage (200 µg/kg body weight) and route licensed for horses. Results show that administering ivermectin orally, at the same dosage of horses, has a pharmacokinetic intermediate behavior between horses and donkeys. This study demonstrates that ivermectin oral paste at horse dosage is effective and safe for the treatment of cyathostomins in mules.

Abstract: Ivermectin (IVM) is an anthelmintic compound commonly used off-label in mules due to its broad-spectrum of activity. Despite the general use of IVM in mules with the same dose and route of administration licensed for horses, significant pharmacokinetic differences might exist between horses and mules, as already observed for donkeys. The aim of the present study was to evaluate the pharmacokinetic profile and anthelmintic efficacy of an oral paste of IVM in mules naturally infected with cyathostomins. Fifteen adult mules with fecal egg counts (FEC) ≥ 200 eggs per gram (EPG), with exclusive presence of cyathostomins, were included in the study. All mules were orally treated with IVM according to the manufacturer's recommended horse dosage (200 µg/kg body weight). FECs were performed before (day-10 and day-3) and after treatment at days 14 and 28 by using a modified McMaster method. The FEC reduction (FECR%) was also calculated. Blood samples were collected from five animals at various times between 0.5 h up to 30 days post treatment to determine pharmacokinetic parameters. The maximum IVM serum concentration (Cmax) was 42.31 ± 10.20 ng/mL and was achieved at 16.80 ± 9.96 h post-treatment (Tmax), area under the curve (AUC) was 135.56 ± 43.71 ng × day/mL. FECR% remained high (>95%) until the 28th day.

Keywords: ivermectin; mule; cyathostomins; pharmacokinetics; anthelmintic efficacy

1. Introduction

The estimated global equine population is 117 million, including 58 million horses, 50 million donkeys and 9 million mules [1]. Although there has been a decrease in mule populations in Europe and the Mediterranean European countries, mules are still widely used as a beast of burden or working equids for draft purposes in developing countries because they are considered to be strong and hardy animals [1,2]. Furthermore, in some countries such as the U.S. there has been a growing interest in using mules also for recreational riding, racing and show purposes [3,4]. The increase in mule use has created the need for more information on how to properly treat and care for them [4].

Little research is available for donkeys but even less is known about mules [5]. In literature, only limited data is available on the dosage, pharmacokinetics and efficacy of drugs in mules [6,7]. The lack of scientific information for mules may be due to the idea that these equids are considered less demanding from a health point of view (mules are considered less susceptible to diseases and fatigue compared to horses) or the fact that they are mostly reared in parts of the world where little routine veterinary care is available [8]. However, mules may be susceptible to the same parasites affecting horses and donkeys. Among all nematodes that may have the greatest negative impact on the wellness of mules, cyathostomins (small strongyles) are of major concern and efficient control based on the appropriate use of anthelmintic drugs is mandatory [9,10].

The three anthelmintic drug classes (e.g., benzimidazoles, pyrimidines and macrocyclic lactones) [11] registered for horses are currently not available for treatment of parasites in mules and only few drugs are specifically licensed for use in other equids like donkeys [12–14]. Therefore, mules and donkeys are usually treated with anthelmintic drugs at the same dosage, route and intervals licensed for horses, despite the lack of scientific reports evaluating pharmacokinetics and efficacy related to their use in these animals [11,15,16].

Among anthelmintic compounds registered for equine species, ivermectin (IVM) is a macrocyclic lactone (ML), commonly used due to its broad-spectrum of activity against both endo- and ectoparasites [11,17]. Ivermectin administered as oral paste, at dosage of 200 µg/kg body weight (BW), is characterized by a highly safe margin and extensively used in the equine industry [18–20]. Ivermectin is a very lipophilic drug and, in general, its pharmacokinetics in domestic species is characterized by a slow phase of absorption, extensive distribution, scant metabolization and slow excretion, principally with feces [21]. In horse species the oral administration of IVM shows a faster absorption compared to subcutaneous injection [22], a mean resident time in the organism that ranges from about four to seven days [20,23,24] and a slow elimination with feces in which the maximum concentration is achieved after two days from administration [25].

Despite the general use of IVM in the mule at the same dose and route indicated for horses, significant pharmacokinetic differences might exist between horses and mules, as already observed between horses and donkeys both for IVM [26] and other drugs [27,28]. Some differences in pharmacokinetics and efficacy have already been observed between mules and horses for antibiotics (i.e., sulfamethoxazole and trimethoprim) [6] and for sedatives (i.e., xylazine) [7].

Anthelmintic activity depends on the interaction between the active drug ingredient and the specific receptors in the parasitic target, but also on drug concentrations obtained at the parasitic site [29,30]. Animal species is one of the main factors affecting the blood drug disposition and thus the concentrations attained for the parasitic targets.

The aim of the present study was to evaluate the blood disposition and the anthelmintic efficacy of an oral formulation of IVM administrated in mules naturally infected by cyathostomins, in terms of percentage of fecal egg count reduction (FECR%), following the international guidelines recommended by the American Association of Equine Practitioners (AAEP) [31].

2. Materials and Methods

2.1. Animals and Experimental Groups

Fifteen adult mules, ten females and five males, with a mean age (±standard deviation) of 11 ± 6 years and a mean weight (±standard deviation) of 511 ± 48 kg were included in the study. Weight was estimated as previously described [32]. Owner's consent was given for all animals included in the study.

The mules were reared in central Italy as working animals and were kept in outdoor paddocks with no shared pasture with other equids. The management was the same for all the mules, animals were allowed to graze for several hours per day and their diet consisted of fresh forage (grass), 8 kg grass hay and 4 kg concentrates per day, and water ad libitum. All the animals enrolled in the study belonged to the same farm and shared the same paddocks from at least 12 months. The mules received treatments for gastrointestinal parasites occasionally in previous years, and no anthelmintic drugs were administrated to the mules in the previous 12 months.

Before starting the study on day −10 (D_{-10}), clinical examinations and hematological investigations were performed on each animal to ensure their health status. Parasitological examinations, consisting of individual fecal egg count (FEC) followed by coproculture, were performed at ten (D_{-10}) and three (D_{-3}) days before anthelmintic administration. All mules used in the study had a FEC ≥ 200 eggs per gram (EPG) with exclusive presence of cyathostomins. A total of fifteen mules were tested for IVM efficacy, in accordance with the guidelines of the World Association for the Advancement of Veterinary Parasitology (WAAVP) [33].

The study was conducted with the favorable consent of the local organization for animal welfare (OPBA, protocol number E81AC.10) of the University of Camerino and the approval of Ministry of Health (protocol number 677/2018-PR) in accordance with directive 2010/63/EU of the European Parliament on the protection of animals used for scientific purpose.

2.2. Treatment Procedures

On day 0 (D_0), a commercially available equine oral paste formulation of IVM (Eqvalan Oral Paste, Merial Italia, Boehringer Ingelheim Animal Health) was administered orally to each mule six hours before feeding (after a fasting period of 12 h) at the horse dosage of 200 μg/kg of BW. All treated animals were observed continuously for the first three hours after drug administration and then weekly until the end of the trial period to detect possible adverse reactions following treatment.

2.3. Sampling Procedures

Individual fecal samples were collected from the rectum of each animal on D_{-10}, D_{-3} and D_0 immediately prior to anthelmintic administration, and then on day 14 (D_{+14}) and day 28 (D_{+28}) post treatment. According to general recommendations proposed by Nielsen et al. (2010) [34], the fecal samples were stored at +4 °C waiting for FEC.

In order to assess IVM pharmacokinetics 5 of the 15 treated mules (females, between 6–21 years), chosen on the basis of a mildest character, and consequently more easily manageable, were enrolled. Whole blood samples (about 10 mL) were collected by jugular vein puncture before drug administration and then at predetermined time-points: 30 minutes (min), 1, 2, 4, 6, 8, 10 and 24 hours (h) and 2, 3, 4, 7, 10, 15, 20, 25 and 30 days (d) after the treatment. Blood samples were centrifuged at 2500 g for 10 min and the serum was collected and stored at −80 °C until estimation of drug concentration.

The sample size as well as the timing of blood sampling was assessed on the basis of previous studies on IVM pharmacokinetics in equine species [20,22–24,26].

2.4. Coprological Examinations

Individual FECs were performed in all mules at D_{-10}, D_{-3}, D_0, D_{+14} and D_{+28}, using a modified McMaster technique with a lower detection limit of 50 eggs per gram (EPG) and a saturated sodium chloride (NaCl) solution with a specific gravity of 1.200 [35]. Pooled fecal cultures were performed at D_{-10} and D_{-3}, while individual fecal cultures were performed for all mules with FEC ≥ 200 EPG at D_0, D_{+14} and D_{+28}. Fecal cultures were incubated at 27 °C for 7–10 days for larval development and 3rd stage larvae were identified according with morphological identification keys [35]. When a coproculture had 100 or less third stage larvae, all were identified; when a coproculture had more than 100 larvae, only 100 larvae were identified.

2.5. Analytical Determination of Ivermectin

One milliliter of serum was added to a 15 mL Falcon tube with 50 μL of a methanolic solution of the internal standard, ivermectin, at 1 μg/mL, 2.5 mL of acetonitrile and 4 g of anhydrous sodium sulphate. After centrifugation, the supernatant was collected in a new tube and the extraction was repeated with 2.5 mL of acetonitrile. The combined extracts were evaporated to dryness and then re-dissolved in 200 μL of water and 2.5 mL of ethyl acetate. The samples were purified through a SPE cartridge (NH2, 100 mg/3 mL, Biotage IST Isolute, Uppsala, Sweden) previously conditioned with 3 mL of methanol and 3 mL of ethyl acetate. Furthermore, 3 mL of ethyl acetate was used to wash the sample tube and transfer onto the SPE cartridge. The eluate was evaporated until dry and the analyte derivatization was executed according to a previously published method [36]. After derivatization, the samples were injected into a high-performance liquid chromatography (HPLC) system.

The HPLC equipment consisted of a Thermo Finnigan Spectrasystem (San Jose, CA, USA) with a P4000 quaternary pump, an AS3000 autosampler and a fluorescence detector (FL3000). The separation was achieved on a Luna C8 (150 × 3.0 mm, 5 μm, Phenomenex, Torrance CA, USA) analytical column equipped with a guard column C8 4 × 2.0 mm (Phenomenex). The mobile phases were acetonitrile (A) and water (B). The flow rate was 1.2 mL/min and the injection volume 50 μL. The gradient profile was as follows (1) 0–3 min, to 85% A; (2) 3–7 min, to 97% A; (3) 7–10 min, 97% A; (4) 10–12 min, to 85% A; and (5) 12–14 min, to 85% A. The total run time per single injection was 14 min. Fluorescence of the derivatized compounds was detected at excitation and emission wavelengths of 364 and 470 nm, respectively. In each analytical batch, a series of five concentration points (2.5, 5, 25, 50 and 100 ng/mL) prepared by spiking blank serum samples were injected for calibration.

Validation parameters were determined in accordance with EMA guidelines on bioanalytical method validation [37]. Five spiked samples were analyzed at four concentrations (5, 50, 100 and 200 ng/mL) on two different days. The observed recoveries ranged from 90% to 106%. Within-run and between-run precisions were in the range 3.5–8.5% and 8.5–13%, respectively. The lower limit of quantification (LLOQ) was 2.5 ng/mL, whereas the upper limit (ULOQ) was 200 ng/mL.

2.6. Pharmacokinetic and Anthelmintic Efficacy

Pharmacokinetic parameters were calculated, for each animal, from serum concentration–time data executing non-compartmental analyses by PK-Solver program [38].

To determine the efficacy of IVM against cyathostomins at each fecal sampling time, an arithmetic mean of EPG was calculated following the American Association of Equine Practitioners (AAEP) parasite control guidelines [31]. For each mule, the percentage efficacy (%) was assessed in terms of fecal egg count reduction (FECR) at different days (D_{+14} and D_{+28}) using the following formula:

FECR (%) = 100 × (Mean EPG pre-treatment − Mean EPG post-treatment/Mean EPG pre-treatment).

2.7. Statistical Analyses

Microsoft Office Excel 2016 software was used for data recording, and FECR, expressed as percentage with 95% confidence intervals, was calculated using the RESO FECRT analysis program, version 4 (http://sydney.edu.au/vetscience/sheepwormcontrol/). The values of the FECR were interpreted according to the AAEP parasite control guidelines [31]. The 95% lower confidence limits (LCL; %) observed for all time points was selected so that resistance would be indicated if the % mean FECR was below 95% and the LCL was below 90% [39].

3. Results

3.1. Parasitological Results

Clinically, none of the animals enrolled in the current study showed any adverse reaction following IVM administration.

The FECs (mean, range and standard deviation) obtained at the different sampling times and the results of the FECR with 95% CI are shown in Table 1. At baseline (D_0), mules included in the study had an average of 1360 EPG, consisting of 100% of cyathostomins; the mean EPG at D_{+14} and D_{+28} were 46.67 and 26.67 EPG, respectively and did not differ statistically ($p > 0.05$) (Supplementary Data). The FECR was 96.57% and 98.04% at D_{+14} and D_{+28}, respectively. The lower 95% CI was over 90% at D_{+14} and D_{+28}.

Table 1. Initial (Day_0) and post-treatment (D_{+14} and D_{+28}) fecal egg counts (FECs) expressed as mean of egg per gram (EPG), relative standard deviation (SD) and percentage of fecal eggs count reduction (FECR%), with their 95% lower and upper confidence intervals (CIs) after an oral treatment of ivermectin (IVM) in mules naturally infected by cyathostomins.

Days	FECs EPG AM	EPG Range	SD	FECR (%)	Lower 95% CI	Upper 95% CI
0	1360.00	200–3550	908.46	-	-	-
14	46.67	0–250	68.26	96.57	90.65	98.74
28	26.67	0–200	57.06	98.04	93.40	99.42

Abbreviations: FECs, fecal egg counts; EPG, egg per gram; AM, arithmetic mean; SD, standard deviation; FECR, fecal egg count reduction; CI, confidence interval.

3.2. Pharmacokinetics Results

Ivermectin was quantifiable in serum 30 min following drug administration in two mules, after 1 h in two mules and after 2 h in one mule. The concentrations of IVM in systemic circulation ranged from 4.2 to 7.4 ng/mL at the first time of appearance in systemic circulation, achieving the Cmax (42.31 ± 10.20 mean ± SD; range 28.3–56.7) between 4 h (only one mule) and 24 h post-treatment. Finally, the serum concentration of IVM decreased progressively, remaining detectable in all animals up to seven days post administration. In one animal, the drug was detectable in serum for 10 days post oral administration, 15 days in another mule and until 20 days in one mule. Figure 1 illustrates the trend of serum IVM concentrations in relation to time. Table 2 shows the main pharmacokinetic parameters of IVM in serum following *per os* administration at dosage of 200 μg/kg BW following non-compartmental analysis.

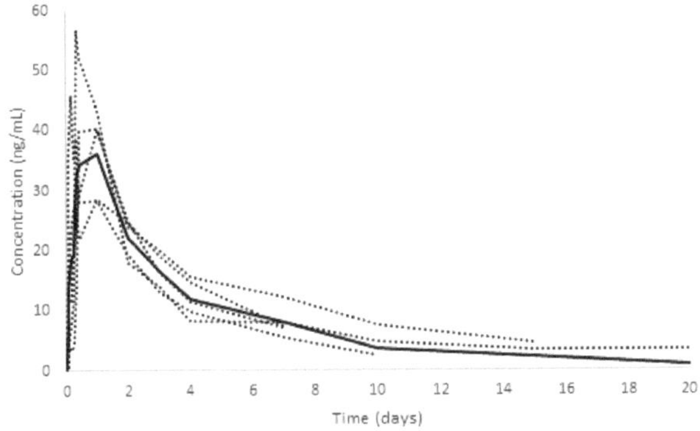

Figure 1. Average (solid line) and single (dot lines) IVM serum concentrations vs. time after oral administration at a dose of 200 μg/kg BW in mules ($n = 5$).

Table 2. Main pharmacokinetic parameters of ivermectin following oral administration at 200 µg/kg BW in mule (present study), horse and donkey (derived from previous published studies).

Species (Number of Animals)	Parameter (Unit)						Reference
	$t_{1/2\lambda z}$ (Day)	T_{max} (h)	C_{max} (ng/mL)	AUC_{0-t} (ng × Day/mL)	$AUC_{0-\infty}$ (ng × Day/mL)	MRT (Day)	
Mule (n = 5)	$2.74^{\S} \pm 2.02^{\dagger}$	16.8 ± 9.96	42.31 ± 10.20	135.56 ± 43.71	163.93 ± 61.82	6.07 ± 4.36	present study
Horse (n = 6)	6.53 ± 0.92	4.08 ± 2.16	61.28 ± 10.73	164.96 ± 30.07	-	7.34 ± 1.30	Gokbulut et al., (2010) [24]
Horse (n = 5)	$4.25^{\S} \pm 0.24$ (a)	9.22 ± 5.71	44.0 ± 23.1	132.7 ± 47.3	-	4.78 ± 0.64	Perez et al., (1999) [23]
Horse (n = 5)	2.93 ± 0.4 (a)	3.60 ± 0.96	51.3 ± 6.1	137.1 ± 35.9	-	4.2 ± 0.4	Perez et al., (2003) [20]
Horse (n = 3)	2.76 ± 0.2	3.3 ±0.7	82.3 ± 12.4	200.92 ± 22.67	-	-	Marriner et al., (1987) [22]
Donkey (n = 3)	7.4 ± 2	24.0 ± 0.0	23.6 ± 4.4	119.3 ± 12.3	-	6.5 ± 0.2	Gokbulut et al., (2005) [26]

(a): terminal half-life resulted from triexponential equation; AUC_{0-t}: area under serum concentration–time curve from zero up to the last concentration ≥ LOQ; $AUC_{0-\infty}$: area under serum concentration–time curve from time zero to infinity; C_{max}: maximum concentration observed; MRT: mean residence time; T_{max}: time of maximum concentration observed (when expressed in days in published studies, it was converted to hours according the formula h = day value × 24); $t_{\frac{1}{2}\lambda z}$: terminal half-life; SD: standard deviation; § harmonic mean; † pseudo standard deviation.

4. Discussion

The current study represents the first study that has explored both pharmacokinetics and efficacy of IVM orally administered in mules. Ivermectin orally administered at 200 μg/kg BW showed intermediate pharmacokinetic parameters between horses and donkeys and seems to be efficacious against cyathostomins.

Ivermectin was the first ML approved in horses as a broad spectrum antiparasitic compound at the recommended dose of 200 μg/kg BW [18]. Since its introduction as an antiparasitic drug in equid species, many authors have investigated the efficacy of IVM on different parasites when administered orally to horses, and also its pharmacokinetics has been well documented [20,22–24,40]. However, there is a paucity of data available in the literature on the pharmacokinetics and antiparasitic efficacy in other equids. Even if donkeys, mules and horses are morphologically similar, they can have considerable differences in pharmacokinetic profiles [6,41,42]. Several studies report a greater capability of drug elimination in donkeys than in horses [13,43,44]. Latzel et al. (2012) [7] observed a significant difference in the half-life of xylazine between horses and mules following IV administration, and at the same time, a less intense sedation in mules. A mule may require up to 50% more of a sedative when compared to a horse to obtain an appropriate level of sedation [45]. Thus, the importance of studying the pharmacokinetics of each drug in the specific target species is evident.

The value of efficacy obtained at D_{+14} in the present study may be considered correctly estimated according to the recommendations by Kaplan and Nielsen (2010) [46], in fact FECR tests have been calculated in fecal samples collected 14 days after treatment and not in the early period post-treatment when ML might cause temporary suppression of the egg shedding and thus leading to a overestimation of the anthelmintic efficacy. However, the obtained FECR values showed an efficacy rate of IVM against cyathostomins lower than those detected in previous studies conducted in horses and donkeys ranging from 96% to 100% [10,47–49]. Since a FECR% ranging between 95–98% should be considered suspected of resistance according to the AAEP guidelines [31], we decided to perform an additional FEC at Day_{+28} to assess if a higher decrease in efficacy may support suspected anthelmintic resistance. ML are characterized by very long ERPs (egg reappearance periods), but recent reports have documented ERPs shortened to 4–5 weeks which may be considered an early indicator of resistance to MLs [31]. The FECR obtained at D_{+28} (98.04%) was higher than D_{+14} (96.57%), which is different from what we would expect; however, the average FECs obtained at two distinct sampling times did not differ statistically. Possible reasons for why an increased percentage in the FECR at D_{+28} was observed, could be related to: (i) few individual variations of egg excretion commonly observed in horses and possibly in mules, (ii) scant efficacy of IVM on the encysted parasites (L3–L4 larval stages) that could conclude their cycle within 28 days increasing the FEC post-treatment and (iii) the analytic sensitivities of the McMaster technique based on a conversion rate, which is however still considered the reference method for the FECR tests.

Consideration for resistances to IVM remained inconclusive because we detected borderline reduced efficacy rates at D_{+14} and D_{+28}. Unfortunately, we were not able to continue to calculate ERP due to behavioral resistances from the animals making it impossible to continue the collection process.

However, borderline reduced efficacy detected in the present study could be related to other factors not associated with resistance. For example, the concern that the sub-optimal dosing of 200 μg/kg of BW may relate to reduced efficacy. We think that it is extremely unlikely that resistant cyathostomin populations infected the sampled mules for a number of reasons: (i) the animals did not share pastures with other equids e.g., horses or donkeys, that could harbor resistant strongyle strains; (ii) any external introduction of animals was done in the last 10 years and (iii) pharmacological pressure within the herd was minimal since the anthelmintic treatments were only occasional.

In this study, the Cmax and AUC observed in mules (42.31 ± 10.20 ng/mL and 135.56 ± 43.71 ng × day/mL, respectively), when treated orally with 200 μg/kg of IVM, are greater than that observed in donkeys (23.4 ± 4.4 ng/mL and 119.3 ± 12.3 ng × day/mL) by Gokbulut et al. (2005) [26] and similar for horses (44.0 ± 23.1 ng/mL and 132.7 ± 47.3 ng × day/mL) [23].

Likewise, Coakley et al., (1999) [50], following administration of flunixin in mules, observed a greater similarity to horses rather than donkeys in drug serum levels.

Nevertheless, it is important to underline that in other studies, after oral IVM administration to horses, greater Cmax and AUCs were seen in respect to the present study [22,24]. The reasons for the differences among these various studies are not clear, but could be due to a different feeding regime. Some authors have speculated that IVM could be absorbed by the particles of digesta causing a low availability of the drug after its oral administration [20,22]. In order to maximize standardization of the treatment conditions in this study, we used a protocol that included fasting for 12 h followed by feed administration 6 h after treatment. Other factors that could lead to a difference in results in above-cited studies could be differences in environment, age, gender and level of fitness as already hypothesized for hematological and blood biochemistry among different equids [4]. Another thought on inconsistent results could be related to a different expression of P-glycoprotein (which has been shown to be involved in the absorption of IVM following its oral administration [21]) in intestinal epithelium of mules compared to other equids. In addition, because IVM is a very lipophilic drug that accumulates in adipose tissue, a different body fat content in mules, compared to horses and donkeys, could be responsible for a different drug distribution and, consequently, for a different profile of drug serum concentrations [21]. Lastly, the different LOQ of the analytical method could have affected the computation of the AUCs [51].

In this study, the Cmax was achieved later than in horses (16.8 h vs. a range, expressed as mean value, between 3.3 and 9.2 h) [20,23,24] and before than in donkeys (24 h) [26]. In three out of five mules, the Tmax was obtained at 24 h; this value was closer to that observed in donkeys compared to that seen in horses [26]. It is important to highlight that the large individual variability observed in serum drug concentrations in the present study, did not allow for generalization of pharmacokinetic behavior of IVM in mules being closer to horses or donkeys. The same large individual variability in systemic drug concentrations was observed in other pharmacokinetic studies on IVM [23,24].

Unfortunately, according to our knowledge and review of literature, there are no PK/PD predictors of efficacy for anthelmintic agents, contrary to antibacterial drugs. Therefore, currently it is not possible to evaluate the potential clinical impact of IVM, orally administered at doses of 200 µg/kg BW in mules, on other parasites. This study may be considered as a first step towards knowledge about pharmacokinetic properties of IVM in mules; further studies are warranted in the future.

5. Conclusions

Studies on correct use of anthelmintic molecules in mules could contribute to improved control of parasites and, consequently a gradual increase in their welfare, supporting a rising of the population worldwide. The present study demonstrated that IVM administered orally at 200 µg/kg BW to mules has intermediate pharmacokinetic parameters between horses and donkeys. Furthermore, IVM seems to be efficacious against cyathostomins in mules even if an optimization of the dosage should be obtained based on further studies.

Supplementary Materials: The following are available online at http://www.mdpi.com/2076-2615/10/6/934/s1, Table S1: Individual FEC at each sampling times and inferential analysis between FECs at Day+14 and Day+28.

Author Contributions: Conceptualization, M.B., F.L., F.V. and B.T.; methodology, M.B., A.D.S., F.V. and F.L.; investigation, M.B., A.D.S., F.V., M.D., R.G. and F.P.; data curation, A.M. and V.V.; writing—original draft preparation, M.B., A.D.S., F.V. and F.L.; writing—review and editing, F.L., B.T. and V.V.; supervision, F.L. All authors have read and approved the submitted version of the manuscript.

Funding: This research received no external funding.

Acknowledgments: Preliminary pharmacokinetic results of this study were presented as oral communication at Donkey Welfare Symposium, University of California, Davis (USA), 26–28 October 2018. This research did not receive any specific grant from funding agencies in the public, commercial or not-for-profit sectors.

Conflicts of Interest: The authors declare no conflict of interest.

References

1. FAO Global Statistical Yearbook. Live Animals (Data). Available online: http://www.fao.org (accessed on 23 March 2020).
2. Barrandeguy, M.E.; Carossino, M. Infectious Diseases in Donkeys and Mules: An Overview and Update. *J. Equine Vet. Sci.* **2018**, *65*, 98–105. [CrossRef]
3. Taylor, T.S.; Matthews, N.S. Donkey and Mule Scenarios: When to stop, think, read or call. In Proceedings of the Annual Convention of the AAEP, Orlando, FL, USA, 4–8 December 2002; pp. 115–116.
4. McLean, A.K.; Wang, W.; Navas-Gonzalez, F.J.; Rodrigues, J.B. Reference intervals for hematological and blood biochemistry reference values in healthy mules and hinnies. *Comp. Clin. Path.* **2016**, *25*, 871–878. [CrossRef]
5. Burnham, S.L. Anatomical Differences of the Donkey and Mule. In Proceedings of the Annual Convention of the AAEP, Orlando, FL, USA, 4–8 December 2002; pp. 102–109.
6. Peck, K.E.; Matthews, N.S.; Taylor, T.S.; Mealey, K.L. Pharmacokinetics of sulfamethoxazole and trimethoprim in donkeys, mules, and horses. *Am. J. Vet. Res.* **2002**, *63*, 349–353. [CrossRef] [PubMed]
7. Latzel, S.T. Subspecies Studies: Pharmacokinetics and Pharmacodynamics of a Single Intravenous Dose of Xylazine in Adult Mules and Adult Haflinger Horses. *J. Equine Vet. Sci.* **2012**, *32*, 816–826. [CrossRef]
8. Baroncini, R. *L'asino, il Mulo e il Bardotto*; Calderini Edagricole: Bologna, Italy, 2001; ISBN 8850600402.
9. Love, S.; Murphy, D.; Mellor, D. Pathogenicity of cyathostome infection. *Vet. Parasitol.* **1999**, *85*, 113–121. [CrossRef]
10. Buono, F.; Roncoroni, C.; Pacifico, L.; Piantedosi, D.; Neola, B.; Barile, V.L.; Fagiolo, A.; Várady, M.; Veneziano, V. Cyathostominae Egg Reappearance Period After Treatment with Major Horse Anthelmintics in Donkeys. *J. Equine Vet. Sci.* **2018**, *65*, 6–11. [CrossRef]
11. Gokbulut, C.; McKellar, Q.A. Anthelmintic drugs used in equine species. *Vet. Parasitol.* **2018**, *261*, 27–52. [CrossRef]
12. McKellar, Q.A.; Scott, E.W. The benzimidazole anthelmintic agents—A review. *J. Vet. Pharmacol. Ther.* **1990**, *13*, 223–247. [CrossRef] [PubMed]
13. Lizarraga, I.; Sumano, H.; Brumbaugh, G.W. Pharmacological and pharmacokinetic differences between donkeys and horses. *Equine Vet. Educ.* **2010**, *16*, 102–112. [CrossRef]
14. Gokbulut, C.; Aksit, D.; Santoro, M.; Roncoroni, C.; Mariani, U.; Buono, F.; Rufrano, D.; Fagiolo, A.; Veneziano, V. Plasma disposition, milk excretion and parasitological efficacy of mebendazole in donkeys naturally infected by Cyathostominae. *Vet. Parasitol.* **2016**, *217*, 95–100. [CrossRef] [PubMed]
15. Veneziano, V.; Di Loria, A.; Masucci, R.; Di Palo, R.; Brianti, E.; Gokbulut, C. Efficacy of eprinomectin pour-on against *Dictyocaulus arnfieldi* infection in donkeys (*Equus asinus*). *Vet. J.* **2011**, *190*, 414–415. [CrossRef] [PubMed]
16. Aupanun, S.; Laus, F.; Poapolathep, A.; Owen, H.; Vullo, C.; Faillace, V.; Giorgi, M. Pharmacokinetic Assessment of the Marker Active Metabolites 4-Methyl-amino-antipyrine and 4-Acetyl-amino-antipyrine After Intravenous and Intramuscular Injection of Metamizole (Dipyrone) in Healthy Donkeys. *J. Equine Vet. Sci.* **2016**, *47*, 55–61. [CrossRef]
17. Lanusse, C.E.; Adrian, L.; Lifschitz, A.; Imperiale, F.A. Macrocyclic Lactones: Endectocide Compounds. In *Veterinary Pharmacology &Therapeutics*; Riviere, M.G., Papich, J., Eds.; Blackwell Publishing: Ames, IA, USA, 2009; pp. 1119–1145.
18. Campbell, W.C.; Leaning, W.H.D.; Seward, R.L. Use of Ivermectin in Horses. In *Ivermectin and Abamectin*; Campbell, W.C., Ed.; Springer: New York, NY, USA, 1989; pp. 234–244.
19. Conder, G.A.; Campbell, W.C. Chemotherapy of nematode infections of veterinary importance, with special reference to drug resistance. *Adv. Parasitol.* **1995**, *35*, 1–70. [PubMed]
20. Pérez, R.; Godoy, C.; Palma, C.; Cabezas, I.; Muñoz, L.; Rubilar, L.; Arboix, M.; Alvinerie, M. Plasma profiles of ivermectin in horses following oral or intramuscular administration. *J. Vet. Med. Ser. A Physiol. Pathol. Clin. Med.* **2003**, *50*, 297–302. [CrossRef]
21. González Canga, A.; Sahagún Prieto, A.M.; José Diez Liébana, M.; Martínez, N.F.; Vega, M.S.; Vieitez, J.J.G. The pharmacokinetics and metabolism of ivermectin in domestic animal species. *Vet. J.* **2009**, *179*, 25–37. [CrossRef]

22. Marriner, S.E.; McKinnon, I.; Bogan, J.A. The pharmacokinetics of ivermectin after oral and subcutaneous administration to sheep and horses. *J. Vet. Pharmacol. Ther.* **1987**, *10*, 175–179. [CrossRef]
23. Pérez, R.; Cabezas, I.; García, M.; Rubilar, L.; Sutra, J.F.; Galtier, P.; Alvinerie, M. Comparison of the pharmacokinetics of moxidectin (Equest®) and ivermectin (Eqvalan®) in horses. *J. Vet. Pharmacol. Ther.* **1999**, *22*, 174–180. [CrossRef]
24. Gokbulut, C.; Cirak, V.Y.; Senlik, B.; Aksit, D.; Durmaz, M.; McKellar, Q.A. Comparative plasma disposition, bioavailability and efficacy of ivermectin following oral and pour-on administrations in horses. *Vet. Parasitol.* **2010**, *170*, 120–126. [CrossRef]
25. Pérez, R.; Cabezas, I.; Sutra, J.F.; Galtier, P.; Alvinerie, M. Faecal excretion profile of moxidectin and ivermectin after oral administration in horses. *Vet. J.* **2001**, *161*, 85–92. [CrossRef]
26. Gokbulut, C.; Boyacioglu, M.; Karademir, U. Plasma pharmacokinetics and faecal excretion of ivermectin (Eqvalan® paste) and doramectin (Dectomax®, 1%) following oral administration in donkeys. *Res. Vet. Sci.* **2005**, *79*, 233–238. [CrossRef]
27. Giorgi, M.; Vullo, C.; De Vito, V.; Catone, G.; Faillace, V.; Laus, F. Pharmacokinetic evaluations of sulpiride after intravenous, intramuscular, and oral single-dose administration in jennies (*Equus asinus*). *J. Equine Vet. Sci.* **2015**, *35*, 13–18. [CrossRef]
28. Giorgi, M.; Laus, F.; de Vito, V.; Owen, H.; Poapolathep, A.; Paggi, E.; Vullo, C. Flupirtine: Preliminary Pharmacokinetics in the Donkey. *J. Equine Vet. Sci.* **2015**, *35*, 309–314. [CrossRef]
29. Lanusse, C.E.; Prichard, R.K. Relationship between pharmacological properties and clinical efficacy of ruminant anthelmintics. *Vet. Parasitol.* **1993**, *49*, 123–158. [CrossRef]
30. Baggot, J.D.; McKellar, Q.A. The absorption, distribution and elimination of anthelmintic drugs: The role of pharmacokinetics. *J. Vet. Pharmacol. Ther.* **1994**, *17*, 409–419. [CrossRef]
31. Nielsen, M.; Mittel, L.; Grice, A.; Erskine, M.; Graves, E.; Vaala, W.; Tully, R.C.; French, D.D.; Bowman, R.; Kaplan, R.M. AAEP Parasite Control Guidelines. 2013. Available online: http://www.aaep.org (accessed on 12 March 2019).
32. Kay, G.; Pearson, R.A.; Ouassat, M. Estimation of the liveweight of working mules in Morocco from their body measurements. *Vet. Rec.* **2004**, *154*, 85–88. [CrossRef]
33. Coles, G.C.; Jackson, F.; Pomroy, W.E.; Prichard, R.K.; von Samson-Himmelstjerna, G.; Silvestre, A.; Taylor, M.A.; Vercruysse, J. The detection of anthelmintic resistance in nematodes of veterinary importance. *Vet. Parasitol.* **2006**, *136*, 167–185. [CrossRef]
34. Nielsen, M.K.; Vidyashankar, A.N.; Andersen, U.V.; DeLisi, K.; Pilegaard, K.; Kaplan, R.M. Effects of fecal collection and storage factors on strongylid egg counts in horses. *Vet. Parasitol.* **2010**, *167*, 55–61. [CrossRef]
35. Ministry of Agriculture Fisheries and Food (MAFF). *Manual of Veterinary Parasitological Techniques*, 3rd ed.; Her Majesty's Stationery Office: London, UK, 1986; ISBN 0112427243.
36. Galarini, R.; Saluti, G.; Moretti, S.; Giusepponi, D.; Dusi, G. Determination of macrocyclic lactones in food and feed. *Food Addit. Contam. Part A Chem. Anal. Control. Expo. Risk Assess.* **2013**, *30*, 1068–1079. [CrossRef]
37. EMA Guideline on Bioanalytical Method Validation EMEA/CHMP/EWP/192217/2009 Rev. 1 Corr. 2**. Available online: https://www.ema.europa.eu (accessed on 23 March 2020).
38. Zhang, Y.; Huo, M.; Zhou, J.; Xie, S. PKSolver: An add in program for pharmacokinetic and pharmacodynamic data analysis in Microsoft Excel. *Comput. Methods Programs Biomed.* **2010**, *99*, 306–314. [CrossRef]
39. Tzelos, T.; Barbeito, J.S.G.; Nielsen, M.K.; Morgan, E.R.; Hodgkinson, J.E.; Matthews, J.B. Strongyle egg reappearance period after moxidectin treatment and its relationship with management factors in UK equine populations. *Vet. Parasitol.* **2017**, *237*, 70–76. [CrossRef]
40. Gokbulut, C.; Nolan, A.M.; McKellar, Q.A. Plasma pharmacokinetics and faecal excretion of ivermectin, doramectin and moxidectin following oral administration in horses. *Equine Vet. J.* **2001**, *33*, 494–498. [CrossRef] [PubMed]
41. Sinclair, M.D.; Mealey, K.L.; Matthews, N.S.; Peck, K.E.; Taylor, T.S.; Bennett, B.S. Comparative pharmacokinetics of meloxicam in clinically normal horses and donkeys. *Am. J. Vet. Res.* **2006**, *67*, 1082–1085. [CrossRef] [PubMed]
42. Grosenbaugh, D.A.; Reinemeyer, C.R.; Figueiredo, M.D. Pharmacology and therapeutics in donkeys. *Equine Vet. Educ.* **2011**, *23*, 523–530. [CrossRef]
43. Matthews, N.S.; Taylor, T.S.; Hartsfield, S.M. Anaesthesia of donkeys and mules. *Equine Vet. Educ.* **1997**, *9*, 198–202. [CrossRef]

44. Toutain, P.L.; Ferran, A.; Bousquet-Mélou, A. Species differences in pharmacokinetics and pharmacodynamics. *Handb. Exp. Pharmacol.* **2010**, *199*, 19–48.
45. Matthews, N.S.; Taylor, T.S. Anesthesia of Donkeys and Mules: How They Differ from Horses. In Proceedings of the Annual Convention of the AAEP, Orlando, FL, USA, 4–8 December 2002; pp. 110–112.
46. Kaplan, R.M.; Nielsen, M.K. An evidence-based approach to equine parasite control: It ain't the 60s anymore. *Equine Vet. Educ.* **2010**, *22*, 306–316. [CrossRef]
47. Baker, N.F.; Miller, J.E.; Madigan, J.E.; Fulton, R.; Seward, R.L. Anthelmintic action of ivermectin, oxibendazole and pyrantel pamoate against thiabendazole-resistant small strongyles of horses. *Equine Pract.* **1984**, *6*, 8–19.
48. Pereira, M.C.; Kohek Júnior, I.; Campos, R.; Lima, S.B.; Foz, R.P.; Campos Pereira, M.; Kohek, I.; Campos, R.; Lima, S.B.; Foz, R.P.P. A field evaluation of anthelmintics for control of cyathostomes of horses in Brazil. *Vet. Parasitol.* **1991**, *38*, 121–129. [CrossRef]
49. Lester, H.E.; Spanton, J.; Stratford, C.H.; Bartley, D.J.; Morgan, E.R.; Hodgkinson, J.E.; Coumbe, K.; Mair, T.; Swan, B.; Lemon, G.; et al. Anthelmintic efficacy against cyathostomins in horses in Southern England. *Vet. Parasitol.* **2013**, *197*, 189–196. [CrossRef]
50. Coakley, M.; Peck, K.E.; Taylor, T.S.; Matthews, N.S.; Mealey, K.L. Pharmacokinetics of flunixin meglumine in donkeys, mules, and horses. *Am. J. Vet. Res.* **1999**, *60*, 1441–1444.
51. Toutain, P.L.; Bousquet-Mélou, A. Bioavailability and its assessment. *J. Vet. Pharmacol. Ther.* **2004**, *27*, 455–466. [CrossRef] [PubMed]

MDPI AG
Grosspeteranlage 5
4052 Basel
Switzerland
Tel.: +41 61 683 77 34

Animals Editorial Office
E-mail: animals@mdpi.com
www.mdpi.com/journal/animals

www.ingramcontent.com/pod-product-compliance
Lightning Source LLC
Chambersburg PA
CBHW041427210926
51850CB00013B/3369